D1639758

C140005530

PHILIP'S ATLAS OF THE

UNIVERSE

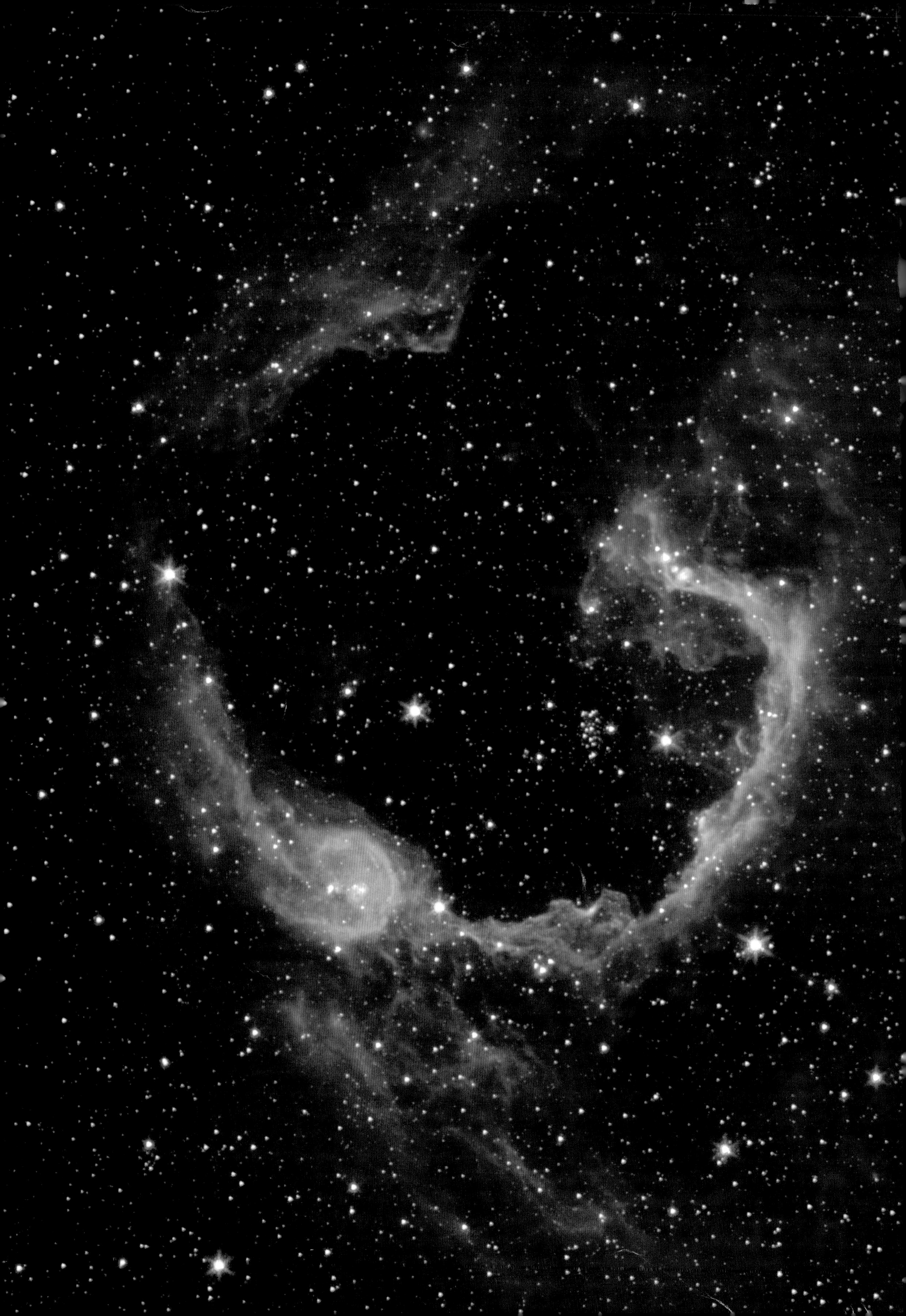

PHILIP'S

ATLAS OF THE UNIVERSE

SIR PATRICK MOORE

NEW EDITION

FOREWORD BY
PIERS SELLERS

Contents

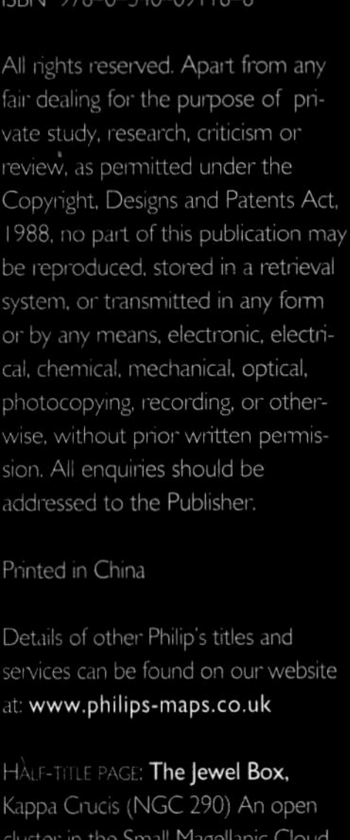

First published in 1994
by Philip's,
a division of Octopus Publishing Group Ltd,
2–4 Heron Quays, London E14 4JP

An Hachette Livre UK Company

This new edition 2007

Editor Robin Rees
Art Editor Mike Brown
Designer Caroline Ohara
Production Sally Banner

A CIP catalogue record for this book is available from the British Library.

ISBN 978-0-540-09118-8

Printed in China

Details of other Philip's titles and services can be found on our website at: **www.philips-maps.co.uk**

HALF-TITLE PAGE: **The Jewel Box,** Kappa Crucis (NGC 290) An open cluster in the Small Magellanic Cloud, 200,000 light years away and about 65 light years across. The contrasting colours of the stars make it a beautiful object even in a small telescope. Image from the Hubble Space Telescope.

OPPOSITE TITLE PAGE: **Star-forming bubble RCW 79** seen from the Spitzer Space Telescope.

STS-121
KELLY
LINDSEY
NOWAK

Foreword

Most of the time, we humans are busy living in the universe at the middle scale - the scale of millimetres out to a few hundred kilometres - which is the scale in which we live and work, can see and otherwise directly experience things around us and thus know the people and places important to our individual lives. Usually, we don't spend much time thinking about what's going on at the very small scale – molecules and atoms – as we have no direct way of experiencing or seeing things that tiny. We have to use our imagination to visualize what goes on in an atomic nucleus and little there corresponds to everyday experience. So our ruminations about atoms are abstract and somehow disconnected from everything else we know.

The universe at the larger scale challenges the imagination in a different way. We can see some of it, with telescopes and imagery, but its size and age are so enormously greater than anything else in our everyday world that it is often hard to attach a sense of reality to it. (Personally, I found it hard to understand that our own world is a relatively small and finite place until I was floating in space above it and saw the Earth as a perfect blue sphere silently rolling below my feet, lit up by a brilliant white sun in a huge black sky.) But with some help from an adept astronomer, it is possible to climb up the ladder of successively larger scales – the area around your home, your country, the Earth, the solar system, the Galaxy - all the way up to the idea of a single place called the universe, and glimpse the true nature of the place where we live. And the universe is full of surprises: the more we look, the more we realize how complex, mysterious, violent and beautiful it is.

Patrick Moore has been studying the heavens for almost his entire life, and his work as a television presenter actually preceded the launch of the first spacecraft, Sputnik, in October 1957 by a few months. He has dedicated his life to both observing the universe as a practising astronomer and to helping the rest of us to understand its mysteries. In this book, he takes the reader on a guided tour of our solar system and the larger cosmos.

Humans are newcomers to the universe, relatively speaking, but we have started the process of studying it, understanding it and slowly moving outwards into it. Who knows where our distant descendents will go? This book should give you an idea of the possibilities.

PIERS SELLERS
August 2007

◀ **Piers Sellers** seen above the Earth on 12 July 2006 on the third spacewalk of the STS-121 mission to the International Space Station, during which he and Michael Fossum tested heat-shield repair techniques for the Space Shuttles.

Introduction

When I wrote the first edition of *The Atlas of the Universe*, in 1970, the great astronomical revolution was just beginning. Electronic devices had started to take over from photographic plates, and computers had become a real force even though they were very crude compared with those of today. Space research was in full swing: men had already landed on the Moon, probes had been sent out to the nearer planets and the first astronomical observatories were in orbit round the Earth.

Since then a great deal has happened. Great new telescopes have been built, allowing us to explore the far reaches of the universe; new theories have forced us to change or even abandon many of the older ideas, even if we have yet to solve fundamental problems such as that of the origin of the universe itself.

The progress of space research has been less smooth. There have been spectacular triumphs, but also some serious setbacks. However, there is one very encouraging note; all nations are working together in space, and the International Space Station now orbiting the Earth really is completely international.

Undoubtedly there will be further problems during the next few decades, but all in all the outlook remains bright. There are still people who question the value of the space programmes, but the cost of a planetary probe does not seem excessive when compared with that of, say, a nuclear submarine, and there are many benefits to mankind: for example, medical research is now closely linked with astronautics.

There is a major difference between this *Atlas* and others. We are used to the superb, highly coloured images produced by the world's greatest telescopes, but in general the colours are added to help in scientific analysis. Obviously I have included some of these false-colour pictures here, but I have concentrated upon things which can actually be seen by an observer who is adequately equipped. This is not always possible, but I have kept to my rule as far as I can.

Since the previous edition, in 2005, much has happened, and I have made many changes. For example, we have included the very latest images from the Spitzer Space Telescope and others, together with the amazing pictures of Mars sent back by the latest spacecraft. Moreover, the practical section has been completely rewritten, and for this I am indebted to Pete Lawrence, whose contributions to astronomy are known world-wide, and whose observatory is absolutely up-to-date. I have also been able to use images by some of the most talented astronomical photographers, notably Damian Peach, Nik Szymanek, Bruce Kingsley, Gordon Rogers, Greg Parker, Dave Tyler and Chris Doherty, and am most grateful to all of them.

Special thanks to John Mason who provided invaluable assistance in the final proof reading, and to Robin Rees, without whom this book would never have appeared.

PATRICK MOORE

◄ **The star-forming region** of NGC1333. This Spitzer Space Telescope image shows two distinct clusters of stars within a nebula. The red area is dust warmed by the stars it surrounds while the blue-green region has numerous stars spewing out jets of materal that are creating the knotty yellow yellow features where they collide with the surrounding cold gas.

Space walk: The picture is a self-portrait taken by astronaut Michael Fossum on 8 July 2006 during a space walk or extravehicular activity while the Discovery orbiter was docked with the International Space Station.

EXPLORING THE UNIVERSE

The Stargazers

The skies are all around us. Even our cave-dwelling ancestors must have looked up and marvelled at what they saw there; the Sun dominated the day, while the Moon ruled the night – were the Sun and Moon gods, or at least the dwelling-places of gods? The first humans did not know, and could not be expected to know, but at least they could make observations, and when writing was developed these observations could be recorded.

It is probably true to say that the earliest true 'stargazers' were the Babylonians, the Chinese and the Egyptians, all of whom divided the stars into groups or constellations, and all of whom drew up calendars; in around 3100 BC the Egyptians gave the length of the year as 365 days, though their understanding of the universe was hampered by their initial mistake of believing that the heavens were formed by the arched body of a goddess, Nut. Certainly the pyramids are astronomically aligned, and the Egyptians also paid great attention to the 'heliacal' rising of the brilliant star Sirius – that is to say, the earliest day in the year when Sirius could be seen in the dawn or morning twilight. This was important because it gave a clue to the time of the annual flooding of the Nile, upon which the Egyptian economy depended.

True astronomy came with the Greeks, who not only looked at the celestial bodies but made real attempts to interpret them. The Greeks knew that there were several 'wandering stars' or planets – now known to us as Mercury, Venus, Mars, Jupiter and Saturn – and that the stars were very remote; at an early stage they showed that the Earth is a globe rather than a flat plane, and they made careful records of unusual phenomena such as eclipses or the appearance of comets. A few of the philosophers, notably Aristarchus (around 230 BC) were bold enough to claim that the Earth is simply a planet moving round the Sun, but the later Greeks reverted to the idea of a central, all-important Earth. Any contrary suggestion was regarded as not only unscientific, but also heretical.

Ptolemy (Claudius Ptolemaeus), last of the great Greek astronomers, lived around the year AD 150. We know nothing about his life, except that his home was in Alexandria, but we owe him a great deal, because he wrote a book which was more or less a summary of ancient science; it has come down to us by way of its Arabic translation (the *Almagest*). He brought the Earth-centred picture of the planetary system to its highest degree of perfection; he completed an excellent star catalogue and he drew the first maps of the known world which were based upon astronomical observation. His maps were not perfect, Scotland is tacked onto England in a sort of back-to-front position.

▼ **Stonehenge** Probably one of the earliest astronomical observatories. The circle is astronomically aligned, though its exact purpose remains a matter of debate. I took the photograph at sunrise from the centre of the circle. In the middle of the picture the Heel Stone can be seen; this is one of the main points of reference. Stonehenge was built by the Beaker people several thousands of years BC, and has nothing to do with Druids ancient or modern.

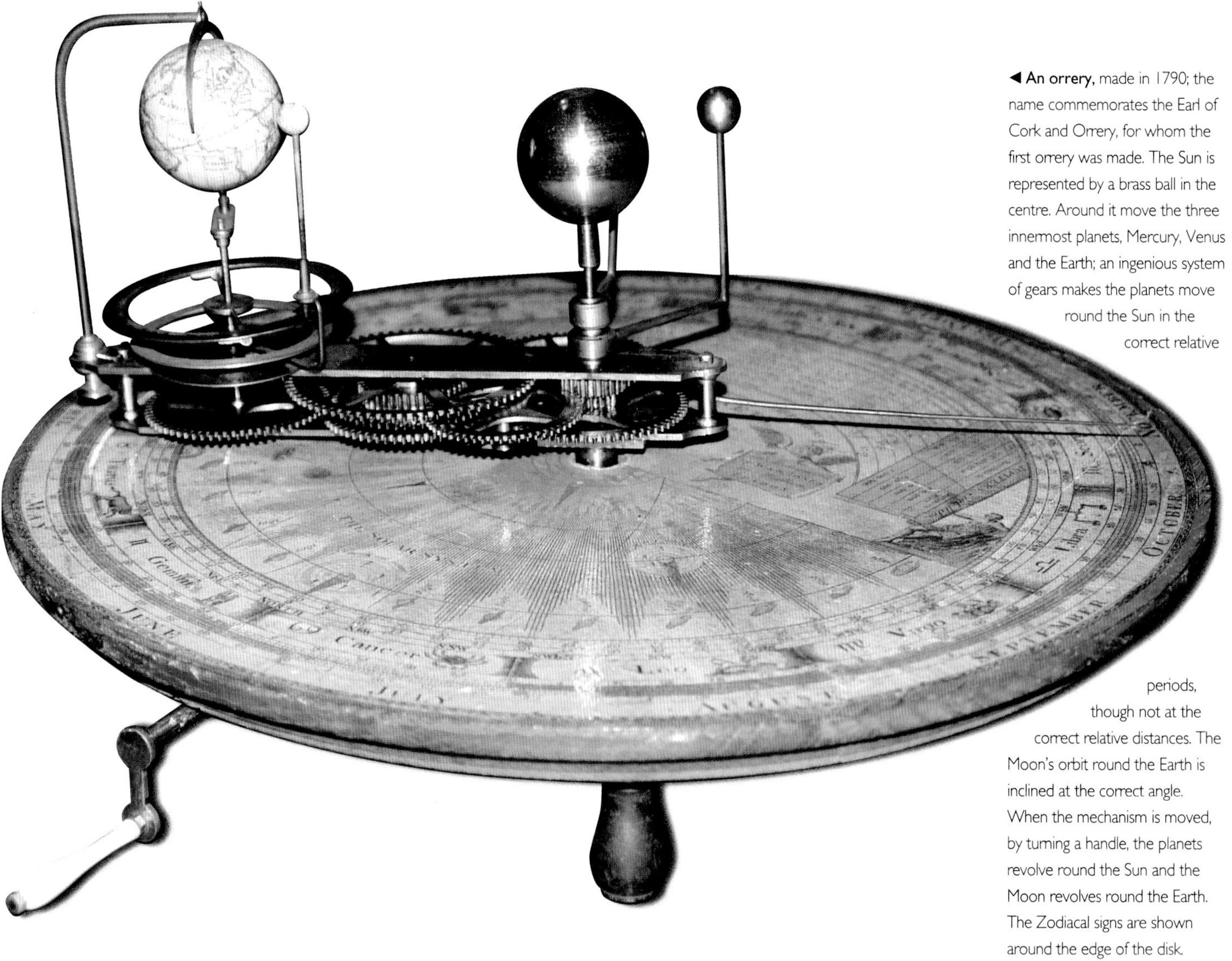

◀ **An orrery,** made in 1790; the name commemorates the Earl of Cork and Orrery, for whom the first orrery was made. The Sun is represented by a brass ball in the centre. Around it move the three innermost planets, Mercury, Venus and the Earth; an ingenious system of gears makes the planets move round the Sun in the correct relative periods, though not at the correct relative distances. The Moon's orbit round the Earth is inclined at the correct angle. When the mechanism is moved, by turning a handle, the planets revolve round the Sun and the Moon revolves round the Earth. The Zodiacal signs are shown around the edge of the disk.

After Ptolemy's time, there were no more astronomical advances for some centuries. When revival came, it was due to the Arabs, the last of whom, Ulugh Beigh, set up an elaborate observatory at his capital, Samarkand in about 1420. Of course he had no telescopes, but he was able to compile star catalogues better than any of his predecessors. With his death, the Arab school of astronomy petered out.

Then came what is called the 'great astronomical revolution'. It began in 1543 with the publication of a book by a Polish churchman, who is always remembered as Nicholas Copernicus. In his book, he removed the Earth from its proud central position in the centre of the Universe, and replaced it with the Sun. He knew that the Church would be bitterly hostile to any such idea, and he prudently held back publication of his book until the very end of his life, but it ushered in what we may call the modern age of astronomy. He was wise to be cautious. In 1600 one of his followers, Giordano Bruno, was burnt at the stake in Rome. One of his crimes in the eyes of the Church was that he believed the Earth to be in orbit aound the Sun.

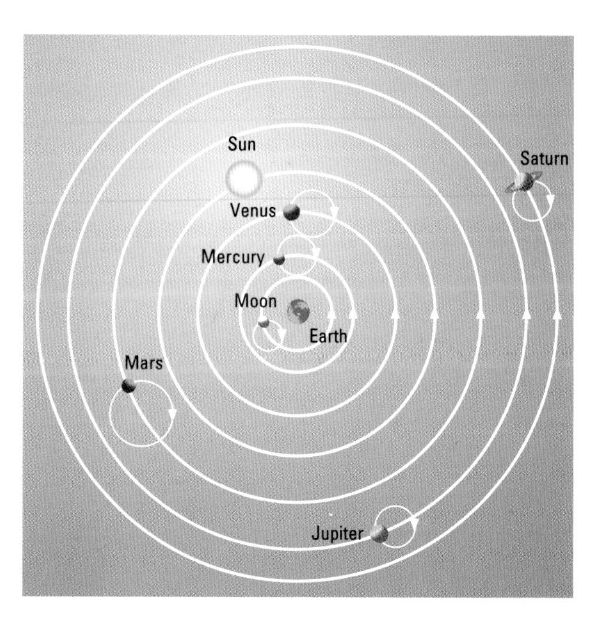

◀ **The Ptolemaic theory** – the Earth lies in the centre of the universe, with the Sun, Moon, planets and stars moving round it in circular orbits. Ptolemy assumed that each planet moved in a small circle or epicycle, the centre of which – the deferent – itself moved round the Earth in a perfect circle.

▶ **The Copernican theory** – placing the Sun in the centre removed many of the difficulties of the Ptolemaic theory, but Copernicus kept the idea of circular orbits, and was even reduced to bringing back epicycles to match observations.

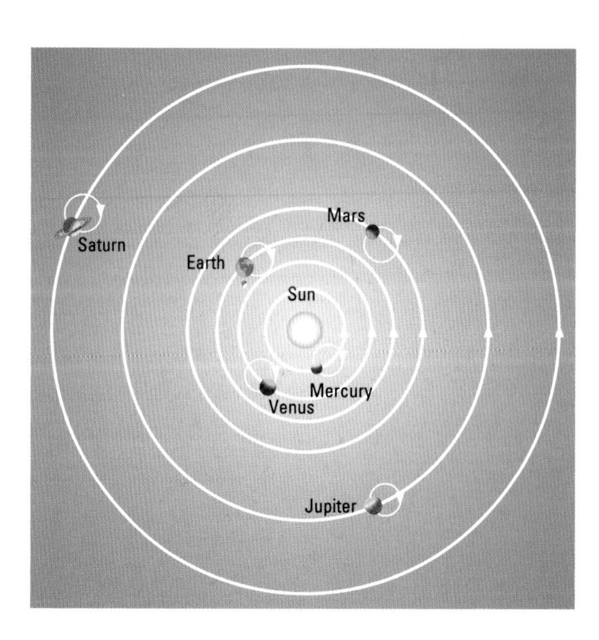

Telescopes and the Stars

▲ **Statue of Tycho Brahe** At the site of his old observatory, now restored, on the Island of Hven in the Baltic Sea.

It is ironical that the first great astronomer after Copernicus – the eccentric Dane, Tycho Brahe – never accepted the heretical idea that the Earth could move round the Sun. Instead, he developed a sort of hybrid system which satisfied few people apart from Tycho himself. Yet he was a brilliant observer, who drew up a star catalogue of amazing accuracy, and also made careful measurements of the movements of the planets, particularly Mars. Tycho died in 1601, and his observations were passed on to his last assistant, a young German named Johannes Kepler, who used them well and was able to show that the planets do indeed move round the Sun – not in perfect circles, but in ellipses. Kepler's three Laws of Planetary Motion have formed the basis of all subsequent work.

A few years after Tycho's death came arguably the greatest development of all – the invention of the telescope. So far as we know, the first telescopes were made in Holland in 1608. One of them came into the possession of Galileo Galilei, in Italy, who made a telescope for himself – 'sparing neither trouble nor expense', as he put it – and in the first weeks of January 1610 he used it to make a series of spectacular discoveries. He saw the craters of the Moon, the phases of Venus, the four large satellites of Jupiter, the 'myriad stars' of the Milky Way and much else. His telescope was a refractor, using a glass lens ('object-glass') to collect and focus the light; the image was then enlarged by a second lens ('eyepiece') which was in effect a magnifying-glass. Galileo's best telescope was not nearly so good as a pair of cheap modern binoculars, but it revolutionized the whole of astronomy – and, indeed, the whole of human thought, because it showed that the Earth could not possibly lie at rest in the centre of the Universe. Had it done so, Venus could not have shown phases from new to full.

Galileo was not tactful. He was an outspoken champion of the Copernican theory, and this brought him into conflict with the Church. He was brought to Rome, put on trial, and condemned as a dangerous heretic; he was forced to 'curse, abjure and detest' the false theory of a moving Earth. Only in 1991 did the Vatican pardon him, and admit that he had been right.

Galileo died in 1642; in the same year Isaac Newton was born in the quiet village of Woolsthorpe in Lincolnshire. He studied at Cambridge University, and was never in any doubt about the truth of the Copernican system, but away from the reach of Rome, he was in no danger of religious persecution; he was free to say and do what he chose, and his life was relatively uneventful. We remember him best for his description of the laws of gravitation, but this was only one of his achievements, and he is widely regarded as the greatest scientist of all time.

In 1665 the University of Cambridge, where he studied, was temporarily closed because of the danger from the plague; Newton retreated to his Woolsthorpe home, where he made several of his major discoveries. One of these was in connection with the nature of light. Light is a wave motion, and the colour depends upon the wavelength – that is to say, the distance between successive wave crests; for visible light red has the longest wavelength and violet the shortest, with orange, yellow, blue and green in between. Newton realized that sunlight is a mixture of all wavelengths, and he set out to prove it. He passed a beam of sunlight through a glass prism, knowing that the different colours would be bent or 'refracted' unequally – the shorter the wavelength the greater the amount of bending. The result was a strip of colours, from red through to violet. This simple experiment produced the first spectrum; today, astronomical spectroscopy is of vital importance.

A glass lens will also bend the colours unequally, and this showed up with telescopes of the kind made by Galileo. A bright object such as a star will be surrounded by gaudy colours which may look beautiful, but are most unwelcome to astronomers. To avoid false colour, Newton produced a completely different kind of telescope, collecting its light by use of a mirror instead of a lens; a mirror reflects all colours equally. The principle of the Newtonian reflector is shown in the diagram on page 295. Newton's first reflector, presented to the Royal Society in London in 1671, had a 25-mm (1-inch) mirror; the largest telescopes of today have enormous light-gathering power, but the principles involved are exactly the same.

One famous story about Newton may well be true. It is said that when he was sitting in his garden he saw an apple fall from a tree; this set off a train of thought in his mind, and he realised that the force pulling the apple was the same as the force which keeps the Moon in its orbit round the Earth. This led on to his Laws of Gravitation, given in his book known generally as the *Principia*, completed in 1611. This book which proved beyond all doubt that far from being all-important, the Earth was merely an ordinary planet. Copernicus began the great revolution; Newton completed it. The whole process had taken just under a 150 years.

▼ **Galileo's telescope** The author holding a replica of Galileo's original telescope of 1610 at the Institute and Museum for the History of Science in Florence, in January 1997.

▶ **Photo-composition** of the Jacobus Kapteyn Telescope dome under the broad sweep of the Milky Way, taken by Nik Szymanek, on the Canary Islands in 1997. The bright planet to the left is Venus.

▼ The Tychonic theory – Tycho Brahe retained the Earth in the central position, but assumed that the other planets moved round the Sun. In effect this was a rather uneasy compromise, which convinced comparatively few people. Tycho adopted it because although he realized that the Ptolemaic theory was unsatisfactory, even he could not bring himself to believe that the Earth was of anything but supreme importance. Tycho's theory found few followers.

▼ Kepler's Laws:
Law 1 A planet moves in an ellipse; the Sun is at one focus, while the other is empty.
Law 2 The radius vector – the line joining the centre of the planet to that of the Sun – sweeps out equal areas in equal times (a planet moves fastest when closest in).
Law 3 For any planet, the square of the revolution period (p) is proportional to the cube of the planet's mean distance from the Sun (a). Once the distance of any planet is known, its period can be calculated, or vice versa. Kepler's Laws make it possible to draw up a scale model of the Solar System; only one absolute distance has to be known, and the rest can then readily be calculated.

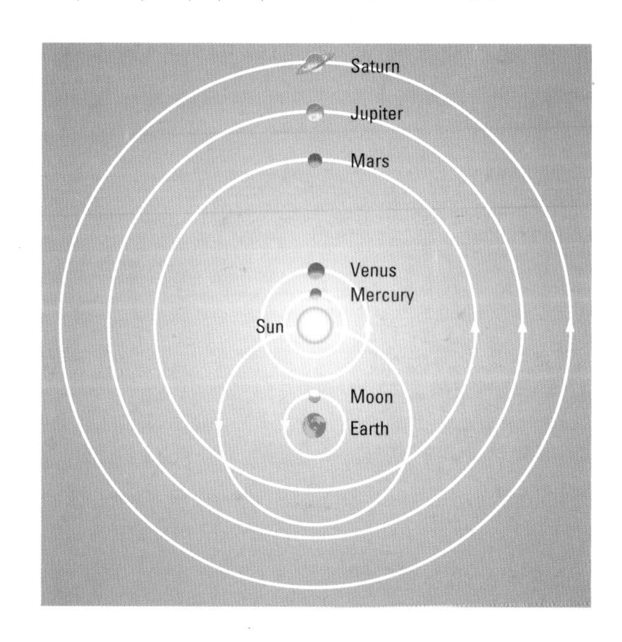

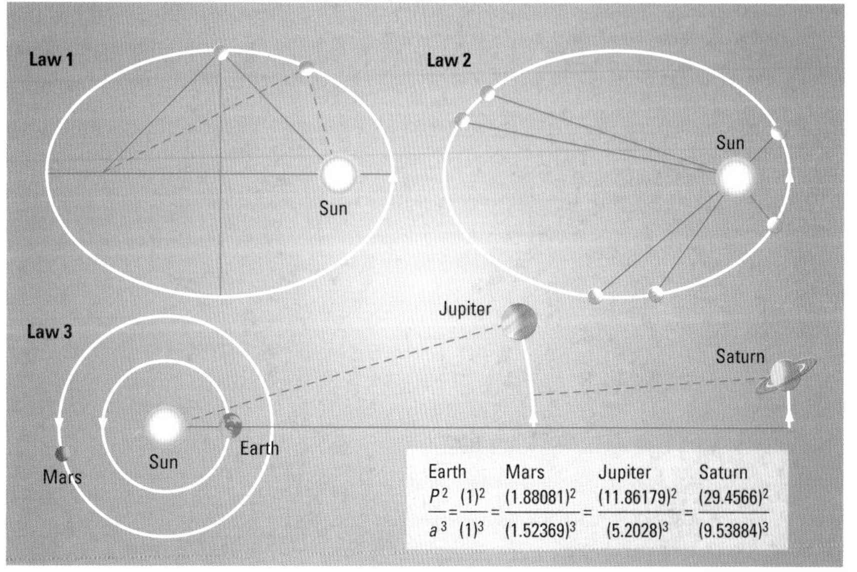

New Horizons

In the story of astronomy there have been a few telescopes of fundamental importance. Galileo's tiny refractor was the first; Newton's equally tiny reflector was the second. Next came the reflector completed in 1789 by William Herschel, a musician-turned astronomer; its mirror was 124 cm (49 inches) across, and it was over 12 metres (40 feet) long. Of course there had been others, and the first major observatories had been founded in Britain. The Royal Greenwich Observatory was set up in 1675 by the express order of King Charles II, because ships out of sight of land needed to know their latitude and longitude, which was done by astronomical observation and required the use of a really accurate star catalogue. At Greenwich the first Astronomer Royal, John Flamsteed, was given the task of compiling such a catalogue, and he succeeded, though it took a long time. (The RGO moved to Herstmonceux in East Sussex in the 1950s, then to Cambridge in 1990 and remained Britain's flagship observatory until its closure in 1998.)

William Herschel was purely an observer. He was born in Hanover, but spent most of his life in England, and became organist in the Octagon Room in Bath. In middle age he took up astronomy and made his own reflecting telescopes; he undertook 'reviews of the heavens' and in 1781 discovered a new planet, now called Uranus, moving far beyond the orbit of Saturn. Herschel became famous and was able to give up his musical career so that he could concentrate on astronomy. It was he who drew up the first reasonable map of our Galaxy, and he discovered hundreds of double stars, star-clusters and nebulae (huge clouds of gas and dust in space).

It was in Herschel's lifetime that astronomical spectroscopy became important, leading to the first ideas about the make-up and nature of the stars. Of course the nearest star, our Sun, is close enough to be examined in detail; the solar spectrum was found to show a rainbow band crossed by narrow dark lines which never changed position or intensity. Later in the 18th century it was found that each line is the trademark of one particular element or group of elements and this tells us 'what the stars are made of'.

Herschel's reflector was outmatched in 1845 by a 183-cm (72-inch) instrument made by the third Earl of Rosse and set up at Birr Castle in central Ireland. It was a strange telescope (happily, still in full working order), and with it Lord Rosse made one discovery of particular importance. In 1781 – the year that Herschel discovered Uranus – the French astronomer Charles Messier had published a catalogue of more than 100 star clusters and nebulae. The nebulae were of two types. Some, such as the Orion Nebula (42 in Messier's catalogue) looked like patches of gas, but others, such as the Great Nebula in Andromeda (M31) were 'starry'. Lord Rosse found that some of the starry nebulae were spiral in form,

▼ **Herschel's '20-ft' reflector**, built on exactly the same pattern as the '40-ft', but with the advantage of being much easier to handle. The optical system used was the Herschelian; there is no flat, and the main mirror is tilted so as to bring the rays of light directly to focus at the upper edge and to one side of the tube – a system which is basically unsatisfactory.

in form, shaped like Catherine wheels. Just under a century later it was found that the spirals are independent galaxies, so far away that in most cases their light, travelling at 2,997,924.58 kilometres per second (186,282.397) miles per second, takes millions of years to reach us.

Astronomical photography began in the mid-19th century, and before long the camera had taken over from the human eye for most astronomical work, as it is less easily deceived and gives a permanent record. From around 1880 the main research telescopes were refractors, smaller in aperture than the Rosse reflector but much more manageable and much more versatile; the largest at the Yerkes Observatory in the USA, has a 102-cm (40-inch) object-glass. But a great change lay close ahead. For the first 20 years of its existence, the Birr telescope was in a class of its own. No other telescope of the time was able to show the spirals and Birr became very much an astronomical centre, particularly as Lord Rosse was always ready to make his astronomical experience and equipment available to all.

▲ **Whirlpool Galaxy M51**
Drawn by Lord Rosse in 1848. The resemblance to a modern photograph is quite uncanny

◄ **The Rosse reflector.**
This telescope was built by the third Earl of Rosse, and completed in 1845. It had a 183-cm (72-inch) metal mirror; the tube was mounted between two massive stone walls, so that it could be swung for only a limited distance to either side of the meridian. This imposed obvious limitations; nevertheless, Lord Rosse used it to make some spectacular discoveries, such as the spiral forms of the galaxies. The telescope has now been fully restored, and by 2001 was again fully operational. This photograph was taken in 1997.

Telescopes of the Mountain-tops

Both of the next telescopes in our list are American; the Hooker 2.5-metre (100-inch) reflector (completed 1917) on Mount Wilson and the Hale 5-metre (200-inch) reflector on Mount Palomar (1948). Each was, for a time, not only the world's largest telescope, but in a class of its own; both were sited at high altitude, well away from the effects of the densest layers of the Earth's atmosphere.

The main architect of the Californian reflectors was George Ellery Hale, an American astronomer who had earned a great reputation for his studies of the Sun. He not only planned large telescopes, but also had the happy knack of persuading friendly millionaires to finance them – something much less difficult a century ago than it is today. Hale master-minded the Yerkes 101-cm (40-inch) refractor, but realized that the real future lay with the reflector. A huge lens is much more troublesome to make than a huge mirror, and it has to be supported round its edge. If it is too large it will become so heavy that it will sag under its own weight, rendering it useless. Hale changed his plans. His first millionaire, John D. Hooker, financed a 125-cm (50-inch) reflector, set up in 1907 on Mount Wilson. It worked well, but Hale was not satisfied, and ten years later the Hooker was ready for use. Instantly it showed that its light-grasp was better than that of any of its predecessors.

Edwin Hubble was one of America's most promising young astronomers. He had the use of the Hooker, and tackled the problem of spirals: were they separate galaxies, or merely features of our own Milky Way system? There was one way to find out. Most stars shine steadily year after year, century after century, but there are some that do not, so that they brighten and fade over short periods. Among these are the Cepheids (named after the prototype star, Delta Cephei in the far north of the sky) which have periods of from a few days to a few weeks, and are as regular as clockwork. It was known that the period of variation was linked with the star's real luminosity; the longer the period, the more powerful the star. Measure the period, at once you know the star's luminosity – and this gives a clue to its distance.

Using the Hooker, Hubble identified Cepheids in some of the spirals, including M31 in Andromeda. They were found to be so powerful that they had to be a long way away – far outside the Milky Way system; our Galaxy was one of many. Hubble gave the distance of the Andromeda Galaxy as 750,000 light-years; almost all the rest were much more remote. Only the Hooker was powerful enough to show the Cepheids beyond the Milky Way. Hubble, together with his assistant Milton Humason, went on to show that apart from those in our local group, all the galaxies are racing away from us – and the farther away they are, the faster they are receding. The whole universe is expanding.

Hale's constant call was for 'More light!' and his final venture was to plan a 5-metre (200-inch) reflector, which would push the observational horizon out even further. Sadly, he did not live to see the completion of the telescope, in 1948, but he would have been proud of the result. The 5-metre, named after Hale, was supreme for several decades. One early discovery made with it, by the German astronomer Walter Baade, was that the Cepheid variables used as 'standard candles' were twice as luminous as had been believed. This meant that their distances had to be doubled, and the observable universe was twice as large as Hubble had expected.

By the middle of the twentieth century, telescopes such as the Mount Wilson and Palomar reflectors could image objects thousands of millions of light years away. Even so, they were limited in one important respect – and this brings us on to what we might call 'invisible astronomy'.

▼ Mount Palomar Observatory The dome of the 5-m (200-in) Hale Telescope which I photographed in 1997. Palomar is at an altitude of approximately 1700 m (5600 ft).

◀ **Mt Wilson** The 2.5-m (100-in) Hooker telescope at Mount Wilson, completed in 1917. It was important in the discovery of Cepheid variables in galaxies other than our own.

▲ **M31 The Andromeda Galaxy.** Photographed with the 122-cm (48-in) Schmidt telescope at Mt Palomar. The two companion galaxies M32 (left) and NGC 205 (right) are well shown.

M31 is a typical spiral galaxy. It considerably larger than our own galaxy, but is unfortunately positioned at such a narrow angle to us so that the full beauty of the spiral is lost. If it were face-on like the Whirlpool Galaxy (M51) – the beautiful spiral in Canes Venatici accompanied by the satellite galaxy NGC 5195 – it would be even more impressive.

The Radio Sky

The next telescope on our list is different from those already described, because it does not produce a visible image, and you certainly cannot look through it. The radio telescope at Jodrell Bank, in Cheshire, uses a 76.2-metre (250-foot) metallic dish to collect radiation which does not affect our eyes. The telescope – named after Sir Bernard Lovell, who masterminded it – came into operation in 1957, and sparked off a complete revolution in astronomy.

Visible light makes up only a very small part of the whole range of wavelengths, or electromagnetic spectrum. It extends from red light, with wavelength of around 700 nanometres, through to violet, at 400 nanometres (one nanometre is equal to one thousand-millionth of a metre). Radiation with wavelengths outside this range does not affect our eyes, so even our largest optical telescopes are of no help. Shorter than violet, we have ultra-violet, X-rays and gamma rays; beyond red we come to infra-red, microwaves and radio waves. It is easy enough to detect infra-red – switch on an electric fire; you will feel the infra-red, in the form of heat, well before the wires start to glow.

Up to less than a century ago, astronomers were limited, just as a pianist would be handicapped if trying to play a Strauss waltz on a piano which lacked all its notes except those of the middle octave. An entirely new kind of instrument had to be devised. The breakthrough was made in 1931, by accident. The American radio engineer Karl Janksy was carrying out investigations into 'static', on behalf of the Bell Telephone Company, when his equipment detected a strange hiss which at first he could not identify. Before long he realized it came from a definite region of the Milky Way, now known to mark the centre of the Galaxy. Jansky never took his investigations much further, but a few years later in 1937 an amateur, Grote Reber, built the first true radio telescope, a metallic 'dish' 9.5 metres (31 feet) across. For some time Reber was the only radio astronomer in the world.

After the end of the war Bernard Lovell, at the University of Manchester, was one of the scientists who saw the potential of radio astronomy. Radio waves can pass through material which is opaque to light waves, and can obtain information from regions such as the dust-surrounded centre of the Galaxy, for example. Lovell planned a 76.2-metre (250-foot) dish, to collect and focus the long wavelength radiation from the sky, bringing it to focus – but the result is not a picture, but a trace on a graph which may look less impressive, but is equally informative.

Radio astronomy observatories are now to be found in many countries. The Lovell telescope is no longer the world's largest, but was the prototype for those which followed, and it is still in the forefront of research. Of course, not all radio telescopes are dishes; some are more like groups of aerials, and all sorts of techniques have been introduced – such as 'aperture synthesis', which is in effect using a number of small telescopes, moveable on rails, to mimic the aperture of a much larger instrument.

Without radio astronomy we would know nothing about bizarre objects such as pulsars; we would be unable to investigate the murky regions where stars are being formed; we would not have detected the cosmic microwave background, which is of fundamental importance in cosmology. But Jodrell Bank showed the way, and in my view it is fair to regard Bernard Lovell as 'the Isaac Newton of radio astronomy'.

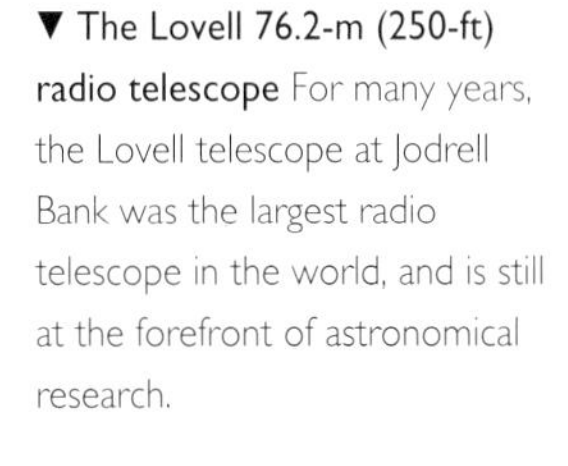

▼ **The Lovell 76.2-m (250-ft) radio telescope** For many years, the Lovell telescope at Jodrell Bank was the largest radio telescope in the world, and is still at the forefront of astronomical research.

◄ **The Very Large Array** (VLA), in New Mexico, is one of the world's premier radio observatories. Its 27 antennae can be arranged into four different Y-shaped configurations. Each antenna is 25 m (82 feet) in diameter, but when the signals are combined electronically the array functions as one giant dish, with the resolution of an antenna 36 km (22 miles) across. In a similar fashion, radio telescopes in different countries, even diferent continents, may be linked together to produce a radio telescope as big as the Earth. This is called Very Long Baseline Interferometry (VLBI).

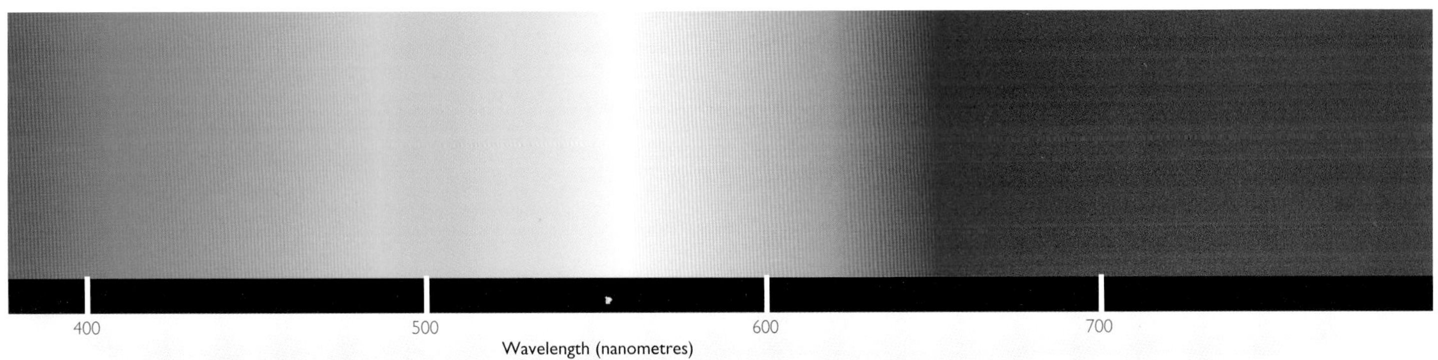

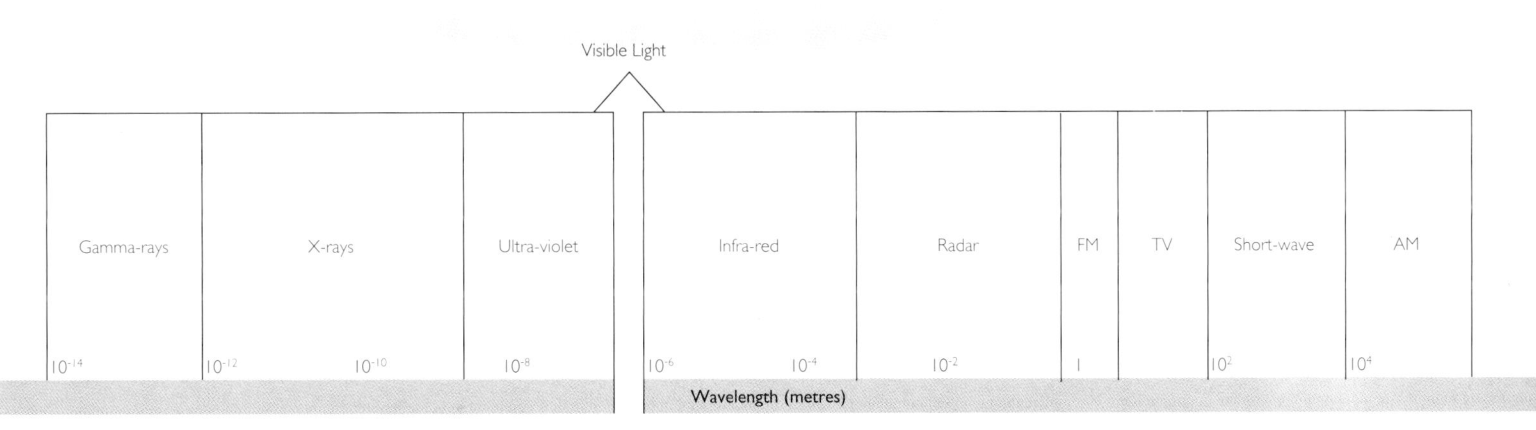

▲ **The electromagnetic spectrum** Astronomy spans the entire electromagnetic spectrum from ultra-short gamma rays to the long radio waves. As can be seen from the diagram, visible light occupies only a very small part of the total range of wavelengths.

Beyond the Atmosphere

▶ **Launch of Ulysses.** Ulysses, a spacecraft designed to survey the poles of the Sun, was launched from Cape Canaveral on 6 October 1990 in the cargo bay of the Space Shuttle Discovery; the probe was made in Europe. The photograph here shows the smoke trail left by the departing Shuttle. I took it with an ordinary camera and was pleased by how well it came out.

We live at the bottom of an ocean of air, which to astronomers is an unmitigated nuisance, partly because it is dirty and unsteady but also because it blocks most of the radiation from space. We are left with only the visible range, plus a few 'windows' – notably in the radio range. Siting observatories at high altitude does help, but it is not a perfect solution. We must send our instruments above the dense layers of atmosphere.

The atmosphere extends upwards for only a limited distance, and above a few hundreds of kilometres we can say that we have reached 'empty space'; admittedly it is not complete vacuum, the density of the material in the space between the worlds of the Solar System is so low that for most purposes it can be ignored. Aircraft cannot fly where there is no air, and the only method is to use rockets, which push against 'themselves' and do not need to have air round them (in fact air is a handicap, because it causes friction and has to be pushed out of the way). Consider a solid-fuel rocket of the kind fired on Bonfire Night (or, in the US, on Independence Day). It consists of a hollow tube filled with gunpowder. When you 'light the blue touch-paper and retire immediately' the gunpowder starts to burn, and hot gas is produced. The gas is forced out through the rocket's exhaust, and propels the tube in the opposite direction, according to what Newton called the Principle of Reaction – 'Every action has an equal and opposite reaction.'

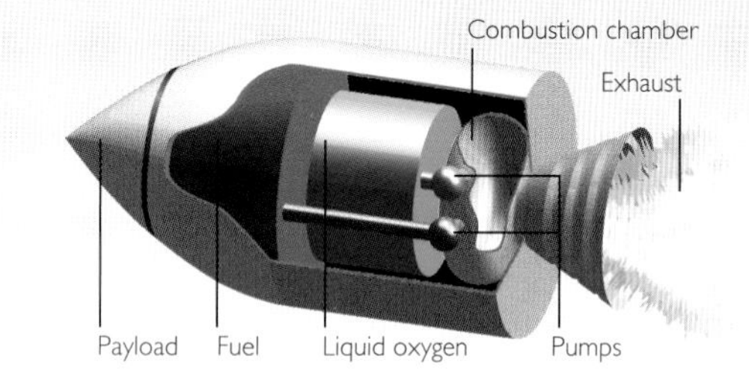

PRINCIPLE OF THE ROCKET
The liquid-propellant rocket uses a 'fuel' and an 'oxidant'; these are forced into a combustion chamber, where they react together, burning the fuel. The gas produced is sent out from the exhaust; and as long as gas continues to stream out, so the rocket will continue to fly. It does not depend upon having atmosphere around it, and is at its best in outer space, where there is no air-resistance to impede its progress.

This is exactly how a space rocket works. Using gunpowder would not be advisable and in a liquid-filled rocket motor two liquids, a fuel and an oxidant, are fed into a combustion chamber, where they ignite and produce the necessary gas. A liquid-fuelled rocket motor is controllable, which is advisable, but a solid-fuelled rocket is not.

The first liquid propellant rocket was launched by the American scientist Robert Hutchings Goddard in 1926. It was a modest affair, rising to only a few tens of feet, but it showed that rockets of this kind were really practicable. During the war, rockets were developed by the Germans, admittedly for military purposes rather than peaceful research, but it has to be admitted that the V2s which bombarded London in 1944 and 1945 were the direct forerunners of the rockets which have taken men to the Moon.

Scientific instruments of all kinds were sent up, and proved to be invaluable. For example, the ultra-short gamma-rays from space cannot pierce our atmosphere, so that gamma-ray astronomy could not begin until detectors were carried aloft. X-ray astronomy was almost as badly handicapped.

A simple rocket can stay up for only a brief period, and can be used only once. Clearly it would be far better to have something which would last for much longer, and in October 1957 the Russians launched the first man-made moon or artificial satellite, Sputnik 1. It was football-sized and carried little apart from a radio transmitter, but as it sped round the Earth, sending out its never-to-be-forgotten 'Bleep, bleep!' signals it marked the start of the true Space Age. It was taken up by rocket and put into an orbit round the Earth. Once there it behaved in exactly the same way that a natural astronomical body would do.

Sputnik 1 was not moving beyond the outer atmosphere, so it was braked by friction and in a few weeks burned away in the upper atmosphere, but since then many hundreds of satellites have been sent up, and today life would seem strange without them. The first transatlantic picture was relayed by a satellite, Telstar, in 1962; without these man-made moons, we in England could never see 'live' a Test Match being played in Australia or a presidential rally being held in New York!

▶ **The V2 weapon.** The V2 was developed during World War II by a German team, headed by Wernher von Braun.

Man in Space

▶ **Saturn V Rocket** The launch of Apollo 11 in 1969 carrying astronauts Armstrong, Aldrin and Collins toward the Moon.

Manned spaceflight began on 12 April 1961, when Major Yuri Gagarin of the Soviet Air Force was launched in the spacecraft Vostok 1 and made a full circuit of the Earth before landing safely in the pre-arranged position. His total flight time was no more than 1 hour 40 minutes, but it was of immense significance, because it showed that true spaceflight could be achieved.

Up to that time nobody was sure about the effects of weightlessness, or zero gravity. Once in orbit, all sensation of weight vanishes, because the astronaut and the spacecraft are in 'free-fall', moving in the same direction at the same rate. (Lie a coin on top of a book, and drop both to the floor; during the descent the coin will not press on the book – with reference to the book, it has become weightless.) In fact, zero gravity did not prove to be uncomfortable. The stage was set for further flights, and these were not long delayed.

The Russians had taken the lead, but the Americans soon followed, with their Mercury programme. All the 'Mercury Seven' made spaceflights (though Deke Slayton had to wait until long after the Mercury missions), and one, Alan Shepard, went to the Moon with Apollo 14 in 1971. Shepard was actually the first American in space; he made a brief sub-orbital 'hop' in 1961.

The first American to orbit the Earth was John Glenn, on 20 February 1962; his flight lasted for 4 hours 55 minutes 23 seconds. His capsule, Friendship 7, was tiny and decidedly cramped. In October 1998 Glenn made his second spaceflight, in the Shuttle Discovery; the contrast between Discovery and Friendship 7 is indeed striking! At the age of 77 Glenn was by far the oldest of all astronauts.

By then there had been many space missions, with elaborate, multi-crewed spacecraft; there had been men and women astronauts from many countries, though the vehicles used were exclusively Russian or American. Inevitably there have been casualties. Two Space Shuttles have been lost; in January 1986 Challenger exploded shortly after launch, and on 1 February 2003 Columbia broke up during re-entry into the atmosphere. Yet, all in all, progress has been amazingly rapid. The first flight in a heavier-than-air machine was made by Orville Wright, in 1903; Yuri Gagarin entered space 58 years later, and only 8 years elapsed between Gagarin's flight and Neil Armstrong's 'one small step' on to the surface of the Moon. It is worth noting that there could have been a meeting between Wright, Gagarin and Armstrong. Their lives overlapped, and I have had the honour of knowing all three.

▼ **Space-walk** Astronaut Piers Sellers in space working on the International Space Station in 2006.

Space Stations

Space stations date back a long way – in fiction – but only in modern times have they become fact. One early post-war design was due to Wernher von Braun, who planned a space wheel; the crew would live in the rim, and rotation of the wheel would simulate gravity for the astronauts. The Von Braun Wheel never progressed beyond the planning stage; it would certainly have been graceful.

The first real space station was the US Skylab, which was manned by three successive crews in 1973–74, and was very successful; a great deal of work was carried out. It remained in orbit until 11 July 1979, when it re-entered the atmosphere and broke up, showering fragments widely over Australia – fortunately without causing any damage or casualties.

Then came Soviet Salyut space stations and later Mir, launched on 20 February 1986. It remained in orbit until 23 March 2001, when it was deliberately brought down into the Pacific. For most of the time between 1986 and 2000 it was inhabited and research of all kinds was undertaken. Problems arose during its last few years, after the end of its planned lifetime, but it was an outstanding success. It was visited by astronauts from many nations (including Britain). Without it, the setting-up of the International Space Station (ISS) would have been far more difficult.

The ISS is being assembled in orbit, more than 350 kilometres (220 miles) above the Earth. In-orbit assembly began on 20 November 1998, with the launch of the Russian-built Zarya (Sunrise) control module; the station was scheduled to be complete by 2004, but it is much delayed. It is truly international, and crew members are changed regularly; flights to and from it, in Russian Soyuz craft and in Space Shuttles, have become routine. Research covers all fields of science, and the ISS has ushered in the new era of space research.

▼ **Skylab.** The first true space station; during 1973–74 it was manned by three successive crews. It continued in orbit until 11 July 1979, when it broke up in the atmosphere.

▲ **The International Space Station.** The latest space station, and the largest manmade object to orbit the Earth. Its construction is not yet complete, but it is now permanently manned.

◄ **Mir in orbit.** The core module, known as the base block, was launched on 20 February 1986. Modules were added to the core in 1987, 1989, 1990 and 1995. It re-entered Earth's atmosphere in March 2001.

Orbiting Telescopes

▼ **Star cluster NGC 602.** Seen by the Hubble Space Telescope near the outskirts of the Small Magellanic Cloud, our neighbouring galaxy, lies this extraordinary region of ridges and swept back shapes that suggest shock waves from the young stars have eroded the dusty material and triggered a progression of star formation.

◀ **Launch of the last of the Great Observatories Program telescopes.** NASA's Spitzer Space Telescope, then called the Spitzer Infra-red Telescope Facility (SIRTF), aboard a Delta II heavy launch vehicle from Cape Canaveral, Florida on 25 August 2003.

Many telescopes have now been sent into orbit. All have had specific tasks; for instance Hipparcos (launched 8 August 1989) was to make very accurate measurements of star positions, while IRAS, the Infra-Red Astronomical Satellite (26 January 1983) searched for cool material and did indeed find clouds of probable planet-forming material around some stars, such as the brilliant Vega. But it was the Hubble Space Telescope (HST), named after the famous American astronomer, which causes the greatest amount of popular interest. It was launched on 24 April 1990, and as I write these words (March 2007) it is still in orbit, sending back magnificent pictures. Its mirror is 2.4 metres (94 inches) across; the telescope moves round Earth at a range of almost 600 kilometres (370 miles) in a period of 94 minutes. It is easily visible to the naked eye, in the guise of a bright star crawling across the sky; in fact it is 13 metres (43 feet) long. I wonder how many millions of people have seen it?

As soon as HST was in orbit, NASA made the devastating discovery that the mirror was faulty. It was slightly too shallow a curve; the error was only 0.002 of a millimetre (1/50 of the thickness of a human hair), but it was enough to ruin the clarity of the images. The only remedy was to send up a Shuttle, so that the astronaut crew could bring the telescope into the payload bay, repair it and then re-launch it. This was managed; there have been several servicing missions, and in the end the telescope's performance was even better than had originally been expected. Seeing conditions are perfect, and moreover the range extends well into the ultra-violet part of the electromagnetic spectrum. Though the HST has a mirror much smaller than some Earth-based telescopes, its pictures have been unrivalled – ranging from surface details on Pluto to galaxies thousands of millions of light-years away, not too far from the boundary of the observable universe.

Of course there have been many other important orbital telescopes, notably XMM-Newton (10 December 1999) which like the Chandra Observatory (23 July 1999) is concerned with X-ray research, and the Spitzer Space Telescope (25 August 2003) which operates in the infra-red and produces images comparable with those of the HST. It is wrong to suggest that orbiting telescopes will supersede Earth-anchored instruments; the two are complementary, but certainly space telescopes are of vital importance in modern astronomy.

It has not been easy to select telescopes which have led to fundamental advances in our knowledge. Many people will disagree with this list, but I give it for what it is worth:

PIONEERING TELESCOPES	
1610	Galileo's first reflector
1671/2	Newton's first reflector
1798	Herschel's '40-ft' reflector
1845	Lord Rosse's 183-cm (72-inch) reflector at Birr Castle
1896	The Yerkes 102-cm (40-in) refractor, in the USA
1917	The Mount Wilson 2.5-m (100-in) reflector
1948	The Palomar 5-m (200-in) reflector
1957	The Lovell 72-m (250-ft) radio telescope
1990	The Hubble Space Telescope
2000	The last mirror of the VLT, in Chile
2003	The Spitzer Space Telescope
2020	Time will tell ?

◀ **Hubble Space Telescope in orbit.** Throughout its long and extraordinarily productive lifespan the HST has produced many beautiful images, but those of the telescope itself floating above our blue planet are as stunning as many of its famous astronomical subjects.

▼ **The Tarantula Nebula (NGC2070, 30 Doradûs)** as imaged by the Spitzer Space Telescope.

Saturn from Cassini, in a spectacularly detailed mosaic of 126 images obtained over two hours on 6 October 2004. The spacecraft was about 6.3 million kilometres (3.9 million miles) from the planet, and the smallest features visible are 38 kilometres (24 miles) across.

THE SOLAR SYSTEM

The Sun's Family

The Solar System is the only part of the universe which we can explore with spacecraft of the kind we can build today. It is made up of one star (the Sun), eight planets (of which the Earth comes third in order of distance from the Sun), and various lesser bodies, such as the satellites, asteroids, comets and meteoroids.

The Sun is a normal star (astronomers even relegate it to the status of a dwarf), but it is the supreme controller of the Solar System, of which all the other members shine by reflected sunlight. It is believed that the planets formed by accretion from a cloud of material which surrounded the youthful Sun; the age of the Earth is known to be about 4.6 thousand million years, and the Solar System itself must be rather older than this.

It is easy to see that the Solar System is divided into two definite parts. There are four rocky planets: Mercury, Venus, the Earth and Mars, beyond which comes a swarm of much smaller bodies known as asteroids. These are followed by the four giants: Jupiter, Saturn, Uranus and Neptune. Beyond Neptune we find another swarm of minor bodies, making up what is known as the Kuiper Belt (in honour of the Dutch astronomer Gerard Kuiper, who suggested its existence). The first Kuiper Belt Object to be discovered was Pluto, found in 1930; it was originally classed as a planet, but we now know that it is not even the largest member of the swarm. Some go out to great distances; Sedna has a period of 12,300 years, and at aphelion is 900 astronomical units (AU) from the Sun.

The four inner planets were formed close to the Sun, so that their light gases were driven away, leaving only their rocky cores; the giants, formed farther out in a colder environment, could retain their light gases – mainly hydrogen – and could grow to immense sizes. Jupiter, by far the largest of them, has a volume 1319 times that of the Earth, and is more massive than all the other planets in the Solar System combined.

The Earth has one satellite: our familiar Moon, which is much the closest natural body in the sky (excluding occasional wandering asteroids). Of the other planets, Mars has two satellites, Jupiter has over 60, Saturn about the same number, Uranus 27 and Neptune 13. However, most of these are very small and probably ex-asteroids; only four planetary satellites (three in Jupiter's system, one in Saturn's) are larger than our Moon.

Comets may be spectacular (as Comet Hale-Bopp was, in 1997), but are of very low mass. The only substantial part of a comet is the nucleus, which has been described as a 'dirty ice-ball'. When a comet nears the Sun the ices begin to evaporate, and the comet may produce a gaseous head (known as a coma), with a long tail. Bright comets have very eccentric orbits, so they come back to the inner part of the Solar System only at intervals of many centuries, and we cannot predict them. There are many short-period comets which return regularly, but all of these are faint; each time a comet passes relatively close to the Sun it loses a certain amount of material, and the short-period comets have to a great extent wasted away.

As a comet moves along it leaves a 'dusty trail' behind it. If the Earth ploughs through one of these trails it collects dust particles, which burn away in the upper air and produce the luminous streaks which we call meteors (or 'shooting stars'). Larger objects, which may survive the fall to the ground, are termed meteorites; they come from the asteroid belt, and are not associated either with comets or with meteors.

Jupiter

Mars

Earth

Venus

Mercury

Asteroid belt

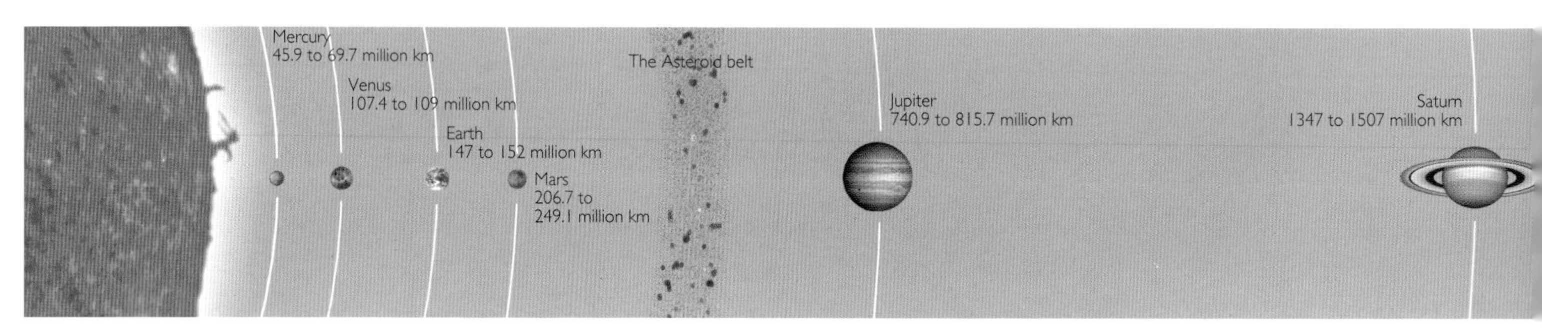

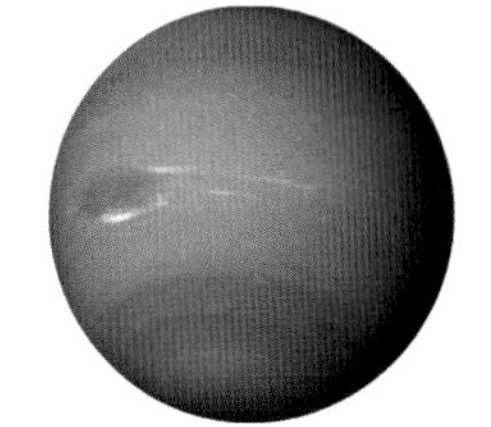

Neptune

How far does the Solar System extend? This is not an easy question to answer. It is possible that there is another planet beyond Neptune and the Kuiper Belt, and it is thought that comets come from a cloud of icy objects orbiting the Sun at a distance of around one to two light years, but we cannot be sure. The nearest star beyond the Sun is just over four light years away, so that if we give the limit of the Solar System as being at a distance of two light years we are probably not very far wrong.

Uranus

Saturn

PLANETARY DATA

		Mercury	Venus	Earth	Mars	Jupiter	Saturn	Uranus	Neptune
Distance from	max.	69.7	109	152	249	816	1507	3004	4537
Sun, millions	mean	57.9	108.2	149.6	227.9	778	1427	2870	4497
of km	min.	45.9	107.4	147	206.7	741	1347	2735	4456
Orbital period		87.97d	224.7d	365.3d	687.0d	11.86y	29.46y	84.01y	164.8y
Synodic period, days		115.9	583.92	—	779.9	398.9	378.1	369.7	367.5
Rotation period (equatorial)		58.646d	243.16d	23h 56m 04s	24h 37m 23s	9h 50m 30s	10h 13m 59s	17h 14m	16h 7m
Orbital eccentricity		0.206	0.007	0.017	0.093	0.048	0.056	0.047	0.009
Orbital inclination, °		7.0	3.4	0	1.8	1.3	2.5	0.8	1.8
Axial inclination, °		2	178	23.4	24.0	3.0	26.4	98	28.8
Escape velocity, km/s		4.25	10.36	11.18	5.03	60.22	32.26	22.5	23.9
Mass, Earth = 1		0.055	0.815	1	0.11	317.9	95.2	14.6	17.2
Volume, Earth = 1		0.056	0.86	1	0.15	1319	744	67	57
Density, water = 1		5.44	5.25	5.52	3.94	1.33	0.71	1.27	1.77
Surface gravity, Earth = 1		0.38	0.90	1	0.38	2.64	1.16	1.17	1.2
Average surface temp., °C		+427	+480	+22	−23	−150*	−180*	−214*	−220*
Albedo		0.06	0.76	0.36	0.16	0.43	0.61	0.35	0.35
Diameter, km (equatorial)		4878	12,104	12,756	6794	143,884	120,536	51,118	50,538
Maximum magnitude		−1.9	−4.4	—	−2.8	−2.6	−0.3	+5.6	+7.7

* For the gas giants, 'average surface temp'. refers to the cloud-top temperature

Uranus
2735 to 3004 million km

Neptune
4456 to 4537 million km

Kuiper Belt

The Earth in the Solar System

▼ **Planet Earth**, seen from the Command Service Module (CSM) of the lunar spacecraft Apollo 10 in May 1969. The Earth is just coming into clear view as the spacecraft moves out from behind the far side of the Moon. The lunar horizon is sharp: this is because there is no atmosphere to cause blurring.

Why do we live on the Earth? The answer must be: 'Because we are suited to it'. There is no other planet in the Solar System which could support Earth-type life except under very artificial conditions. Our world has the right sort of temperature, the right sort of atmosphere, a plentiful supply of water and a climate which is to all intents and purposes stable – and has been so for a very long time.

The Earth's path round the Sun does not depart much from the circular form, and the seasons are due to the tilt of the rotational axis, which is 23.4° to the perpendicular. We are actually closer to the Sun in January, when it is winter in the northern hemisphere, than in July – but the difference in distance is not really significant, and the greater amount of water south of the equator tends to stabilize the temperature.

The axial inclination varies to some extent, because the Earth is not a perfect sphere; the equatorial diameter is 12,756 kilometres (7927 miles), the polar diameter only 12,714 kilometres (7901 miles) – in fact, the equator bulges out slightly. The Sun and Moon pull on this bulge, and the result is that over a period of 25,800 years the axis sweeps out a cone of angular radius about 23.4° around the perpendicular to the plane of the Earth's orbit. Because of this effect – termed precession – the positions of the celestial poles change. At the time when the Egyptian pyramids were built, the north pole star was Thuban in the constellation of Draco; today we have Polaris in Ursa Minor and in 12,000 years from now the pole star of the northern hemisphere will be the brilliant Vega, in Lyra.

We have found out a great deal about the history of the Earth. Its original atmosphere was stripped away, and was replaced by a secondary atmosphere which leaked out from inside the globe. At first this new atmosphere contained much more carbon dioxide and much less free oxygen than it does now, so we would have been quite unable to breathe it. Life began in the sea; and by the time plants spread on to the lands around 430 million years ago, much of the carbon dioxide had already been removed from the atmosphere, and plants on the land continued this process.

Life was slow to develop, as we know from studies of fossils; we can build up a more or less complete geological record, and it has been found that there were several great 'extinctions', when many life-forms died out. One of these occurred about 65 million years ago, when the dinosaurs died out – for reasons which are still not clear, though it has been suggested that the cause was a major climatic change caused by the impact of a large asteroid. In any case, humans are newcomers to the terrestrial scene. If we give a time-scale in which the total age of the Earth is represented by one year, the first true people will not appear until 11 pm on 31 December.

Throughout the Earth's history there have been various cold spells or Ice Ages, the last of which ended only 10,000 years ago. In fact, the last Ice Age was not a period of continuous glaciation; there were several cold spells interrupted by warmer periods, or 'interglacials', and it is by no means certain that we are not at the moment simply in the middle of an interglacial. The reasons for the Ice Ages is not definitely known, and may be somewhat complex, but we have to remember that even though the Sun is a steady, well-behaved star its output is not absolutely constant; in historical times there have been marked fluctuations – for example, the so-called 'little ice age' between 1645 and 1715, when the Sun was almost free of spots and Europe, at least, was decidedly colder than it is at the present moment. At present (2007) we are experiencing a period of global warming, as has happened often enough in the past; the Sun is, to a minor extent, a variable star. Warming will probably continue for several decades yet, and will be followed by a period of cooling.

Neither can the Earth exist for ever. Eventually the Sun will change; it will swell out to become a giant star, and the Earth will certainly be destroyed. Luckily there is no immediate cause for alarm. The crisis will not be upon us for several thousands of millions of years yet, and it is probably true to say that the main danger to the continued existence of life on Earth comes from ourselves.

▲ **The Blue Marble** NASA's detailed true-colour image is compiled from a number of satellite based observations. The global weather systems are beautifully shown.

◄ **Image taken from the Mir space station;** in the foreground, seen against the Earth, is the nose of the Space Shuttle Atlantis.

The Earth as a Planet

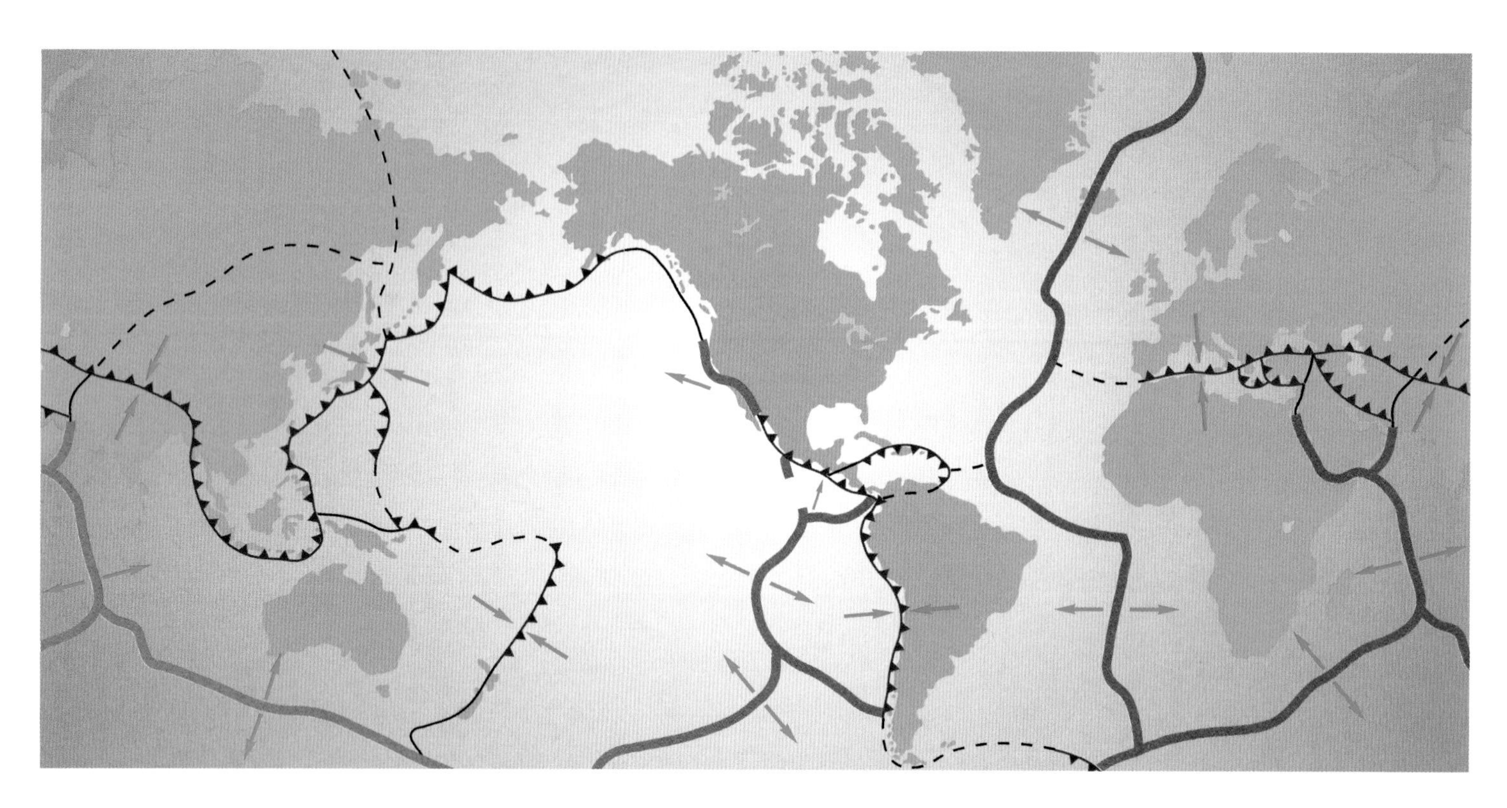

Divergent plate boundaries

Convergent plate boundaries

Uncertain plate boundaries

Direction of plate movement

▲ **The Earth's crust** is divided into six large tectonic plates and a number of minor ones. They are separated by mid-ocean ridges, deep-sea trenches, active mountain belts and fault zones. Volcanic eruptions and earthquakes are largely confined to the areas where plates meet. During the geological history of the Earth, these plates have moved around, creating and re-creating continents.

The Earth's crust, on which we live, does not extend down very far – some 10 kilometres (6 miles) below the oceans and 50 kilometres (30 miles) below the continents. Temperature increases with depth, and at the bottom of the world's deepest mines, those in South Africa, the temperature rises to 55°C. Below the crust we come to the mantle, where the solid rocks behave as though plastic. The mantle extends down to 2900 kilometres (1800 miles), and then we come to the iron-rich liquid outer core, which accounts for c.30 per cent of the Earth's mass. Inside this is the solid core, which accounts for only 1.7 per cent of the Earth's mass and has been said to 'float' in the liquid. The central temperature is thought to be 4000–5000°C.

A glance at a world map shows that if the continents could be cut out in the manner of a jigsaw puzzle, they would fit neatly together. For example, the bulge on the east coast of South America fits into the hollow of west Africa. This led the Austrian scientist Alfred Wegener to suggest that the continents were once joined together and have now drifted apart. His ideas were ridiculed for many years, but the concept of 'continental drift' was accepted years ago, and has led to the science of plate tectonics, which explains the mechanism.

The Earth's lithosphere (the crust and the upper part of the mantle) is divided into well-marked plates. When plates are moving apart, hot mantle material rises up between them to form new oceanic crust. When plates collide, one plate may be forced beneath another – a process known as subduction – or they may buckle and force up mountain ranges. Regions where the tectonic plates meet are subject to earthquakes and volcanic activity, and it is from earthquake waves that we have drawn much of our knowledge of the Earth's internal constitution.

The point on the Earth's surface vertically above the origin or 'focus' of an earthquake is termed the epicentre. Several types of waves are set up in the globe. First there are the P or primary waves, which are waves of compression and are often termed 'push' waves; there are also S or secondary waves, which are also called 'shake-waves' – they may be likened to the transverse waves set up in a mat when it is shaken by one end and cause more d damage. Finally there are the L or long waves, which travel round the Earth's surface. The P waves can travel through liquid, but the S waves cannot, and by studying how they are transmitted through the Earth it has been possible to measure the size of the Earth's liquid core.

If earthquakes can be destructive, then so can volcanoes. The mantle, below the crust, contains pockets of magma (hot, fluid rock), and above a weak point in the crust the magma may force its way through, building up a volcano. When the magma reaches the surface it solidifies and cools, to become lava. Hawaii provides perhaps the best example of long-continued vulcanism. On the main island there are two massive shield volcanoes, Mauna Kea and Mauna Loa, which are actually loftier than Everest, though they do not rise so high above sea level because, instead of rising above the land, they have their roots deep in the ocean-bed. Because the crust is shifting over the mantle, Mauna Kea has moved away from the 'hot spot' and has become extinct – at least, one hopes so, because one of the world's major observatories has been built upon its summit. Mauna Loa now stands over the 'hot spot', and is very active indeed, though in time it too will be carried away and will cease to erupt.

Other volcanoes, such as Vesuvius in Italy, are cone-shaped. The magma forces its way up through a vent, and if this vent is blocked the pressure may build up until there is a violent explosion – as happened in AD 79, when the Roman towns of Pompeii and Herculaneum were destroyed. There have been many devastating volcanic eruptions, one of the latest being that of Mount Pinatubo in the Philippines, which sent vast quantities of dust and ash into the upper atmosphere in 1991. The world's southernmost active volcano is Mount Erebus, in Antarctica.

The Earth is not the only volcanic world in the Solar System. There are constant eruptions upon Io, one of the satellites of Jupiter; there are probably active volcanoes on Venus, and we cannot be certain that all the Martian volcanoes are extinct. However, it does not seem that plate tectonics can operate upon any other planet or satellite, so that in this respect the Earth is unique in our experience.

◀ **Division of the plates.** Aerial photograph over Iceland, near Reykjavik. Magma does not just reach the Earth's surface where plates are converging. Iceland is on part of the mid-Atlantic Ridge, where the sea-floor is being pulled apart by the movement of the plates. Lava emerges from long fissures in the ground to form new crust. A rift forms as the ground is pulled farther apart and eventually the crust will weaken enough for another eruption to take place. Photograph by Christine Wheeler, 2007.

▼ **Seismic activity** has allowed scientists to study the inner structure of the Earth. The crust is only on average 10 km (6 mi) thick beneath the oceans and 50 km (30 mi) thick beneath the land. Below is the 2900-km (1800-mi) thick mantle of hot, plastic rock. Inside that is an outer liquid core, 2100 km (1300 mi) thick, with a solid core inside it, 2700 km (1700 mi) in diameter.

VOLCANOES

Volcanoes form where tectonic plates meet. Pockets of magma force themselves up from the mantle through weak points in the crust. The molten magma may bubble inside the crater or give off clouds of ash and gas.

Magma may also find its way to the surface via side vents. A volcano may be inactive for a considerable time, allowing the magma to solidify near the surface. Huge pressure can then build up beneath it, often with devastating results.

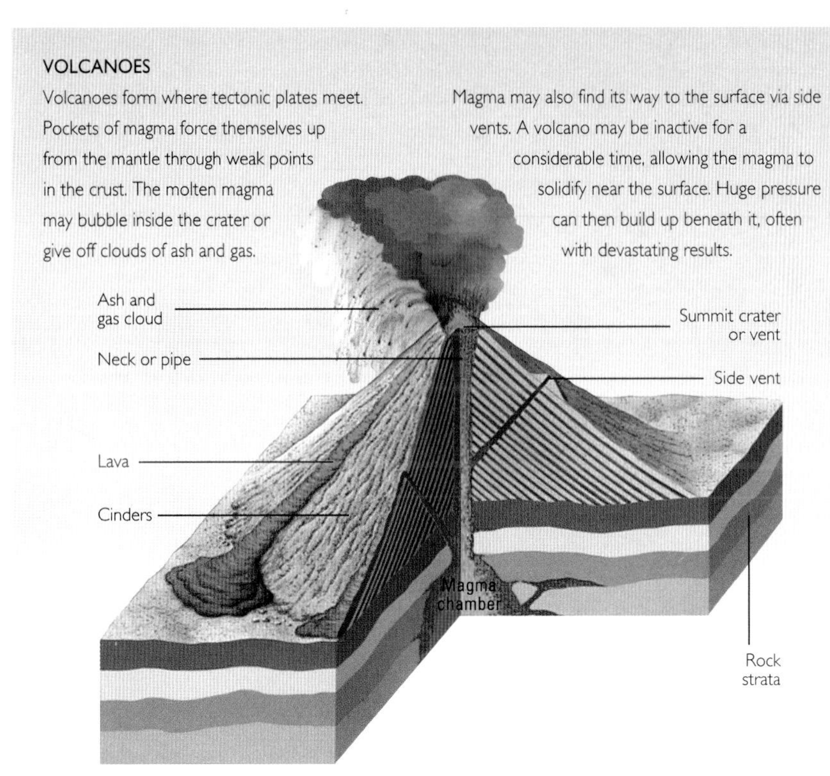

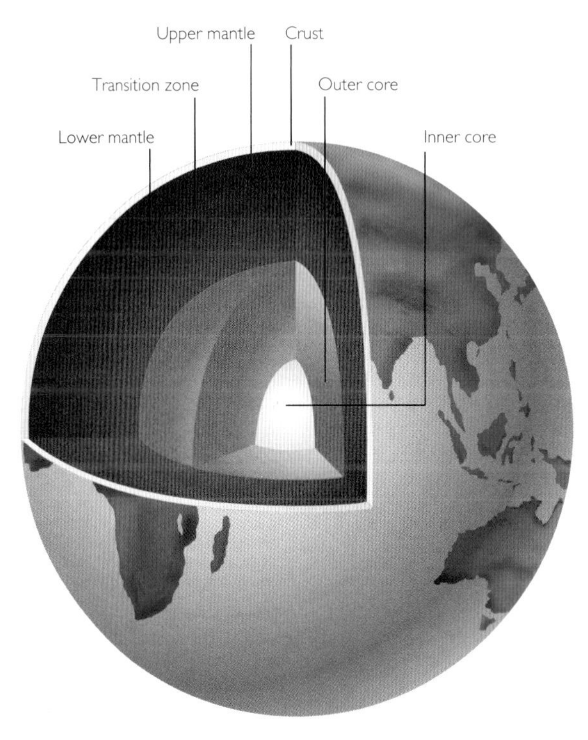

The Earth's Atmosphere and Magnetosphere

As seen from space the Earth is truly magnificent, as we have been told by all the astronauts – particularly those who have observed it from the Moon, although it contrary to some claims, is quite impossible to see features such as the Great Wall of China! The outlines of the seas and continents show up clearly, and there are also clouds in the atmosphere, some of which cover wide areas.

The science of meteorology has benefited greatly from space research methods, because we can now study whole weather systems instead of having to rely upon reports from scattered stations. The atmosphere is made up chiefly of nitrogen (78 per cent) and oxygen (21 per cent), which does not leave much room for anything else; there is some argon, a little carbon dioxide, and traces of gases such as krypton and xenon, together with a variable amount of water vapour.

The atmosphere is divided into layers. The lowest of these, the troposphere, extends upwards between about 8 kilometres (5 miles) and more than 17 kilometres (over 10 miles) – it is deepest over the equator. It is here that we find clouds and weather. The temperature falls with increasing height, and at the top of the layer has dropped to –55°C; the density is, of course, very low.

Above the troposphere comes the stratosphere, which extends up to about 50 kilometres (30 miles). Surprisingly, the temperature does not continue to fall; indeed it actually rises, reaching –10°C at the top of the layer. This is because of the presence of ozone, the molecule of which is made up of three oxygen atoms instead of the usual two in molecular oxygen; ozone is warmed by short-wave radiation from the Sun. However, the rise in temperature does not mean increased heat. Scientifically, temperature is defined by the rate at which the atoms and molecules fly around; the greater the speeds, the higher the temperature. In the stratosphere, there are so few molecules that 'heat' is negligible. The 'ozone layer' shields us from harmful radiations coming from space. Whether it is being damaged by our own activities is a matter for debate, but the situation needs to be watched.

Above the stratosphere comes the mesosphere and the thermosphere which together comprise the ionosphere, which extends from about 50 to 600 kilometres (30 to 370 miles); it is here that some radio waves are reflected back to the ground, making long-range communication possible. In the ionosphere we find the lovely noctilucent clouds, which are quite unlike ordinary clouds and may possibly be caused

▼ **Aurora** A bright aurora visible over the city of Tromsø. The green colour indicates the presence of excited oxygen atoms. Pete Lawrence.

by water droplets condensing as ice on to meteoritic particles; their average height is around 80 kilometres (50 miles). The ionosphere is often divided into the mesosphere, up to 80 kilometres (50 miles), and the thermosphere, up to 200 kilometres (125 miles). Beyond comes the exosphere, which has no definite boundary, but simply thins out until the density is no more than that of the interplanetary medium. There is also the Earth's geocorona, a halo of hydrogen gas which extends out to about 95,000 kilometres (60,000 miles).

Aurorae, or polar lights – aurora borealis in the northern hemisphere and aurora australis in the southern – are also found in the ionosphere; the usual limits are from 100 to 700 kilometres (60 to 440 miles), though these limits may sometimes be exceeded. Aurorae are seen in various forms: glows, rays, bands, draperies, curtains and 'flaming patches'. They change very rapidly and can be extremely brilliant. They are caused by electrified particles from space, mainly originating in the Sun, which collide with atoms and molecules in the upper atmosphere and make them glow. Because the particles are electrically charged, they tend to cascade down around the magnetic poles, so aurorae are best seen from high latitudes. They are very common in places such as Alaska, northern Norway, northern Scotland and Antarctica, but are much rarer from lower latitudes such as those of southern England, though there are occasional brilliant displays – as in 1989, 2001 and 2003. From the equator they are hardly ever seen. Auroral activity is more or less permanent around the so-called auroral ovals, which are 'rings' positioned asymmetrically round the magnetic poles. When there are violent disturbances in the Sun, producing high-speed particles, the ovals broaden and expand, producing displays farther from the main regions. Aurorae have been known for many centuries. The Roman emperor Tiberius, who reigned from AD 14 to 37, once dispatched fire engines to the port of Ostia because a brilliant red aurora led him to believe that the whole town was ablaze.

The Earth has a strong magnetic field. The region over which this field is dominant is called the magnetosphere; it is shaped rather like a tear-drop, with its tail pointing away from the Sun. On the sunward side of the Earth, it extends to about 65,000 kilometres (40,000 miles), but on the night side it spreads out much farther.

Inside the magnetosphere there are two zones of strong radiation; they were detected by the first successful American satellite, Explorer 1 in February 1958, and are known as the Van Allen zones, in honour of the scientist who designed the equipment. There are two main zones, one with its lower limit at just under 8000 kilometres (5000 miles) and the other reaching out to 37,000 kilometres (23,000 miles). The inner belt, composed chiefly of protons, dips down towards the Earth's surface over the South Atlantic, because the Earth's magnetic field is offset from the planet's axis of rotation, and this 'South Atlantic Anomaly' presents a distinct hazard to sensitive instruments carried in artificial satellites.

It cannot be said that we understand the Earth's magnetic field completely, and there is evidence of periodic reversals, as well as changes in intensity. At least it is certain that the field is caused by currents in the iron-rich liquid core. Incidentally, it is worth noting that the Moon and Venus have no detectable magnetic fields and that of Mars is extremely weak, though that of Mercury is stronger than might have been expected. Magnetically, Earth is quite unlike the other inner planets.

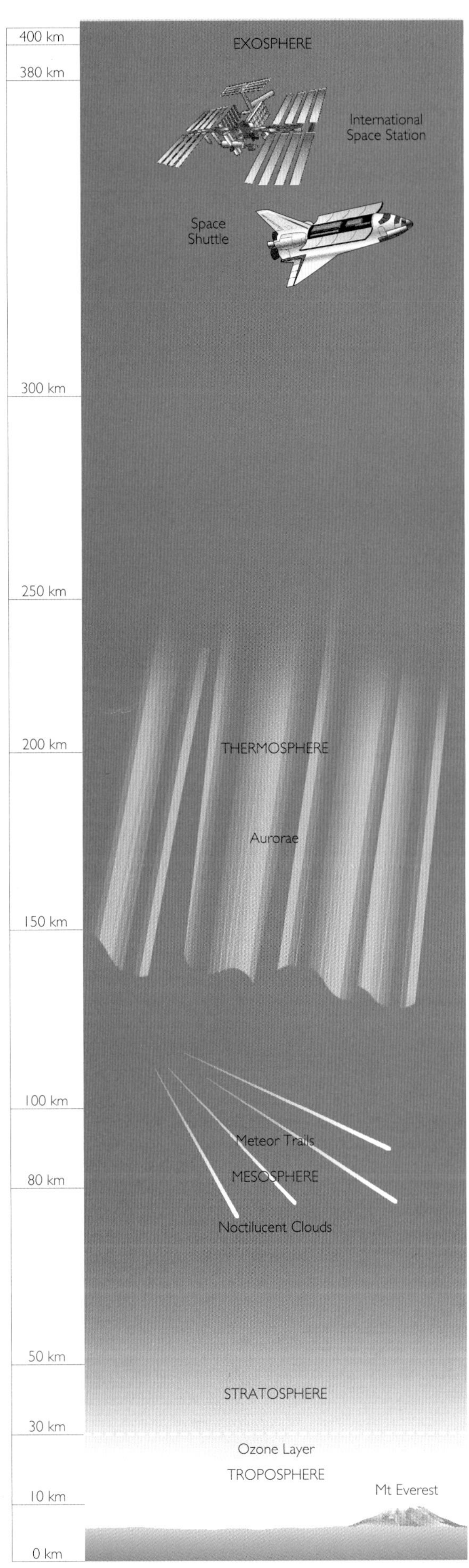

◀ **The Earth's atmosphere** consists of the troposphere, extending from ground level to a height of between 8 and 17 km (5–10 mi); the stratosphere extends up to around 50 km (30 mi); the mesosphere, between 50 and around 80 km (30 to 50 mi); the thermosphere from around 80 up to 200 km (50 to 125 mi); beyond this height lies the exosphere.

The Earth–Moon System

▼ **Earth and Moon** in a single frame, the first of its kind ever taken by a spacecraft – Voyager 1 on September 7, 1977 from a distance of 11.66 million km (7.25 million mi).

The Moon is officially classed as the Earth's satellite, but in many ways it may be better to regard the Earth-Moon system as a double planet; the mass ratio is 81 to 1, whereas for example Titan, the largest satellite of Saturn, has a mass only $\frac{1}{4150}$ that of Saturn itself – even though Titan is considerably larger than our Moon.

We are by no means certain about the origin of the Moon. The attractive old theory according to which it simply broke away from the Earth, leaving the hollow now filled by the Pacific Ocean, has long been discounted. It may be that the Earth and the Moon were formed together from the solar nebula, but there is increasing support for the idea that the origin of the Moon lies in a collision between the Earth and a large wandering body, so that the cores of the Earth and the impactor merged, and debris from the Earth's mantle, ejected during the collision, formed a temporary ring round the Earth from which the Moon subsequently built up. The Earth's mantle is much less massive than its core, and this theory would explain why the Moon is not so dense as the Earth; moreover, analyses of the lunar rocks show that the Moon and the Earth are of about the same age.

It is often said that 'the Moon goes round the Earth'. In a way this is true. To be strictly accurate, however, the two bodies move together round their common centre of mass, or barycentre; however, since the barycentre lies deep inside the Earth's globe, the simple statement is good enough for most purposes.

The orbital period is 27.3 days, and everyone is familiar with the phases, or apparent changes of shape, from new to full. When the Moon is in the crescent stage, the 'dark' side may often be seen shining faintly. There is no mystery about this; it is caused by light reflected on to the Moon from the Earth, and is therefore known as earthshine. It can be quite conspicuous. Note, incidentally, that because the Earth and the Moon are moving together round the Sun, the synodic period (that is to say, the interval between one New Moon and the next) is not 27.3 days, but 29.5 days.

The Moon's axial rotation period is equal to its orbital period. This is due to tidal friction over the ages. During its early history, the Moon was much closer to the Earth than it is now, and the Earth's rotation period was shorter; even today the 'day' is becoming longer, while the Moon is being driven outwards from the Earth. However, these effects are very slight. The Moon's distance is increasing at a rate of less than 4 centimetres (1.5 inches) per year.

▼ **The New Moon** (1 and 9) occurs when the Moon is closest to the Sun. In the Crescent Moon (2), Mare Crisium is prominent between the eastern limb and the terminator. Earthshine is often seen.

The Half Moon, First Quarter (3) reveals Mare Serenitatis with the great chain of craters near the central meridian. Since the Sun is still low over the area that can be seen, the features are well defined.

The Gibbous Moon (4) reveals the great ray-craters Tycho and Copernicus. Although the craters are well illuminated and readily identifiable, their spectacular rays are not yet as striking as they will soon become.

The Full Moon (5). There are no shadows, and the rays from Tycho and Copernicus are so prominent that crater identification becomes difficult. The lunar maria take on a decidedly dark hue against the brilliant rays.

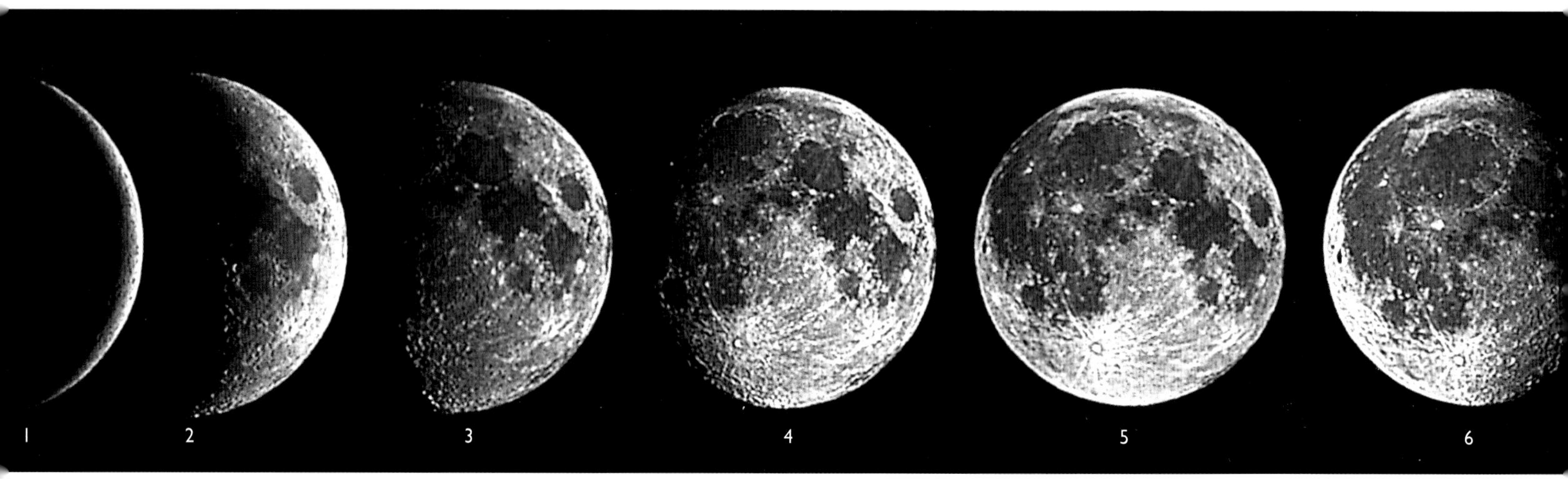

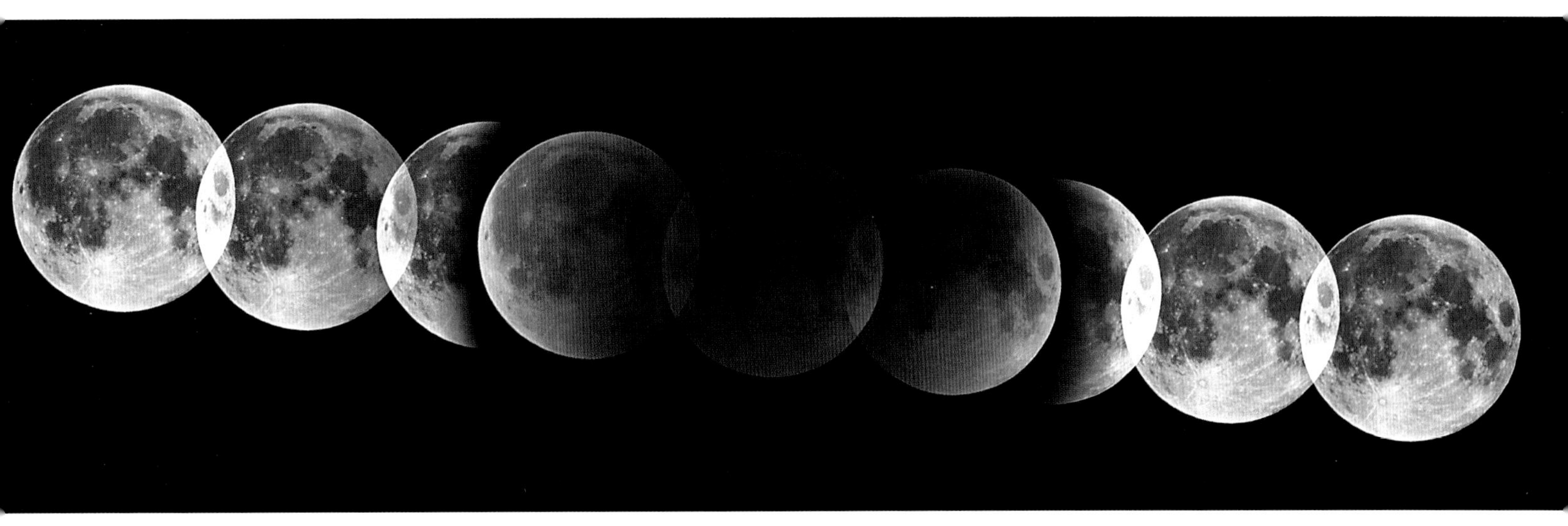

The captured or 'synchronous' rotation of the Moon means that one part of it is always turned away from the Earth, so until 1959, when the Soviets sent their probe Lunik 3 on a 'round trip', we knew nothing definite about it. In fact it has proved to be basically similar in appearance to the region we have always known, though there are some significant differences; for example, the far side has only one major 'sea', the Mare Orientale, and there are no mountains comparable to the Apennines. The rotation has been synchronous since a fairly early stage in the history of the Earth-Moon system.

The Moon has a crust, a mantle and a core. There is a loose upper layer, termed the regolith, from 1 to 20 metres (3 to 65 feet) deep; below comes a layer of shattered bedrock about 1 kilometre (0.6 miles) thick, and then a layer of more solid rock going down to about 25 kilometres (15 miles). Next comes the mantle, and finally the core, which is metal-rich and is probably between 1000 and 1500 kilometres (600 to 930 miles) in diameter. The core is hot enough to be molten, though the central temperature is much less than that of the Earth.

The Moon's low escape velocity means that it retains little atmosphere. A trace remains, and was detected by instruments taken to the Moon by Apollo 17 in 1972; the main constituents are helium (due to the solar wind) and argon (seeping out from below the crust). The atmosphere seems to be in the form of a collisionless gas, and the total weight of the entire atmosphere can be no more than about 30 tons. The density is of the order of around 10^{-14} that of the Earth's.

▲ **The Moon** does not generally disappear during eclipses, for some sunlight is refracted on to it by the Earth's atmosphere. Some eclipses are 'bright', with beautiful colours, others are dark, and records exist of the Moon vanishing completely, usually because of dust or volcanic ash in the Earth's upper air. A bright eclipse is shown in this multi-exposure sequence by Akira Fujii.

LUNAR DATA

Distance from Earth, centre to centre:	
max. (apogee)	406,697 km (252,681 mi)
mean	384,400 km (238,828 mi)
min. (perigee)	356,410 km (221,438 mi)
Orbital period	27.321661 days
Axial rotation period	27.321661 days
Synodic period (interval between successive New Moons)	29d 12h 44m 3s
Mean orbital velocity	3680 km/h (2286 mi/h)
Orbital inclination	5° 9'
Apparent diameter:	max. 33' 31"
	mean 31' 6"
	min. 29' 22"
Density, water = 1	3.34
Mass, Earth = 1	0.012
Volume, Earth = 1	0.020
Escape velocity	2.38 km/s (1.48 mi/s)
Surface gravity, Earth = 1	0.165
Albedo	0.07
Mean magnitude at Full:	−12.7
Diameter	3476.6 km (2160 mi)

The Waning Moon (6). This is not as brilliant as the waxing Gibbous Moon. More of the dark maria which were once thought to be seas are illuminated. They are, in fact, gigantic plains of volcanic lava.

The Half Moon, Last Quarter (7). The rays are less striking; shadows inside the large craters are increasing. The Old Moon (8) occurs just before the New, seen in the dawn sky. Earthshine may often be seen.

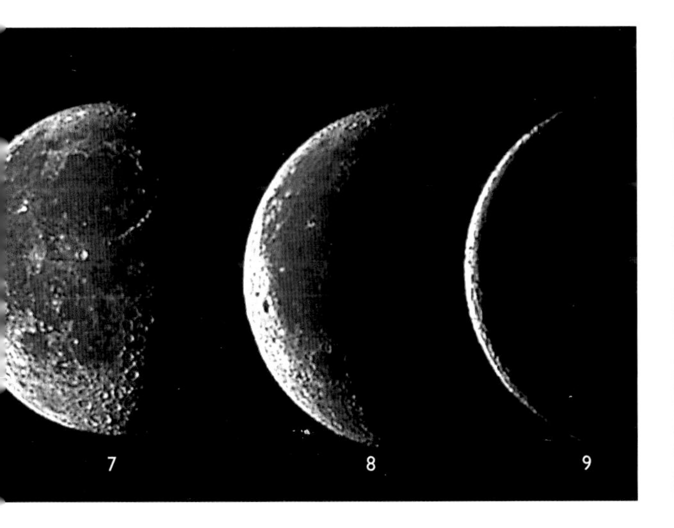

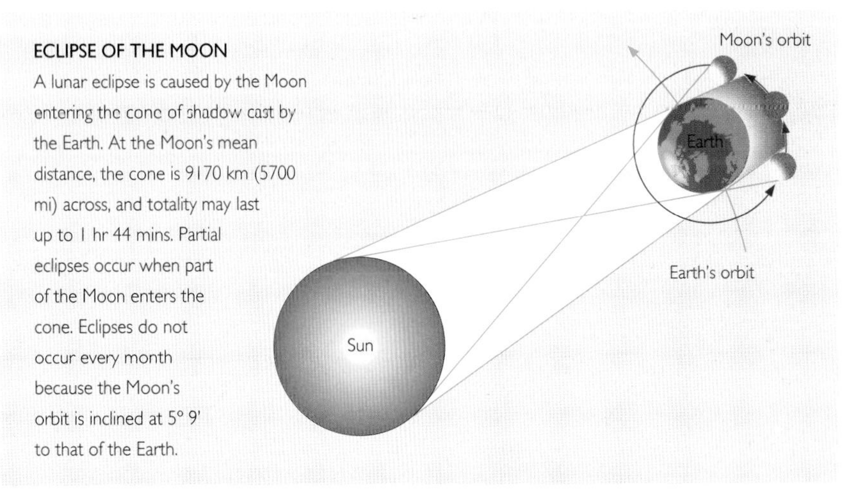

ECLIPSE OF THE MOON

A lunar eclipse is caused by the Moon entering the cone of shadow cast by the Earth. At the Moon's mean distance, the cone is 9170 km (5700 mi) across, and totality may last up to 1 hr 44 mins. Partial eclipses occur when part of the Moon enters the cone. Eclipses do not occur every month because the Moon's orbit is inclined at 5° 9' to that of the Earth.

Features of the Moon

The Moon is much the most spectacular object in the sky to the user of a small telescope. There is an immense amount of detail to be seen, and the appearance changes dramatically from one night to the next because of the changing angle of solar illumination. A crater which is imposing when close to the terminator – the boundary between the daylight and night hemispheres – may be almost impossible to identify near Full Moon, when there are virtually no shadows.

The most obvious features are the wide dark plains known as seas or maria (sing. mare). For centuries now it has been known that there is no water in them (and never has been!), but they retain their romantic names such as the Mare Imbrium (Sea of Showers), Sinus Iridum (Bay of Rainbows) and Oceanus Procellarum (Ocean of Storms). They are of various types. Some, such as the Mare Imbrium, are essentially circular in outline, with mountainous borders; the diameter of the Mare Imbrium is 1300 kilometres (800 miles). Other seas, such as the vast Oceanus Procellarum, are irregular and patchy, so that they give the impression of being lava 'overflows'. There are bays, such as the Sinus Iridum which leads off the Mare Imbrium and is a superb sight when the Sun is rising or setting over it, catching the mountain-tops while the floor is still in shadow and producing the appearance often nicknamed the 'Jewelled Handle'.

Most of the major maria form a connected system. There is, however, one exception: the isolated, well-formed Mare Crisium, near the Moon's north-east limb, which is easily visible as a separate object with the naked eye. It appears elongated in a north-south direction, but this is because of the effect of foreshortening; the north-south diameter is 460 kilometres (285 miles), while the east-west diameter is 590 kilometres (370 miles). Maria still closer to the limb are so foreshortened that they can be made out only under favourable conditions.

NAMED NEAR-SIDE SEAS (MARIA)	
Sinus Aestuum	The Bay of Heats
Mare Australe	The Southern Sea
Mare Crisium	The Sea of Crises
Palus Epidemiarum	The Marsh of Epidemics
Mare Foecunditatis	The Sea of Fertility
Mare Frigoris	The Sea of Cold
Mare Humboldtianum	Humboldt's Sea
Mare Humorum	The Sea of Humours
Mare Imbrium	The Sea of Showers
Sinus Iridum	The Bay of Rainbows
Mare Marginis	The Marginal Sea
Sinus Medii	The Central Bay
Lacus Mortis	The Lake of Death
Palus Nebularum	The Marsh of Mists
Mare Nectaris	The Sea of Nectar
Mare Nubium	The Sea of Clouds
Mare Orientale	The Eastern Sea
Oceanus Procellarum	The Ocean of Storms
Palus Putredinis	The Marsh of Decay
Sinus Roris	The Bay of Dews
Mare Serenitatis	The Sea of Serenity
Mare Smythii	Smyth's Sea
Palus Somnii	The Marsh of Sleep
Lacus Somniorum	The Lake of the Dreamers
Mare Spumans	The Foaming Sea
Mare Tranquillitatis	The Sea of Tranquility
Mare Undarum	The Sea of Waves
Mare Vaporum	The Sea of Vapours

The whole lunar scene is dominated by the craters, which range from vast enclosures such as Bailly, 293 kilometres (182 miles) in diameter, down to tiny pits. No part of the Moon is free from them; they cluster thickly in the uplands, but are also to be found on the floors of the maria and on the flanks and crests of mountains. They break into each other, sometimes distorting each other so completely the original forms are hard to trace; some have had their walls so reduced by lava flows that they have become 'ghosts', and some craters have had their 'seaward' walls so breached that they have become bays. Fracastorius, at the edge of the Mare Nectaris, is a good example of this.

Riccioli, a Jesuit astronomer who drew a lunar map in 1651, named the main craters after various personalities, usually scientists. His system has been followed up to the present time, though it has been modified and extended, and later astronomers such as Newton have come off second-best. Some unexpected names are found. Julius Caesar has his own crater, though this was for his association with calendar reform rather than his military prowess.

Central peaks and groups of peaks, are common, and the walls may be massive and terraced. Yet in profile a crater is not in the least like a steep-sided mine shaft. The walls rise to only a modest height above the outer surface, while the floor is sunken; the central peaks never rise as high as the outer ramparts, so that in theory a lid could be dropped over the crater! Some formations, such as Plato in the region of the Alps and Grimaldi near the western limb, have floors dark enough to make them identifiable under any conditions of illumination; Aristarchus, in the Oceanus Procellarum, is only 37 kilometres (23 miles) across, but has walls and a central peak so brilliant that when lit only by earthshine it has sometimes been mistaken for a volcano in eruption. One crater, Wargentin in the south-west limb area, has been filled with lava to its brim, so that it has taken on the form of a plateau. It is almost 90 kilometres (55 miles) across.

The most striking of all the craters are Tycho, in the southern uplands, and Copernicus in the Mare Nubium. Under high light they are seen to be the centres of systems of bright rays, which spread out for hundreds of kilometres. They are surface features, casting no shadows, so that they are well seen only when the Sun is reasonably high over them; near Full Moon they are so prominent that they drown most other features. Interestingly, the rays around Tycho do not come from the centre of the crater, but are tangential to the walls. There are many other minor ray-centres, such as Kepler in the Oceanus Procellarum and Anaxagoras in the north polar area.

The main mountain ranges form the borders of the regular maria; thus the Mare Imbrium is bordered by the Alps, Apennines and Carpathians. Isolated peaks and hills abound, and there are also domes, which are low swellings often crowned by summit craterlets. One feature of special interest is the Straight Wall, in the Mare Nubium – which is neither straight, nor a wall! The land to the west drops by about 300 metres (1000 feet), so that the 'wall' is simply a fault in the surface. Before Full Moon its shadow causes it to appear as a black line; after this it reappears as a bright line, with the Sun's rays shining on its inclined face. It is by no means sheer, and the gradient seems to be no more than 40°. In the future it will no doubt become a lunar tourist attraction.

◀ **The Full Moon:** photograph by Bruce Kingsley using a 28-cm (11-in) reflector. As is usual in images captured using a telescope, South is at the top. Mare Crisium, to the left (east) is well shown, under favourable libration. The tiny dot above right, is the planet Saturn. This shows how small Saturn looks compared with the Moon!

Valleys are found here and there, notably the great gash cutting through the Alps. The so-called 'Rheita Valley' in the south-eastern uplands is really a chain of craters which have merged, and crater-chains are very common on the Moon, sometimes resembling strings of beads. There are also rills – alternatively known as rilles or clefts – which are crack-like collapse features. Some of these, too, prove to be crater-chains either wholly or in part. The most celebrated rills are those of Hyginus and Ariadaeus, in the region of the Mare Vaporum, but there are intricate rill systems on the floors of some of the large craters, such as Gassendi, near the northern boundary of the Mare Humorum, and Alphonsus, the central member of a chain of great walled plains, near the centre of the Moon's disk, of which the flat-floored, 148-kilometre (92-mile) Ptolemaeus is the largest crater.

Many of the maria are crossed by ridges, which are low, snaking elevations of considerable length. Ridges on the maria are often the walls of ghost craters which have been so completely inundated by lava that they are barely recognizable.

The craters were produced by a violent meteoritic bombardment which began at least 4500 million years ago and ended about 3850 million years ago. There followed widespread vulcanism, with magma pouring out from below and flooding the basins. The lava flows ended rather suddenly about 3200 million years ago, and since then the Moon has shown little activity apart from the formation of an occasional impact crater. It has been claimed that the ray-craters Tycho and Copernicus may be no more than a thousand million years old, though even this is ancient by terrestrial standards. Even the youngest craters on the Moon are far older than any still existing on Earth.

There is little activity now; occasional localized glows and obscurations, known as Transient Lunar Phenomena (TLP), are thought to be caused by the release of gas from below the crust, but the Moon today is essentially changeless.

MAIN MOUNTAIN RANGES	
Alps	Northern border of Imbrium
Altai Scarp	South-west of Nectaris, from Piccolomini
Apennines	Bordering Imbrium
Carpathians	Bordering Imbrium to the south
Caucasus	Separating Serenitatis and Nebularum
Cordillera	Limb range, near Grimaldi
Haemus	Southern border of Serenitatis
Harbinger	Clumps of peaks in Imbrium, near Aristarchus
Jura	Bordering Iridum
Percy	North-west border of Humorum; not a major range
Pyrenees	Clumps of hills bordering Nectaris to the east
Riphaeans	Short range in Nubium
Rook	Limb range, associated with Orientale
Spitzbergen	Mountain clump in Iridum, north of Archimedes
Straight Range	In Imbrium, near Plato; very regular
Taurus	Mountain clumps east of Serenitatis
Tenerife	Mountain clumps in Imbrium, south of Plato
Ural	Extension of the Riphaeans

Lunar Landscapes

Lunar photographs taken with even small telescopes can show a surprising amount of detail, and there always appears to be something new to see. It is not hard to compile one's own lunar photographic atlas.

► **The Ptolemaeus chain** (Bruce Kingsley, 25-cm (10-in) reflector). Ptolemaeus, in the north, is 148 km (92 mi) across, and has a fairly smooth floor, with a deep crater, Ammonius. Alphonsus has a central peak and rills; Arzachel is deeper, with a high central peak.

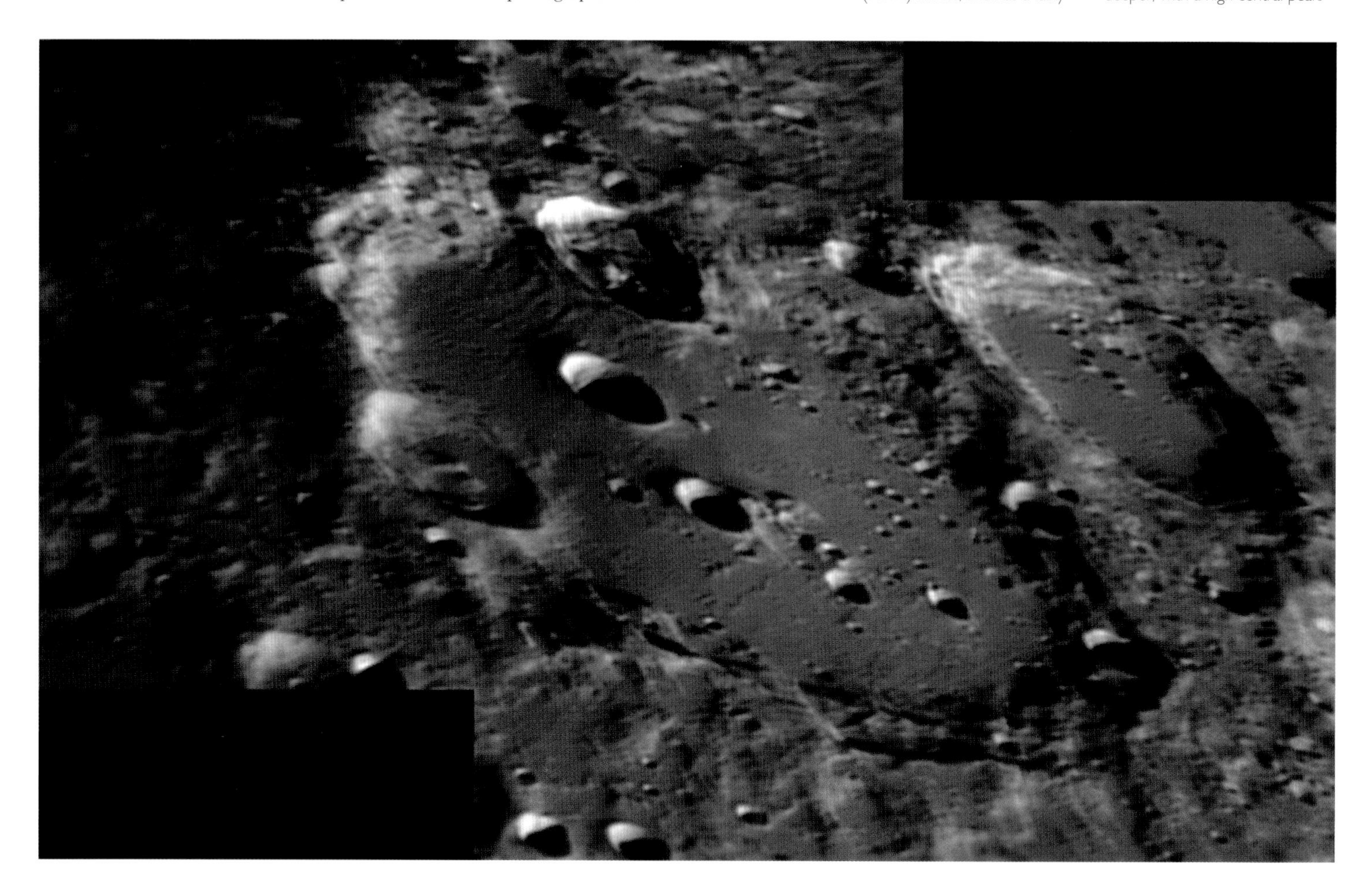

▲ **Clavius** The diameter is 233 km (145 mi); its walls rise to over 3660 m (12,000 ft). The north-east wall is broken by a large crater, Porter, and there is an arc of craters across the floor, of which the largest is Rutherford. When on the terminator, Clavius can be distinguished with the naked eye.

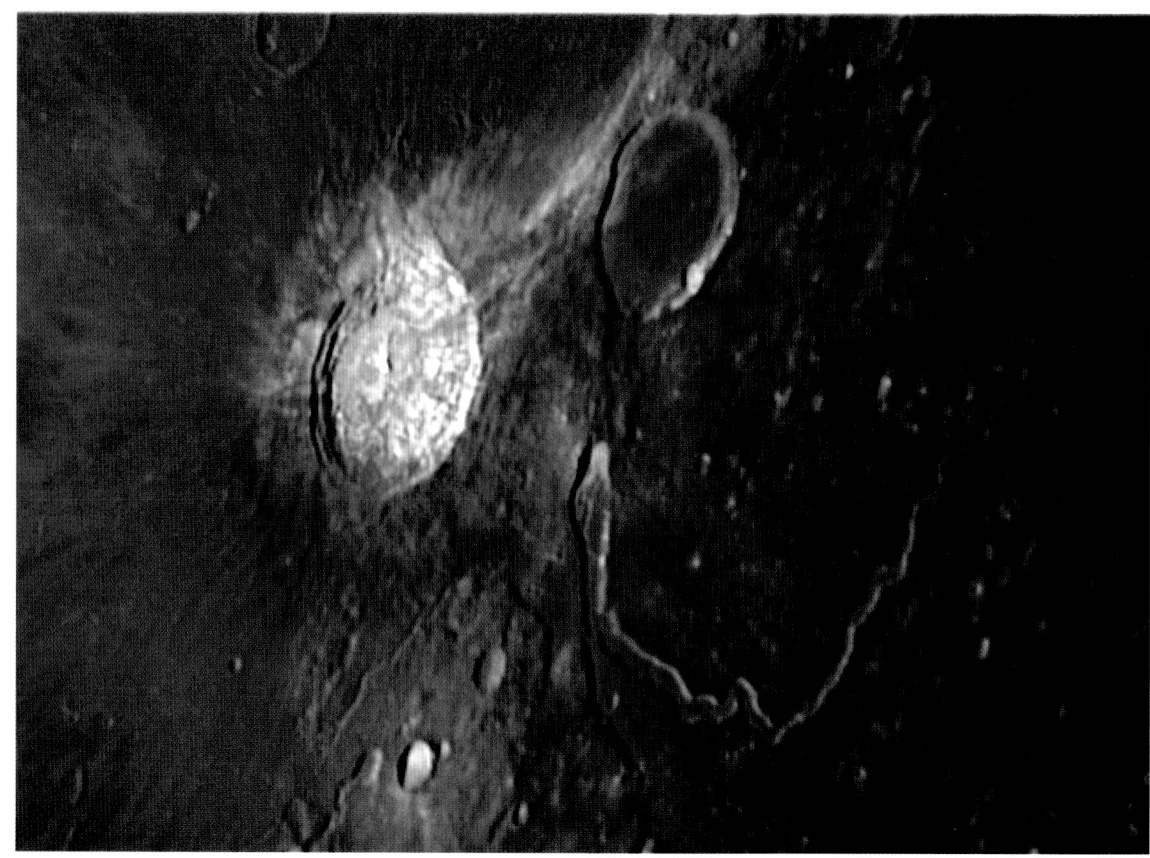

► **Aristarchus** The brightest crater on the entire Moon. Note the strange dark bands running up the wall, and the central peak. The crater adjoining it, Herodotus, has a darkish floor, and issuing from it is the long, winding valley, known as Schröter's valley, although the crater named after Schröter himself is nowhere near it! Aristarchus is obvious whenever it is in sunlight and can also be seen by earthlight. Unwary observers have been known to mistake it for an erupting volcano. Image by Dave Tyler.

The Far Side of the Moon

► **The far side of the Moon** (part), imaged by the Galileo spacecraft. The Mare Orientale is almost central, while the dark area to the right is the Oceanus Procellarum; the small bright spot is the crater Aristarchus.

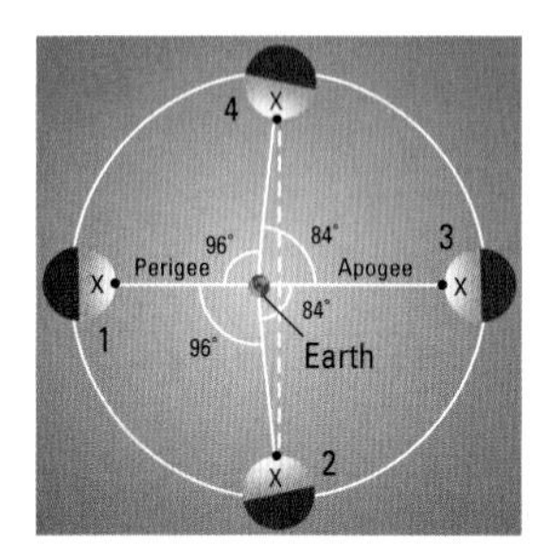

▲ **Libration in longitude.** X is the centre of the Moon's disk, as seen from Earth. At position 1 the Moon is at perigee. After a quarter of its orbit it has reached position 2; but since it has travelled from perigee it has moved slightly faster than its mean rate, and has covered 96° instead of 90°. As seen from Earth, X lies slightly east of the apparent centre of the disk, and a small portion of the far side has come into view in the west. After a further quarter-month the Moon has reached position 3. It is now at apogee, and X is again central. A further 84° is covered between positions 3 and 4, and X is displaced towards the west, so that an area beyond the mean eastern limb is uncovered. At the end of one orbit the Moon has arrived back at 1, and X is once more central on the Moon's disk as seen from Earth.

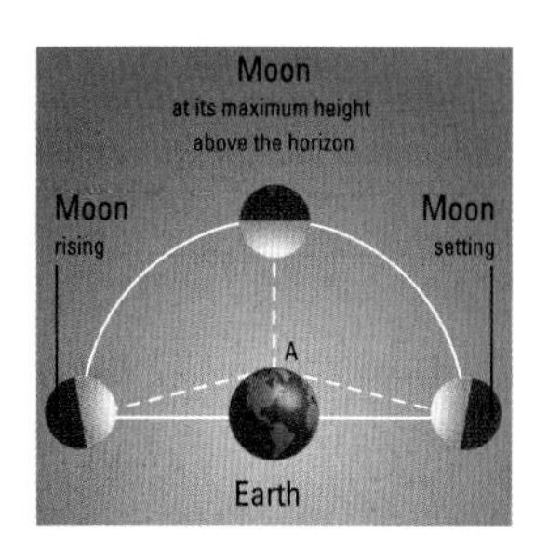

▲ **Diurnal libration.** We are observing from the Earth's surface at A, not its centre, so that we can see a little way alternately round the northern and southern limbs.

Look at the Moon, even with the naked eye, and you will see the obvious features such as the principal maria. The positions of these features on the disk are always much the same, because of the synchronous rotation. Yet there are slight shifts, due to the effects known as librations. All in all we can examine a grand total of 59 per cent of the lunar surface, and only 41 per cent is permanently averted, though of course we can never see more than 50 per cent at any one time.

The most important libration – the libration in longitude – is due to the fact that the Moon's path round the Earth is elliptical rather than circular, and it moves at its fastest when closest to us (perigee). However, the rate of axial rotation does not change, so that the position in orbit and the amount of axial spin become periodically 'out of step'; we can see a little way round alternate mean limbs. There is also a libration in latitude, because the Moon's orbit is inclined by over 5 degrees, and we can see for some distance beyond the northern and southern limbs. Finally there is a diurnal or daily libration, because we are observing from the surface, not the centre of the Earth.

All these effects mean that the 'libration regions' are carried in and out of view. They are so foreshortened that it is often difficult to distinguish between a crater and a ridge, and before 1959 our maps of them were very imperfect. About the permanently hidden regions nothing definite was known. It was reasonable to assume that they were basically similar to the familiar areas – though some strange ideas had been put forward from time to time. (The 19th-century Danish astronomer Andreas Hansen once proposed that all the Moon's air and water had been drawn round to the far side, which might well be inhabited!) The first pictures of the far side were obtained in October 1959 by the Soviet space probe Lunik 3 (also known as Luna 3). It went right round the Moon, taking pictures of the far side and later sending them back by television techniques. The pictures are very blurred and lacking in detail, but they were good enough to show that, as expected, the far side is just as barren and just as crater-scarred as the areas we have always known. Later spacecraft, both manned and automatic, have enabled us to draw up very complete maps of the entire lunar surface.

There is a definite difference between the near and the far sides, no doubt because the Moon's rotation has been synchronous since a fairly early stage in the evolution of the Earth-Moon system; the crust is thickest on the far side. One major sea, the Mare Orientale, lies mainly on the hidden regions; only a small part of it can be seen from Earth, and then only under conditions of favourable libration. Pictures from spacecraft have shown it to be a vast, multi-ringed structure which is probably the youngest of all the lunar seas. Otherwise there are no large maria on the far side, and this is the main difference between the two hemispheres.

One very interesting feature is Tsiolkovskii, 240 kilometres (150 miles) in diameter. It has a dark floor which in many

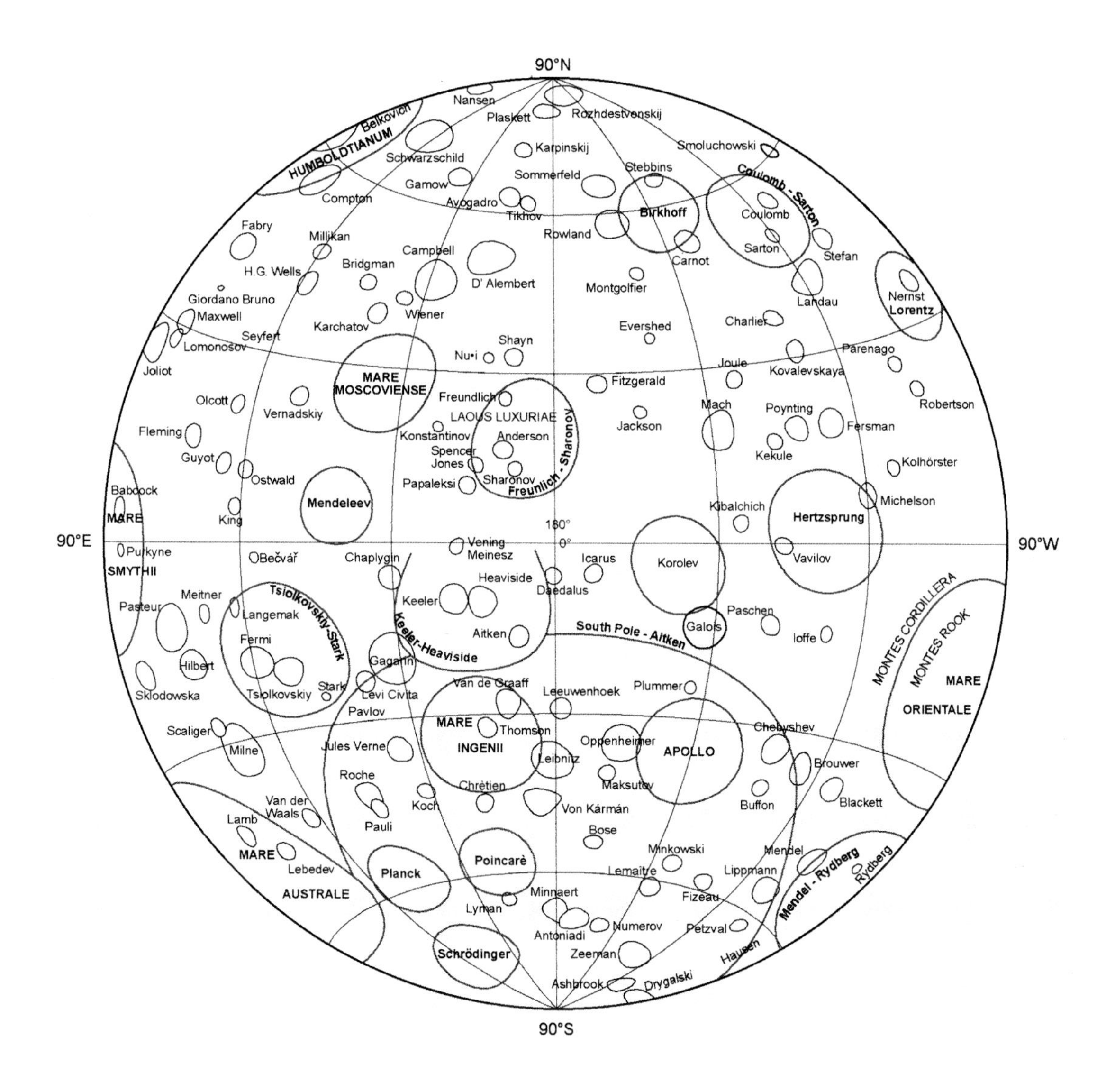

◀ The far side of the Moon – first recorded from the Soviet space probe Lunik 3 in 1959, and now fully mapped.

▼ Tsiolkovskii is exceptional in many ways. It is 240 km (150 mi) in diameter, with terraced walls and a massive central mountain structure. The darkness is caused by lava; in fact Tsiolkovskii seems to be intermediate in type of lunar feature, falling somewhere between a crater and a mare, or sea.

photographs gives the impression of being shadowed, though the real cause of the darkness is the hue of the floor itself; there is no doubt that we are seeing a lake of solidified lava, from which a central peak rises. In many ways Tsiolkovskii seems to be a sort of link between a crater and a mare. It intrudes into a larger but less regular basin, Fermi, which has the usual light-coloured interior.

Many of the familiar types of features are seen on the far side, and the distribution of the craters is equally non-random; when one formation breaks into another, it is always the smaller crater which is the intruder. Valleys, peaks and ray systems exist. Though the Moon has no overall magnetic field that we can detect, there are regions of localized magnetism here and there; one of these lies near the rather irregular far-side crater Van de Graaff. It has been suggested that the Moon used to have a definite magnetic field which has now died away.

On the original Lunik 3 picture a long, bright feature running for hundreds of kilometres was shown, and was thought to be a mountain range which was promptly named in honour of the Soviet Union. Alas, it was later found that the feature is nothing more than a surface ray, and the Soviet Mountains were tactfully deleted from the maps. However, it was surely right to name the most imposing far-side feature in honour of Konstantin Eduardovich Tsiolkovskii, the great pioneer who was writing about spaceflight more than 100 years ago.

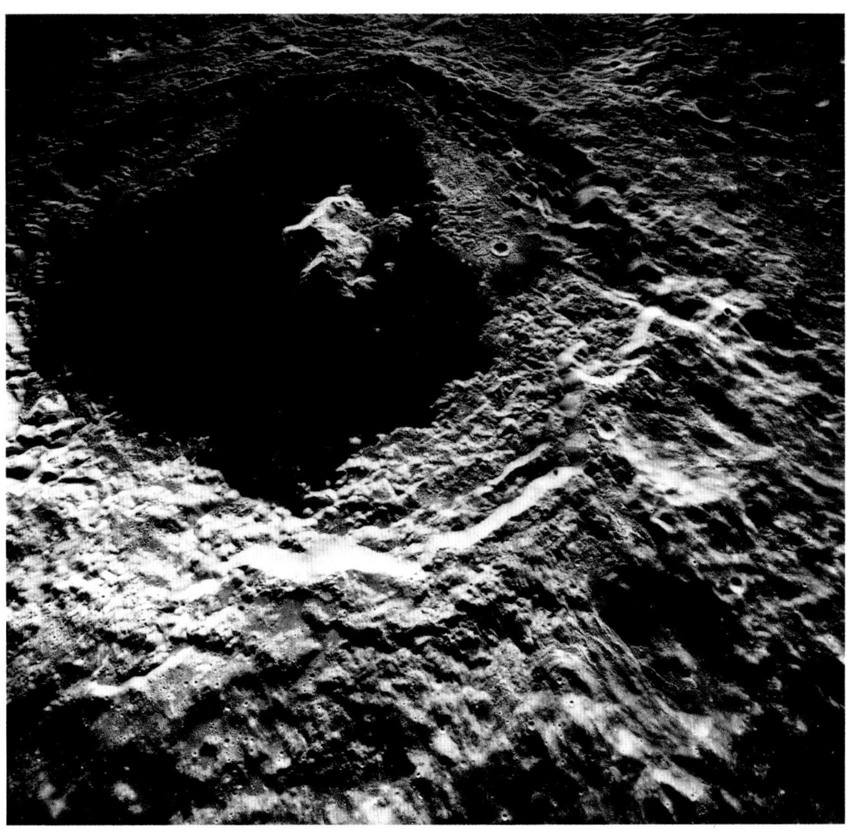

Missions to the Moon

▼ **Apollo 15.** The first mission in which a 'Moon Car', or Lunar Roving Vehicle, was taken to the Moon, enabling the astronauts to explore much greater areas. Astronaut Irwin stands by the rover, with one of the peaks of the Apennines in the background. The electrically powered rovers performed faultlessly. The peak is much further from the rover than may be thought; distances on the Moon are notoriously difficult to estimate.

The Russians took the lead in exploring the Moon with spacecraft. Their Luniks contacted the Moon in 1959, and they were also the first to make a controlled landing with an automatic probe. Luna 9 came gently down in the Oceanus Procellarum, on 3 February 1966, and finally disposed of a curious theory according to which the lunar seas were coated with deep layers of soft dust. Later, the Russians were also able to send vehicles to the Moon, collect samples of lunar material, and bring them back to Earth. It is now known that they had planned a manned landing there in the late 1960s, but had to abandon the idea when it became painfully clear that their rockets were not sufficiently reliable. By 1970 the 'Race to the Moon' was definitely over.

American progress had been smoother. The Ranger vehicles crash-landed on the Moon, sending back data and pictures before being destroyed; the Surveyors made soft landings, obtaining a tremendous amount of information; and between 1966 and 1968 the five Orbiters went round and round the Moon, providing very detailed and accurate maps of virtually the entire surface. Meanwhile, the Apollo programme was gathering momentum.

By Christmas 1968 the crew of Apollo 8 were able to go round the Moon, paving the way for a landing. Apollo 9 was an Earth-orbiter, used to test the lunar module which would go down on to the Moon's surface. Apollo 10, the final rehearsal, was another lunar orbiter; and then, in July 1969, first Neil Armstrong, then Buzz Aldrin, stepped out on to the bleak rocks of the Mare Tranquillitatis from the Eagle, the lunar module of Apollo 11. Millions of people on Earth watched Armstrong make his immortal 'one small step' on to the surface of the Moon. The gap between our world and another had at last been bridged.

Apollo 11 was a preliminary mission. The two astronauts spent more than two hours outside their module, setting up the first ALSEP (Apollo Lunar Surface Experimental Package), which included various instruments – for example a seismometer, to detect possible 'moon-quakes'; a device for making a final search for any trace of lunar atmosphere, and an instrument designed to collect particles from the solar wind. Once their work was completed (interrupted only briefly by a telephone call from President Nixon) the astronauts went back into the lunar module; subsequently they lifted off, and rejoined Michael Collins, the third member of the expedition, who had remained in lunar orbit. The lower part of the lunar module was used as a launching pad, and was left behind, where it will remain until it is collected and removed to a lunar museum. The return journey to Earth was flawless.

Apollo 12 (November 1969) was also a success; astronauts Conrad and Bean were even able to walk over to an old Surveyor probe, which had been on the Moon ever since 1967, and bring parts of it home. Apollo 13 (April 1970) was a near-disaster; there was an explosion during the outward journey, and the lunar landing had to be abandoned. With Apollo 14 (January 1971) astronauts Shepard and Mitchell took a 'lunar cart' to carry their equipment, and with the three final missions, Apollo 15 (July 1971), 16 (April 1972) and 17 (December 1972) a Lunar Roving Vehicle (LRV) was used, which increased the range of exploration very considerably. One of the Apollo 17 astronauts, Dr Harrison Schmitt, was a professional geologist who had been given training specially for the mission.

◀ **Apollo 16.** The Lunar Module ascends from the Moon's surface towards the Command Service Module (CSM). Mare Fecunditatis can be seen in the background. Astronauts Duke and Young explored a wide area, and set up a number of scientific experiments in the Apollo Lunar Surface Experimental Package (ALSEP). The Lunar Module was designed to make a landing on the Moon and return the astronauts to orbit. The upper section has one ascent engine only, and there can be no second chance. The photograph was taken from the orbiting CSM.

The Apollo programme has increased our knowledge of the Moon beyond all recognition - and yet in a way it was limited; the missions were really in the nature of reconnaissances. The various ALSEPs continued operating for some years, until they were eventually switched off mainly on financial grounds.

No men have been to the Moon since 1972, but if all goes well this will soon change. There are firm plans to return there during the next decade, and to set up a fully-fledged Lunar Base at some time between 2020 and 2030. The Base must be truly international; Earthmen must go into space together.

► **The scene from Apollo 17.** The Lunar Rover is well shown. The sky is of course jet-black; one is reminded of Buzz Aldrin's description of the Moon as 'magnificent desolation'.

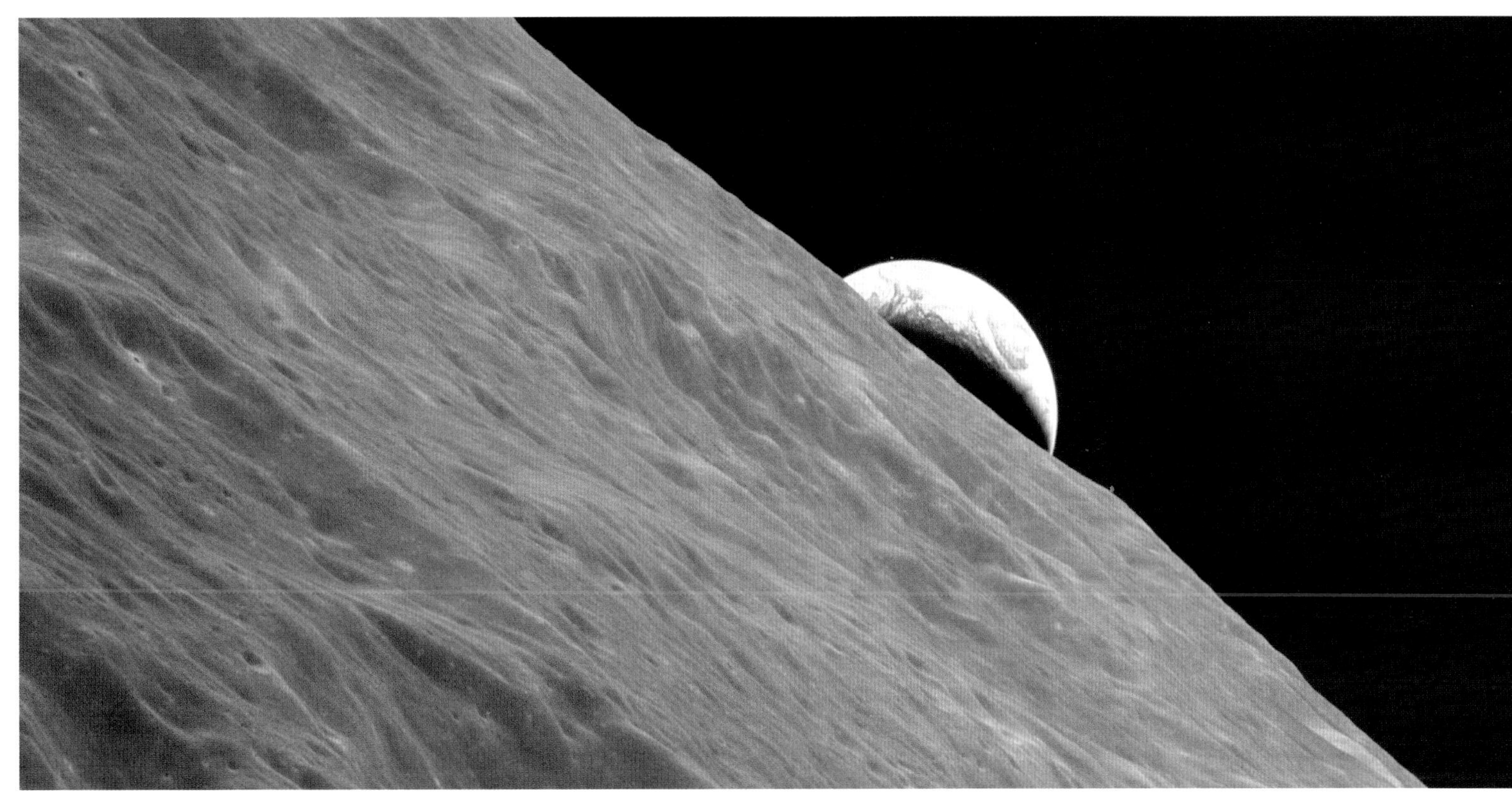

▲ **Earthrise.** This picture was taken from Apollo 17, the last manned lunar mission. It shows the crescent Earth rising over the limb of the Moon; when the photograph was taken, Apollo 17 was in lunar orbit.

► **Apollo 17.** During one of the Moon walks, Dr Schmitt, the geologist, suddenly called attention to what seemed to be orange soil inside a small crater, unofficially named Shorty. At first it was thought to indicate recent fumarole activity, but the colour was due to very small, very ancient glassy 'beads'.

Clementine, Prospector and SMART-1

▶ **Mayer** Image of craters W. Bond marked by two smaller craters and the narrow ditch running off the left-hand side of the image, and C. Mayer, the large isolated crater at lower right. This area is northern central region near Mare Frigoris.

The 1994 mission to the Moon, Clementine, was named after an old American mining song – because, after the lunar part of its programme, the probe was scheduled to go on to an asteroid, and it has been suggested that in the future it may be possible to carry out mining operations on asteroids.

Clementine was funded partly by NASA and partly by the US Department of Defense. The military authorities were anxious to test instruments and techniques capable of locating hostile ballistic missiles, and the only way to circumvent the strict regulations about this sort of activity was to go to the Moon. Therefore, the Department could test its anti-ballistic missile system and do some useful scientific work as well.

Clementine was launched on 24 January 1994 from the Vandenburg Air Force Base in California, and began its Earth programme. It weighed 140 kilograms (300 lb) and carried an array of advanced sensors. After completing this part of its mission several manoeuvres were carried out, and Clementine entered lunar orbit on 21 February 1995. For 71 days it orbited the Moon in a highly inclined path, which took it from 415 kilometres (260 miles) to 2940 kilometres (1830 miles) from the Moon; a full research programme was successfully completed.

Clementine surveyed the whole of the Moon. Many gravity measurements were made, and superb images obtained; the inclined orbit of the probe meant that the polar regions could be mapped more accurately than ever before. For example, there were detailed views of the vast South Pole-Aitken Basin, which is 2250 kilometres (1400 miles) in diameter and 12 kilometres (7 miles) deep. There was also the Mendel-Rydberg Basin, 630 kilometres (390 miles) across, which is less prominent because it lies under a thick blanket of débris from the adjacent Mare Orientale.

It was claimed that Clementine had detected indications of ice inside some of the polar craters, whose floors are always in shadow – but ice did not seem at all probable on a world such as the Moon. Clementine left lunar orbit on 3 May 1995; it had been hoped to rendezvous with a small asteroid, Geographos, but a programming error ruled this out.

The next lunar probe, Prospector, was launched on 3 January 1998, and carried out an extensive mapping survey. On 31 July 1999 it was deliberately crashed into a polar crater, in the hope that water might be detected in the débris, but no signs of water were found and the idea of lunar ice has been generally abandoned.

The European probe SMART-1 was launched on 27 September 2003, on an Ariane 5 rocket. It used ion propulsion, with xenon propellant, and took 13½ months to reach the Moon; it entered lunar orbit on 15 November 2004, and began a 3-year programme of general research, plus photography. It was extremely successful (SMART stands for Small Missions for Advanced Research in Technology). At the end of its career, on 3 September 2006, it was deliberately crashed on to the surface. As before, the ejected débris showed no sign of icy material.

▼ **The South Polar Region** of the Moon as imaged from Clementine, showing some of the huge craters.

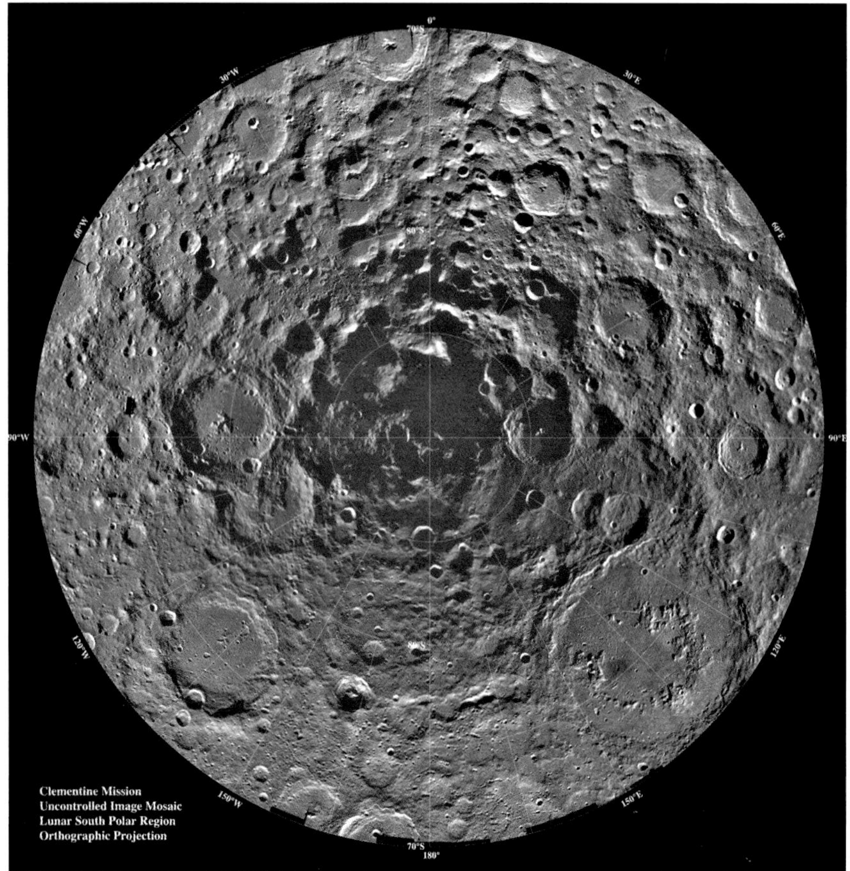

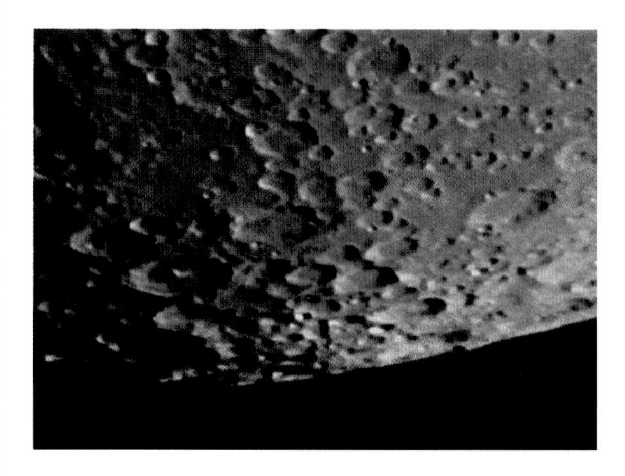

▲ **The impact site** of Prospector, in the lunar south polar region near the craters Moretus and Short.

▶ **Lift-off of SMART-1**, 27 September 2003, by Ariane 5 rocket from the European spaceport in Kourou.

arianespace
esa
esa
cnes
cnes
ariane

The Moon: *First Quadrant (North-east)*

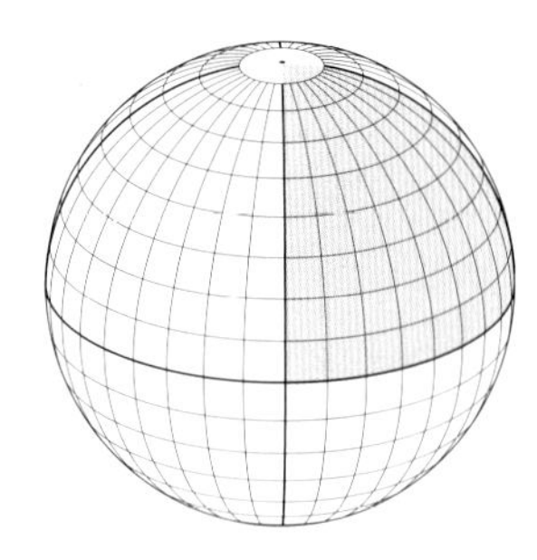

The First Quadrant is occupied largely by maria. The whole of the Mare Serenitatis and Mare Crisium are included, and most of the Mare Tranquillitatis and the darkish Mare Vaporum, with parts of the Mare Frigoris and Mare Foecunditatis. There are also some small seas close to the limb (Smythii, Marginis and Humboldtianum) which are never easy to observe because they are so foreshortened. There are also large walled plains close to the limb, such as Neper and Gauss.

In the south, the Mare Serenitatis is bordered by the Haemus Mountains (Montes Haemus), which rise to 2400 metres (7800 feet). The Alps (Montes Alpes) run along the southern border of the Mare Frigoris, and here we find the magnificent Alpine Valley (Vallis Alpes), which is 130 kilometres (80 miles) long and is much the finest formation of this type on the Moon; a delicate rill runs along its floor, and there are obscure parallel and transverse valleys. Mont Blanc (Mons Blanc), in the Alps, rises to 3500 metres (11,500 feet). Part of the Apennine range (Montes Apenninus) extends into this quadrant, with lofty peaks such as Mount Bradley and Mount Hadley (Mons Bradley and Mons Hadley), both of which are over 4000 metres (13,000 feet) high. There are several major rill systems (Ariadaeus, Hyginus, Triesnecker, Ukert and Bürg) and an area near Arago which is rich in domes. The Apollo 11 astronauts landed in the Mare Tranquillitatis, not far from Maskelyne, and Apollo 17 came down in the area of Littrow and the clumps of hills which are called the Taurus Mountains (Montes Taurus).

Agrippa A fine crater with a central peak and terraced walls. It forms a notable pair with its slightly smaller neighbour **Godin**.

Arago A well-formed crater, with the smaller, bright **Manners** to the south-east. Close to Arago is a whole collection of domes – some of the finest on the Moon; many of them have summit craterlets.

Archytas The most prominent crater on the irregular Mare Frigoris. It has bright walls and a central peak.

Ariadaeus A small crater associated with a major rill system. The main rill is almost 250 kilometres (150 miles) long, and has various branches, one of which connects the system with that of **Hyginus** – which is curved, and is mainly a craterlet-chain. Another complex rill system is associated with **Triesnecker** and **Ukert**. All these features are visible with a small telescope under good conditions.

Aristillus This makes up a group together with **Archimedes** (which is shown on the map of the Second Quadrant) and **Autolycus**. All three are very prominent. Under high illumination Autolycus is also seen to be the centre of a minor ray-system.

Aristoteles This and **Eudoxus** form a prominent pair of walled plains. Aristoteles has walls rising to 3300 metres (11,000 feet) above the floor. **Atlas** and **Hercules** form another imposing pair. Atlas has complex floor detail, while inside Hercules there is one very bright crater.

Bessel The main formation on the Mare Serenitatis; a well-formed crater close to a long ray which crosses the mare and seems to belong to the Tycho system.

Bürg A crater with a concave floor; the very large central peak is crowned by a craterlet (**Rømer** is another example of this.) Bürg stands on the edge of a dark plain which is riddled with rills.

Challis This and **Main** form a pair of 'Siamese twins' – a phenomenon also found elsewhere, as with Steinheil and Watt in the Fourth Quadrant.

Cleomedes A magnificent enclosure north of the Mare Crisium. The wall is interrupted by one very deep crater, **Tralles**.

Dionysius One of several very brilliant small craterlets in the rough region between the Mare Tranquillitatis and the Mare Vaporum; others are **Cayley**, **Whewell** and **Silberschlag**.

Endymion A large enclosure with a darkish floor. It joins the larger but very deformed **De la Rue**.

Gioja The north polar crater – not easy to examine from Earth. It is well formed, and intrudes into a larger but low-walled formation.

Julius Caesar This and **Boscovich** are low-walled, irregular formations, notable because of their very dark floors.

Le Monnier A fine example of a bay, leading off the Mare Serenitatis. Only a few mounds of its 'seaward' wall remain.

Linné A famous formation. It was once suspected of having changed from a craterlet into a white spot at some time between 1838 and 1866, but this is certainly untrue. It is a small, bowl-shaped crater standing on a white patch.

Manilius A fine crater near the Mare Vaporum, with brilliant walls; it is very prominent around the time of Full Moon. So too is **Menelaus**, in the Haemus Mountains.

Picard This and **Peirce** are the only prominent craters in the Mare Crisium.

Plinius A superb crater 'standing sentinel' on the strait between the Mare Serenitatis and the Mare Tranquillitatis. It has high, terraced walls; the central structure takes the form of a twin crater.

Posidonius A walled plain with low, narrow walls and a floor crowded with detail. It forms a pair with its smaller neighbour **Chacornac**.

Proclus One of the most brilliant craters on the Moon. It is the centre of an asymmetrical ray system; two rays border the **Palus Somnii**, which has a curiously distinctive tone.

Sabine This and **Ritter** make up a pair of almost perfect twins – one of many such pairs on the Moon.

Taruntius A fine example of a concentric crater. There is a central mountain with a summit pit, and a complete inner ring on the floor. This sort of arrangement is difficult to explain by random impact.

Thales A crater near De la Rue, prominent near Full Moon because it is a ray-centre.

Vitruvius On the Mare Tranquillitatis, near the peak of Mount Argaeus. It has bright walls, with a darkish floor and a central peak.

Note: north is up.

SELECTED CRATERS: FIRST QUADRANT

Crater	Diameter, km	Lat. °N	Long. °E	Crater	Diameter, km	Lat. °N	Long. °E
Agrippa	48	4	11	Hyginus	6	8	6
Apollonius	48	5	61	Jansen	26	14	29
Arago	29	6	21	Julius Caesar	71	9	15
Archytas	34	59	5	Le Monnier	55	26	31
Ariadaeus	15	5	17	Linné	11	28	12
Aristillus	56	34	1	Littrow	35	22	31
Aristoteles	97	50	18	Macrobius	68	21	46
Atlas	69	47	44	Main	48	81	9
Autolycus	36	31	1	Manilius	36	15	9
Bessel	19	22	18	Manners	16	5	20
Bond, W.C.	160	64	3	Maskelyne	24	2	30
Boscovich	43	10	11	Mason	31	43	30
Bürg	48	45	28	Menelaus	32	16	16
Cassini	58	40	5	Messala	128	39	60
Cauchy	13	10	39	Neper	113	7	83
Cayley	13	4	15	Peirce	19	18	53
Challis	56	78	9	Picard	34	15	55
Chacornac	48	30	32	Plana	39	42	28
Cleomedes	126	27	55	Plinius	48	15	24
Condorcet	72	12	70	Posidonius	96	32	30
De la Rue	160	67	56	Proclus	29	16	47
Democritus	37	62	35	Rømer	37	25	37
Dionysius	19	3	17	Ritter	32	2	19
Endymion	117	55	55	Sabine	31	2	20
Eudoxus	64	44	16	Sulpicius Gallus	13	20	12
Firmicus	56	7	64	Taquet	10	17	19
Gärtner	101	60	34	Taruntius	60	6	48
Gauss	136	36	80	Thales	39	59	41
Geminus	90	36	57	Theaetetus	26	37	6
Gioja	35	North polar		Tralles	48	28	53
Godin	43	2	10	Triesnecker	23	4	4
Hercules	72	46	39	Ukert	23	8	1
Hooke	43	41	55	Vitruvius	31	18	31

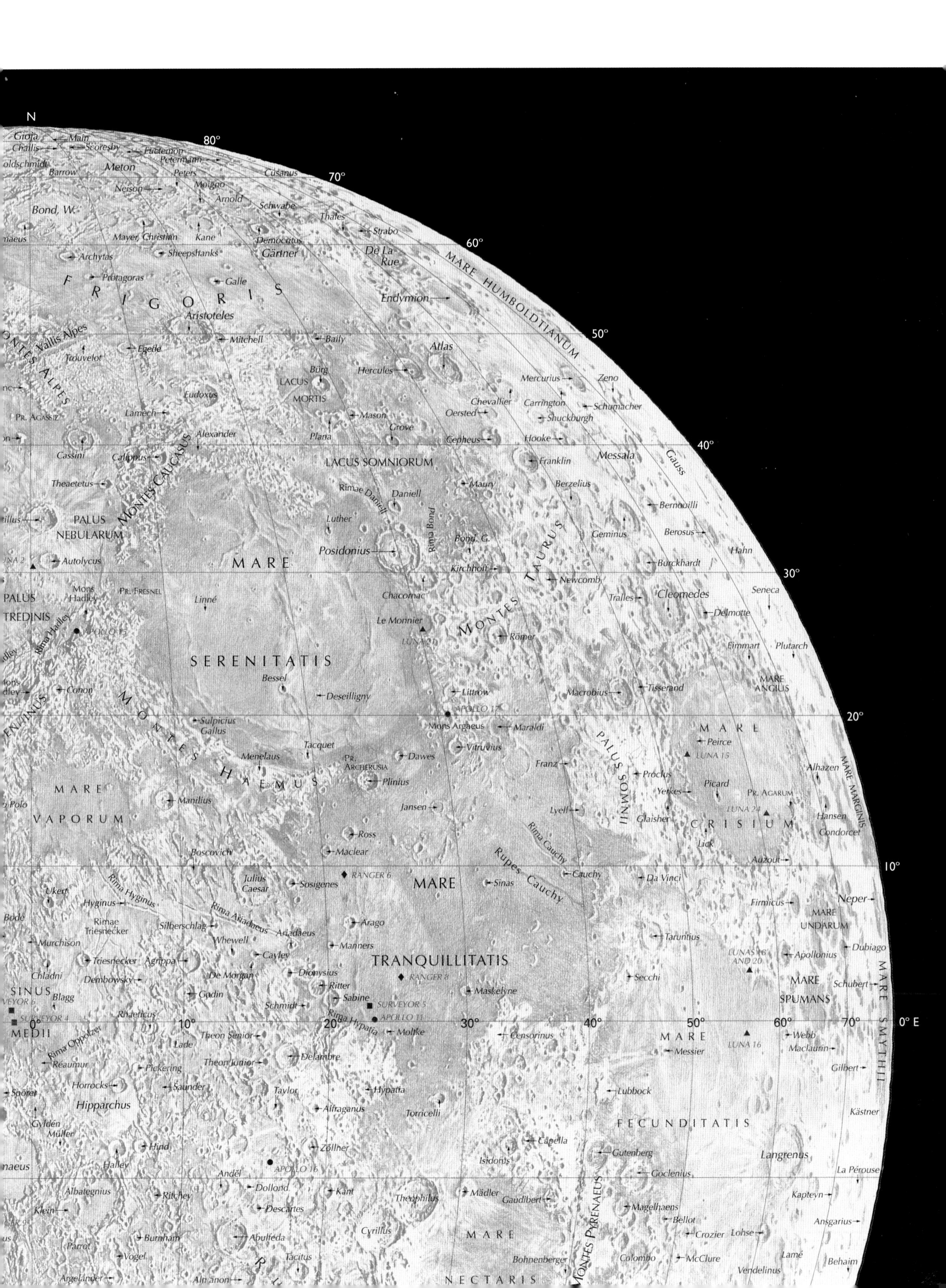

N
80°
70°
60°
50°
40°
30°
20°
10°
0°
10°
20°
30°
40°
50°
60°
70°
0° E
Gioja
Main
Challis
Scoresby
Euctemon
Petermann
Barrow
Meton
Peters
Cusanus
Neison
Moigno
Arnold
Schwabe
Bond, W.
Thales
Strabo
Mayer, Christian
Kane
Democritus
Archytas
Sheepshanks
Gärtner
De La Rue
MARE HUMBOLDTIANUM
Protagoras
Galle
FRIGORIS
Endymion
Aristoteles
Vallis Alpes
Mitchell
Baily
Trouvelot
Egede
Atlas
MONTES ALPES
Bürg
Hercules
LACUS MORTIS
Mercurius
Zeno
Eudoxus
Chevallier
Carrington
Schumacher
Lamech
Mason
Oersted
Shuckburgh
PR. AGASSIZ
Plana
Grove
Alexander
Cepheus
Hooke
Cassini
Calippus
MONTES CAUCASUS
Franklin
Messala
Gauss
LACUS SOMNIORUM
Theaetetus
Maury
Berzelius
Rimae Daniell
Daniell
Bernoulli
PALUS NEBULARUM
Luther
Rima Bond
Geminus
Berosus
Bond, G.
Posidonius
Hahn
Autolycus
MARE
Kirchhoff
Burckhardt
Newcomb
MONTES TAURUS
Mons Hadley
PR. FRESNEL
Chacornac
Tralles
Cleomedes
Seneca
PALUS PUTREDINIS
Linné
Delmotte
Rima Hadley
Le Monnier
APOLLO 15
LUNA 21
MONTES
Römer
Eimmart
Plutarch
SERENITATIS
MARE ANGUIS
Bessel
Conon
Littrow
Macrobius
Tisserand
Deseilligny
APOLLO 17
MONTES HAEMUS
Sulpicius Gallus
Mons Argaeus
Maraldi
MARE
PALUS SOMNII
Peirce
Tacquet
LUNA 15
Menelaus
Dawes
Vitruvius
PR. ARCHERUSIA
Franz
Proclus
Alhazen
MARE MARGINIS
Plinius
Yerkes
Picard
PR. AGARUM
MARE
Manilius
Jansen
Lyell
LUNA 24
Hansen
VAPORUM
Glaisher
CRISIUM
Condorcet
Rima Cauchy
Ross
Lick
Maclear
Boscovich
Rupes Cauchy
Auzout
Julius Caesar
RANGER 6
Cauchy
Da Vinci
Ukert
Sosigenes
MARE
Sinas
Rima Hyginus
Hyginus
Firmicus
Neper
Bode
Rimae Triesnecker
Rima Ariadaeus
Arago
MARE UNDARUM
Silberschlag
Ariadaeus
Whewell
Taruntius
Murchison
Manners
Dubiago
Cayley
LUNAS 18 AND 20
Apollonius
Triesnecker
Agrippa
TRANQUILLITATIS
Chladni
De Morgan
Dionysius
Secchi
MARE SPUMANS
Dembowski
RANGER 8
Schubert
SINUS MEDII
Ritter
Blagg
Godin
Sabine
Maskelyne
SURVEYOR 5
MARE SMYTHII
Schmidt
SURVEYOR 4
Rhaeticus
Rima Hypatia
APOLLO 11
Theon Senior
Moltke
Censorinus
Webb
Rima Oppolzer
Lade
MARE
LUNA 16
Maclaurin
Delambre
Messier
Réaumur
Theon Junior
Gilbert
Pickering
Spörer
Horrocks
Saunder
Lubbock
Hipparchus
Taylor
Hypatia
Alfraganus
Kästner
Torricelli
Gyldén
Müller
FECUNDITATIS
Capella
Hind
Zöllner
Gutenberg
Langrenus
Halley
Isidorus
APOLLO 16
La Pérouse
Andĕl
Goclenius
Dollond
Albategnius
Kant
Theophilus
Mädler
Gaudibert
Ritchey
Kapteyn
MONTES PYRENAEUS
Magelhaens
Klein
Descartes
Bellot
Ansgarius
Burnham
Abulfeda
Cyrillus
Crozier
Lohse
MARE
Parrot
Vogel
Tacitus
Bohnenberger
Colombo
McClure
Lamé
Behaim
Vendelinus
Argelander
Almanon
NECTARIS

The Moon: Second Quadrant (North-west)

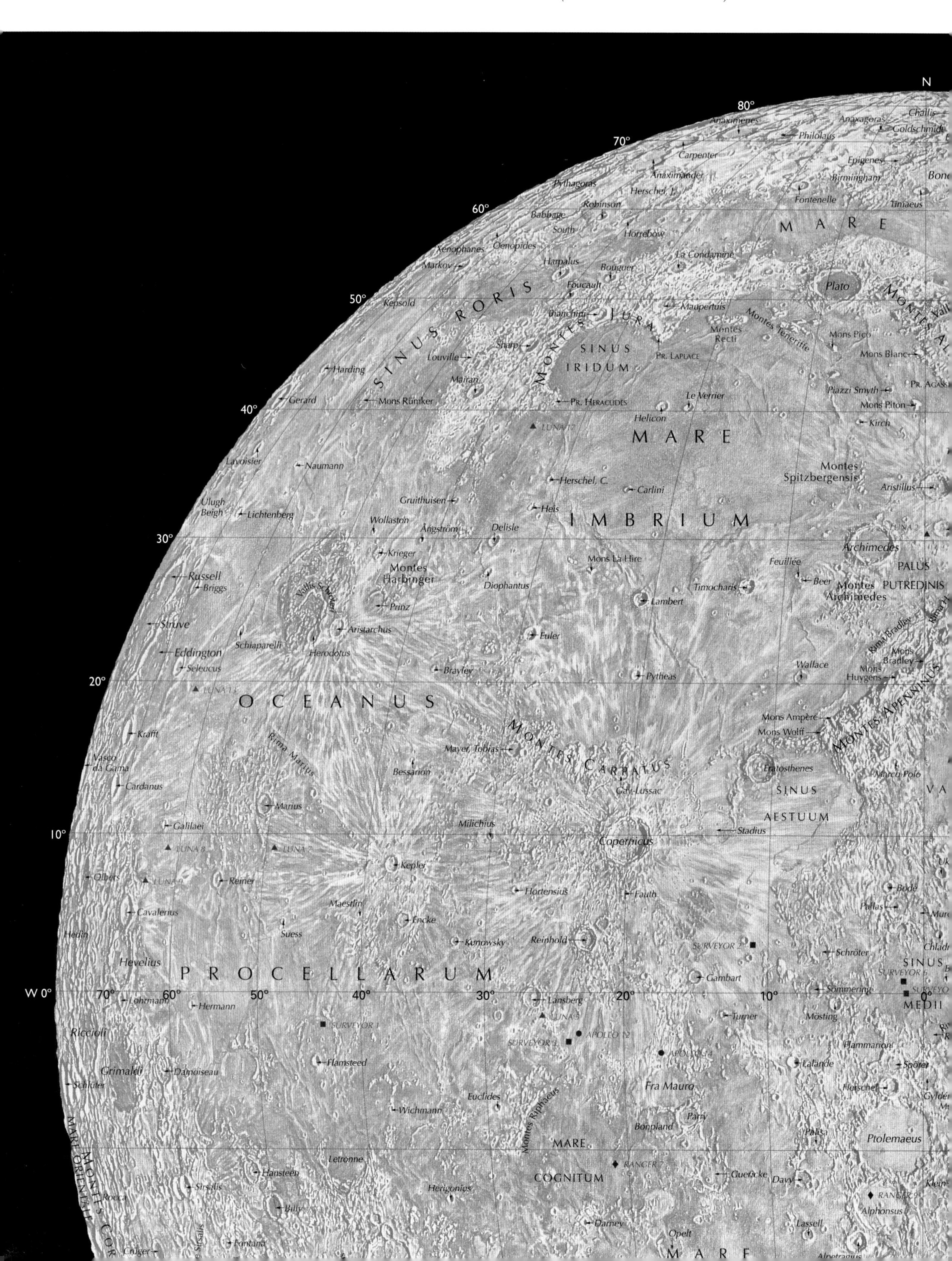

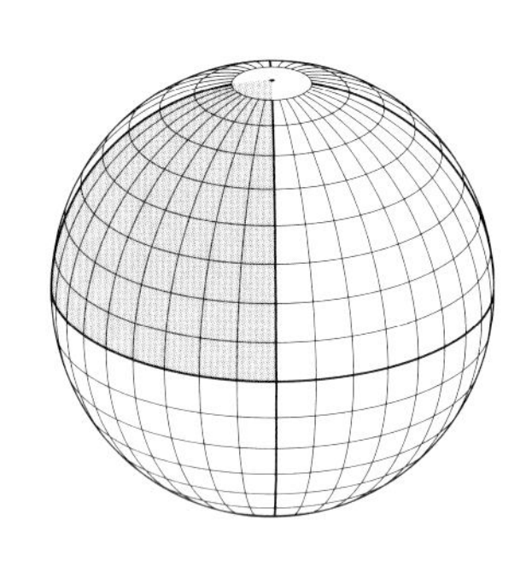

This is another 'marine quadrant', containing virtually the whole of the Mare Imbrium and most of the Oceanus Procellarum, as well as the Sinus Aestuum, Sinus Roris, a small part of the Sinus Medii and a section of the narrow, irregular Mare Frigoris. The Sinus Iridum, leading off the Mare Imbrium, is perhaps the most beautiful object on the entire Moon when observed at sunrise or sunset, when the solar rays catch the top of the Jura Mountains (Montes Jura) which border it. There are two prominent capes, Pr Laplace and Pr Heraclides; the seaward wall of the bay has been virtually levelled. It was in this area that the first Soviet 'crawler', Lunokhod 1, came down in 1970.

The Apennines (Montes Apenninus) make up the most conspicuous mountain range on the Moon; with the lower Carpathians (Montes Carpatus) in the south, they make up much of the border of the Mare Imbrium. The Straight Range (Montes Recti), in its north, is made up of a remarkable line of peaks rising to over 1500 metres (5000 feet); the range is curiously regular, and there is nothing else quite like it on the Moon. The Harbinger Mountains (Montes Harbinger), in the Aristarchus area, are made up of irregular clumps of hills. Isolated peaks include Mons Pico and Piton, in the Mare Imbrium. Pico is very conspicuous and is 2400 metres (7900 feet) high; the area between it and Plato is occupied by a ghost ring which was once called Newton, though the name has now been transferred to a deep formation in the southern uplands and the ghost has been relegated to anonymity.

Anaxagoras A well-formed crater with high walls and central peak. It is very bright, and is the centre of a major ray-system, so that it is easy to find under all conditions of illumination.

Archimedes One of the best-known of all walled plains; regular, with a relatively smooth floor. It forms a splendid trio with Aristillus and Autolycus, which lie in the First Quadrant.

Aristarchus The brightest crater on the Moon. Its brilliant walls and central peak make it prominent even when lit only by earthshine; there are strange darkish bands running from the central peak to the walls. Close by is **Herodotus**, of similar size but normal brightness. This is the area of the great **Schröter Valley**, (Vallis Schröteri) which begins in a 6-kilometre (4-mile) crater outside Herodotus; broadens to 10 kilometres (6 miles), producing the feature nicknamed the Cobra-Head, and then winds its way across the plain. The total length is 160 kilometres (100 miles), and the maximum depth 1000 metres (3300 feet). It was discovered by the German astronomer Johann Schröter, and is called after him, though Schröter's own crater is a long way away in the area of Sinus Medii and Sinus Aestuum. Many TLP have been recorded in this area.

Beer This and **Feuillé** are nearly identical twins – one of the most obvious craterlet-pairs on the Moon.

Birmingham Named not after the city, but after an Irish astronomer. It is low-walled and broken, and one of several formations of similar type in the far north; others are **Babbage**, **South** and **John Herschel**.

Carlini One of a number of small, bright-walled craterlets in the Mare Imbrium. Others are **Caroline Herschel**, **Diophantus**, **De l'Isle** and **Gruithuisen**.

Copernicus The 'Monarch of the Moon', with high, terraced walls and a complex central mountain group. Its ray-system is second only to that of Tycho; at or near Full Moon it dominates the entire area.

Einstein A great formation in the limb region, beyond the low, double and very reduced **Struve**. Einstein has a large central crater. It is visible only under favourable libration (and is not on the map) – as when I discovered it in 1945, using my 30-centimetre (12-inch) reflector.

Eratosthenes A magnificent crater, with massive walls and a high central peak; it marks one end of the Apennines, and is very like Copernicus except that it is smaller and lacks a comparable ray system. West of it is **Stadius**, a typical ghost ring; it has a diameter of 70 kilometres (44 miles), but its walls have been so reduced that they are barely traceable. Probably the walls can be nowhere more than about 10 metres (33 feet) high.

Hevelius One of the great chain which includes Grimaldi and Riccioli (in the Third Quadrant) and **Cavalerius**. Hevelius has a convex floor and a low central peak; a system of rills lies on the floor. West of Hevelius is **Hedin**, visible only under conditions of extreme libration; it is 98 kilometres (61 miles) in diameter, with irregular, broken walls.

Kepler A bright crater, and the centre of a major ray-system. Its southern neighbour, **Encke**, is of about the same size, but is much less bright and has no comparable ray-system.

Le Verrier This and **Helicon** make up a prominent crater-pair in the Mare Imbrium, near Sinus Iridum.

Lichtenberg A small crater which glows against the dark mare surface. Unusual coloration effects have been reported here.

Plato A large walled plain with fairly low walls, and an iron-grey floor which makes it readily identifiable under any conditions of illumination. There are a few craterlets on the floor, some of which can be 'missed' when they ought logically to be visible. Plato is perfectly circular, though as seen from Earth it is foreshortened into an oval.

Pythagoras Were it farther on the disk, Pythagoras would be truly magnificent, with its high, terraced walls and massive central peak. Farther along the limb, to the south, is the smaller, similar but still very imposing **Xenophanes**.

Timocharis A well-marked formation with a central crater (a peculiarity which it shares with **Lambert**). Timocharis is the centre of a rather obscure system of rays.

Note: north is up.

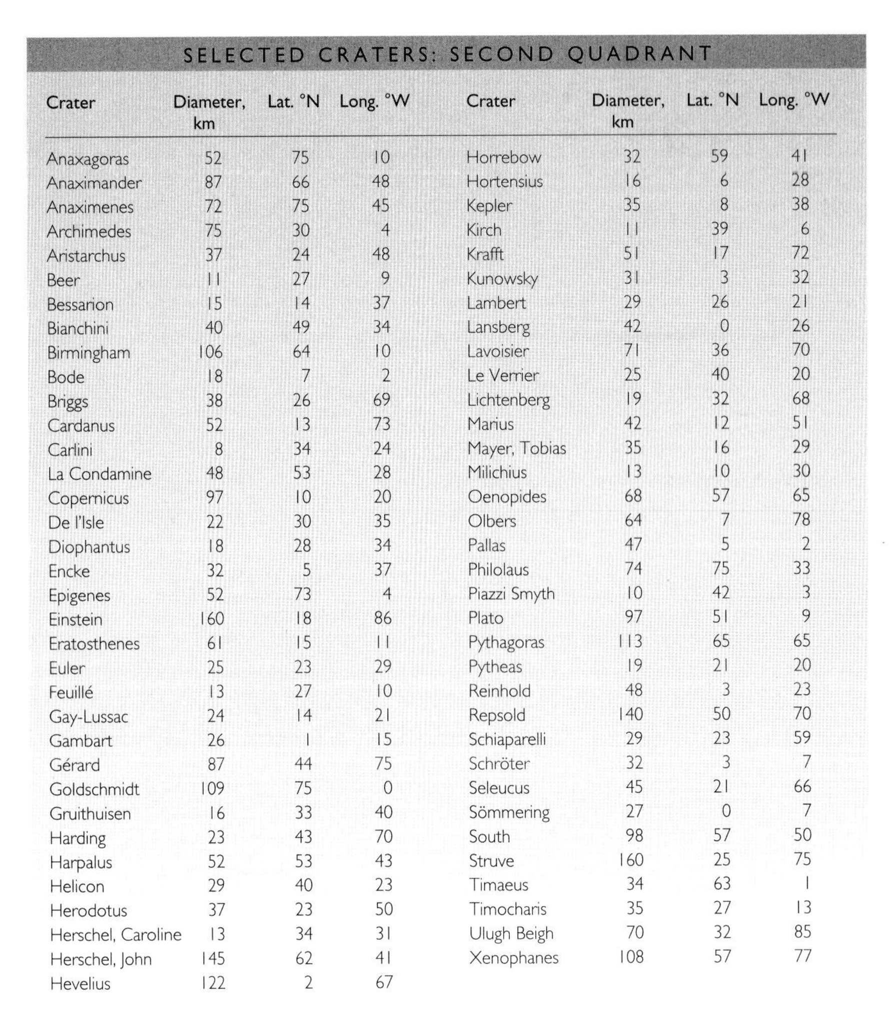

SELECTED CRATERS: SECOND QUADRANT

Crater	Diameter, km	Lat. °N	Long. °W	Crater	Diameter, km	Lat. °N	Long. °W
Anaxagoras	52	75	10	Horrebow	32	59	41
Anaximander	87	66	48	Hortensius	16	6	28
Anaximenes	72	75	45	Kepler	35	8	38
Archimedes	75	30	4	Kirch	11	39	6
Aristarchus	37	24	48	Krafft	51	17	72
Beer	11	27	9	Kunowsky	31	3	32
Bessarion	15	14	37	Lambert	29	26	21
Bianchini	40	49	34	Lansberg	42	0	26
Birmingham	106	64	10	Lavoisier	71	36	70
Bode	18	7	2	Le Verrier	25	40	20
Briggs	38	26	69	Lichtenberg	19	32	68
Cardanus	52	13	73	Marius	42	12	51
Carlini	8	34	24	Mayer, Tobias	35	16	29
La Condamine	48	53	28	Milichius	13	10	30
Copernicus	97	10	20	Oenopides	68	57	65
De l'Isle	22	30	35	Olbers	64	7	78
Diophantus	18	28	34	Pallas	47	5	2
Encke	32	5	37	Philolaus	74	75	33
Epigenes	52	73	4	Piazzi Smyth	10	42	3
Einstein	160	18	86	Plato	97	51	9
Eratosthenes	61	15	11	Pythagoras	113	65	65
Euler	25	23	29	Pytheas	19	21	20
Feuillé	13	27	10	Reinhold	48	3	23
Gay-Lussac	24	14	21	Repsold	140	50	70
Gambart	26	1	15	Schiaparelli	29	23	59
Gérard	87	44	75	Schröter	32	3	7
Goldschmidt	109	75	0	Seleucus	45	21	66
Gruithuisen	16	33	40	Sömmering	27	0	7
Harding	23	43	70	South	98	57	50
Harpalus	52	53	43	Struve	160	25	75
Helicon	29	40	23	Timaeus	34	63	1
Herodotus	37	23	50	Timocharis	35	27	13
Herschel, Caroline	13	34	31	Ulugh Beigh	70	32	85
Herschel, John	145	62	41	Xenophanes	108	57	77
Hevelius	122	2	67				

The Moon: Third Quadrant (South-west)

Highlands occupy a large part of the Third Quadrant, though part of the huge Mare Nubium is included together with the whole of the Mare Humorum. There are some high mountains on the limb, and a very small part of the Mare Orientale can be seen under really favourable libration; otherwise the main mountains are those of the small but prominent Riphaean range, on the Mare Nubium. Of course the most prominent crater is Tycho, whose rays dominate the entire surface around the time of Full Moon. This quadrant also includes two of the most prominent chains of walled plains, those of Ptolemaeus and Walter; the dark-floored Grimaldi and Riccioli; the celebrated plateau Wargentin, and the inappropriately named Straight Wall (Rupes Recta). The most important rill systems (rimae) are those of Sirsalis, Ramsden, Hippalus and Mersenius.

Bailly One of the largest walled plains on the Moon, but unfortunately very foreshortened. It has complex floor detail, and has been described as 'a field of ruins'.

Billy This and **Crüger** are well-formed, and notable because of their very dark floors, which make them easily identifiable.

Bullialdus A particularly fine crater, with massive walls and central peak. It is not unlike Copernicus in structure, though it is not a ray-centre.

Capuanus A well-formed crater, with a darkish floor upon which there is a whole collection of domes.

Clavius A vast walled plain, with walls rising to over 4000 metres (13,000 feet). The north-western walls are broken by a large crater, **Porter**, and there is a chain of craters arranged in an arc across the floor. Near the terminator, Clavius can be seen with the naked eye.

Euclides A small crater near the Riphæan Mountains, easy to find because it is surrounded by a bright nimbus.

Fra Mauro One of a group of low-walled, reduced formations on the Mare Nubium (the others are **Bonpland**, **Parry** and **Guericke**). Apollo 14 landed near here.

Gassendi A grand crater on the north border of the Mare Humorum. The wall has been reduced in places, and is broken in the north by a large crater. There is a rill-system on the floor, and TLP have been seen here. North of Gassendi is a large bay, **Letronne**.

Grimaldi The darkest formation on the Moon. The walls are discontinuous, but contain peaks rising to 2500 metres (8000 feet). Adjoining it is **Riccioli**, which is less regular but has one patch on its floor almost as dark as any part of Grimaldi.

Hippalus A fine bay in the Mare Humorum, associated with a system of rills. Like another similar bay, **Doppelmayer**, Hippalus has the remnant of a central peak.

Kies A low-walled crater on the Mare Nubium, with a flooded floor. Near it lies a large dome with a summit craterlet.

Maginus A very large formation with irregular walls; other large walled plains of the same type in the area are **Longomontanus** and **Wilhelm I**. Maginus is curiously obscure around the time of Full Moon.

Mercator This and **Campanus** form a notable pair. They are alike in form and shape, but Mercator has the darker floor.

Mersenius A prominent walled plain closely west of the Mare Humorum, associated with a fine system of rills.

Moretus A very deep formation in the southern uplands, with a particularly fine central peak.

Newton One of the deepest formations on the Moon, but never well seen because it is so close to the limb.

Pitatus This has been described as a 'lagoon' on the coast of the Mare Nubium. It has a dark floor and a low central peak. A pass connects it with its neighbour **Hesiodus**, which is associated with a long rill extending south-westwards.

Ptolemaeus The largest member of the most imposing line of walled plains on the Moon. Ptolemaeus has a flattish floor with one large crater, **Ammonius**; **Alphonsus** has a central peak and rills on its floor; **Arzachel** is smaller, with higher walls and a more developed central peak. Several TLP have been seen in Alphonsus. Nearby is **Alpetragius**, with regular walls and a central peak crowned by a craterlet.

Purbach One of a line of three major walled plains on the edge of the Mare Nubium. The other members are **Walter**, which has fairly regular walls, and **Regiomontanus**, which gives the impression of having been squashed between Walter to the south and Purbach to the north.

Scheiner This and **Blancanus** are two large, important walled plains close to Clavius.

Schickard One of the major walled plains on the Moon. The walls are rather low and irregular; the floor contains some darkish patches as well as various hills and craterlets.

Schiller A compound formation, produced by the fusion of two old rings.

Sirsalis One of 'Siamese twins' with its neighbour. It is associated with a long and very prominent rill.

Thebit Near the Straight Wall, the crater is broken by Thebit A, which is in turn broken by Thebit F.

Tycho The great ray-crater. Its bright walls make it prominent even in low illumination. Near Full Moon it is clear the rays come tangentially from the walls rather than from the centre.

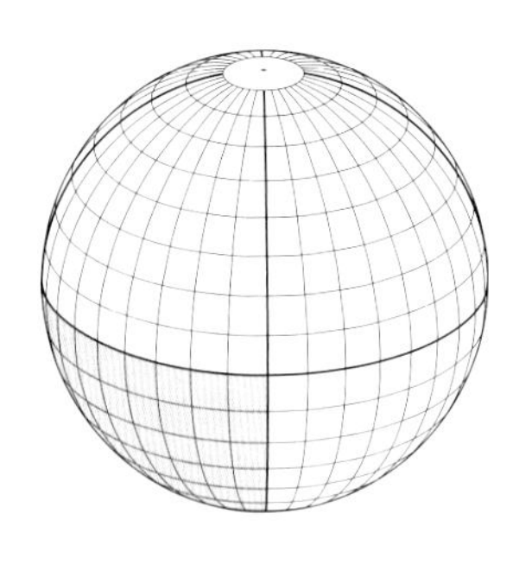

Note: north is up.

SELECTED CRATERS: THIRD QUADRANT

Crater	Diameter, km	Lat. °S	Long. °W	Crater	Diameter, km	Lat. °S	Long. °W
Agatharchides	48	20	31	Lalande	24	4	8
Alpetragius	43	16	4	Lassell	23	16	8
Alphonsus	129	13	3	Le Gentil	140	73	80
Arzachel	97	18	2	Letronne	113	10	43
Bailly	294	66	65	Lexell	63	36	4
Bayer	52	51	35	Lohrmann	45	1	67
Bettinus	66	63	45	Longomontanus	145	50	21
Billy	42	14	50	Maginus	177	50	6
Birt	18	22	9	Mercator	38	29	26
Blancanus	92	64	21	Mersenius	72	21	49
Bonpland	58	8	17	Moretus	105	70	8
Bullialdus	50	21	22	Mösting	26	1	6
Byrgius	64	25	65	Nasireddin	48	41	0
Campanus	38	28	28	Newton	113	78	20
Capuanus	56	34	26	Nicollet	15	22	12
Casatus	104	75	35	Orontius	84	40	4
Clavius	232	56	14	Parry	42	8	16
Crüger	48	17	67	Phocylides	97	54	58
Cysatus	47	66	7	Piazzi	90	36	68
Damoiseau	35	5	61	Pictet	48	43	7
Darwin	130	20	69	Pitatus	86	30	14
Davy	32	12	8	Ptolemaeus	148	9	3
Deslandres	186	32	6	Purbach	120	25	2
Doppelmayer	68	28	41	Regiomantanus	129 × 105	28	0
Euclides	12	7	29	Riccioli	160	3	75
Flammarion	72	3	4	Rocca	97	15	72
Flamsteed	19	5	44	Saussure	50	43	4
Fra Mauro	81	6	17	Scheiner	113	60	28
Gassendi	89	18	40	Schickard	202	44	54
Gauricus	64	34	12	Schiller	180 × 97	52	39
Grimaldi	193	6	68	Segner	74	59	48
Gruemberger	87	68	10	Short	70	76	5
Guericke	53	12	14	Sirsalis	32	13	60
Hainzel	97	41	34	Thebit	60	22	4
Hansteen	36	11	52	Tycho	84	43	11
Heinsius	72	39	18	Vieta	52	29	57
Hell	31	32	8	Vitello	38	30	38
Herigonius	16	13	34	Walter	129	33	1
Herschel	45	6	2	Wargentin	89	50	60
Hesiodus	45	29	16	Weigel	55	58	39
Hippalus	61	25	30	Wichmann	13	8	38
Inghirami	97	48	70	Wilhelm I	97	43	20
Kies	42	26	23	Wilson	74	69	33
Kircher	74	67	45	Wurzelbauer	80	34	16
Klaproth	119	70	26	Zucchius	63	61	50
Lagrange	165	33	72	Zupus	26	17	52

The Moon: *Fourth Quadrant (South-east)*

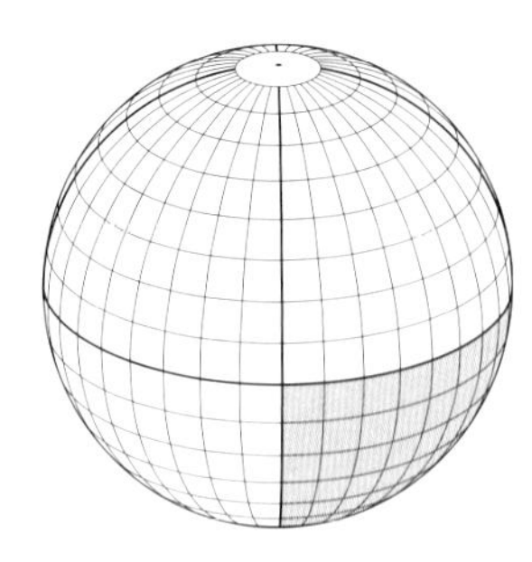

The Fourth Quadrant is made up mainly of highlands, though it does contain the Mare Nectaris, part of the Mare Fecunditatis and the irregular limb-sea Mare Australe. There are large ruined enclosures such as Janssen and Hipparchus, and three imposing formations in a group – Theophilus, Cyrillus and Catharina. We also find four members of the great Eastern Chain: Furnerius, Petavius, Vendelinus and Langrenus. There are two crater-valleys, those of Rheita and Reichenbach, plus the fascinating little Messier, which was once (wrongly) suspected of recent structural change. The feature once called the Altai Mountain range is now known as the Altai Scarp (Rupes Altai), which is certainly a better name for it; it is concentric with the border of the Mare Nectaris, and runs north-west from the prominent crater Piccolomini.

Alfraganus A small, very bright crater; minor ray centre.

Aliacensis This crater and its neighbour **Werner** are very regular. There are several rather similar crater-pairs in the Fourth Quadrant; others are **Abenezra-Azophi** and **Almanon-Abulfeda**.

Capella Crater cut by a valley, with a particularly large central peak with summit pit. Adjoins **Isidorus**, of similar size.

Fracastorius A great bay opening out of the Mare Nectaris. Its seaward wall has been virtually destroyed. Between it and Theophilus is a smaller bay, **Beaumont**.

SELECTED CRATERS: FOURTH QUADRANT

Crater	Diameter, km	Lat. °S	Long. °E	Crater	Diameter, km	Lat. °S	Long. °E
Abenezra	43	21	12	La Pérouse	72	10	78
Abulfeda	64	14	14	Legendre	74	29	70
Airy	35	18	6	Licetus	74	47	6
Albategnius	129	12	4	Lilius	52	54	6
Alfraganus	19	6	19	Lindenau	56	32	25
Aliacensis	84	31	5	Lockyer	48	46	37
Apianus	63	27	8	Maclaurin	45	2	68
Azophi	43	22	13	Mädler	32	11	30
Barocius	80	45	17	Magelhaens	40	12	44
Beaumont	48	18	29	Manzinus	90	68	25
Blanchinus	53	25	3	Marinus	48	40	75
Boguslawsky	97	75	45	Messier	13	2	48
Bohnenberger	35	16	40	Metius	81	40	44
Brisbane	47	50	65	Mutus	81	63	30
Buch	48	39	18	Neander	48	31	40
Büsching	58	38	20	Nearch	61	58	39
Capella	48	8	36	Oken	80	44	78
Catharina	89	18	24	Palitzsch	41	28	64
Cyrillus	97	13	24	Parrot	64	15	3
Delambre	52	2	18	Petavius	170	25	61
Demonax	121	85	35	Phillips	120	26	78
Donati	35	21	5	Piccolomini	80	30	32
Fabricius	89	43	42	Pitiscus	80	51	31
Faraday	64	42	18	Playfair	43	23	9
Faye	35	21	4	Pons	32	25	22
Fermat	40	23	20	Pontécoulant	97	59	65
Fernelius	64	38	5	Rabbi Levi	80	35	24
Fracastorius	97	21	33	Réaumur	45	2	1
Furnerius	129	36	60	Reichenbach	48	30	48
Goclenius	52	10	45	Rheita	68	37	47
Gutenberg	72	8	41	Riccius	80	37	26
Hagecius	81	60	46	Rosse	16	18	35
Halley	35	8	6	Sacrobosco	84	24	17
Hecataeus	180	23	84	Steinheil	70	50	48
Helmholtz	97	72	78	Stevinus	70	33	54
Hind	26	8	7	Stöfler	145	41	6
Hipparchus	145	6	5	Tacitus	40	16	19
Hommel	121	54	33	Theon Junior	16	2	16
Horrocks	29	4	6	Theon Senior	17	1	15
Humboldt, Wilhelm	193	27	81	Theophilus	101	12	26
				Torricelli	19	5	29
Isidorus	48	8	33	Vendelinus	165	16	62
Janssen	170	46	40	Vlacq	90	53	39
Kant	30	11	20	Watt	72	50	51
La Caille	53	24	1	Webb	26	1	60
Langrenus	137	9	61	Werner	66	28	3

Goclenius. A fairly regular crater, making up a group with less perfect **Gutenberg** and deformed **Magelhaens**.

Hipparchus A very large enclosure not far from Ptolemaeus. It is very broken, but under low light is still impressive. It adjoins **Albategnius**, which is rather better preserved and has a low central peak.

Humboldt, Wilhelm A huge formation, too foreshortened to be well seen – though space probe pictures show that it has considerable floor detail, including a system of rills. It adjoins the smaller formation of **Phillips**, which is of similar type.

Janssen A vast enclosure, but in a very poor state of repair. Its walls are broken in the north by **Fabricius** and in the south by the bright-walled **Lockyer**.

Langrenus One of the great Eastern Chain. It has high, terraced walls, rising to over 3000 metres (10,000 feet), and a bright twin-peaked central elevation. Near Full Moon, Langrenus appears as a bright patch. Disturbances inside Langrenus have been photographed by A. Dollfus – the best proof to date of TLPs.

Mädler A prominent though irregular crater on the Mare Nectaris. It is crossed by a ridge.

Messier This and its twin, Messier A (formerly known as W. J. Pickering) lie on the Mare Fecunditatis. They show remarkable changes in appearance over a lunation, though there has certainly been no real change in historic times. The unique 'comet' ray extends to the west.

Metius A well-formed walled plain near Janssen.

Oken A crater along the limb from the Mare Australe, easy to identify because of its darkish floor.

Petavius A magnificent crater – one of the finest on the Moon. Its walls rise to over 3500 metres (11,500 feet) in places; the slightly convex floor contains a complex central mountain group, and a prominent rill runs from the centre to the south-west wall. Oddly enough, Petavius is none too easy to identify at full moon. Immediately outside it is **Palitzsch**, once described as a 'gorge'. In fact, it is a crater-chain – several major rings which have coalesced.

Piccolomini The prominent, high-walled crater at the arc of the Altai Scarp.

Rheita A deep crater with sharp walls. Associated with it is the so-called 'Valley' (Rupes Rheita) over 180 kilometres (110 miles) long and in places up to 25 kilometres (15 miles) broad; it is not a true valley, but is made up of craterlets. Not far away is the Reichenbach valley, which is of similar type but is not so conspicuous or so well-formed.

Steinheil This and its neighbour **Watt** make up a pair of 'Siamese twins'.

Stöfler A grand enclosure, with an iron-grey floor which makes it easy to find. Part of the rampart has been destroyed by the intrusion of **Faraday**.

Theon Senior and **Theon Junior**. Very bright craterlets near the regular, conspicuous **Delambre**. In many ways they resemble Alfraganus.

Theophilus One of the most superb features of the Moon, and in every way the equal of Copernicus except that it is not a ray-centre. It is very deep, with peaks rising to 4400 metres (14,400 feet) above the floor. There is a magnificent central mountain group. It adjoins **Cyrillus**, which is less regular and in turn adjoins very rough-floored **Catharina**.

Vendelinus A member of the Eastern Chain, but less regular than Langrenus or Petavius, and presumably older. It has no central peak, and in places the walls are broken.

Vlacq A deep, well-formed crater with a central peak; it is a member of a rather complex group, of which other members are **Hommel** and **Hagecius**.

Webb A crater very near the lunar equator, with a darkish floor and a central hill; centre of system of short, faint rays.

Note: North is up.

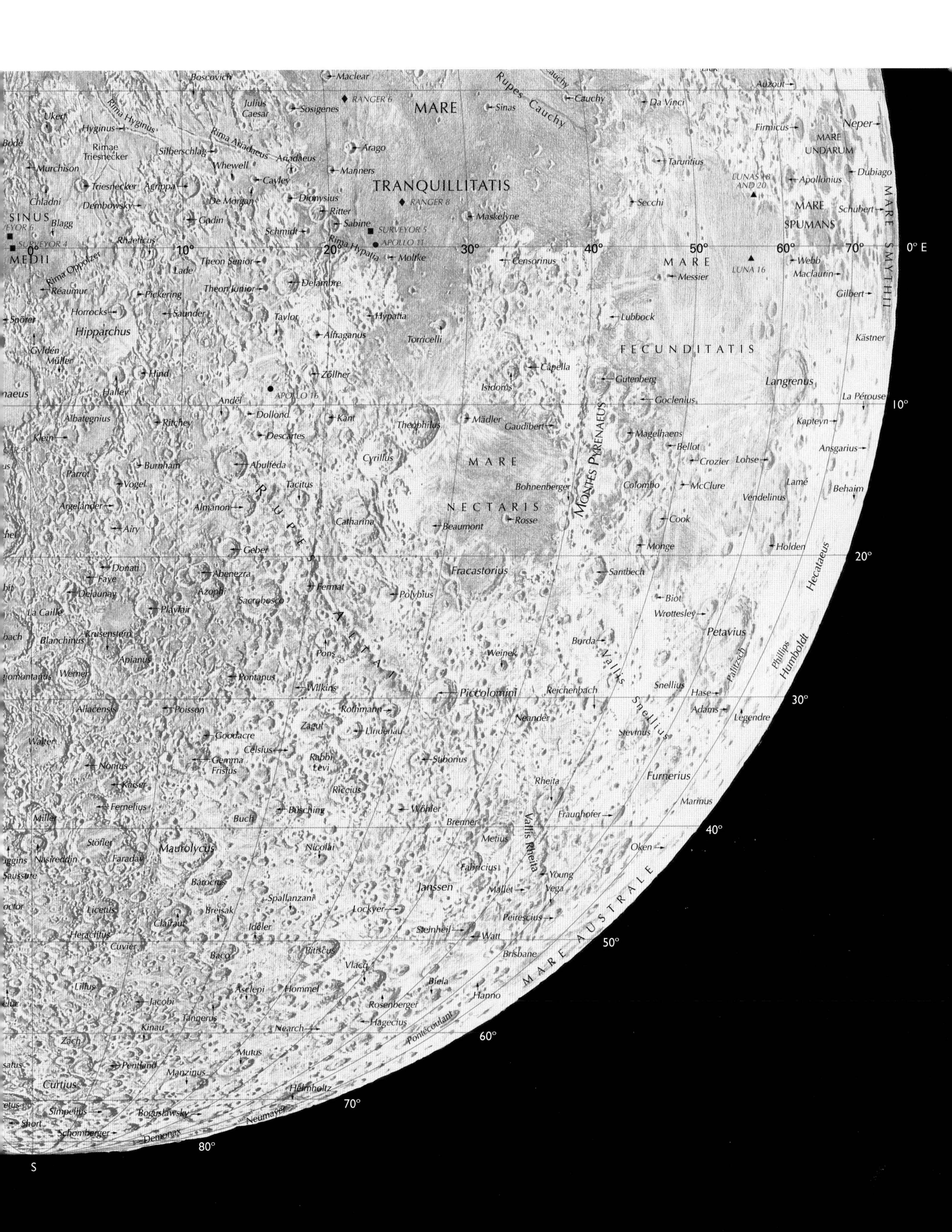
Boscovich
Maclear
Rupes Cauchy
Cauchy
Auzout
Julius Caesar
Sosigenes
RANGER 6
MARE
Sinas
Da Vinci
Ukert
Rima Hyginus
Hyginus
Rima Ariadaeus
Firmicus
Neper
MARE UNDARUM
Bode
Rimae Triesnecker
Silberschlag
Arago
Ariadaeus
Taruntius
Murchison
Whewell
Manners
Dubiago
Cayley
LUNAS 18 AND 20
Apollonius
Triesnecker
Agrippa
TRANQUILLITATIS
Chladni
De Morgan
Dionysius
RANGER 8
Secchi
MARE SPUMANS
Schubert
MARE SMYTHII
SINUS MEDII
Dembowsky
Ritter
Maskelyne
Blagg
Godin
Sabine
SURVEYOR 5
Schmidt
SURVEYOR 4
Rhaeticus
Rima Hypatia
APOLLO 11
0°
10°
20°
30°
40°
50°
60°
70°
0° E
Rima Oppolzer
Theon Senior
Moltke
Censorinus
MARE
LUNA 16
Webb
Lade
Messier
Maclaurin
Delambre
Réaumur
Theon Junior
Gilbert
Pickering
Horrocks
Saunder
Taylor
Hypatia
Lubbock
Spörer
Hipparchus
Alfraganus
Torricelli
Kästner
FECUNDITATIS
Gyldén
Müller
Capella
Hind
Zöllner
Gutenberg
Langrenus
Halley
Isidorus
APOLLO 16
Anděl
Goclenius
La Pérouse
Albategnius
Dollond
10°
Kant
Mädler
Ritchey
Theophilus
Gaudibert
Kapteyn
MONTES PYRENAEUS
Magelhaens
Klein
Descartes
Bellot
Ansgarius
Cyrillus
Burnham
Abulfeda
Crozier
Lohse
MARE
Parrot
Vogel
Tacitus
Bohnenberger
Colombo
McClure
Lamé
Behaim
Vendelinus
Argelander
Almanon
NECTARIS
Catharina
Rosse
Cook
Beaumont
Airy
RUPES ALTAI
Monge
Holden
Geber
Donati
Hecataeus
20°
Fracastorius
Santbech
Faye
Abenezra
Azophi
Fermat
Delaunay
Polybius
Biot
Sacrobosco
La Caille
Playfair
Wrottesley
Krusenstern
Petavius
Blanchinus
Borda
Phillips
Humboldt
Pons
Apianus
Weinek
Palitzsch
Vallis Snellius
Werner
Pontanus
Snellius
Wilkins
Piccolomini
Reichenbach
Hase
30°
Aliacensis
Poisson
Rothmann
Adams
Legendre
Neander
Zagut
Lindenau
Stevinus
Goodacre
Walter
Celsius
Gemma Frisius
Rabbi Levi
Nonius
Suborius
Furnerius
Rheita
Kaiser
Riccius
Marinus
Fernelius
Büsching
Wöhler
Fraunhofer
Miller
Buch
Brenner
40°
Vallis Rheita
Metius
Stöfler
Oken
Nasireddin
Maurolycus
Faraday
Nicolai
Fabricius
Young
Barocius
MARE AUSTRALE
Janssen
Mallet
Vega
Spallanzani
Licetus
Breisak
Lockyer
Peirescius
Clairaut
Ideler
Steinheil
Heraclitus
Watt
Cuvier
50°
Bitiscus
Brisbane
Baco
Vlacq
Lilius
Biela
Asclepi
Hommel
Jacobi
Hanno
Rosenberger
Tannerus
Kinau
Nearch
Hagecius
Pontécoulant
60°
Zach
Mutus
Pentland
Manzinus
Curtius
Helmholtz
70°
Simpelius
Boguslawsky
Neumayer
Short
Schomberger
Demonax
80°
S

The Planets seen from Earth

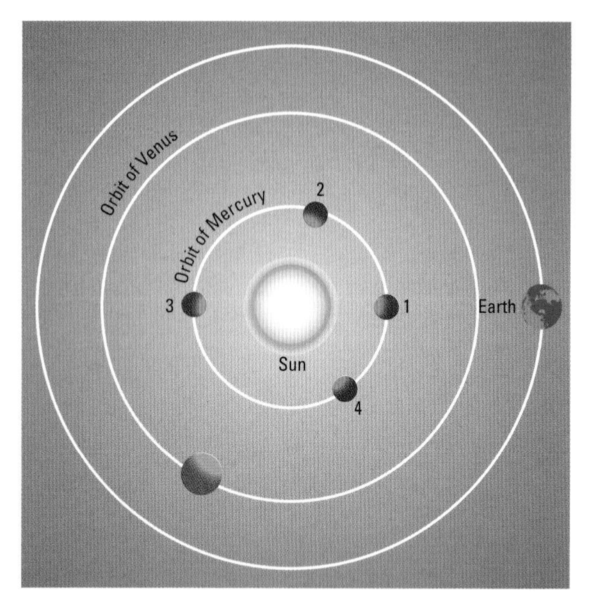

◀ Phases of Mercury. (1) New. (2) Dichotomy (half-phase). (3) Full. (4) Dichotomy. For the sake of clarity, I have not taken the Earth's movement round the Sun into account in this diagram.

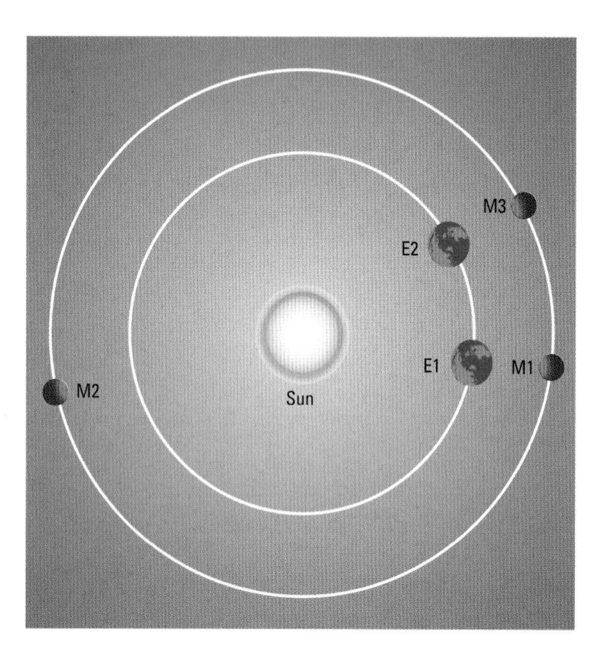

◀ Movements of Mars. With the Earth at E1 and Mars at M1, Mars is in opposition. A year later the Earth has come back to E1, but Mars has only reached M2. The Earth has to catch Mars up before there is another opposition, with the Earth at E2 and Mars at M3.

▶ Retrograde motion of Mars. As the Earth catches Mars up and passes it, the movement will seem to be retrograde, so that between 3 and 5 Mars will appear to go backwards in the sky – east to west, against the stars, instead of west to east.

The word 'planet' really means 'wanderer', and the planets were first identified in ancient times by their movements against the starry background. Because their orbits are not greatly inclined to that of the Earth – less than 4° for all the planets apart from Mercury – they seem to keep to a well-defined band around the sky, termed the Zodiac. There are twelve official Zodiacal constellations, though a thirteenth, Ophiuchus (the Serpent-bearer) does cross the zone for some distance.

The 'inferior' planets, Mercury and Venus, are closer to the Sun than we are, and have their own way of behaving. They seem to stay in the same general area of the sky as the Sun, which makes them awkward to observe – particularly in the case of Mercury, where the greatest elongation from the Sun can never be as much as 30°. They show phases similar to those of the Moon, from new to full, but there are marked differences. At new phase, the dark side of the planet is turned towards us, and we cannot see it at all unless the alignment is perfect, when the planet will appear in transit as a dark disk crossing the face of the Sun. This does not happen very often; Venus was last in transit in 2004, and will be again on 6 June 2012, and then not for another 105 years when there will be another pair of transits 8 years apart. Transits of Mercury are less uncommon; the last was on 8 November 2006; followed by that of 9 May 2016.

When an inferior planet is full, it is on the far side of the Sun, and is to all intents and purposes out of view. At other times (top left diagram) the phase may be crescent, half (dichotomy), or gibbous (between half and full). At new, the planet is at inferior conjunction; when full, it is at superior conjunction. These movements mean that the inferior planets are best seen either in the west after sunset, or in the east before sunrise. They never remain above the horizon throughout a night.

The superior planets, the orbits of which lie beyond that of the Earth in the Solar System, can reach superior conjunction – though for obvious reasons they can never pass through inferior conjunction. When seen at right angles to the Sun, they are said to be at quadrature. When near quadrature Mars can show an appreciable phase – down to 85 per cent – so that when viewed through a telescope its shape resembles that of the Moon a day or two from full. The giant planets are so far away that their phases are inappreciable.

When the Sun, the Earth and a planet are lined up, with the Earth in the mid position, the planet is at opposition; it is exactly opposite to the Sun in the sky, and is best placed for

▲ **Transit of Venus,** 8 June 2004. Photograph by Jamie Cooper, from Selsey in Sussex. Venus is jet-black; unfortunately there were no large sunspots with which to compare it.

observation. The interval between one opposition and the next is known as the synodic period.

The movements of Mars are shown in the bottom left and right-hand diagrams opposite. It is clear that oppositions do not occur every year; the Earth has to 'catch Mars up', and the mean synodic period is 780 days. Oppositions of Mars occur in 2007 and 2010, but not in 2008 or 2009. As the Earth 'passes' Mars, there is a period when the planet will move against the stars in an east-to-west or retrograde direction. The giant planets are so much farther away, and move so much more slowly, that they come to opposition every year. Jupiter's synodic period is 399 days, but that of Neptune is only 367.5 days, so that it comes to opposition less than two days later every year.

There should be no trouble in identifying Jupiter and Venus, because they are always so brilliant; Mercury is unlikely to be seen unless deliberately looked for, while Uranus is on the fringe of naked-eye visibility and Neptune is much fainter. Mars at its best can actually outshine all the planets apart from Venus, but when at its faintest it is little brighter than the Pole Star, though its strong red colour will usually betray it. Saturn is brighter than most stars, and because it takes almost 30 years to complete one journey round the Zodiac it can be found without difficulty once initially identified.

Planets can pass behind the Moon, and be occulted. The planets themselves may occult stars, and these events are interesting to watch, but they do not happen very often, and an occultation of a bright star by a planet is very rare.

Mercury

Mercury, the innermost planet, is never easy to study from Earth. It is small, with a diameter of only 4878 kilometres (3030 miles); it always stays in the same region of the sky as the Sun, and it never comes much within 80 million kilometres (50 million miles) of us. Moreover, when it is at its nearest it is new, and cannot be seen at all except during the comparatively rare transits.

Mercury has a low escape velocity, and it has always been clear that it can have little in the way of atmosphere. The orbital period is 88 days. It was once assumed that this was also the length of the axial rotation period, in which case Mercury would always keep the same face turned towards the Sun, just as the Moon does with respect to the Earth; there would be an area of permanent day, a region of everlasting night, and a narrow 'twilight zone' in between, over which the Sun would bob up and down over the horizon – because the orbit of Mercury is decidedly eccentric, and there would be marked libration effects. However, this has been shown to be wrong. The real rotation period is 58.6 days, or two-thirds of a Mercurian year, and this leads to a very curious calendar indeed. To an observer on the planet's surface, the interval between sunrise and sunset would be 88 Earth-days and the length of day (sunrise to sunrise) is 176 days.

PLANETARY DATA – MERCURY	
Sidereal period	87.969 days
Rotation period	58.6461 days
Mean orbital velocity	47.87 km/s (29.76 mi/s)
Orbital inclination	7° 00' 15".5
Orbital eccentricity	0.206
Apparent diameter	max. 12".9, min. 4".5
Reciprocal mass, Sun = 1	6,000,000
Density, water = 1	5.44
Mass, Earth = 1	0.055
Volume, Earth = 1	0.056
Escape velocity	4.3 km/s (2.7 mi/s)
Surface gravity, Earth = 1	0.38
Mean surface temperature	350°C (day); –170°C (night)
Oblateness	Negligible
Albedo	0.06
Maximum magnitude	–1.9
Diameter	4878 km (3030 mi)

Earth

▼ **Mariner 10.** So far, this is the only spacecraft to have by-passed Mercury; it was also the first to use the gravity-assist technique. It has provided us with our only good maps of the surface, and has shown that the Earth-based maps (even Antoniadi's) were very inaccurate. Even so, it was able to image less than half the surface, so that our knowledge of the topography of Mercury is still very incomplete.

The orbital eccentricity makes matters even stranger, because the heat received at perihelion is 2½ times greater than at aphelion. At a 'hot pole', where the Sun is overhead at perihelion, the temperature rises to +427°C, but at night a thermometer would register –183°C. Mercury has an extremely uncomfortable climate.

To an observer situated at a hot pole, the Sun will rise when Mercury is at aphelion and the solar disk will be at its smallest. As the Sun nears the zenith, it will grow in size, but for a while the orbital angular velocity will be greater than the constant spin angular velocity; our observer will see the Sun pass the zenith, stop and move backwards in the sky for eight Earth-days before resuming its original direction of motion. There are two hot poles, one or the other of which will always receive the full blast of solar radiation when Mercury is at perihelion. An observer 90° away will have a different experience; the Sun will rise at perihelion, so that after first coming into view it will sink again before starting its climb to the zenith. At sunset it will disappear, and then rise again briefly before finally departing, not to rise again for another 88 Earth-days.

Mercury is denser than any planet apart from the Earth. There seems to be an iron-rich core about 3600 kilometres (2250 miles) in diameter (larger than the whole of the Moon), containing about 80 per cent of the total mass; by weight Mercury is 70 per cent iron and only 30 per cent rocky material. The core is presumably molten, and above it comes a 600-kilometre (370-mile) mantle and crust composed of silicates.

Most of our detailed knowledge of Mercury has been obtained from one probe, Mariner 10. It was launched on 3 November 1973, and after by-passing the Moon made rendezvous with Venus on 5 February 1974. The gravity field of Venus was used to send Mariner in towards an encounter with Mercury, and altogether there were three active passes: on 29 March and 21 September 1974, and 16 March 1975, by which time the equipment was starting to fail. The last messages were received on 24 March 1975. A new mission, Messenger, was launched in 2004, but will not enter orbit round Mercury until 2011, but will make close pssses of the planet in January and October 2008 and September 2009.

As expected, the atmosphere proved to be almost non-existent. The ground pressure is about 10^{-9} mb ($\frac{1}{1000}$ million) of a millibar, and the main constituent is helium, presumably drawn from the solar wind. A magnetic field was detected, with a surface value about one per cent of the Earth's field; there are two magnetic poles of opposite polarity, inclined by 11 degrees to the rotational axis. The polarity of the field is the same as ours; that is to say, a compass needle would point north. The field is just strong enough to deflect the solar wind away from the planet's surface.

It has to be admitted that Mercury is not a rewarding telescopic object and little will be seen apart from the characteristic phase. Any form of life there seems out of the question.

▲ **Mercury from Mariner 10.** Six hours after its closest approach, Mariner 10 took this series of 18 images of Mercury's surface, which have been combined to make a photomosaic.
Note that, in general, the arrangement of the craters follows the lunar pattern; small craters break into larger ones, not vice versa. There are also ray centres. The north pole is at the top.

◀ **The latest probe, Messenger, blasts off from Cape Caneveral atop a Delta II rocket on 3 August 2004.**

Venus

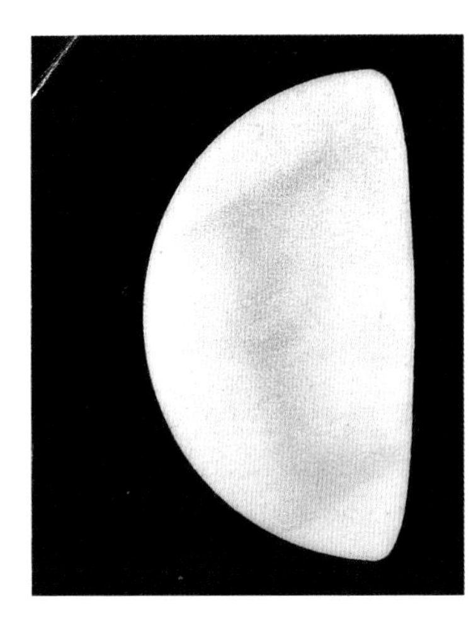

▲ **Venus** drawn with my 31-cm (12½-inch) reflector. All that could be seen were very vague, cloudy shadings which are necessarily rather exaggerated in the sketch, together with slightly brighter areas near the cusps. The terminator appeared essentially smooth. The internal structure of Venus may not be too unlike that of the Earth, but with a thicker crust and an iron-rich core which is smaller both relatively and absolutely. There is no detectable magnetic field and, like Mercury, Venus has no satellite.

Venus, the second planet in order of distance from the Sun, is as different from Mercury as it could possibly be. It is far brighter than any other star or planet, and can cast strong shadows; very keen-sighted people can see the phase with the naked eye during the crescent stage, and binoculars show it easily. Yet telescopically Venus is a disappointment. Little can be seen, and generally the disk appears blank. We are looking not at a solid surface, but at the top of a layer of cloud which never clears. Before the Space Age, we knew very little about Venus as a world.

We knew the size and mass; Venus is only very slightly inferior to the Earth, so that the two are near-twins. The orbital period is 224.7 days, and the orbit round the Sun is almost circular. Estimates of the rotation period ranged from less than 24 hours up to many months, but the favoured value was about a month. The vague shadings sometimes visible on the disk were much too indefinite to give any reliable results. There was also the Ashen Light, or dim visibility of the 'night' side, when Venus was in the crescent phase. It seemed to be real, but few people agreed with the 19th-century astronomer Franz von Paula Gruithuisen that it might be due to illuminations on the planet's surface lit by the local inhabitants to celebrate the accession of a new emperor! A recent theory attributes the Ashen Light to the very hot surface of the planet shining dimly through the atmosphere. This certainly seems more plausible than Gruithuisen's revellers.

It was suggested that Venus might be in the condition of the Earth during the Carboniferous period, with swamps and luxuriant vegetation of the fern and horsetail type; as recently as the early 1960s many astronomers were confident that the surface was mainly covered with water, though it was also thought possible that the surface temperature was high enough to turn Venus into a raging dust-desert. It had been established that the upper part of the atmosphere, at least, was made up mainly of carbon dioxide, which tends to shut in the Sun's heat.

The first positive information came in December 1962, when the American spacecraft Mariner 2 passed by Venus at a range of less than 35,000 kilometres (21,800 miles) and sent back data which at once disposed of the attractive 'ocean' theory. In 1970 the Soviets managed to make a controlled landing with Venera 7, which transmitted for 23 minutes before being put out of action, and on 21 October 1975 another Soviet probe, Venera 9, sent back the first picture direct from the surface. It showed a forbidding, rock-strewn landscape, and although the rocks are grey they appear orange by reflection from the clouds above. The atmospheric pressure was found to be around 90 times that of the Earth's air at sea level, and the temperature is over 480°C.

Radar measurements have shown that the rotation period is 243.2 days – longer than Venus' 'year'; moreover, the planet rotates from east to west, in a sense opposite to that of the Earth. If it were possible to see the Sun from the surface of Venus, it would rise in the west and set in the east and the following sunrise would be 118 Earth-days later, so that in its way the calendar of Venus is every bit as strange as that of Mercury. The reason for this retrograde rotation is not known. According to one theory, Venus was hit by a massive body early in its evolution and literally knocked over. This does not sound very plausible, but it is not easy to think of anything better.

It has been found that the top of the atmosphere lies around 400 kilometres (250 miles) above the surface, and that the upper clouds have a rotation period of only 4 days. The upper clouds lie at an altitude of 70 kilometres (44 miles), and there are several definite cloud layers, though below 30 kilometres (19 miles) the atmosphere is relatively clear and calm. The atmosphere's main constituent is indeed carbon dioxide, accounting for over 96 per cent of the whole; most of the rest is nitrogen. The clouds are rich in sulphuric acid; at some levels there must be sulphuric acid 'rain' which evaporates before reaching ground level.

There was a time – not so long ago – when Venus was regarded as much more friendly than Mars, and was the main planetary target of spacecraft. A manned flight there did not seem to be out of the question. We know better now, a mission to Venus may be possible in the future, but certainly not for a very long time yet. Once again Mars has become the main target and Venus has been relegated to second place.

▼ **Four photographs of Venus** taken on the same scale. As the phase shrinks, the apparent diameter increases.

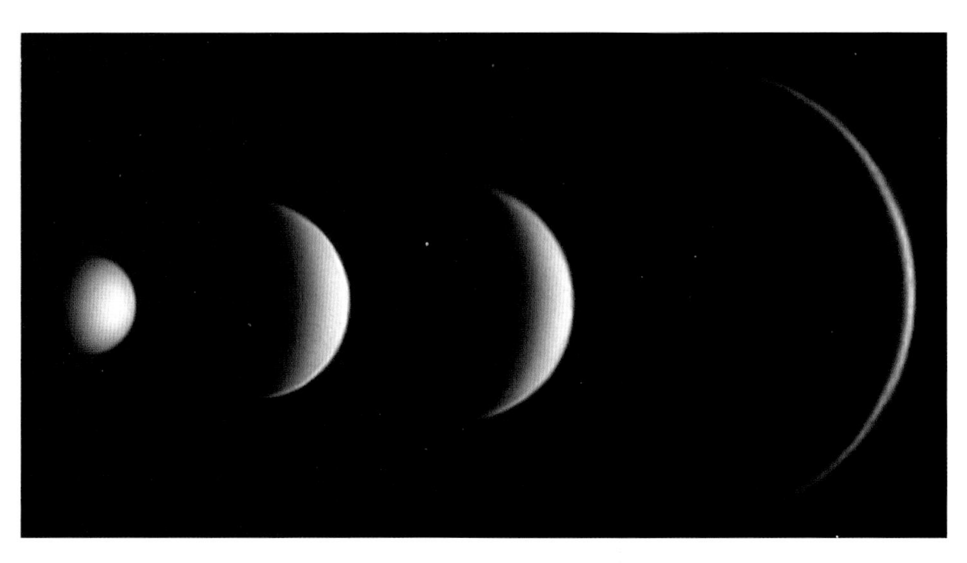

PLANETARY DATA – VENUS	
Sidereal period	224.701 days
Rotation period	243.16 days
Mean orbital velocity	35.02 km/s (21.76 mi/s)
Orbital inclination	3°23' 39".8
Orbital eccentricity	0.007
Apparent diameter	max. 65".2 min. 9".5 mean 37".3
Reciprocal mass, Sun = 1	408,520
Density, water = 1	5.25
Mass, Earth = 1	0.815
Volume, Earth = 1	0.86
Escape velocity	10.36 km/s (6.43 mi/s)
Surface gravity, Earth = 1	0.903
Mean surface temperature	cloud-tops 33°C surface +480°C
Oblateness	0
Albedo	0.76
Maximum magnitude	−4.4
Diameter	12,104 km (7523 mi)

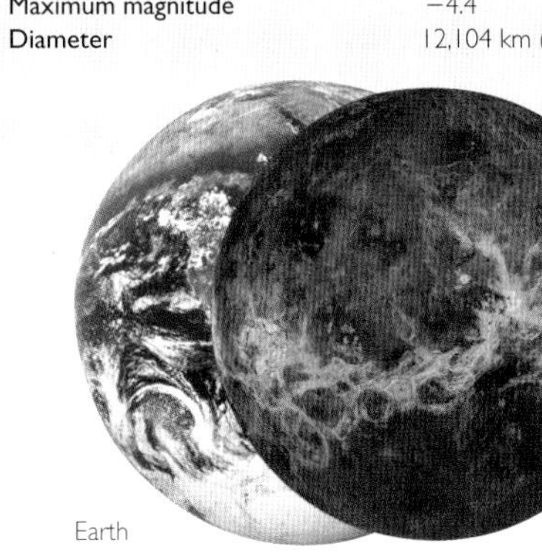

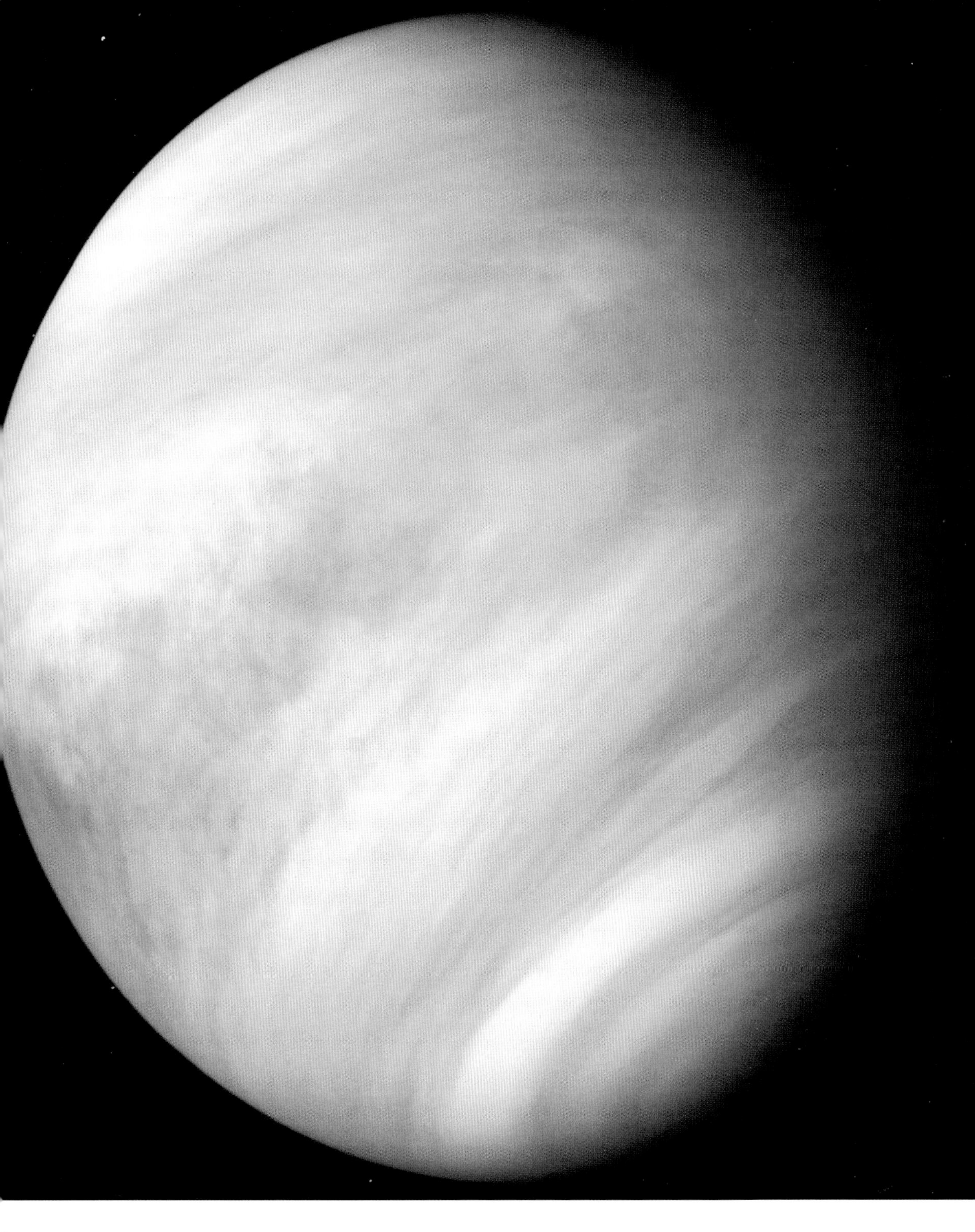

◀ **The clouds of Venus** imaged by the Pioneer Venus Orbiter spacecraft. The upper clouds race around the planet in just 4 days. The well-known 'D'-shape cloud feature is well shown.

▼ **Venera 13** on the surface of Venus in March 1982. Part of the spacecraft is shown in this picture; the temperature was measured at 457°C and the pressure at 89 atmospheres. The rock was reddish-brown, and the sky brilliant orange.

Mapping Venus

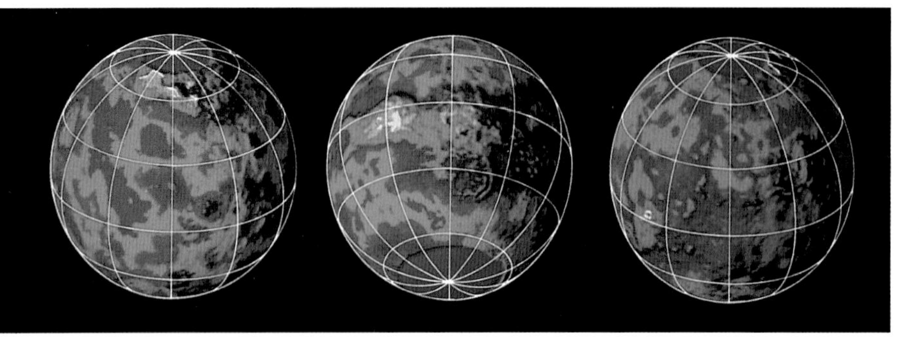

▲ Topographic globes of Venus. NASA's Pioneer Venus Orbiter mapped the planet using radar. The map was compiled as a false-colour representation with blue indicating low levels and yellow and red higher areas. Ishtar Terra and Aphrodite Terra stand out very clearly.

Because we can never see the surface of Venus, the only way to map it is by radar. It has been found that Venus is a world of plains, highlands and lowlands; a huge rolling plain covers 65 per cent of the surface, with lowlands accounting for 27 per cent and highlands for only 8 per cent. The higher regions tend to be rougher than the lowlands, and this means that in radar they are brighter (in a radar image, brightness means roughness).

There are two main upland areas, Ishtar Terra and Aphrodite Terra. Ishtar, in the northern hemisphere, is 2900 kilometres (1800 miles) in diameter; the western part, Lakshmi Planum, is a high, smooth, lava-covered plateau. At its eastern end are the Maxwell Montes, the highest peaks on Venus, which rise to 11 kilometres (nearly 7 miles) above the mean radius and 8.2 kilometres (5 miles) above the adjoining plateau. Aphrodite straddles the equator; it measures 9700 × 3200 kilometres (6000 × 2000 miles), and is made up of several volcanic massifs, separated by fractures. Diana Chasma, the deepest point on Venus, adjoins Aphrodite.

(En passant, it has been decreed that all names of features in Venus must be female. The only exception is that of the Maxwell Montes. The name of the Scottish mathematician James Clerk Maxwell had been assigned to them before the official edict was passed!)

A smaller highland area, Beta Regio, includes the shield volcano, Rhea Mons and the rifted mountain Theia Mons. Beta, which is cut by a huge rift valley rather like the Earth's East African Rift Valley, is of great interest. It is likely that Rhea is still active, and there can be no doubt that the whole surface of Venus is dominated by vulcanism. Venus' thick crust will not slide over the mantle in the same way as that of the Earth, so that plate tectonics do no apply; when a volcano forms over a hot spot it will remain there for a very long period. Lava flows are found over the whole of the surface.

Craters are plentiful, some of them irregular in shape while others are basically circular. The largest, Mead, has a

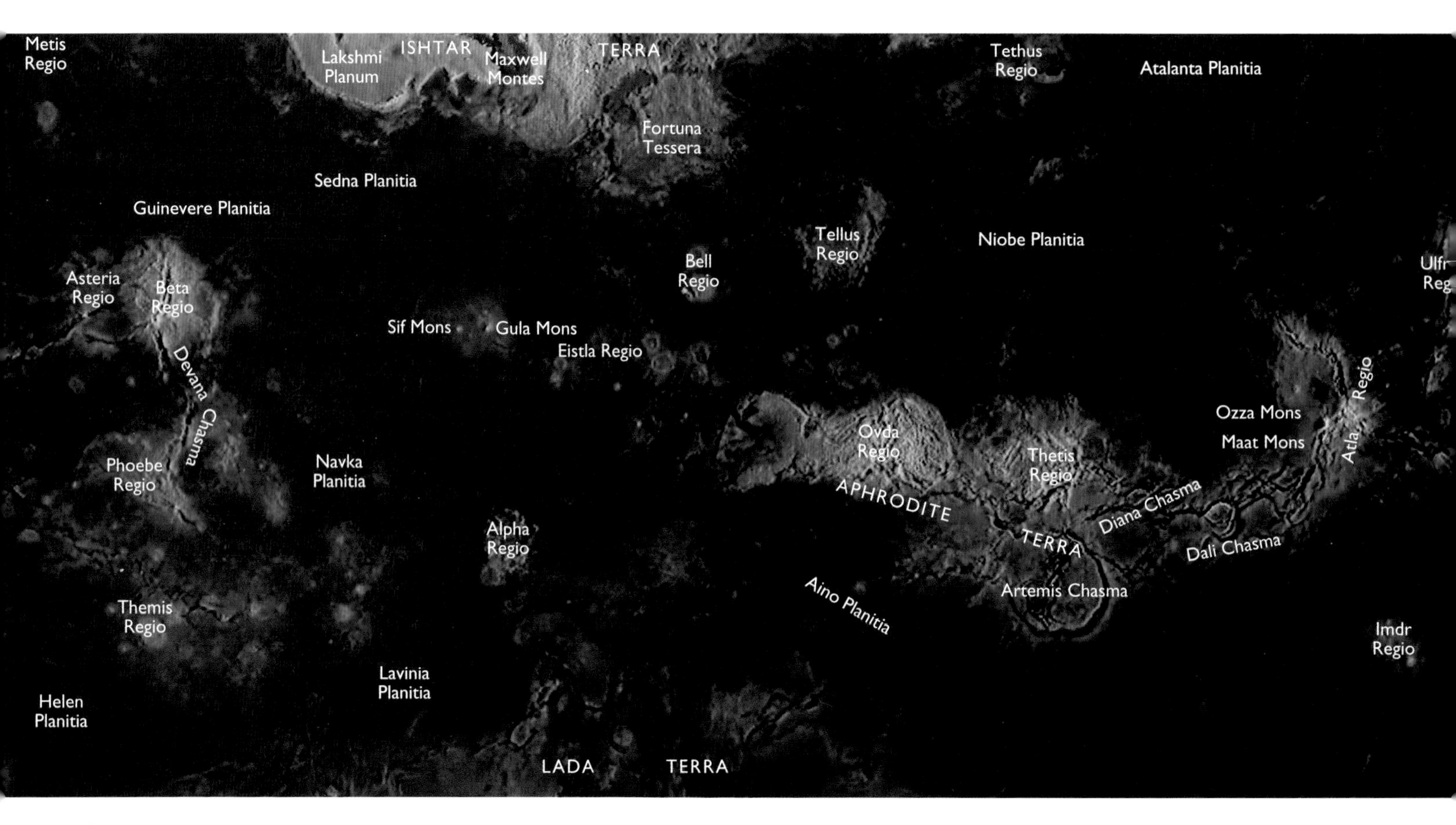

▼ Map of Venus, from radar measurements. Low-lying areas are shown in blue, intermediate in green, higher areas in yellow, and the highest in red.

◀ ▶ The topography of Venus. These images were obtained by the Magellan radar altimeter during its 24 months of systematic mapping of the surface of Venus. Colour is used to code elevation, and simulated shading to emphasize relief. Red corresponds to the highest elevations and blue to the lowest. At left are the two polar regions in orthographic projection. The image at far left is centred on the North Pole, and that at near left on the South Pole. The four images at right are hemispheric views centred on (from top to bottom) 0°E, 90°E, 180°E and 270°E. North is at the top. The resolution of detail on the surface is about 3 km (2 mi). A mosaic of the Magellan images forms the base for the maps; gaps in the coverage were filled with images from the Earth-based Arecibo radar, with extra elevation data from the Venera spacecraft and the US Pioneer Venus missions.

diameter of 280 kilometres (175 miles), though small craters are less common than on Mercury, Mars or the Moon because of the dense shielding atmosphere.

There are circular lowland areas, such as Atalanta Planitia, east of Ishtar; there are systems of faults, and there are regions now called tesserae – high, rugged tracts extending for thousands of square kilometres and characterized by intersecting ridges and grooves. Tesserae used to be called 'parquet terrain', but although the term was graphic it was abandoned as being insufficiently scientific.

Venus has been contacted by fly-by probes, radar-carrying orbiters and soft landers; in 1985 the two Soviet probes en route for Halley's Comet dispatched two balloons into the upper atmosphere of the planet, so that information could be sent back from various levels as the balloons drifted around. The Magellan probe (launched 1989) surveyed the surface with its radar in much more detail, and Venus Express (2005) has confirmed and extended the earlier findings.

FEATURES ON VENUS – SELECTED LIST

	Feature	Lat.°	Long.°
TERRAE	Aphrodite	40 S–5 N	140–000
	Ishtar	52–75 N	080–305
REGIONES	Alpha	29–32 S	000
	Asteria	18–30 N	228–270
	Beta	20–38 N	292–272
	Metis	72 N	245–255
	Phoebe	10–20 N	275–300
	Tellus	35 N	080
	Thetis	02–15 S	118–140
PLANITIA	Atalanta	54 N	162
	Lakshmi Planum	60 N	330
	Lavinia	45 S	350
	Leda	45 N	065
	Niobe	138 N–10 S	132–185
	Sedna	40 N	335
CHASMA	Artemis	30–42 S	121–145
	Devana	00	289
	Diana	15 S	150
	Heng-O	00–10 N	350–000
	Juno	32 S	102–120
CRATERS	Colette	65 N	322
	Lise Meitner	55 S	322
	Pavlova	14 N	040
	Sacajewa	63 N	335
	Sappho	13 N	027
VOLCANO	Rhea Mons	31 N	285
MOUNTAIN	Theia Mons	29 N	285

Magellan and Venus Express

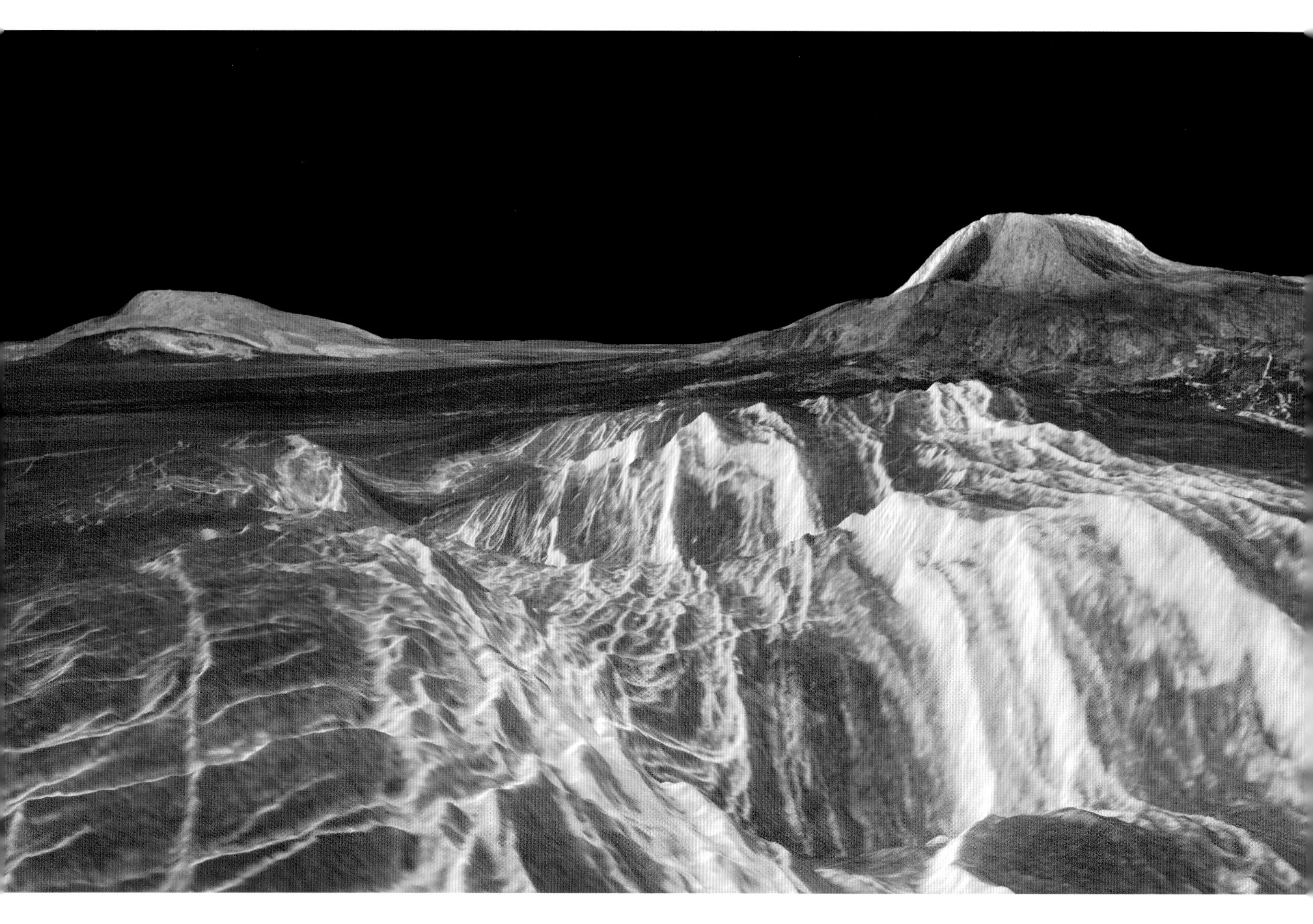

▲ **Eistla Regio.** This false-colour perspective of the western part of the Eistla Regio region of Venus depicts the view looking north-west from a point 700 km (440 mi) from the summit crater of Gula Mons, the mountain seen at top right which stands 3 km (2 mi) above the surrounding plain. The foreground is dominated by a large rift valley.

The various American and Soviet spacecraft to Venus, launched between 1961 and 1984, had provided a great deal of information about the planet, but there was still need for better radar coverage. This was the purpose of the Magellan probe, launched from the Shuttle Atlantis on 4 May 1989. It was hoped that the resolution would be far better than anything achieved by the earlier missions, and so it proved. Radar mapping began in September 1990, and by 1993 over 98 per cent of the planet's surface had been covered. Cycle 4 ended on 24 May of that year. One cycle is 243 Earth-days, during which Venus rotates completely beneath the spacecraft's orbital plane. When Magellan first went into orbit round Venus, the period was 3.2 hours, and the minimum distance was 289 kilometres (180 miles), though in September 1992 this was reduced to 184 kilometres (115 miles).

Magellan could resolve features down to 120 metres (400 feet). The main dish, 3.7 metres (12 feet) across, sent down a pulse at an oblique angle to the spacecraft, striking the surface below much as a beam of sunlight will do on Earth. The surface rocks modify the pulse before it is reflected back to the antenna; rough areas are radar bright, smooth areas are radar-dark. A smaller antenna sends down a vertical pulse, and the time-lapse between transmission and return gives the altitude of the surface below to an accuracy of 10 metres (33 feet). Magellan has shown fine details on the volcanic surface. There are for example multiple lava flows, with varying radar reflectivity indicating rocky and smoother areas. There are flows which have clearly been made by very liquid lava, and even show river-like meandering. The features known as tesserae are high, rugged tracts extending for several thousands of kilometres; one of these is Alpha Regio, shown on the facing page. Magellan showed many 'coronae', caused by plumes of hot material rising from below the surface. Arachnoids, so far found only on Venus, are so named because of their superficial resemblance to spiders' webs; they are circular to ovoid in shape, with concentric rings and intricate outward-extending features. They are similar in form to the coronae – circular volcanic structures surrounded by ridges, grooves and radial lines. There are strange looking objects which have been nicknamed 'pancakes'; these too are volcanic in origin. There are strong indications of explosive vulcanism here and there. The crater Cleopatra on the eastern slopes of the Maxwell Montes is about 100 kilometres (62 miles) in diameter.

The scale and colours of the images shown here are products of computer processing; for example, the vertical scale in the image of Gula Mons seen at top right in the main picture has been deliberately increased to accentuate its features. The colours are not as they would be seen by an observer on the planet – even assuming that he or she could get there, which would be by no means easy! For example, the bright patches representing lava flows would not appear as such to the naked eye.

Magellan did all that had been expected of it; at the end of its career it was burned away in Venus' atmosphere on 11

October 1994. The next mission to the planet was the European probe Venus Express, launched on a Soyuz Fregat carrier from Baikonur (Kazakhstan) on 9 November 2005. It reached its final orbit round Venus on 7 May 2006 – a highly eccentric path with an orbital period of 24 hours. The distance from Venus ranges between 250 and 66,000 kilometres (155 and 41,000 miles), and the orbit is polar.

One of the early images showed a huge vortex in the atmosphere almost directly over the south pole, corresponding to a similar cloud structure over the north pole. New temperature measurements of the surface were obtained, identifying 'hot spots' which may be linked with present-day active vulcanism.

Venus has given up many of its secrets, and is no longer 'the planet of mystery', but we have to admit that the more we learn about it, the more inhospitable it seems to be. The chances of finding any life there appear to be effectively nil.

Maxwell Montes. The Maxwell Montes are seen as a large bright patch below centre in this image. They are the highest mountains on the planet, with peaks extending more than 7 km (4.5 mi) above the surface.

Atalanta Planitia. This vast plain can be seen to the right of the fault lines which radiate from near the central region.

Lakshmi Planum. This plateau lies just to the left of the Maxwell Montes. It stands 2.5–4 km (1.5–2.5 mi) above the surface and is covered by lava.

North Pole. The north pole of Venus lies at the very centre of the image. Longitude zero is to the right. There is major faulting above the pole.

▲ **The northern hemisphere of Venus.** This false-colour projection of the surface was created from data gathered during the three cycles of Magellan's radar mapping observations and supplemented with earlier Pioneer Venus data, and the general colour hue comes from the Soviet Venera lander's images taken whilst on the surface of the planet in 1972.

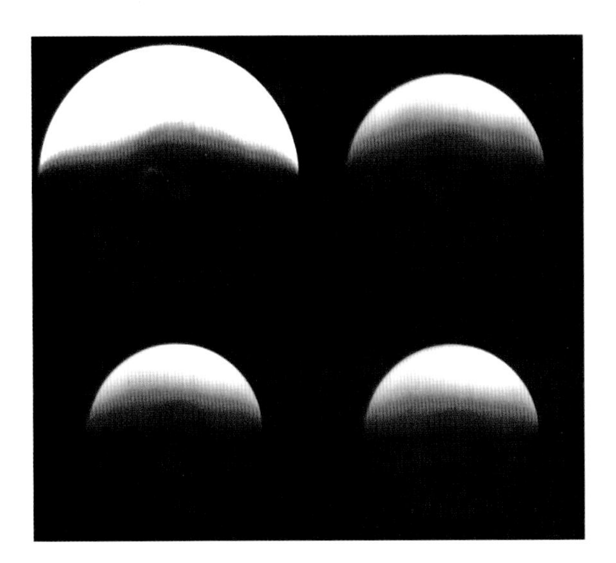

▲ **Venus double vortex.** Infra-red images from Venus Express' Ultraviolet/Visible/Near/Infra-red spectrometer (VIRTIS) showing a double vortex in motion over the south pole. The images were obtained at different distances from the planet.

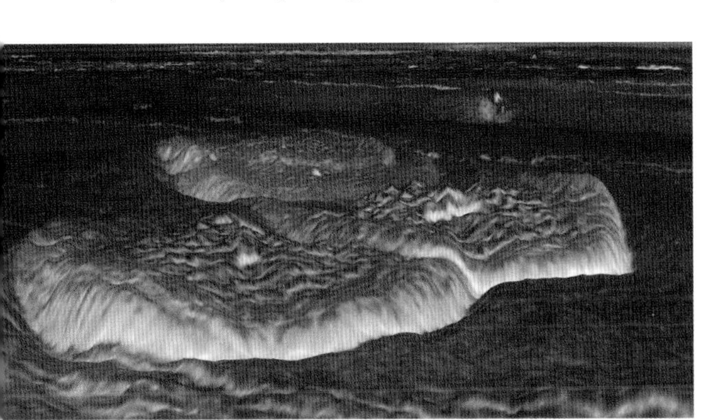

▲ **Alpha Regio.** This mosaic of radar images shows part of the eastern edge of Alpha Regio. The area contains seven dome-like hills, three of which are visible here, averaging 25 km (16 mi) in diameter and 750 m (2400 ft) in height. They may have been formed by successive lava eruptions.

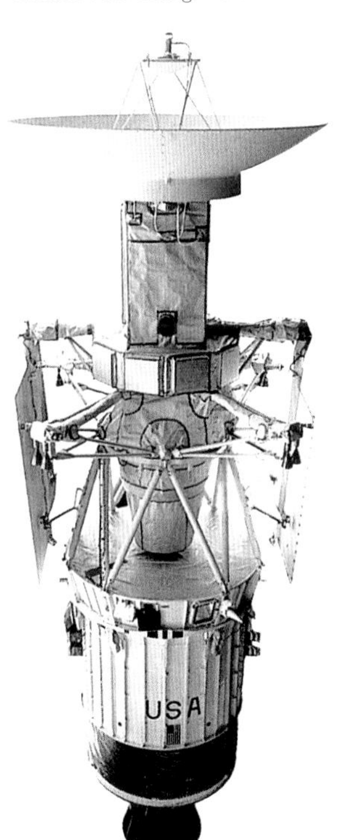

▶ **Magellan.** Released from the cargo bay of the space Shuttle Atlantis on 4 May 1989, the spacecraft reached Venus on 10 August 1990. After 37 silent minutes while it swung around the back of the planet, it emerged in a perfect orbit. Its radar mapping programme was completed by September 1993.

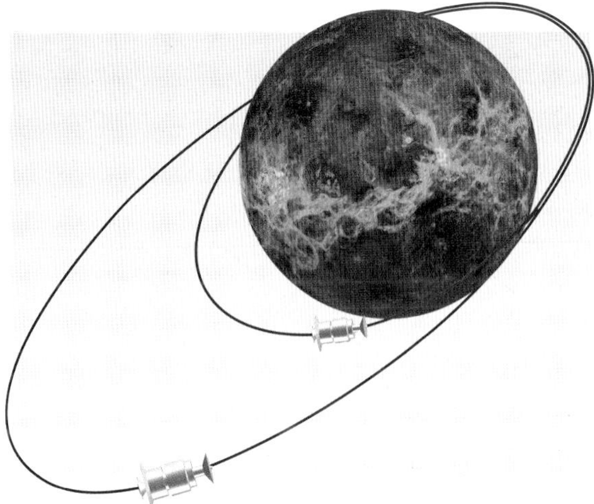

MOVEMENTS OF MAGELLAN

The Magellan space probe was put into an orbit around Venus, and mapped the surface in a series of 20-km (12.5-mi) swathes. When this programme was completed, the probe was put into a more elliptical orbit. When the whole of the surface had been mapped, in 1993, Magellan entered a circular orbit to undertake gravitational studies of the planet.

Mars

Mars, the first planet beyond the orbit of the Earth, has always been of special interest, because until relatively recently it was thought that advanced life forms might exist there. Less than a century ago, there was even a prize (the Guzman Prize) offered in France to be given to the first man to establish contact with beings on another world – Mars being specifically excluded as being too easy!

Mars is considerably smaller and less dense than the Earth, and in size it is intermediate between the Earth and the Moon. The escape velocity of 5 kilometres per second (3.1 miles per second) is high enough to hold down a thin atmosphere, but even before the Space Age it had become clear that the atmosphere is not dense enough to support advanced Earth-type life; neither could oceans exist today on the surface. The axial tilt is much the same as ours, so that the seasons are of similar type even though they are much longer. Mars' orbital period is 687 days. The axial rotation period, easily measured from observations of the surface markings, is 24 hours 37 minutes 22.6 seconds, so that a Martian 'year' contains 668 Martian days or 'sols'.

The orbit of Mars is decidedly eccentric. The distance from the Sun ranges between 249 million and 207 million kilometres (between 155 million and 129 million miles), and this has a definite effect upon Martian climate. As with Earth, perihelion occurs during southern summer, so that on Mars the southern summers are shorter and warmer than those of the north, while the winters are longer and colder.

At its nearest, Mars may come within 59 million kilometres (36 million miles) of the Earth, closer than any other planet apart from Venus. Small telescopes will then show considerable surface detail. First there are the polar ice-caps, which vary with the seasons; at its greatest extent the southern cap may extend down to latitude 50°, though at minimum it becomes very small. Because of the more extreme climate in the southern hemisphere, the variations in the size of the ice cap are greater than those in the north.

PLANETARY DATA – MARS	
Sidereal period	686.980 days
Rotation period	24h 37m 22s.6
Mean orbital velocity	24.1 km/s (15 mi/s)
Orbital inclination	1° 50' 59".4
Orbital eccentricity	0.093
Apparent diameter	max. 25".7, min. 3".5
Reciprocal mass, Sun = 1	3,098,700
Density, water = 1	3.94
Mass, Earth = 1	0.107
Volume, Earth = 1	0.150
Escape velocity	5.03 km/s (3.1 mi/s)
Surface gravity, Earth = 1	0.380
Mean surface temperature	−23°C
Oblateness	0.009
Albedo	0.16
Maximum magnitude	−2.8
Diameter (equatorial)	6794 km (4222 mi)

Earth

▼ **Mars in 2005,** Damian Peach, 11–23 October (23.5-cm (9.25-in) Schmidt-Cassegrain). The V-shaped Syrtis Major is prominent, with Hellas and Cimmeria. A dust cloud is visible across Argyre, with smaller dust cones across Chryse and Erythraeum. The dust activity is best seen in the last three images. As is usual with visual observers, South is up.

The dark areas are permanent, though minor variations occur; as long ago as 1659 the most conspicuous dark feature, the rather V-shaped patch now known as the Syrtis Major, was recorded by the Dutch astronomer Christiaan Huygens. Originally it was assumed that the dark areas were seas, while the ochre tracts which cover the rest of the planet represented dry land. When it was found that the atmospheric pressure is too low for liquid water, it was believed that the dark areas were old sea-beds filled with vegetation. This view was generally accepted up to the time of the first fly-by made by Mariner 4, in 1965.

There are various bright areas, of which the most prominent is Hellas, in the southern part of the planet. At times it is so bright that it has been mistaken for an extra polar cap, and it was once thought to be a snow-covered plateau, though it is now known to be a deep basin.

In general the Martian atmosphere is transparent, but clouds can be seen in it, and there are occasional dust-storms which may spread over most of the planet, hiding the surface features completely. What apparently happens is that if the windspeed exceeds 50 to 100 metres per second (160 to 320 feet per second), tiny grains of surface material are whipped up and given a 'skipping' motion, known technically as saltation. When they strike the surface they force still smaller grains into the atmosphere, where they remain suspended for weeks. Widespread dust-storms are commonest when Mars is near perihelion, and the surface winds are at their strongest.

The first reasonably reliable maps of Mars date back to the 1860s. The various features were named, mainly after astronomers; the old maps show Mädler Land, Lassell Land, Beer Continent and so on. (The latter name honoured

Wilhelm Beer, a German pioneer of lunar and planetary observation.) Then, in 1877, G. V. Schiaparelli produced a more detailed map and renamed the features, so that, for example, the most prominent dark marking on Mars, the V-shaped feature drawn by Huygens so long ago, and originally known as the Kaiser Sea, was renamed Syrtis Major. It is Schiaparelli's nomenclature, modified and extended, which we use today.

Schiaparelli also drew strange, artificial-looking lines across the ochre deserts, which he called canali or channels; inevitably this was translated as 'canals' and the suggestion was made that the features might be artificial waterways. This view was championed by Percival Lowell, who built the great observatory at Flagstaff in Arizona mainly to study Mars, and equipped it with an excellent 61-centimetre (24-inch) refractor. Lowell believed that the canals represented a planet-wide irrigation system, built by the local inhabitants to pump water from the ice-caps at the poles through to the equator. Disappointingly, it has now been proved that the canals do not exist; they were merely tricks of the eye, and Lowell's Martians have been banished to the realm of science fiction.

The best pre-Space Age maps of Mars were those drawn up by E. M. Antoniadi in the 1920s and early 1930s. The telescope used was the Meudon 83-centimetre (33-inch) refractor, and Antoniadi's charts proved to be amazingly accurate, but the real 'breakthrough' came with Mariner 4 in 1965.

Basically, Antoniardi's nomenclature has been retained; for example his Solis Lacus is not a lake, and his Mare Cimmerium is not a sea. Before the Space Age, Mars was thought to be a world with comparatively little surface relief.

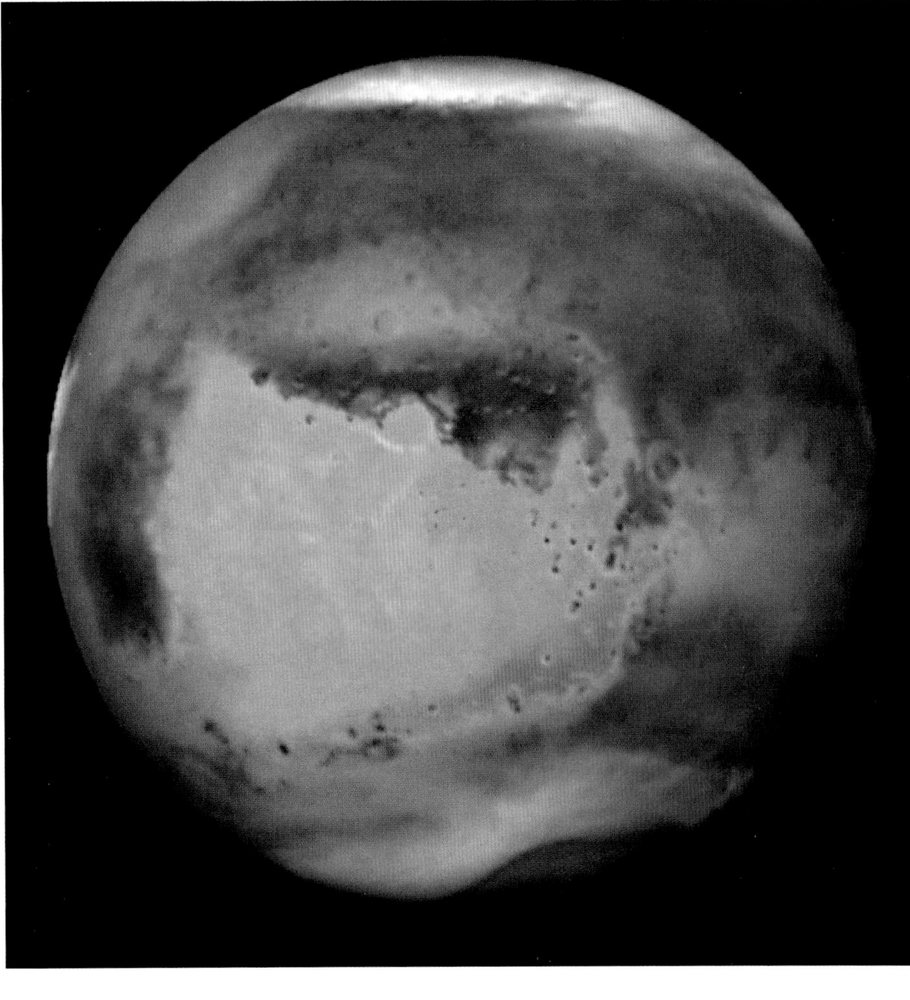

▲ **Mars: Hubble Space Telescope.** South is at the top. The south polar ice is evident, as is frost in the north. Syrtis Major is seen to the right, Acidalia Planitia lower left, Sabaeus near the centre of the image. Hellas is visible below Syrtis Major, but is not cloud-filled. A large dust storm is churning high above the North polar cap (bottom). A smaller dust storm can be seen nearby, and another large dust storm is spilling out of the Hellas basin (top left).

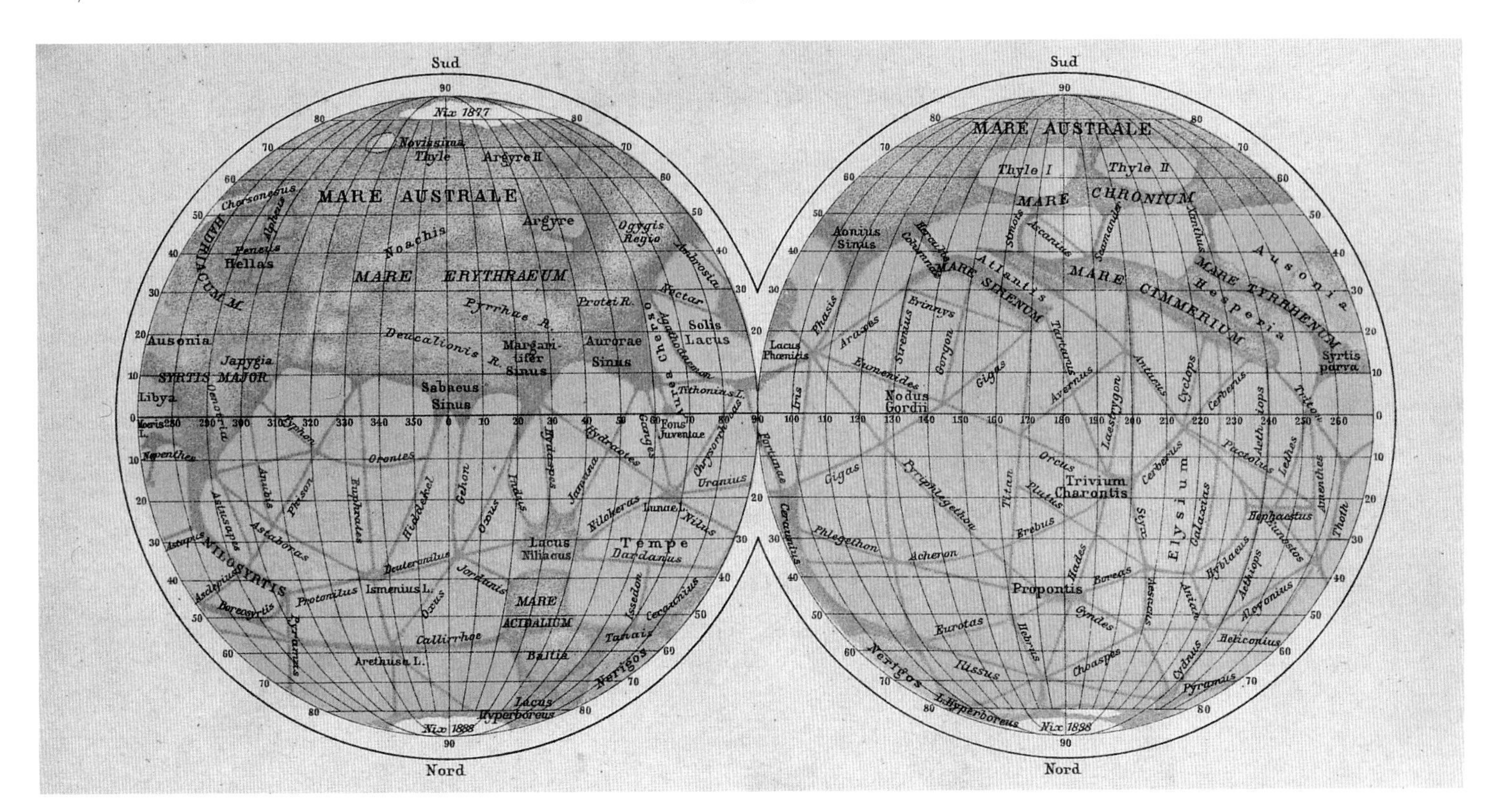

▼ **Schiaparelli's charts of Mars,** compiled from observations made between 1877 and 1888. The main dark features are clearly shown – but so too are the 'canals', which are now known to be non-existent! Schiaparelli's map uses the nomenclature which he introduced in 1877.

Map of Mars

The photographic map of Mars given here shows many features which can be seen with an adequate telescope, such as a 30-centimetre (12-inch) reflector, under good conditions, though others, such as craters, are out of range. The map is drawn with south at the top, as this is the normal telescopic view. The most prominent feature is the triangular Syrtis Major, once thought to be an old sea-bed filled with vegetation but now known to be a plateau. A band of dark markings (the 'Great Diaphragm') runs round the planet rather south of the equator, and there are some features of special interest, such as the Solis Planum, which shows variations in shape and intensity. There are two large basins, Hellas and Argyre, which can sometimes be very bright, Hellas particularly so; this is the deepest basin on Mars, and at the bottom the atmospheric pressure is 8.9 millibars, though this is still not high enough to allow liquid water to exist. In the north the main feature is the wedge-shaped Acidalia Planitia. Observers sometimes continue to use the pre-Space Age names, so that Solis Planum is 'Solis Lacus' and Acidalia Planitia is 'Mare Acidalium'.

▶ **Caidera of Olympus Mons:** Mars Express, 21 January 2004. The image was obtained from a range of 270 km (170 mi); the whole image is 103 (64 mi) across. The caldera is about 3 km (1.9 mi) deep, while the volcano itself rises to 24 km (15 mi) above the surrounding terrain. Violent eruptions ceased in the recent past, relatively speaking, but a certain amount of residual activity may linger on even now.

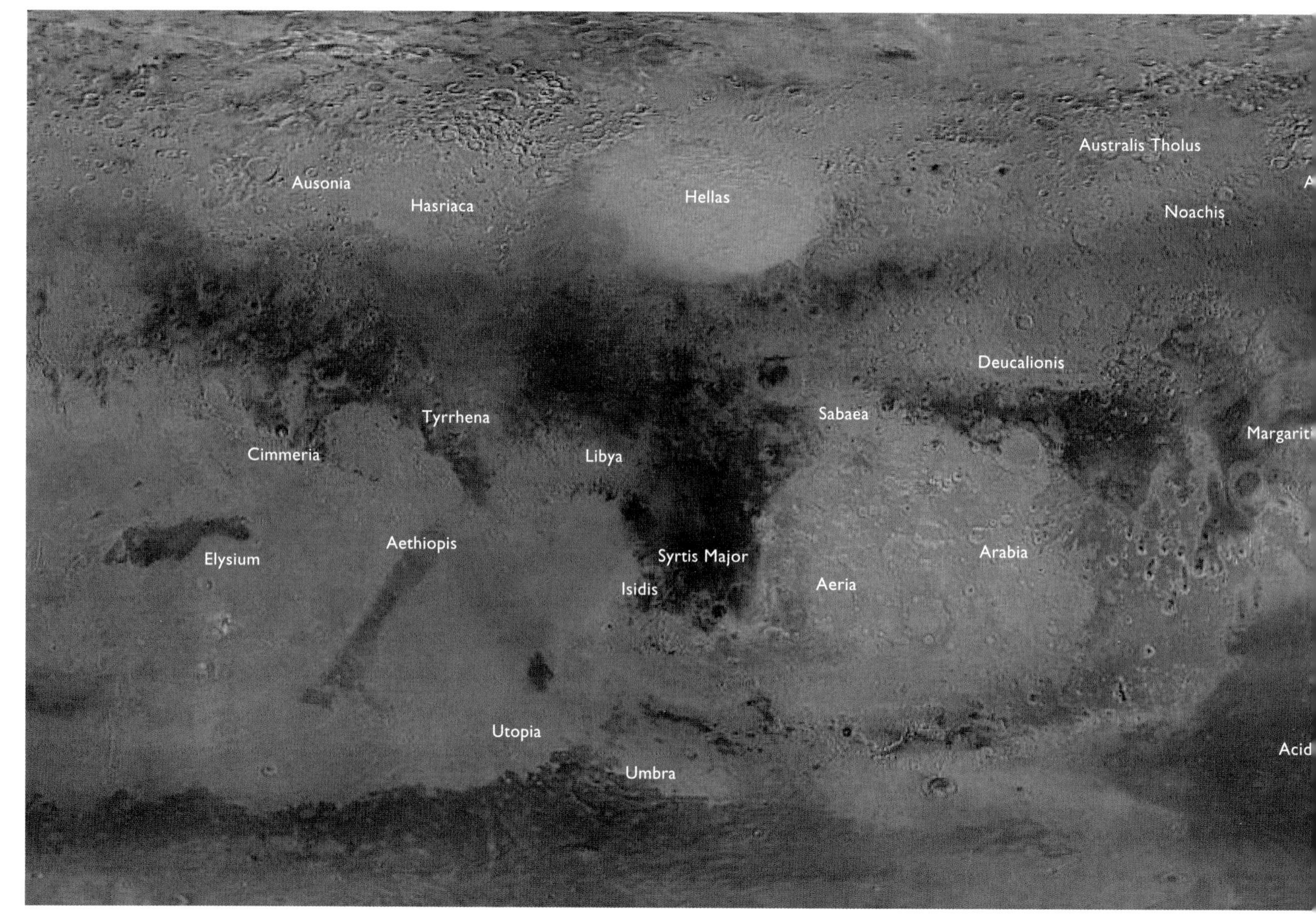

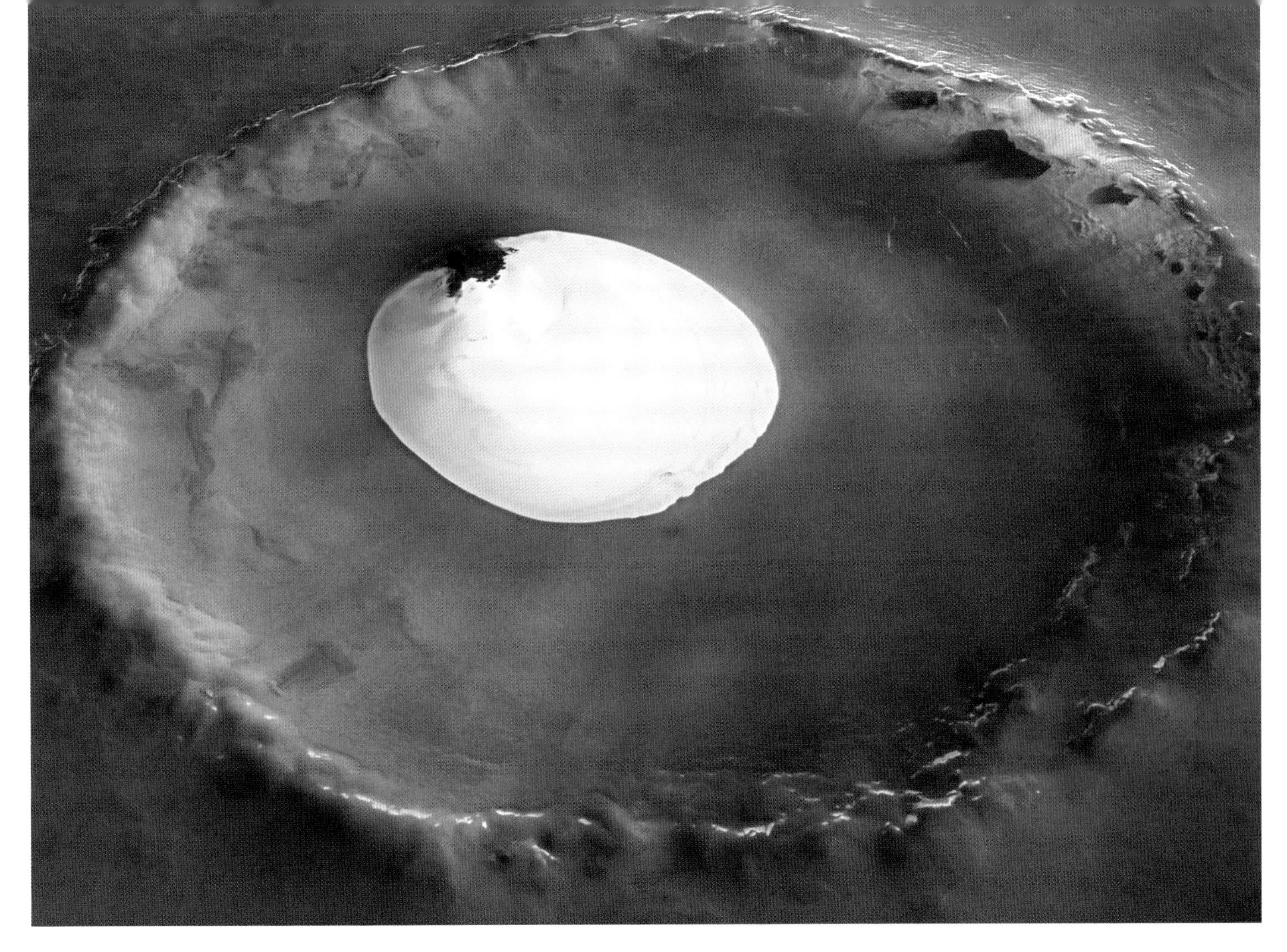

▲ **Residual ice in a crater** in the Vastitas Borealis. The crater is 35 km (22 mi) across; sunlight is blocked by its 2-km (6500-ft) wall. The ice may be up to 180 m (600 ft) deep.

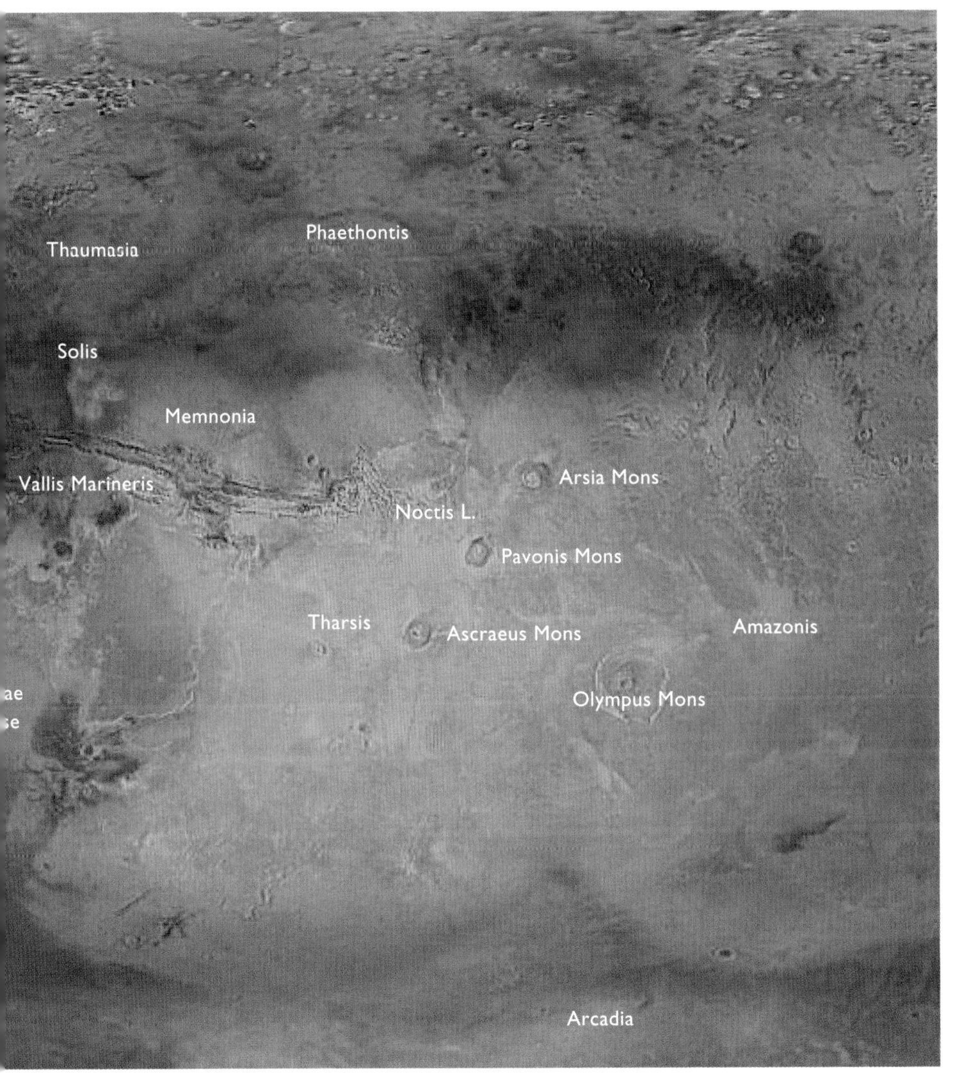

◀ **Map of Mars.** The map given here is shown with south at the top, as is usual for visual observers. The two prominent features are the v-shaped Syrtis Major, which crosses the equator, and Acidalia Planitia in the north. The basins of Hellas and Argyre can often be striking. The great volcanoes of the Tharsis bulge can be seen as small patches in moderate-sized telescopes, and were formerly believed to be lakes. Craters are beyond the range of ordinary instruments.

MAIN TYPES OF FEATURES

Catena – line or chain of craters, e.g. Tithonia Catena.
Chaos – areas of broken terrain, e.g. Aromatum Chaos.
Chasma – very large linear chain, e.g. Capri Chasma.
Colles – hills, e.g. Deuteronilus Colles.
Dorsum – ridge, e.g. Solis Dorsum.
Fossa – ditch: long, shallow, narrow depression, e.g. Claritas Fossae.
Labyrinthus – canyon complex. Noctis Labyrinthus is the only really major example.
Mensa – small plateau or table-land, e.g. Nilosyrtis Mensae.
Mons – mountain or volcano, e.g. Olympus Mons.
Patera – shallow dome-like volcanic structure, e.g. Alba Patera.
Planitia – smooth, low-lying plain, e.g. Hellas Planitia.
Planum – plateau; smooth high area, e.g. Hesperia Planum.
Rupes – cliff, e.g. Ogygis Rupes.
Terrae – lands, names often given to classical albedo features, e.g. Sirenum Terra, formerly known as Mare Sirenum.
Tholus – domed hill, e.g. Uranius Tholus.
Vallis – valley, e.g. Vallis Marineris.
Vastitas – extensive plain. Vastitas Borealis is the main example.

Missions to Mars

Many spacecraft have now been sent to Mars. Some have been triumphantly successful whilst others have failed for no apparent reason – and, ironically, the Russians have had little luck at all, because their probes have either lost contact, crash-landed without sending back any useful information or missed the Red Planet altogether. This is curious in view of their excellent results from Venus, which is a much more difficult target.

It was left to NASA to take the lead, which they did in 1964 with Mariners 3 and 4. The first of these was a failure, but Mariner 4, launched on 28 November, flew past Mars on 14 July 1965 at a range of 9789 kilometres (6080 miles), sending back over 20 images which caused us to change all our ideas about Mars. (The probe itself went on its way round the Sun; it was finally lost in December 1967, and no doubt is still in solar orbit.) To the surprise of many astronomers – not all – Mars showed craters, and seemed outwardly to be more like the Moon than the Earth. The canals were finally consigned to history, and it was fairly obvious that the dark areas were not evidence of any kind of vegetation. Next came Mariners 6 and 7, launched in early 1969; they by-passed Mars on 31 July and 4 August respectively, sending back clear images from a range of over 3400 kilometres (2000 miles). Again the pictures showed craters and peaks. By ill chance, all these early Mariners surveyed the least interesting parts of the Martian surface.

Mariner 9, launched from Cape Canaveral on 30 May 1971, was more ambitious; it was scheduled to enter a closed orbit round the planet, and did so on 13 November. Its orbit took it to within 1700 kilometres (1050 miles) of the planet, and altogether it returned 7329 images before losing contact in October 1972. It was Mariner 9 which gave us our first views of the great volcanoes, including the towering Olympus Mons, which is three times as high as our Everest and is crowned by an 85-kilometre (53-mile)caldera. It was found that there are two marked bulges in the crust, Tharsis and Elysium, and it is here that we find the main volcanoes. Along the Tharsis bulge lie Ascraeus Mons, Pavonis Mons and Arsia Mons, all of which had often been seen by Earth-based observers though they had no way of finding out what they were (Pavonis Mons was believed to be a lake, while Olympus Mons was called Nix Olympica, the Olympic Snow). The two hemispheres of Mars are unalike. In general the southern part of the planet is the higher, though it does contain two large, deep basins, Hellas and Argyre. The northern hemisphere is lower and less cratered, though it does contain part of the Tharsis ridge.

There were great valleys, notably the Vallis Marineris (Mariner Valley), just south of the equator, which runs for more than 4500 kilometres (2800 miles), with a maximum width of 600 kilometres (370 miles); the deepest part is 7 kilometres (4 miles) below the rim. There are complex systems such as Noctis Labyrinthus (once taken for a lake, and called Noctis Lacus), with canyons up to 20 kilometres (12 miles) wide, making up the pattern which has been called the Chandelier. There are what can only be dry riverbeds, with 'islands', and there are craters ranging from vast enclosures down to tiny pits. The crater named in honor of Schiaparelli has a diameter of 461 kilometres (287 miles).

Mariner established that the polar caps are mainly composed of water ice, with a seasonal outer layer of carbon diox-

ide ice. Disappointingly, the atmosphere was found to be much more rarefied than had been expected. Instead of having a ground pressure of around 86 millibars, the value generally accepted up to that time, it was found that the presure is below 10 millibars everywhere – and instead of being composed of nitrogen, it is almost pure carbon dioxide. During southern winter, when carbon dioxide from the atmosphere condeses out on to the large polar cap, the air pressure drops perceptively.

After a series of Soviet failures, NASA turned back to Mars in 1975, launching two Viking probes. Each was made up of an orbiter and a lander. The orbiters continued the surveys started by Mariner 9, and also relayed data from the landers, both of which touched down gently; the first on 19 June 1976 in the 'Golden Plain', Chryse (22.4°N, 47.5°W) and the second on the following 7 August, in Utopia (48°N, 226°W). They were braked during descent partly by parachutes and partly by retro-rockets, landing at no more than 9.6 km/h (less than 6 mph), mercifully avoiding the large rocks strewn on the surface. Both used 'grabs' to collect samples of the surface, which were then drawn back into the probes themselves and carefully analysed. The results were somewhat perplexing, but there was no definite evidence of biological activity. Neither were there strong 'Marsquakes'. The pictures from the surface showed barren, rocky deserts, with a thin veneer of red material. The sky was yellowish pink; winds were light and temperatures ranged between −96°C after dawn to a maximum of −31°C near noon. When the Vikings finally 'went silent', most astronomers tended to dismiss Mars as sterile and inert. The best was yet to come!

▲ **Mars from Viking 2.** This photo of Mars was taken by Viking 2 from 419,000 km (260,000 mi) away as the spacecraft approached the planet on 5 August 1976. Water-cloud plumes extend north-west from the western flank of Ascraeus Mons, northernmost of the three great volcanoes that line Tharsis Ridge. The middle volcano, Pavonis Mons, is just visible on the dawn terminator, below and west of Ascraeus Mons. Vallis Marineris, the great system of rift canyons, extends from the centre of the picture at the terminator downwards to the east. The bright basin near the bottom of the photo is the Argyre Basin, one of the largest basins on Mars. The ancient crater lies near the south pole (not visible in this photo) and is brightened by the icy frosts and fogs that lie in the bottom of the basin and are characteristic of the near-polar regions of Mars when each is experiencing its winter season.

◀ **Coprates Chasma** This image, taken by the High Resolution Stereo Camera on ESA's Mars Express spacecraft, shows Coprates Chasma, a major trough located roughly in the centre of the Vallis Marineris canyon system, and Coprates Catena, the three parallel troughs to the left of the image.

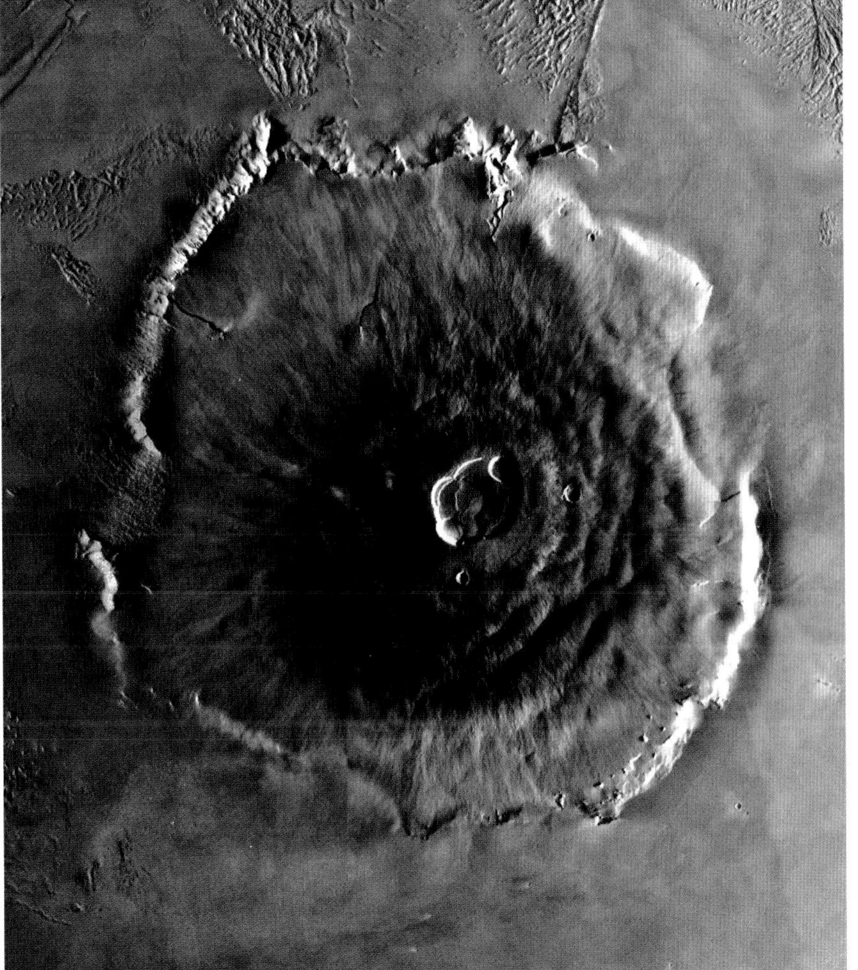

▶ **Olympus Mons,** the loftiest peak on Mars – three times the height of our Everest. It is a typical shield volcano, crowned by a complex caldera.

The Search for Life on Mars

Despite the negative results from the Vikings, the quest for life on Mars was not given up; nobody expected to find anything as advanced as a beetle, or even a blade of grass, but it seemed reasonable to expect very lowly organisms. On 2 December 1996 a new probe was launched toward the Red Planet – Pathfinder, which carried a small rover, Sojourner. This time there was to be no controlled landing. Pathfinder was encased in tough airbags and was designed to come down at high speed, bouncing more than a dozen times before coming to rest. Once Pathfinder had settled down and assumed an upright position, 'petals' in its hull would open, so that Sojourner could emerge and crawl down a ramp on to the Martian surface. It was an ambitious project, and Sojourner itself was only about the size of a household microwave, but it was in fact a highly sophisticated probe, capable of carrying out on-the-spot analyses of the Martian rocks. It could move around, controlled by NASA operators over 190 million kilometres (120 million miles) away.

Everything went well. Transmissions began immediately, and the results were of great interest. Pathfinder had come down near the end of what was known as the Ares Vallis; the valley had once been filled with a raging torrent of water, and had been expected to bring rocks of all kinds down on to the floodplain – which did indeed turn out to be the case. Sedimentary rocks were identified, showing that the site had been water-filled; the rocks were volcanic, which again was no surprise. Wind speeds were light, but there were numerous 'dust devils' (miniature tornadoes). The ground temperature rose to −13°C at noon, though it plummeted to −75°C during the night. Altogether the lander sent back 16,000 images, and 550 were received from Sojourner before contact was finally lost in October 1997.

▼ The search for life. Viking I drew in material from the red 'desert', analysed it, and sent back the results. It had been expected that signs of organic activity would be detected, but this proved not to be the case. If Martian life exists it must be very primitive. Both the Viking landers have long since ceased to operate, so that for a final decision we must await the results from new spacecraft. There are already plans to send a probe to Mars, collect material, and return it to Earth for analysis.

▲ First colour picture from Viking I. The lander of Viking I came down in the 'golden plain' of Chryse, at 22.4°N, 47.5°W. The picture shows a red, rock-strewn landscape; the atmospheric pressure at the time was approximately 7 millibars. The first analysis of the surface material was made from Viking: I–44 per cent silica, 5.5 per cent alumina, 18 per cent iron, 0.9 per cent titanium and 0.3 per cent potassium. In Viking's ay, colour correction techniques were by no means perfect and the images appear redder than they should be.

Note that Pathfinder and Sojourner were not making deliberate searches for life. That would be a task reserved for later missions to the planet. Their main aim was to find the areas where life might be expected – and in particular to show that Mars had once been wetter, warmer and more welcoming than it is today.

Mars was not neglected in the closing years of the twentieth century and the opening decade of the twenty-first. Between 1996 and 2006 NASA launched six spacecraft and Europe and Japan one each. There were three failures – Japan's Nozomi, which never managed to rendezvous with Mars, and NASA's Mars Climate Orbiter and Mars Polar Lander; Mars Climate Orbiter burned up in the Martian atmosphere, following confusion over the use of Imperial and metric units and Mars Polar Lander probably crashed on the surface. The rest did all that had been expected of them, and Mars Global Surveyor, launched in November 1996, sent back data for almost ten years; the last signals were received from it on 2 November 2006.

The successful craft were Mars Global Surveyor, Mars Odyssey (launched 7 April 2001), Mars Express (2 June 2003) and Mars Reconnaissance Orbiter (12 April 2005), plus the two Mars Exploration Rovers, Spirit and Opportunity. Mars Express was of special interest because it carried a small British lander, Beagle 2, designed to make a gentle touch-down and look for any signs of life, past or present. It would not be able to move around, but it carried equipment enabling it to analyse the surface materials and even probe down to a few inches. It was released from its carrier on schedule and should have landed in the Isidis Planitia region of Mars at 02.54 GMT on Christmas Day, 2003. Alas, no signal was ever received from it. Either it crashed on to the surface, or else its radio failed for some reason or other. One day it – or its wreckage – may be identified.

If living orgainisms are found on Mars, we will be confident that life will appear wherever conditions for it are tolerable, and will evolve as far as the environment permits. It follows that life in the universe is likely to be widespread. Mars has much to tell us, but we cannot hope to learn the full story until we have obtained samples and analysed them in our laboratories. Only then will we be able to answer the question which has been asked so many time, 'Is there life on Mars?'

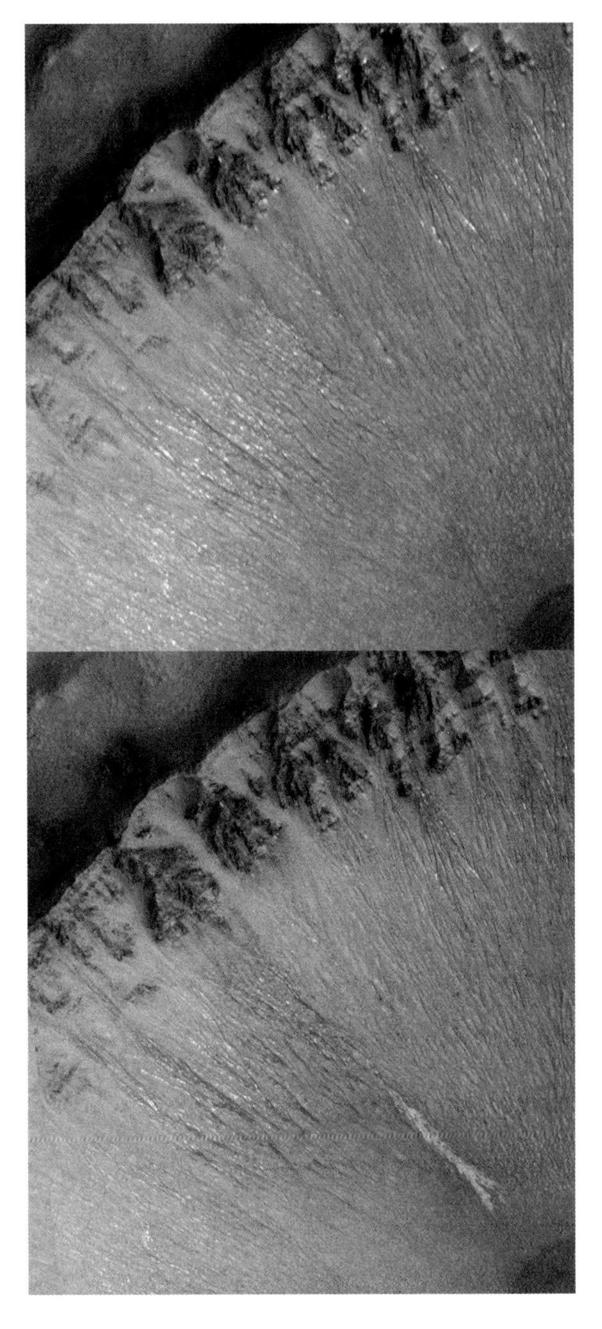

◀ **Water flowing on Mars?** These two images, of the same area, indicate that water gushed out from below the ground and flowed down the crater before evaporating or freezing, leaving the deposit seen on the second picture (September 2005), but not on the first (August 1999).

▼ **Pathfinder on Mars.** The lander was renamed the Sagan Memorial Station in honour of Carl Sagan. This panoramic view was obtained by the Mars Pathfinder Imager. Lander petals, ramps and deflated air bags are clearly visible, as is the Sojourner rover, which is using its Alpha Proton X-Ray Spectrometer (APXS) to analyse the rock at right, nicknamed 'Yogi' because it looked like a bear.

Exploring Mars

By the start of 2007 Mars had been surveyed in more detail than ever before. Mars Global Surveyor, Mars Odyssey, Mars Express and Mars Reconnaissance Orbiter had done their work well, and all except Mars Global Surveyor were still active (it was a pity about Climate Orbiter and Polar Lander). The Martian landscape was amazingly varied. First and foremost, of course, were the towering volcanoes; plate tectonics do not operate on Mars, so that when a volcano is born over a 'hot spot' it stays there – Olympus Mons, a straightforward shield volcano, is three times as high as Everest and is crowned by a complex caldera.

▼ The face on Mars (Cydonia). The first picture was taken by Viking 1 in 1976, and caused great excitement among those who believed in Martian life. Alas, the second image – Mars Express, 21 September 2006 – shows the structure to be an ordinary massif. Martians, farewell!

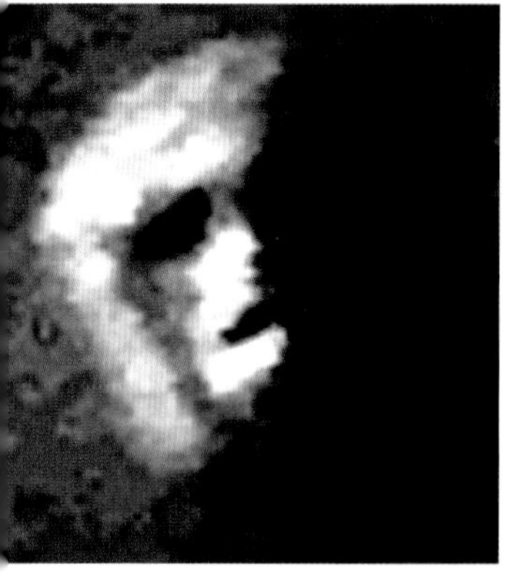

Images sent back by spacecraft show smooth areas inside the caldera; has there been any recent activity? Major eruptions belong to the past, but it is not safe to say that the volcanoes are completely extinct, and the surface of the Olympus Mons caldera may have been resurfaced as recently as two million years ago. The volcano lies 1500 kilometres (930 miles) west of the main Tharsis chain consisting of Arsia Mons, Pavonis Mons and Astreaus Mons. These too are shield volcanoes, and here we find splendid examples of lava tubes, particularly in the region of the 12 kilometres (7.4 miles) high Pavonis Mons. Lava tubes are produced when lava on top of a lava flow cools and forms a crust, while the subsurface lava remains molten. The longest tube shown by Mars Express images stretches for 60 kilometres (37 miles).

Impact craters are found all over Mars. One, 35 kilometres (22 miles) across and so far unnamed, was imaged by Mars Express in February 2005; inside it is a layer of residual ice, where incoming rays from the Sun are blocked by the crater wall. The ice may well be several thousand metres thick. There are images of frost-covered craters, such as Lowell. A particularly good picture was obtained of the crater Victoria on the Meridiani Planum, not far from the equator; over 730 metres (2400 feet) across, with sharply scalloped walls. On the floor there is a beautiful dune field, well shown on an image obtained in November 2006 (see page 83). Some of the old riverbeds show features which can only be islands which diverted the surviving flood waters; there are the long valleys characteristic of Mars – the Kasei Vallis network has a total length of over 2200 kilometres (1300 miles).

There was one amusing diversion. In the region of Cydonia not far from Chryse, where the first lander had come down, Viking Orbiter pictures showed a structure which really did look like a human face. Flying Saucer enthusiasts were quick to claim that it must be artificial, and NASA felt compelled to make it a target for the cameras of Global Surveyor and Express; predictably, the 'Face' proved to be nothing more dramatic than an ordinary mountain group. There was also the crater Galle, nicknamed the 'Happy Smiling Face' for obvious reasons, but I doubt whether anybody seriously suggested that it had been designed by Martian architects!

The Mars Express orbiter carried MARSIS, the Mars Advanced Radar for Subsurface and Ionospheric Sounding, used to locate buried structures; for example an almost circular feature, 260 kilometres (155 miles) in diameter shallowly buried under the surface layer in Chryse. But the most significant discovery of all was made from studying Global Surveyor pictures of small craters in the Terra Sirenum region. One of these craters was imaged in August 1999 and again in September 2005. The 2005 picture showed a deposit thought to be due to water bursting out from a spring, and flowing downhill, remaining liquid for long enough to leave traces. One NASA investigator commented, 'if you were standing in one of the gullies and saw this torrent of water sweeping down towards you, you'd probably want to get out of the way.'

More observations are needed; not all authorities are convinced about the presence of water just below the surface of Mars, but the evidence in favour is undeniably strong. Mars is not a dead world, and there have even been suggestions that the surface oceans may eventually return.

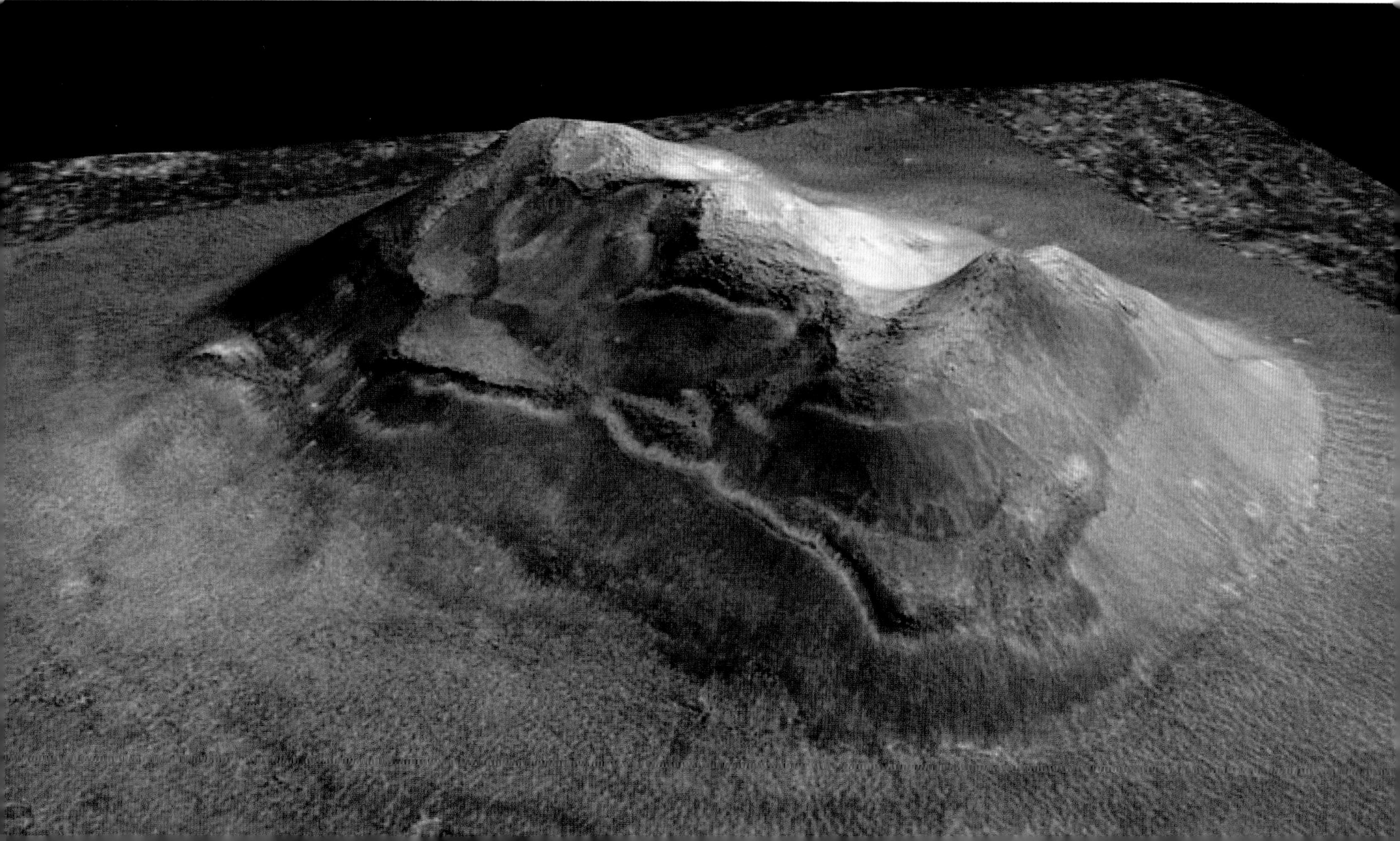

▲ Mars' north pole in spring. This striking image mosaic was obtained by the Mars Orbiter Camera aboard the Mars Global Surveyor in May 2002. Mars' north polar cap is the bright white region at top. The temperature difference between this cold region and the warmer ground surrounding it results in swirling winds and the clouds of several dust storms be seen clearly.

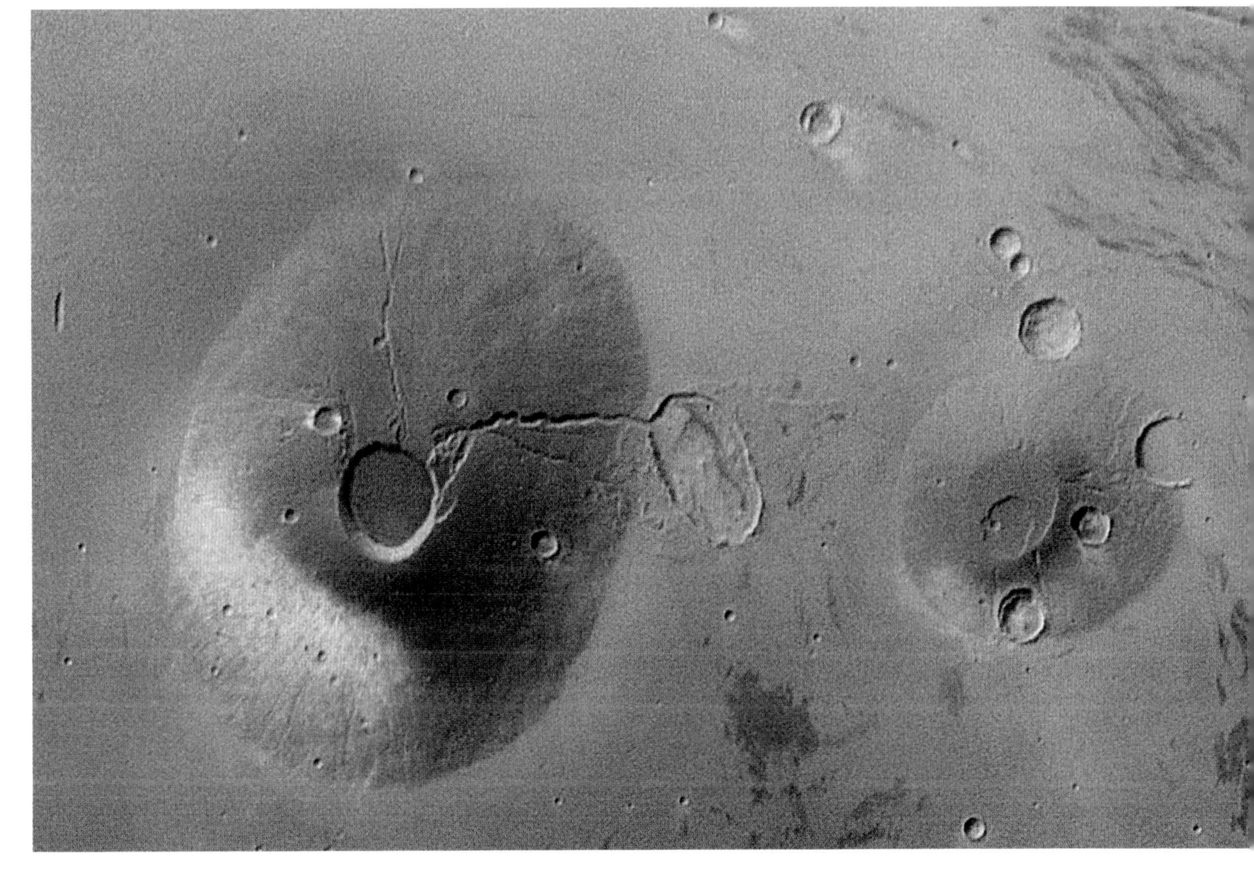

▲ Volcanoes on Mars. This Mars Global Surveyor image was obtained by the Mars Orbiter Camera in March 2002. Visible are two volcanoes – Ceraunius Tholus (left) and Uranius Tholus (right). Craters can be seen on both these volcanoes, indicating that they are old and no longer active. Dust from the global dust storm of 2001 lies on the slopes of Ceraunius Tholus, towards the left in this image. The caldera of this volcano is about 25 km (16 mi) wide.

Mars: Spirit and Opportunity

The two Mars Exploration Rovers, MER-A and MER-B – better known as Spirit and Opportunity – were launched in 2003, on 1 June and 7 July respectively. They touched down on Mars using the 'airbag' method, Spirit on 4 January 2004 and Opportunity on the following 25 January. They were expected to remain active for 90 Martian days (sols). In fact both had passed their thousandth day by November 2006, and were still functioning excellently in 2007. At one stage dust accumulating on the upper part of Spirit threatened to interfere with transmissions. An obliging Martian dust devil blew the dust away; Nature can sometimes be co-operative.

The two vehicles are to all intents and purposes identical. Each is 2.7 metres (8.7 feet) in diameter and 1.6 metres (5.2 feet) tall, including the entry vehicle. The main body is made of aluminium with an outer ring covered by the solar panels. There are six wheels mounted on a rocker-bogie suspension which means that the rovers can negotiate rough ground, and even climb over small rocks.

▼ Sunset over Gusev crater. On 19 May 2005 this view was obtained by Spirit as the Sun sank toward the crater rim. From Mars, the Sun looks considerably smaller than it does from Earth.

The landing sites had been carefully chosen. Spirit was aimed at Gusev crater, 14.5°S, 175.4°E; it is 170 km (106 miles) in diameter, and is thought to be between three and four thousand million years old. It was chosen because a long channel system, Ma'adim Vallis, drains into it, indicating that the crater must once have been a lake. Opportunity was sent to Meridiani Planum, on the opposite side of Mars, where observations from the orbiters had indicated the presence of haematite, a mineral usually associated with water. Both landings were perfect. Spirit came down just where it had been expected to do; Opportunity plumped down in a 30-metre (100-foot) crater which was promptly christened Eagle. The NASA planners were delighted about this, because Eagle contained rocky outcroppings which indicated that Meridiani Planum had once been a small sea. The haematite detected from orbit was found to be encased in spherules which were nicknamed 'blueberries'.

Spirit and Opportunity were well equipped. There were of course cameras able to send both panoramic and close-range views; spectrometers, used for close-up examination of the minerology of the rocks; a Rock Abrasion Tool (RAT) for removing dusty and weathered rock surfaces and exposing the fresh material below; an X-ray spectrometer, a microscopic imager and much else. Equipment and communications gave surprisingly little trouble. Of course there were problems; one of Spirit's wheels jammed and had to be dragged along, while Opportunity spent some time stuck in a drift, but it is fair to say that both rovers exceeded the wildest hopes of the NASA planners.

Probably their main achievement has been to prove that Mars did once have oceans, lakes and rivers; Steve Squyres,

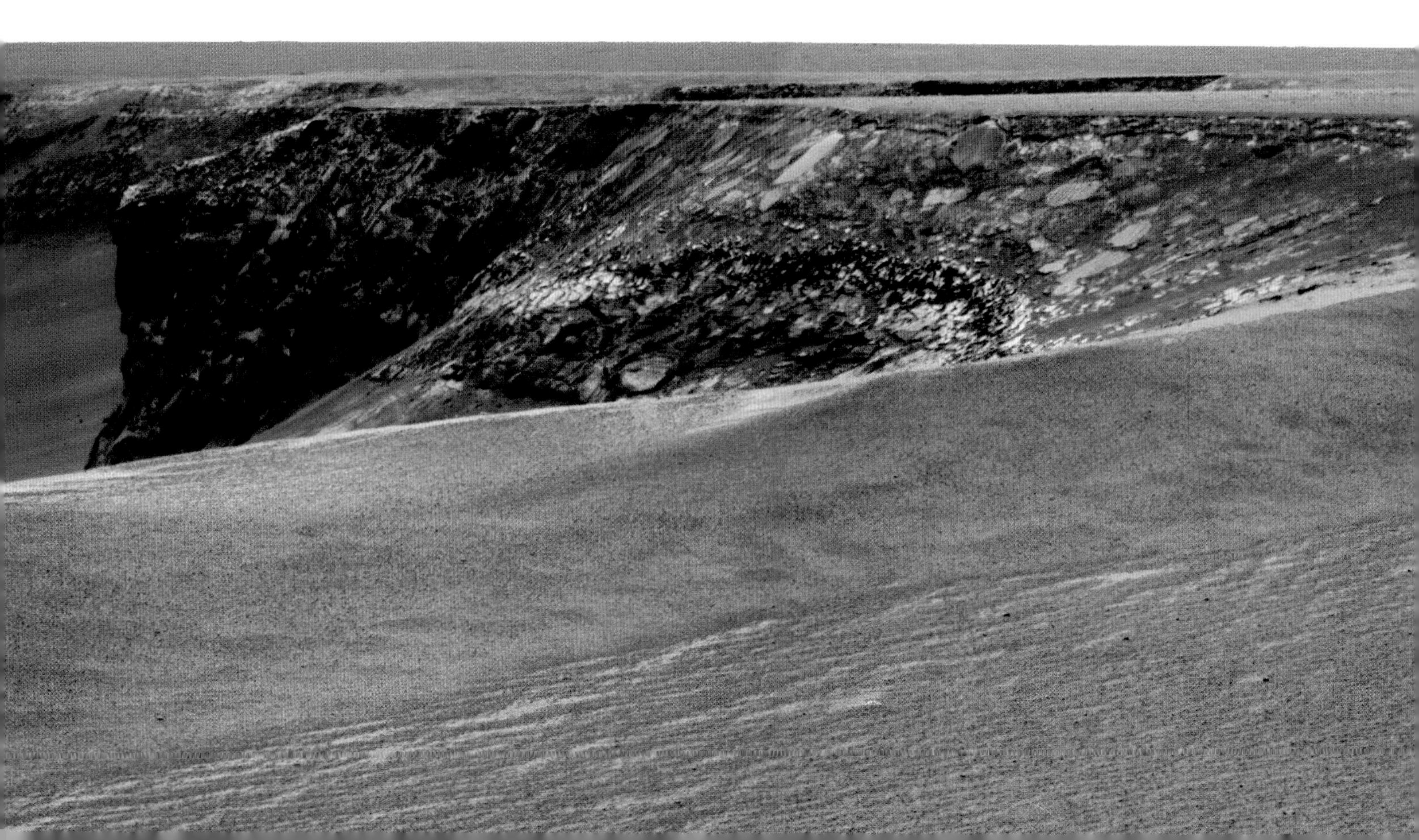

Principal Investigator for scientific instruments, commented that Meridiani Planum, where Opportunity landed, had been 'drenched' with water. Mars was once capable of supporting life. Whether it actually did so, we do not know.

The views of Mars were varied – and spectacular by any standards. For example Opportunity visited Victoria Crater, and was actually imaged by Mars Reconnaissance Orbiter; Opportunity's tracks were also shown. The rover made its way to the top of a large promontory over the crater, christened Cape Verde, and was able to take a true-colour mosaic of the crater below.

I suppose that everyone will have a special favourite among the many pictures so far sent back by the rovers. I have to make a choice between two, both from Spirit. One is the panorama of the Columbia Hills complex, inside Gusev; Spirit actually climbed to the top of Husband Hill. But in my view pride of place must go to the haunting picture of sunset, taken on 19 May 2005 as the Sun was about to sink below the rim of Gusev crater – on the 419th sol on the Martian surface. The Sun is much smaller than as seen from Earth; the foreground terrain is a rock outcrop, and the crater wall is 80 km (50 miles) away.

Up to January 2007 the rovers could not plan more than a limited distance ahead, but they have now been provided with new software which enables them to plan a route to a spot 50 metres (164 feet) away, evading obstacles which they would not be able to surmount. As I write these words (July 2007) both Spirit and Opportunity are still sending back data, although a major dust storm has hindered operations of Opportunity and to a lesser extent Spirit.

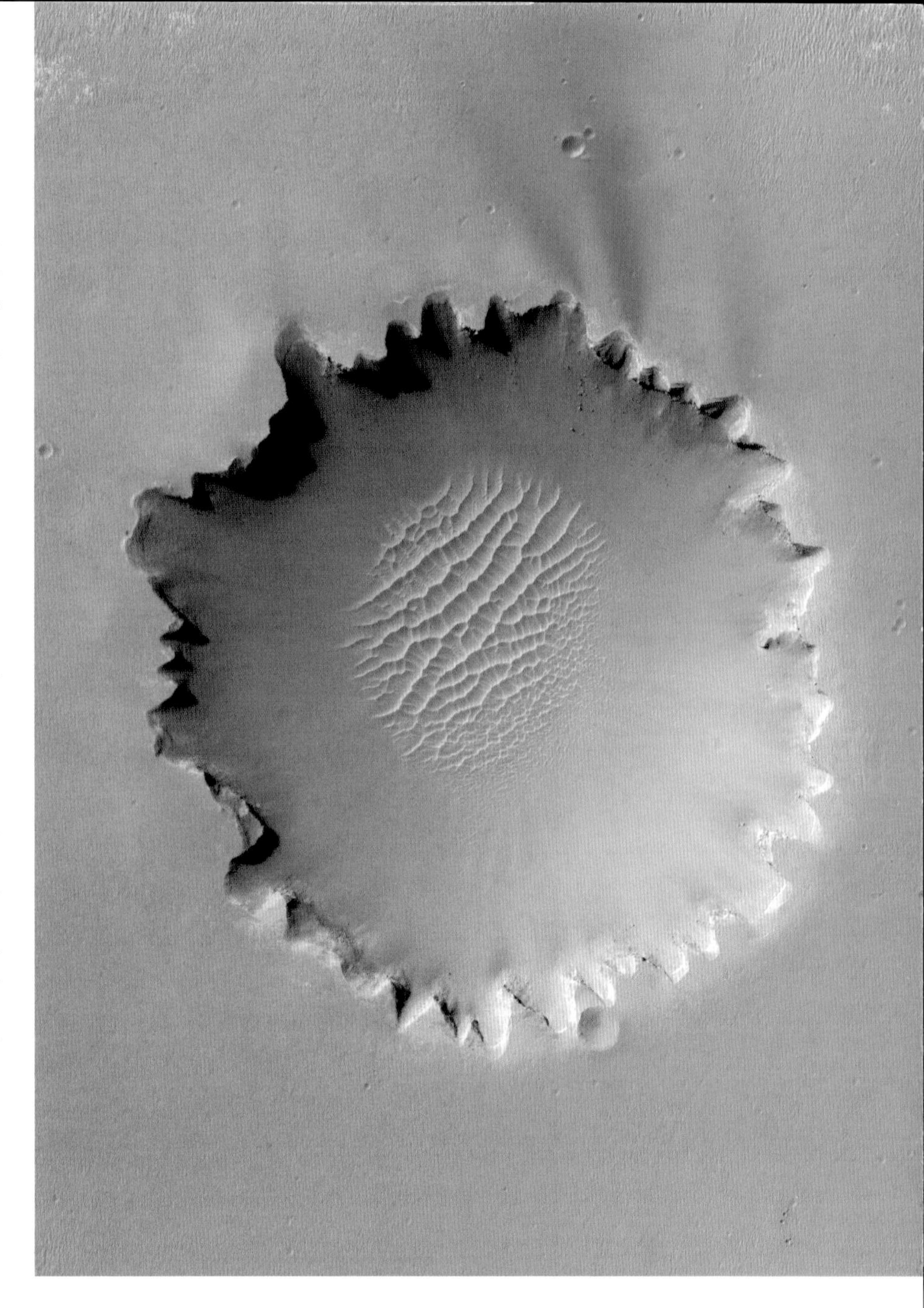

▶ **Victoria Crater** in Meridiani Planum, 730 m (2400 ft) across; pictured here by the MRO it was visited by Opportunity in September 2006.

▼ **Inside Victoria Crater** (Opportunity), looking from 'Duck Bay' to the dramatic promontory 'Cabro Frio' 200 m (600 ft) away.

Jupiter

Jupiter, giant of the Sun's family, lies well beyond the main asteroid zone. It is more massive than all the other planets combined, even though it has only $\frac{1}{1047}$ of the mass of the Sun. Despite its distance, it shines more brightly than any other planet apart from Venus and, very occasionally, Mars.

A casual look through a telescope is enough to show that Jupiter is quite unlike the Earth or Mars. Its surface is made up of gas, and the disk is obviously flattened; even though Jupiter is so large it has a rotation period of less than 10 hours, and this makes the equator bulge out. Moreover, Jupiter does not spin in the way that a solid body would do. The equatorial region has a period about five minutes less than that of the rest of the planet, and discrete features have periods of their own, so that they drift around in longitude.

It used to be thought that Jupiter was a miniature Sun, warming its satellite system, but this is very far from being the case; the upper clouds are bitterly cold. According to the latest models, there is a silicate core at a high temperature – perhaps 30,000°C – overlaid by layers of hydrogen; the lower layer is so compressed that it takes on the characteristics of a metal. Above comes the 'atmosphere', made up of 86 per cent hydrogen and 14 per cent helium, with small amounts of ammonia, methane and water vapour. The main markings on the disk are the cloud belts, several of which are always very much in evidence – notably the two Equatorial Belts, one to either side of the equator. There are spots, wisps and festoons; Jupiter shows vivid colours, and its surface is always changing. Wind speeds are very high; the outer atmosphere consists of a thick lower deck and a much more tenuous upper layer. There are no sharp lines of demarcation, so that it is not easy to decide just where the atmosphere merges into the true body of the planet.

Jupiter has an immensely powerful magnetic field; its magnetosphere may extend as far as the orbit of Saturn. There are zones of radiation which would instantly kill any astronaut incautious enough to venture inside them; Jupiter is a world to be viewed from a respectful distance. In addition, the planet is a strong radio emitter.

Amateur observers have always paid great attention to Jupiter, and have carried out very valuable work. Today, the amateur with a modest backyard telescope can take images far better than any which could have been obtained at any professional observatory a few decades ago. And Jupiter is a fascinating world; one must always be prepared for the unexpected!

▼ **Jupiter,** 14 April 2004; Dave Tyler, 20-cm (8-in) reflector. South is up. The Red Spot is prominent, and the belts are rich in detail. This image is far better than any obtainable 20 years ago with the best professional equipment. Three satellites are shown.

PLANETARY DATA – JUPITER	
Sidereal period	4332.59 days/11.86 years
Rotation period (equatorial)	9h 50m 30s
Mean orbital velocity	13.06 km/s (81 mi/s)
Orbital inclination	1° 18' 15".8
Orbital eccentricity	0.048
Apparent diameter	max. 50".1, min. 30".4
Reciprocal mass, Sun = 1	1047.4
Density, water = 1	1.33
Mass, Earth = 1	317.89
Volume, Earth = 1	1318.7
Escape velocity	60.22 km/s (37.42 mi/s)
Surface gravity, Earth = 1	2.64
Mean surface temperature	−150°C (at cloud tops)
Oblateness	0.06
Albedo	0.43
Maximum magnitude	−2.6
Diameter (equatorial)	143,884 km (89,424 mi)
Diameter (polar)	133,700 km (83,100 mi)

▲ **North Polar Region**
Lat. +90° to +55° approx.
Usually dusky in appearance and variable in extent.
The whole region is often featureless. The North Polar Current has a mean period of 9hrs 55mins 42secs.

North North North Temperate Belt
Mean Lat. +45°
An ephemeral feature often indistinguishable from the NPR.

North North Temperate Zone
Mean Lat. +41°
Often hard to distinguish from the overall polar duskiness.

North North Temperate Belt
Mean Lat. +37°
Occasionally prominent, sometimes fading altogether, as in 1924.

North Temperate Zone
Mean Lat. +33°
Very variable, both in width and brightness.

North Temperate Belt
Mean Lat. +31° to +24°
Usually visible, with a maximum extent of about 8° latitude. Dark spots at southern edge of the North Temperate Belt are not uncommon.

North Tropical Zone
Mean Lat. +24° to +20°
At times very bright. The North Tropical Current, which overlaps the North Equatorial Belt, has a period of 9hrs 55mins 20secs.

North Equatorial Belt
Mean Lat. +20° to +7°
The most prominent of all the Jovian belts. This region is extremely active and has a large amount of detail.

Equatorial Zone
Mean Lat. +7° to −7°
Covering about one-eighth of the entire surface of Jupiter, the EZ exhibits much visible detail.

Equatorial Band
Mean Lat. −0.4°
At times the EZ appears divided into two components by a narrow belt, the EB, at or near to the equator of Jupiter.

South Equatorial Belt
Mean Lat. −7° to −21°
The most variable belt. It is often broader than the NEB and is generally divided into two components by an intermediate zone. The southern component contains the Red Spot Hollow (RSH).

South Tropical Zone
Mean Lat. −21° to −26°
Contains the famous Great Red Spot. The STrZ was the site of the long-lived South Tropical Disturbance.

Great Red Spot
Mean Lat. −22°
Although there are other spots visible on Jupiter's surface, both red and white, the Great Red Spot is much the most prominent. It rotates in an anticlockwise direction.

South Temperate Belt
Mean Lat. −26° to −34°
Very variable in width and intensity; at times it appears double.

South Temperate Zones
Mean Lat. −38°
Often wide; may be extremely bright. Spots are common.

South South Temperate Belt
Mean Lat. −44°
Variable, with occasional small white spots.

South South South Temperate Belt
Mean Lat. −56°

South Polar Region
Lat. −58° to −90° approx.
Like the NPR, very variable in extent.

▼ **Jupiter** This image by Dave Tyler shows a general view of Jupiter. South is up. The satellite is Ganymede and the shadow appears as a black spot.

Missions to Jupiter

Several spacecraft have now passed Jupiter. First there were Pioneer 10 (December 1973) and Pioneer 11 (December 1974), which carried out preliminary surveys; Pioneer 11 was subsequently sent on to a brief encounter with Saturn. Next came the much more sophisticated Voyagers, No 1 (March 1979) and No 2 (July of the same year). Both of these then went on to carry out detailed studies of Saturn and its satellite system; Voyager 2 went on to encounter Uranus and Neptune as well. All of these four early probes are now on their way out of the Solar System permanently. In February 1992 the Ulysses solar polar probe passed near to Jupiter, mainly to use its gravitational pull to send Ulysses soaring far out of the plane of the ecliptic, but observations of Jupiter were also made – the opportunity was too good to be missed. The most recent probe specifically targeting Jupiter, Galileo, was dispatched in October 1990 and reached its target in December 1995. Two more spacecraft – Cassini and New Horizons – passed the planet on their way to Saturn and Pluto, respectively, and obtained images of cloud structure, the rings and some satellites.

Very important results came from the Voyagers. Particular attention was paid to the magnetosphere, which is very extensive. It is not spherical, but has a long 'magnetotail' stretching away from the direction of the Sun and extends out to well over 700 million kilometres (over 400 million miles), so that at times it may even engulf the planet Saturn. There are zones of radiation ten thousand times stronger than the Van Allen zones of the Earth, so that any astronaut foolish enough to venture into them would quickly die from radiation poisoning. Indeed, the unexpectedly high level of radiation almost crippled the equipment in Pioneer 10, the first probe to pass the planet, and subsequent spacecraft were aimed so as to pass quickly over the equatorial region, where the danger is at its worst. The Voyagers were carefully constructed so as to tolerate twice the anticipated of radiation; only minor effects were noted when Voyager 1 approached within 350,000 kilometres (220,000 miles) of Jupiter, but Voyager 2, passing at 650,000 kilometres (400,000 miles), found the level to be three times stronger. Evidently the zones are very variable. The magnetic field is extremely complicated, and is reversed relative to that of the Earth, so that a compass needle would point south; the magnetic axis is inclined to the rotational axis by 10°.

An obscure ring system was discovered and superb pictures of the planet's surface were obtained, showing the turbulent, vividly coloured clouds and spots. Aurorae and lightning flashes were recorded on the night side, and observations of all kinds were made. It was seen that marked changes had occurred between the Pioneer and Voyager passes, and even in the interval between the Voyager 1 and 2 encounters, the shape of the Red Spot was different, for example.

Galileo was launched in October 1989, and after a somewhat roundabout journey reached Jupiter in December 1995. It was made up of two parts: an entry vehicle, and an orbiter, which separated from each other well before arrival. The entry probe was scheduled to plunge into the Jovian clouds,

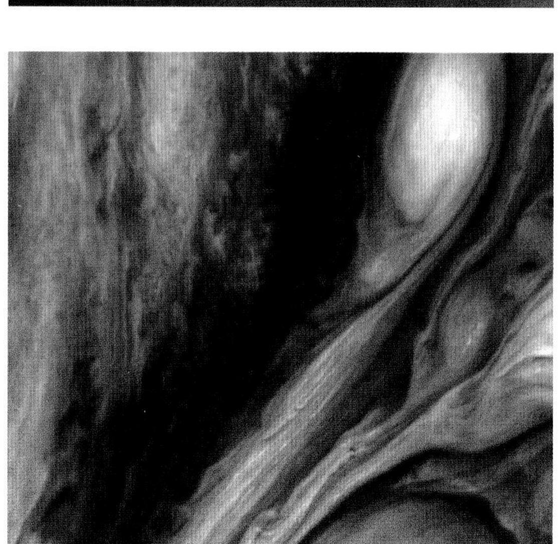

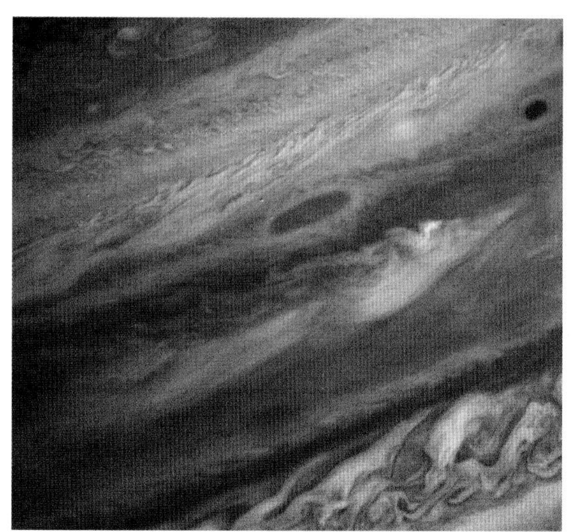

▶ **Voyager views of Jupiter.** The photograph at top left shows the Great Red Spot from Voyager 1 on 1 March 1979. Long subject to observation, the spot forms a hollow in the adjoining belt, and though it periodically disappears for a while, it always returns. Top right shows the North Equatorial Belt from Voyager 2, taken on 28 June 1979; the colours have been deliberately enhanced to bring out more detail. The wisp-like plumes of the equatorial zone can be seen across the middle of the image. Bottom left shows the region just east of the Great Red Spot in a view from Voyager 1. The colours have again been exaggerated to bring out subtle variations in shading. The view at bottom right was taken by Voyager 2 on 29 June 1979. It extends from latitude 40°S to 40°N, and shows an equatorial zone similar to that at top right. A region of turbulence is visible in the bottom righthand corner, just west of the Great Red Spot, where western and eastern winds combine.

and transmit data for over an hour until it was destroyed; the orbiter would orbit Jupiter for several years, sending back images of the planet and its satellites.

The high-gain antenna, a particularly important part of the communications link, failed to unfurl; some data were lost, but much of the planned programme could be carried out. The entry probe plunged into the clouds on schedule, and continued to transmit data for 75 minutes, by which time it had penetrated to a depth of 160 kilometres (100 miles).

Some of the results were unexpected. For instance, it had been thought that the strong Jovian winds would be confined to the outer clouds and would slacken with increasing depth, but this did not happen; by the time contact was lost, the winds were just as strong as they had been at the surface, indicating that the driving force was not the Sun, but heat radiating from the interior. There was much less lightning activity than had been expected, and, surprisingly, the Jovian atmosphere was dry, so it was thought that the water observed after the Shoemaker-Levy impact came from either the dying comet or from deeper in the Jovian atmosphere. However, later analysis showed that the Galileo entry probe had plunged into the clouds in an unusually 'dry' area of the planet – an equivalent of a Jovian desert, so the lack of water was not typical. One mystery at least was solved, but it cannot be claimed that we have as yet anything like a complete knowledge of the interior of Jupiter. We can hardly hope for another cometary impact, so that presumably we must wait for another deliberate entry into Jupiter's cloud layer.

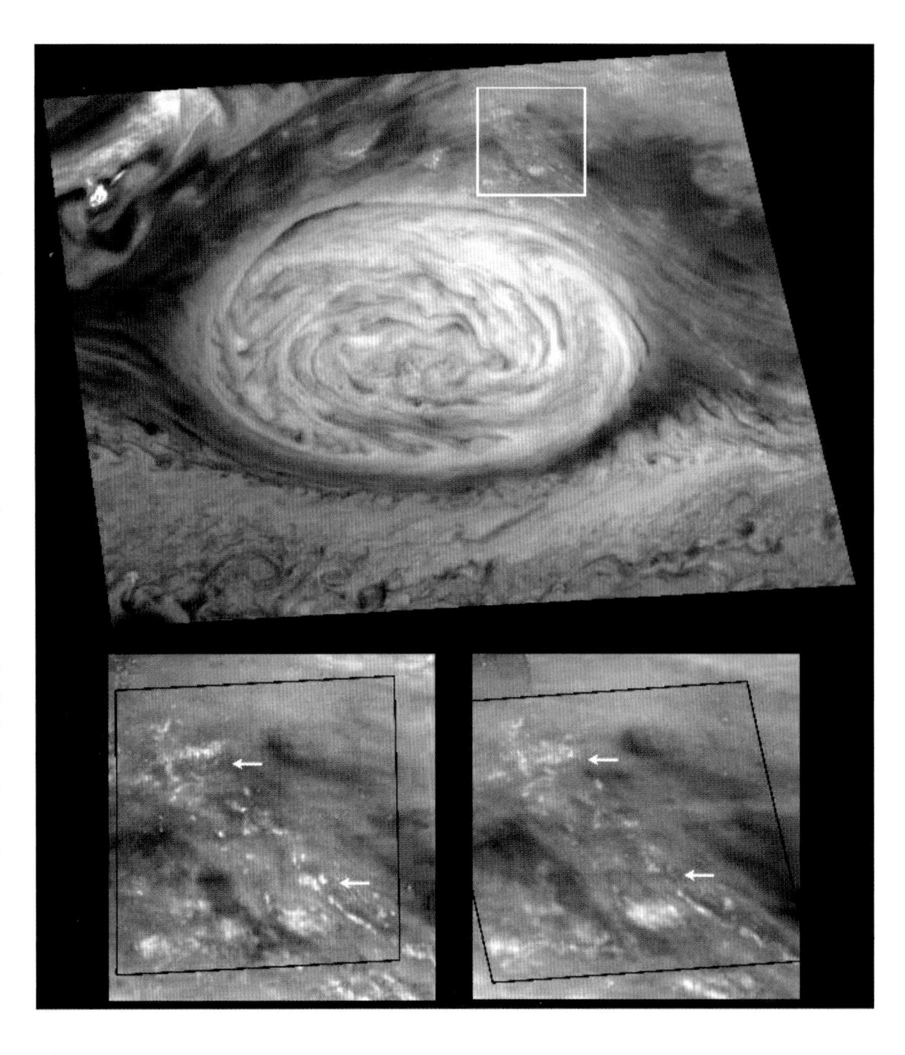

► **Thunderheads on Jupiter;** bright, white cumulus clouds. In the upper image, they edge the Great Red Spot. Imaged from Galileo 26 July 2006, SSI camera, from 1,384,000 km (860,000 mi).

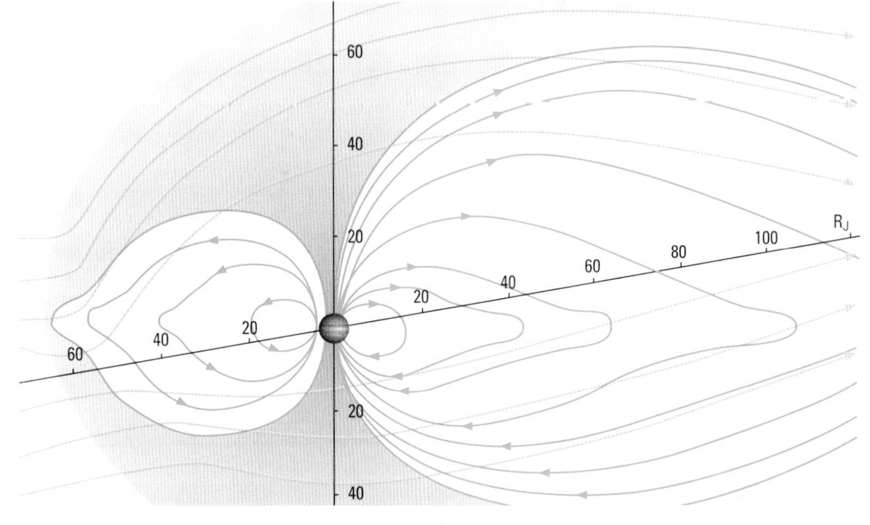

► **The magnetosphere of Jupiter.** Solar wind particles approaching from the left collide with the structurally complex magnetosphere. Inside the bow shock lies the magnetopause. The whole magnetically active region is enveloped by the 'magnetosheath'.

▼ **Jupiter's ring.** This picture of the western portion of Jupiter's main ring was made by the Galileo spacecraft on 9 November 1996.

Comet Shoemaker–Levy 9

▶ **Comet collision** – a Hubble Space Telescope image showing numerous Comet Shoemaker–Levy 9 impact sites on Jupiter shortly after the collision of the last cometary fragment. The impact sites appear as dark 'smudges' lined up across the middle of Jupiter's southern hemisphere and are easily mistaken for 'holes' in the giant planet's atmosphere. In reality, they are the chemical debris 'cooked' in the tremendous fireballs that exploded in Jupiter's atmosphere as each fragment impacted the planet. This material was then ejected high above the bright multicoloured cloud tops where it was caught in the winds of the upper atmosphere and eventually dispersed around the planet.

There have been two recent impacts on Jupiter – one natural, one man-made. Each has provided us with a great deal of new information.

In March 1993 three American comet-hunters, Eugene and Carolyn Shoemaker and David Levy, discovered what they described as a 'squashed comet'; it was their ninth discovery, so the comet became known as Shoemaker-Levy 9 (SL9). It was unlike anything previously seen. It was orbiting Jupiter, and had been doing so for at least 20 years; calculation showed that on 7 July 1992 it had skimmed over the Jovian cloud tops at a mere 21,000 kilometres (13,000 miles) and the nucleus had been torn apart, so that it had been transformed into a sort of string-of-pearls arrangement. Over 20 fragments were identified, and were conveniently lettered from A to W.

It was soon found that the comet was on a collision course, and that the fragments would hit Jupiter in July 1994. The first fragment, A, impacted on 16 July, just on the side of Jupiter turned away from Earth, but the planet's quick spin soon brought the impact site into view. The other fragments followed during the next few days, and produced dramatic effects; there were huge scars, visible with a very small telescope. The cometary fragments were only a few kilometres in diameter, but were travelling at tremendous speed. Spectacular pictures were obtained from ground observatories and from the Hubble Space Telescope; the effects of the impacts were detectable for months.

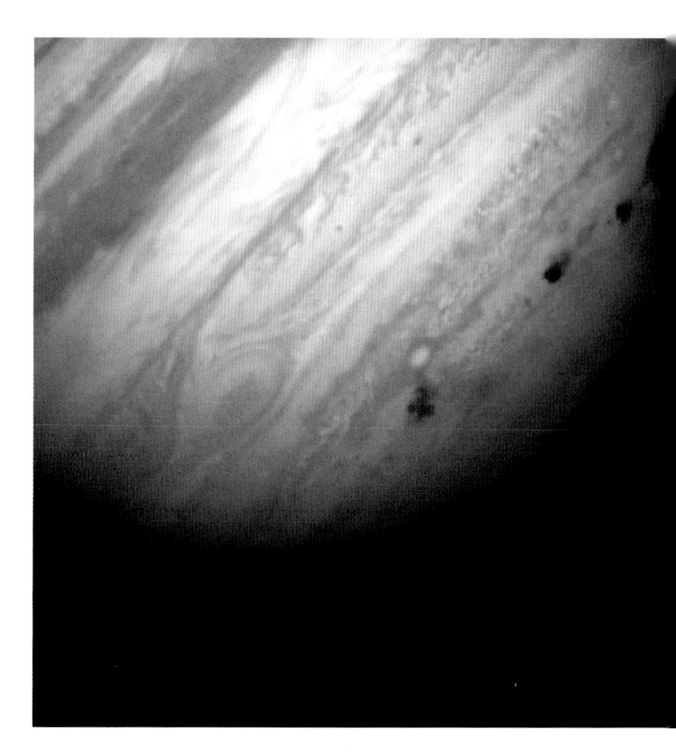

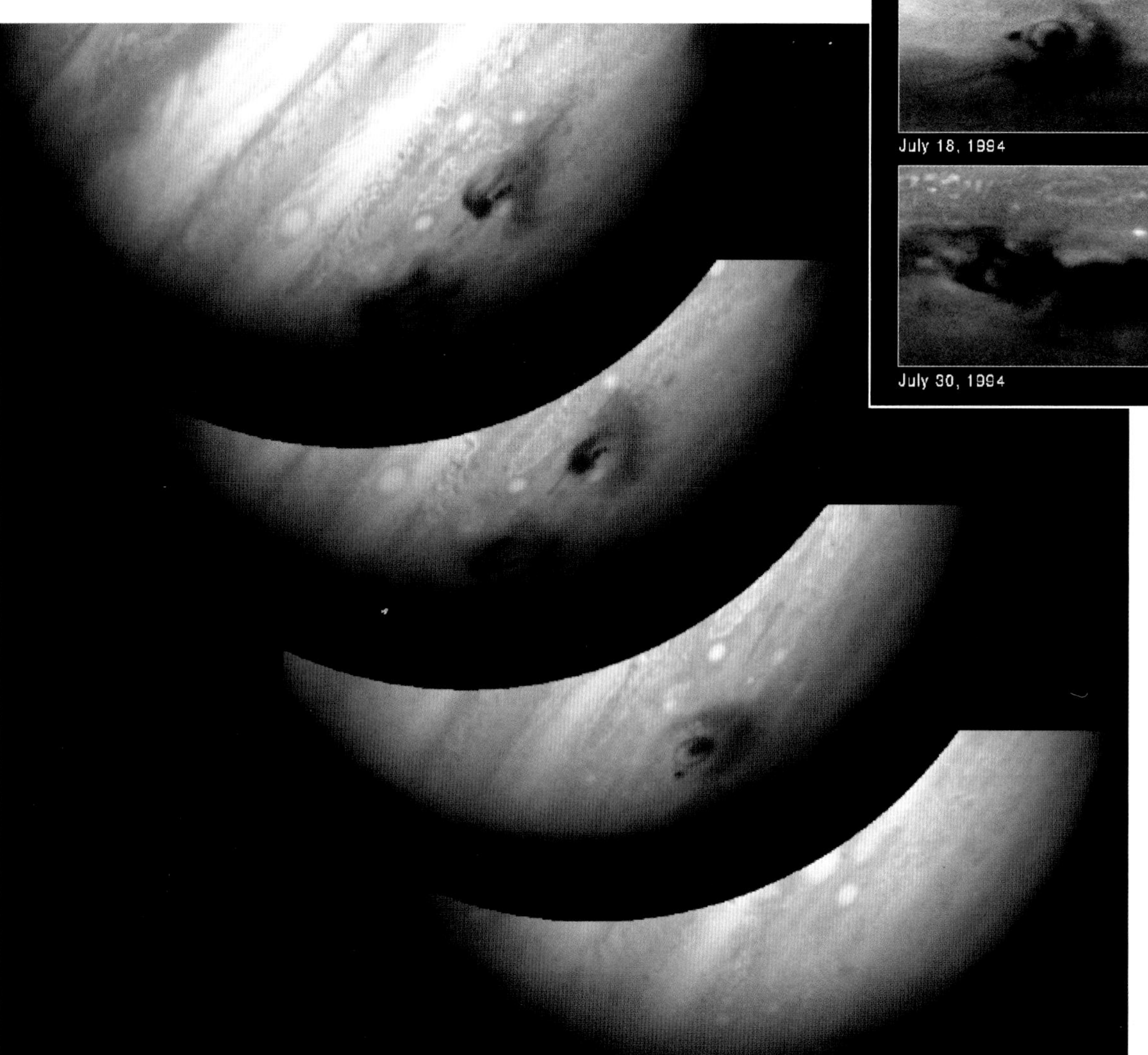

◀ **Impact of Shoemaker-Levy 9.** This mosaic of images taken with the HST's WFPC2 shows the early evolution of the site of the impact of fragment G, the largest fragment. From bottom to top: 5 minutes after impact, the plume can be seen over the planet's limb; after 90 minutes the site had rotated into view; after 3 days the dark stain is spreading out like a bruise (the second impact site is that of fragment L) and after 5 days, the debris is beginning to be smeared out by the planet's winds.

◄▼ **Comet Shoemaker–Levy 9.** A composite HST image taken in visible light showing the evolution of the brightest region of the comet. In this false-colour representation, different shades of red are used to display different intensities of light:
1 July 1993 – data taken prior to the HST servicing mission. The separation of the two brightest fragments is only 0.3", so ground-based telescopes could not resolve this pair. The other two fragments just to the right of the closely spaced pair are only barely detectable because of HST's spherical aberration.
24 January 1994 – the first HST observation after the servicing mission. The two brightest fragments are now about 1" apart, and the two fainter fragments are much more clearly seen. The light near the faintest fragment is not as concentrated as the light from the others and is elongated in the direction of the comet's tail.
30 March 1994 – the latest HST observation shows that the faintest fragment has become a barely discernible 'puff'. Also, the second faintest fragment has clearly split into two distinct fragments by March.

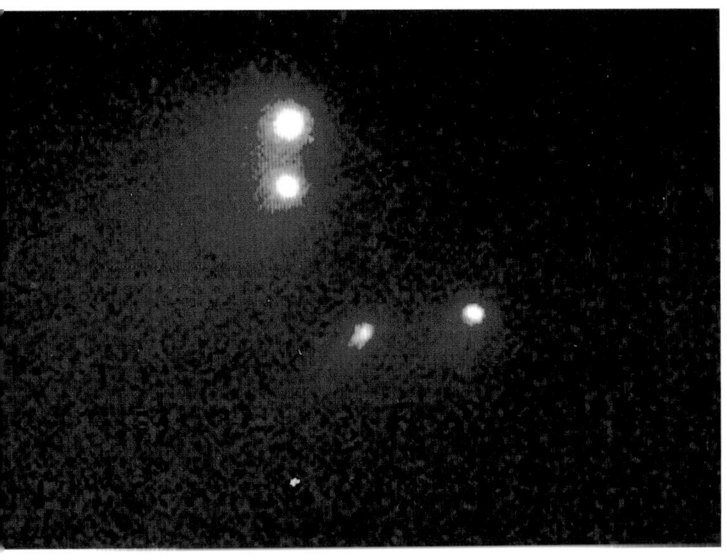

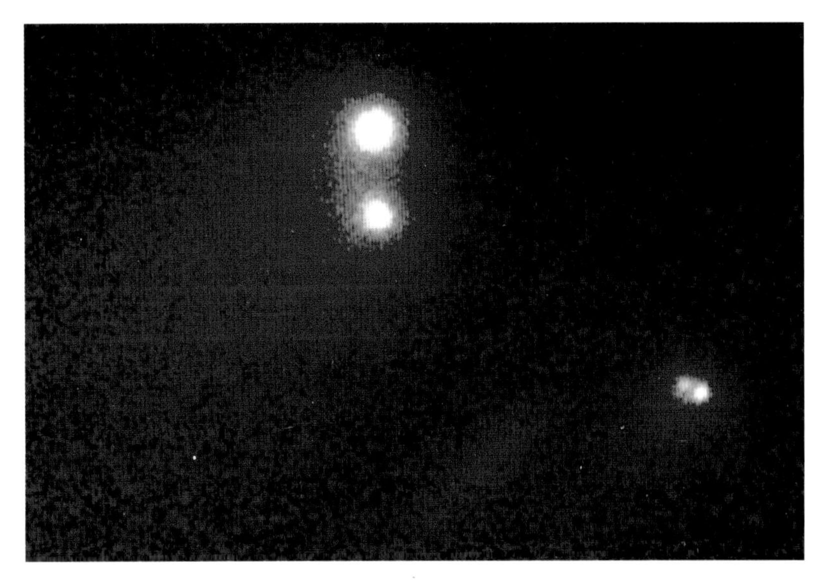

◄ **Further evolution of fragment G's impact site.** In the near infra-red, the impact site of fragment D is engulfed by that of fragments G and S. More than a month after impact, the dark stain is still clearly visible.

▼ **The disruption of Comet Shoemaker-Levy 9.** This HST image taken in May 1994 – some two months before impact – shows the comet broken into over 20 pieces.

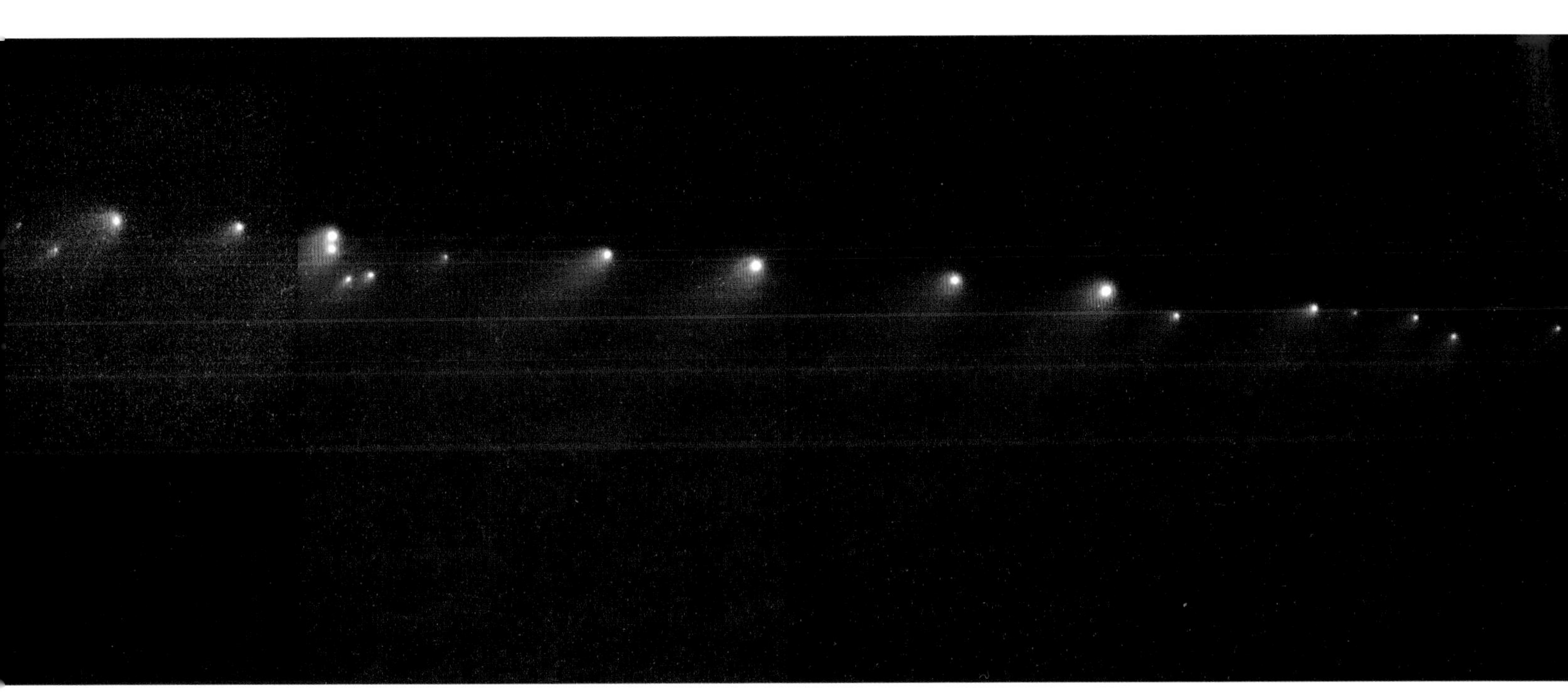

Jupiter: Rings and Satellites

Jupiter has a ring system, first detected in 1979 by the Voyager 1 spacecraft, but there is no comparison with the glorious ring-system of Saturn. Jupiter's rings are dark and obscure, so there is no chance of observing them with an ordinary Earth-based telescope. There are three main components, the Halo, Main and Gossamer rings.

Name	Distance from centre of Jupiter, km	Width, km
Halo	89,400–123,000	33,600
Main	123,000–128,940	5940
Gossamer	128,940–280,000	151,060

The rings seem to be composed of small particles knocked off the small inner satellites (Metis, Adrastea, Amalthea and Thebe) by meteoric impacts. The Halo and Gossamer rings are extremely tenuous. In addition there is a distant, very faint ring of unknown origin.

Jupiter has a wealth of satellites. Four – Europa, Callisto, Io and Ganymede – are of planet-size; Europa is slightly smaller than our Moon, Io slightly larger and Ganymede and Callisto much larger; Ganymede is larger than Mercury, though less massive. All can be seen with any small telescope, and it is fascinating to follow their movements from night to night; they may be occulted by Jupiter, pass across the Jovian disk in transit, or be eclipsed by Jupiter's shadow. During transits, their own shadows cast on the Jovian clouds are very prominent. The satellites were discovered in 1610 by Galileo, using his first tiny telescope, and are always known collectively as the Galieans (they were discovered independently at the same time by Simon Marius, and it is possible that a naked-eye record of Ganymede was made by a Chinese astronomer, Gan De, in 364 BC). Europa is icy and smooth, Ganymede and Callisto icy and cratered, and Io violently volcanic, with major eruptions going on all the time. It has been said, 'there is no such thing as an uninteresting Galilean.'

The other satellites are small, and only Amalthea is as much as 250 kilometres (230 miles) in diameter. To date (February 2007) the grand total of known attendants is 63. In addition to the Galileans, only a few were known before the Voyager missions: Himalia, Lysithea, Leda, Elara, Ananke, Carme and Pasiphaë. The dwarf outer satellites are captured asteroids; some have retrograde motion and are so far from Jupiter that their orbits are not even approximately circular – no two successive orbits are alike. No doubt many more tiny asteroidal satellites await discovery.

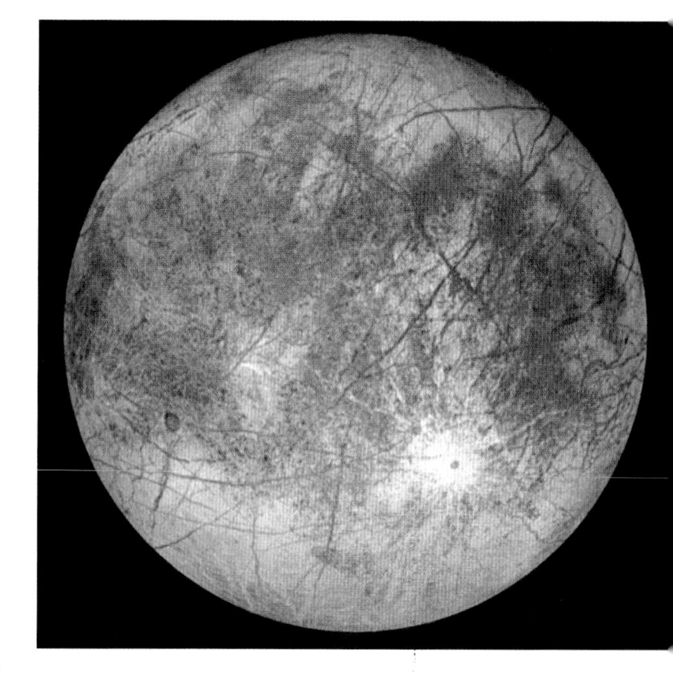

◄► **The four Galilean satellites** of Jupiter as imaged by the Galileo spacecraft: Io (top left) on 3 July 1999, Europa (top right) on 7 September 1996, Ganymede (bottom left) on 26 June 1996, and Callisto (bottom right) in May 2001. The two brightest satellites, Io and Europa, apparently have surfaces of very different composition. Io is thought to be covered with sulphur and salts, and its surface shows evidence of ongoing volcanic activity. Europa's surface is mainly water ice with some rocky areas; long fractures can been seen in the crust. Ganymede has both ice and rock exposed on its surface, and the bright spots are relatively recent impact craters. Callisto's surface is primarily rock-covered ice, covered with craters. These surface properties contrast sharply with the interiors of the satellites: Io has a hot, molten interior of silicate material, but no water or ice. Europa, Ganymede and Callisto may all have ice/water in the interior. and Europa have rocky interiors, which contain large amounts of water or ice. All the images were obtained by the Solid State Imaging (SSI) system on NASA's Galileo orbiter.

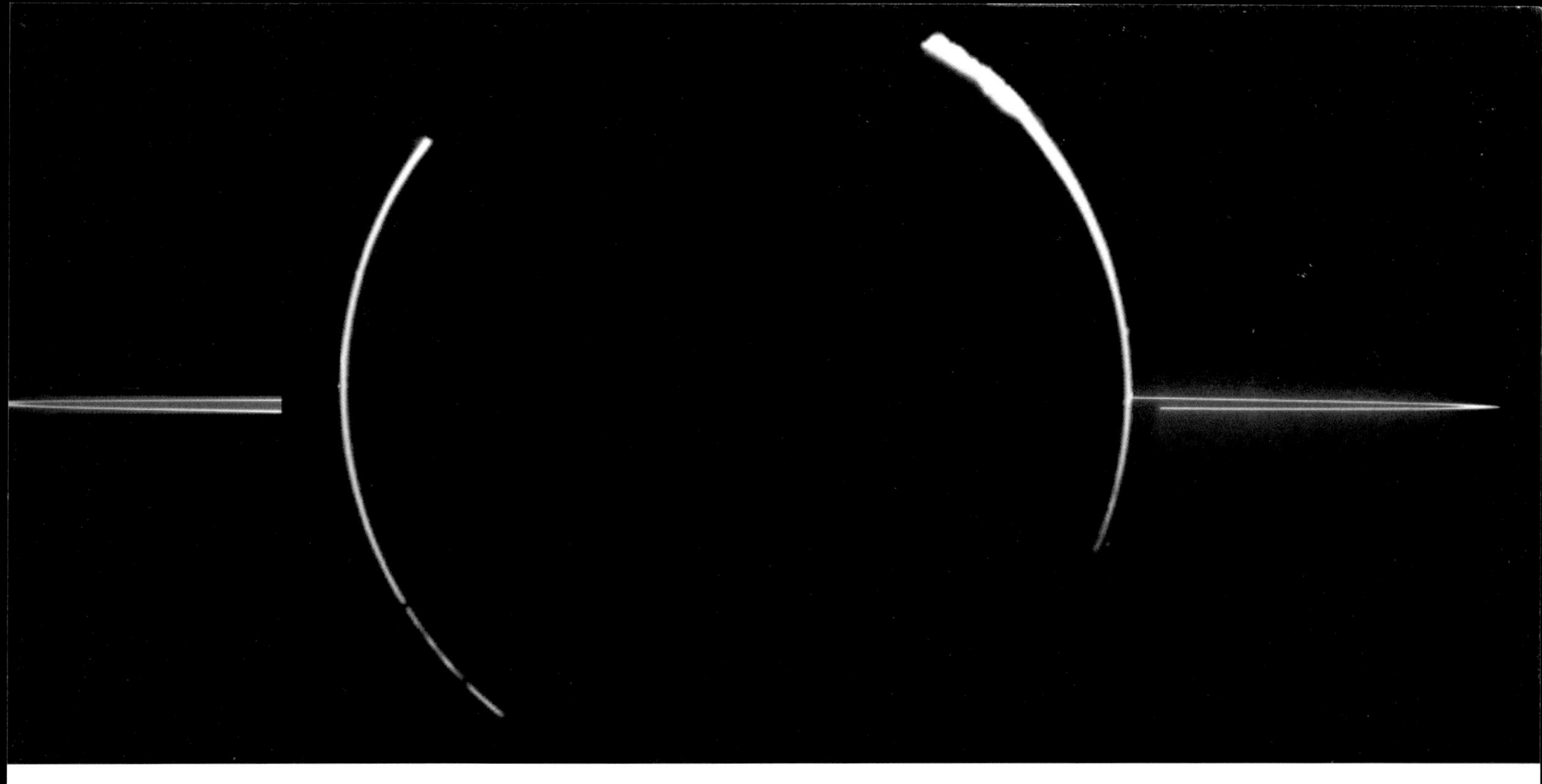

▲ **View of jupiter's ring system,** acquired by Galileo when the Sun was behind the planet, and the spacecraft was in Jupiter's shadow.

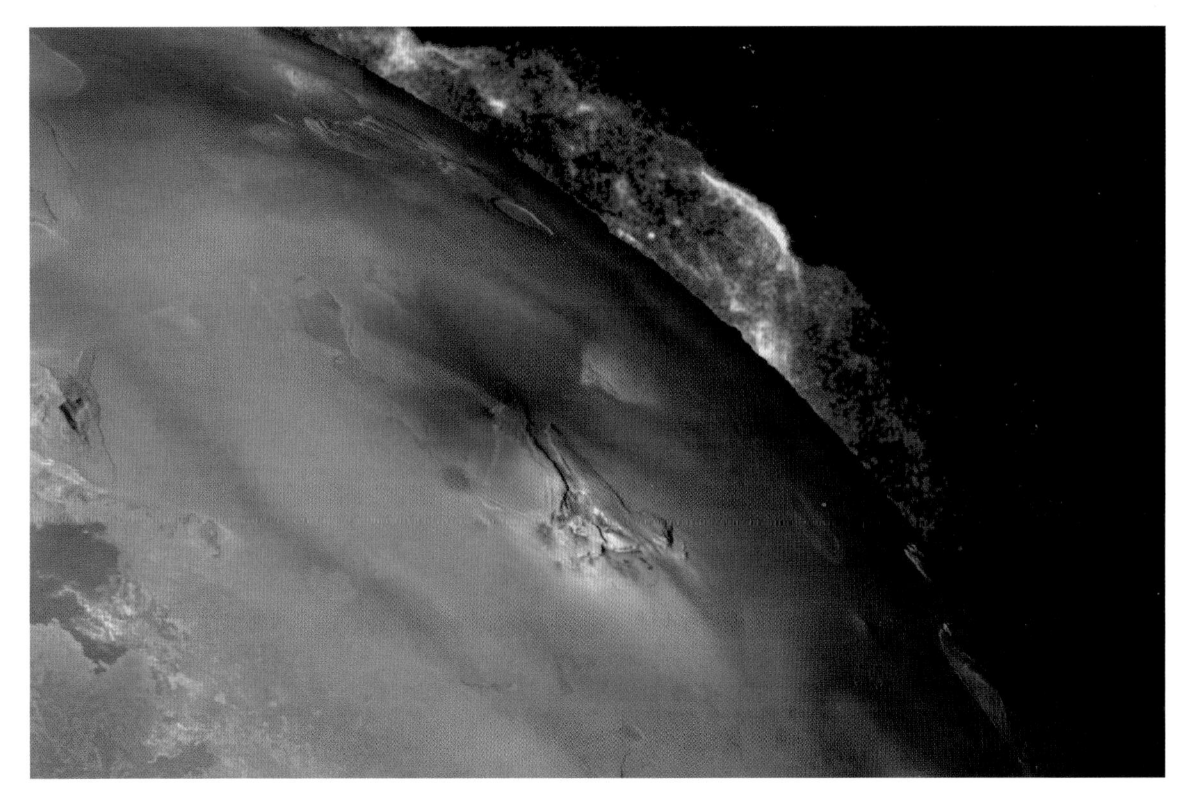

◄ **The volcanoes of Io** from Voyager. Activity, seen along the limb, seems to be constant; the crust is certainly unstable and material is sent up from the volcanoes to a height of hundreds of kilometres above the surface.

SATELLITES OF JUPITER

Name	Distance from Jupiter, km	Orbital period, days	Orbital Incl., °	Orbital ecc.	Diameter, km	Density, water = 1	Escape vel., km/s	Mean opp., mag.
Metis	127,900	0.290	0	0	40	3?	0.02?	17.4
Adrastea	128,980	0.298	0	0	26 × 20 × 16	3?	0.01?	18.9
Amalthea	181,300	0.498	0.45	0.003	262 × 146 × 143	3?	0.16?	14.1
Thebe	221,900	0.675	0.9	0.013	110 × 90	3?	0.8?	15.5
Io	421,600	1.769	0.04		3660 × 3637 × 3631	3.55	2.56	5.0
Europa	670,900	3.551	0.47	0.009	3130	3.04	2.10	5.3
Ganymede	1,070,00	7.155	0.21	0.002	5268	1.93	2.78	4.6
Callisto	1,880,000	16.689	0.51	0.007	4806	1.81	2.43	5.6
Leda	11,094,000	238.7	26.1	0.148	8	3?	0.1?	20.2
Himalia	11,480,000	250.6	27.6	0.158	186	3?	0.1?	14.8
Lysithea	11,720,000	259.2	29.0	0.107	36	3?	0.01?	18.4
Elara	11,737,000	259.7	24.8	0.207	76	3?	0.05?	16.7
Ananke	21,200,000	631*	147	0.17	30	3?	0.01?	18.9
Carme	22,600,000	692*	164	0.21	40	3?	0.02	18.0
Pasiphaë	23,500,000	735*	145	0.38	40	3?	0.02	17.7
Sinope	23,700,000	758*	153	0.28	35	3?	0.01?	18.3

(* = retrograde)

Jupiter: Io

▶ **Io imaged from Galileo.** Io is the most volcanically active world in the Solar System, and there are few impact craters. It is pocked with dark volcanic vents and layers of colourful sulphurous deposits.

▼ **Changes on Io** in close-up: three views of the Pillan Patera region from Galileo, taken in April 1997 (left), September 1997 (middle) and July 1999 (right), Pele (centre left) and Pillan (above right). Between April and September 1997 the volcano Pillan erupted, producing the dark deposit that can be sent above right centre. By 1999 the red material from Pele (at left centre) has started to cover the dark material from Pillan. It also appears that a further volcano, to the right of Pillan, has erupted since the 1997 images. The pictures cover an area approximately 1650 km wide and 1750 km high (1025 and 1090 mi).

The Galileo orbiter has sent back superb images of the Galilean satellites. Ganymede and Callisto are icy and cratered; the former's magnetic field was a great surprise. The most spectacular images are those of Io and Europa.

The Ionian scene has changed markedly since the Voyager passes, while the existence of an underground ocean inside Europa is a real possibility. Some surface details on the satellites can be seen with the Hubble Space Telescope, and the events on Io can be monitored, but of course the clarity of the Galileo pictures is unrivalled.

Io is the most volcanically active world in the Solar System; eruptions are going on all the time. It is strange that there is so great a difference between Io and the inert, ice-coated Europa. Io, is also the only satellite to show activity apart from Saturn's Enceladus and Neptune's Triton, with their ice geysers.

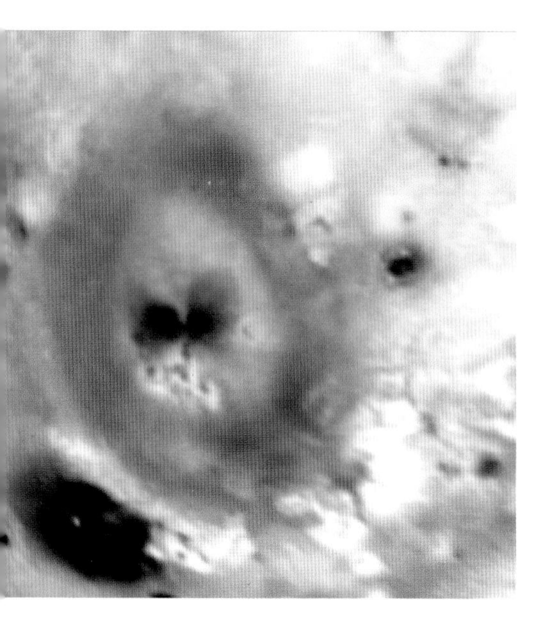

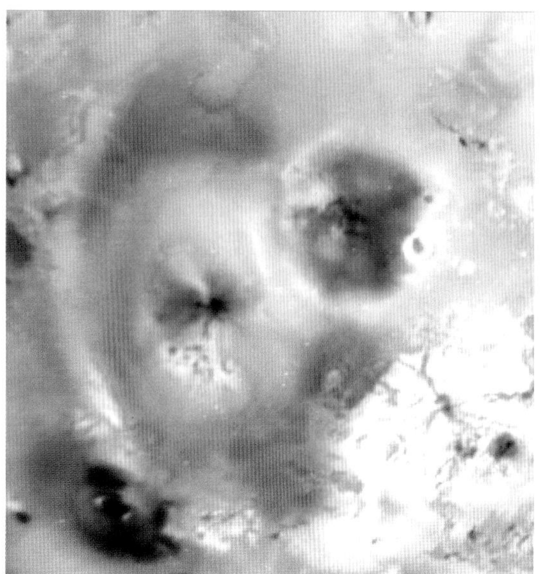

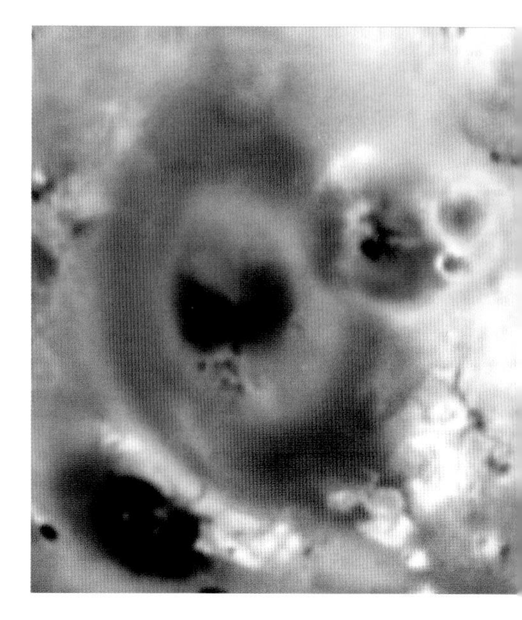

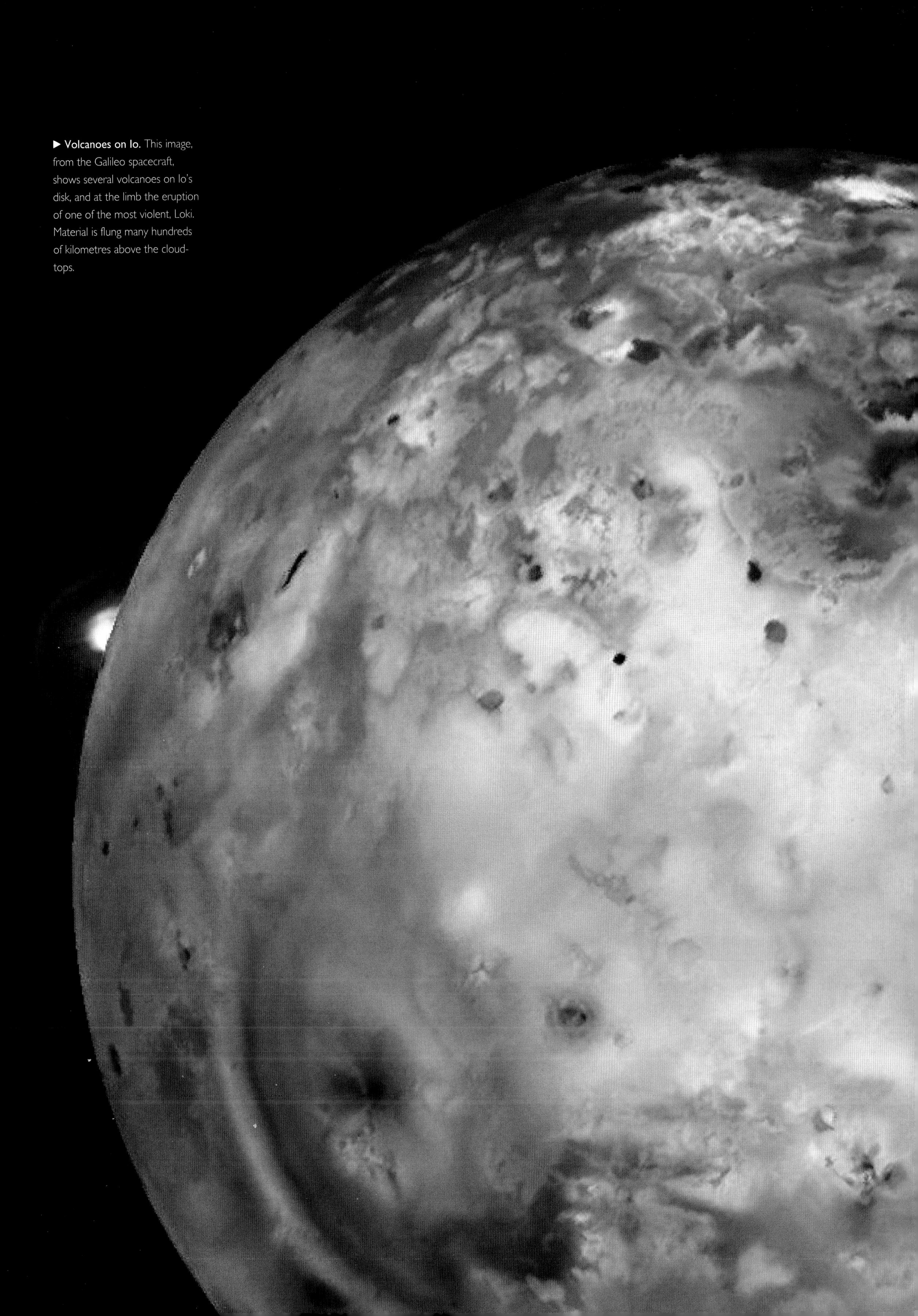

▶ **Volcanoes on Io.** This image, from the Galileo spacecraft, shows several volcanoes on Io's disk, and at the limb the eruption of one of the most violent, Loki. Material is flung many hundreds of kilometres above the cloud-tops.

Jupiter: Europa, Ganymede and Callisto

Europa's surface had already been revealed by Voyager 2 to be most interesting and results from Galileo were eagerly awaited. Images show what appear to be giant blocks of ice that have broken apart and settled into new positions. The icy plains subdivide into units with patches where contaminants are present and fractures and ridges. The surface is much younger than that of other icy satellites, to judge from the paucity of impact craters which shows that Europa experiences a rapid rate of resurfacing, though it is not so rapid as on Io. It has been suggested that beneath Europa's icy crust there may be an ocean of liquid water, perhaps even containing life. However, there is no proof that an ocean exists, and the discovery of life there would indeed be surprising.

Callisto is the outermost and darkest of the Galilean satellites and is heavily cratered. It is the only one of the group not to show any signs of current or past geological activity.

Ganymede is the largest satellite in the Solar System and has an old crater-pitted surface. It is predominantly made of two types of terrain, with older dark areas broken into by younger, lighter, ridged sections.

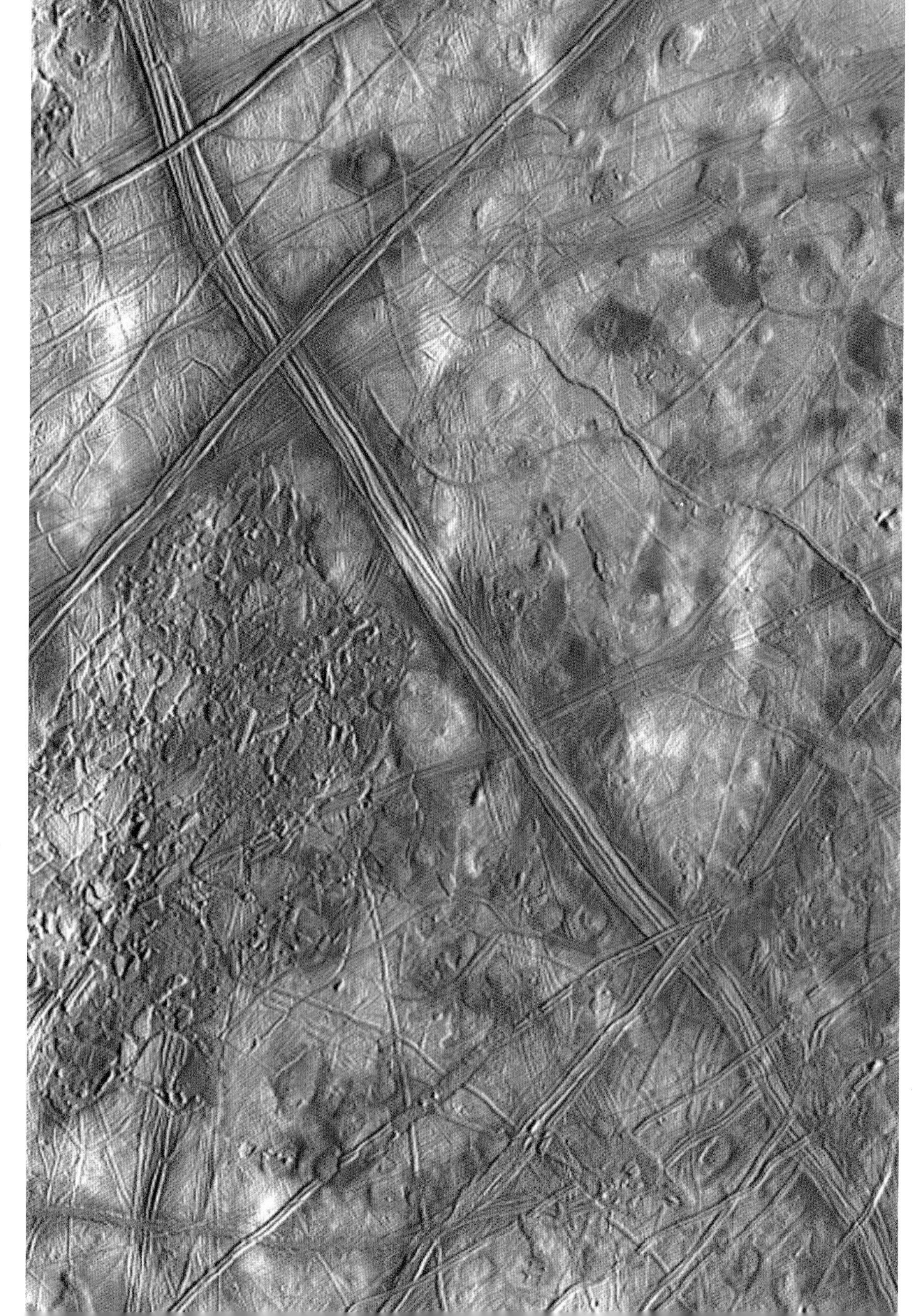

► **Europa, from Galileo.** False colour has been used to enhance the visibility of certain features in this composite of three images of Europa's crust. It shows what appear to be blocks that have broken apart and settled in new locations. The icy plains, shown here in bluish hues, subdivide into units with different albedos at infra-red wavelengths, probably because of differences in the grain size of the ice. The brown and reddish hues represent regions where contaminants are present in the ice. The lines are fractures or ridges in the crust. The composite was produced by Galileo imaging team scientists at the University of Arizona. The images were obtained during September 1996, December 1996 and February 1997 from 677,000 km (417,489 mi).

▲ **Ganymede.** Jupiter's largest satellite; a mosaic created from Galileo images. This image of southern Galileo Regio shows impact craters in various stages of degradation. Almost all of the craters appear flat. The two prominent light coloured craters are almost completely erased by the flow in the icy crust.

▼ **Callisto.** Outermost of the Galilean satellites; it is heavily cratered, and seems to be inert. This image shows Valhalla, one of the two huge ringed basins.

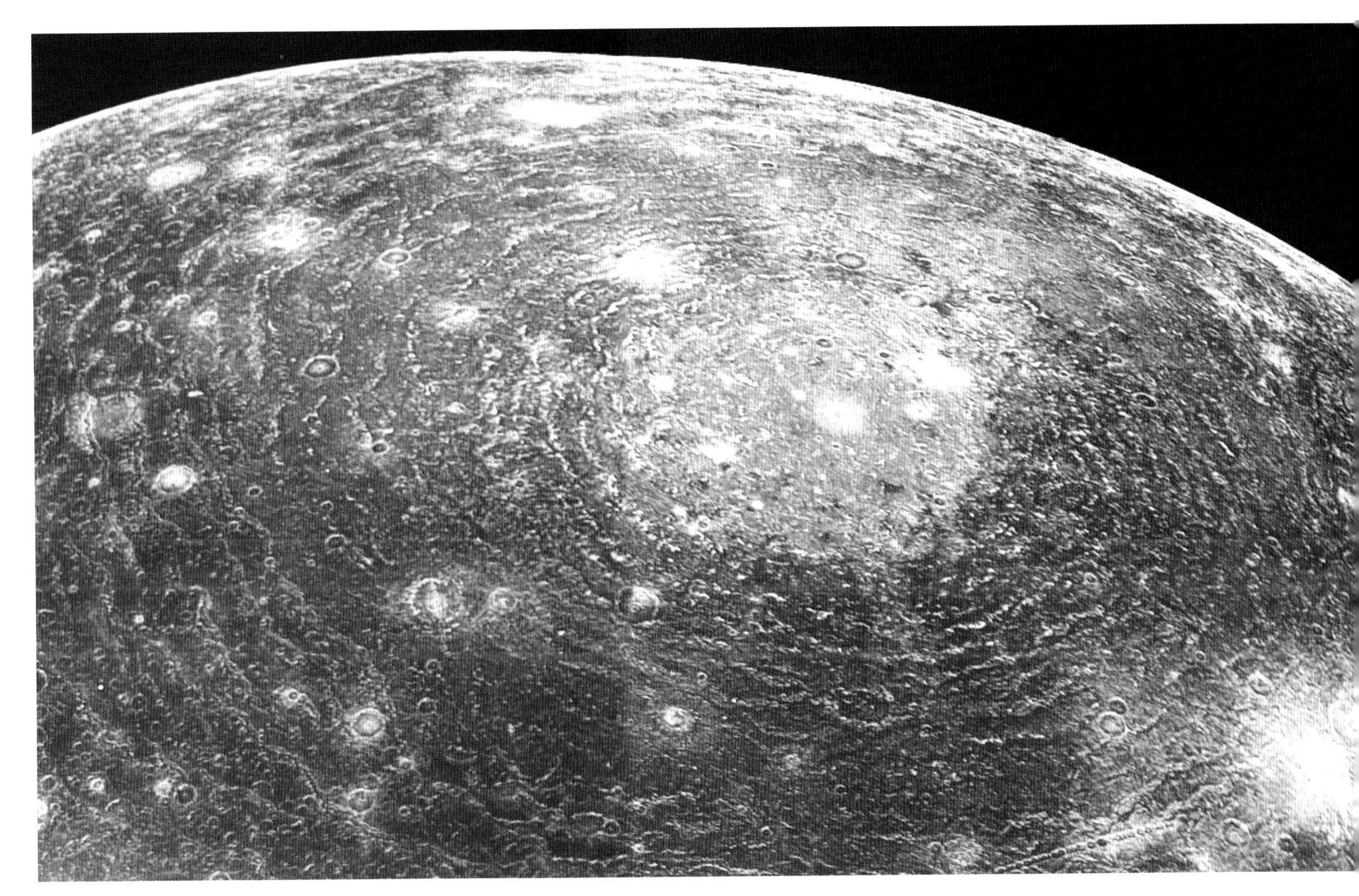

Rings of Saturn

The ring system of Saturn is unique, quite unlike the dark, obscure rings of Jupiter, Uranus and Neptune. Saturn's rings were first seen in the 17th century, and explained by Christiaan Huygens, in 1659 in *Systema saturnium*; previously Saturn had even been regarded as a triple planet.

There are two bright rings (A and B) and a fainter inner ring (C) which is usually known as the Crêpe or Dusky Ring because it is semi-transparent. The bright rings are separated by a gap known as the Cassini Division in honour of G. D. Cassini, who discovered it in 1675. Various fainter rings, both inside and outside the main system, had been reported before the Space Age, but there had been no definite confirmation. The main system is relatively close to the planet, and lies well within the Roche limit – that is to say, the minimum distance at which a fragile body can survive without being gravitationally disrupted; the outer edge of the A Ring lies at 135,200 kilometres (84,000 miles) from Saturn's centre, while Mimas, the innermost of the satellites known before the space missions, is much farther out at 185,600 kilometres (116,000 miles).

The rings extend from an inner radius of 67,000 kilometres (42,000 miles) to an outer radius of 480,000 kilometres (380,000 miles) and are a few hundred metres (up to 1000 feet) thick, with an overall diameter of 960,000 kilometres (600,000 miles). The ratio of thickness to diameter is about 1 : 1 million. When the rings are edgewise-on to us they almost disappear. Edgewise presentations occur at intervals of 13 years 9 months and 15 years 9 months alternately, as in 1966, 1980, 1995, 2009 and 2025. This inequality is due to Saturn's orbital eccentricity.

During the shorter interval, the south pole is tilted sunwards – in other words, it is summer in the southern hemisphere – and part of the northern hemisphere is covered up by the rings; during this time Saturn passes through perihelion, and is moving at its fastest. During the longer interval the north pole is turned sunwards, so that parts of the southern hemisphere are covered up; Saturn passes through aphelion, and is moving at its slowest.

The rings are at their most obscure when the Earth is passing through the main plane or when the Sun is doing so. It is wrong to claim that they vanish completely; they can be followed at all times with powerful telescopes, but they cannot be seen with smaller instruments, and at best they look like very thin, faint lines of light.

No solid or liquid ring could exist so close to Saturn (if, indeed, such a ring could ever be formed in the first place). It has long been known that the rings are made up of small particles, all moving round the planet in the manner of tiny moons. There is no mystery about their composition; they are made up of a mixture of rock and water ice.

► **The edgewise-on rings.** These pictures were taken with the Hubble Space Telescope in 1995, when the rings were edge-on to the Earth. (Top) Titan, to the left, is casting its shadow on to Saturn's disk; four of the icy satellite are shown – from left to right Mimas, Tethys, Janus and Enceladus. (Lower) The rings are now slightly tilted; Dione (lower right) casts a long, thin shadow across the ring system. Tethys is at the upper left.

▼ **Aspects of the rings** as seen from the HST annually from 1994 to 2000.

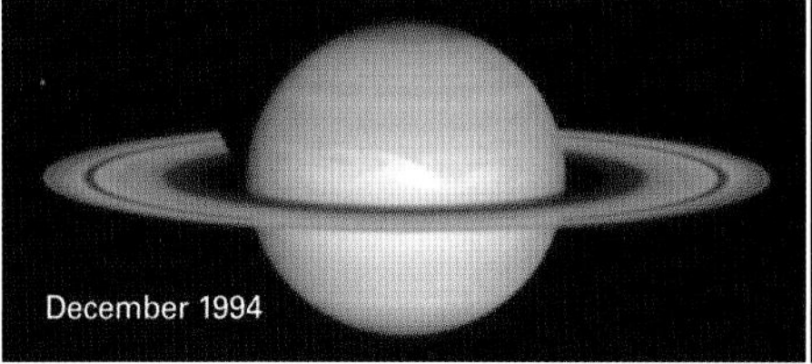

Of the two main rings, B is the brighter. The Cassini division is very conspicuous when the system is favourably tilted to the Earth, and even before the Pioneer and Voyager missions several minor divisions had been reported, though only one (the A Ring's Encke division) had been confirmed, and it was believed that the other divisions were mere 'ripples' in an otherwise fairly regular and homogeneous flat ring.

In 1850, W. C. and G. P. Bond observed a dark band across Saturn inside the B Ring and C. W. Tuttle surmised that it might be caused by a dusky ring. In 1907 the French observer G. Fournier announced the discovery of a dark ring inside the C Ring; though at that time confirmation was lacking, and it became known as the D Ring. Voyager confirmed its existence and found the E, F and G rings.

The Cassini Division was thought to caused mainly by the gravitational perturbations of the 400-kilometre (250-mile) satellite Mimas, which had been discovered by William Herschel as long ago as 1789. A particle moving in the division would have an orbital period exactly half that of Mimas, and cumulative perturbations would drive it away from the 'forbidden zone'. No doubt there is some substance in this, though the Voyager revelations showed that there must be other effects involved as well. The rings turned out to be completely different from anything which had been expected before the space missions.

DISTANCES AND PERIODS OF RINGS AND INNER SATELLITES

	Distance from centre of Saturn, km	Period, h
Cloud-tops	60,330	10.66
Inner edge of D 'Ring'	67,000	4.91
Inner edge of C Ring	73,200	5.61
Inner edge of B Ring	92,200	7.93
Outer edge of B Ring	117,500	11.41
Middle of Cassini Division	119,000	11.75
Inner edge of A Ring	121,000	11.92
Encke Division	133,500	13.82
Pan	133,600	14
Outer edge of A Ring	135,200	14.14
Atlas	137,670	14.61
Prometheus	139,350	14.71
F Ring	140,600	14.94
Pandora	141,700	15.07
Epimetheus	151,420	16.65
Janus	151,420	16.68
Inner edge of G Ring	165,800	18
Outer edge of G Ring	173,800	21
Inner edge of E Ring	180,000	22
Mimas	185,540	22.60
Enceladus	238,040	32.88
Tethys	294,760	32.89
Dione	377,420	2.74
Outer edge of E Ring	480,000	4
Rhea	527,040	4.52

▼ **Saturn,** imaged by Damian Peach on 13 January 2005, with a 24-cm (9¼-in) reflecting telescope. The minor divisions in the rings are clearly shown. Note also the belts on the disk, and the disk's polar 'hood'.

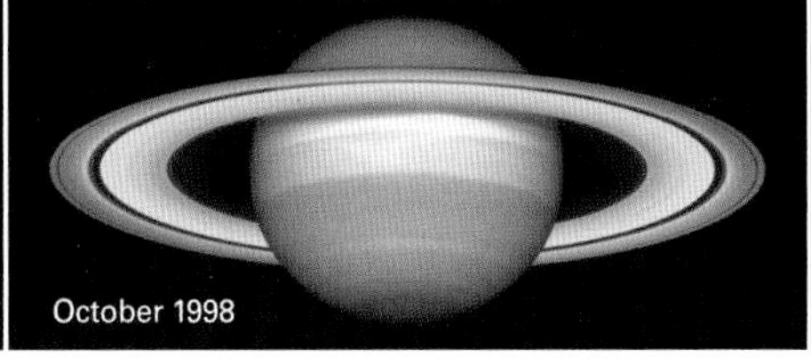
October 1998

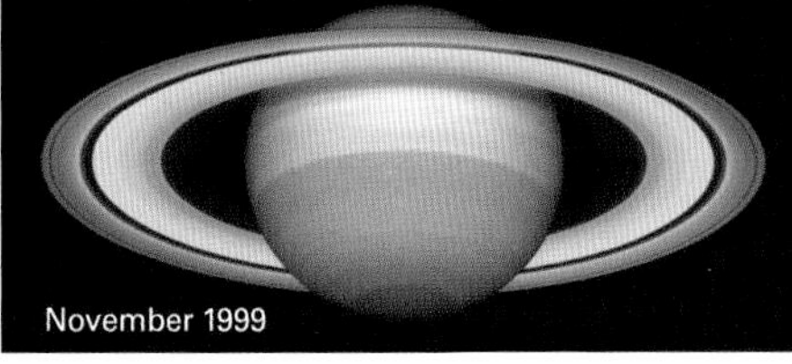
November 1999

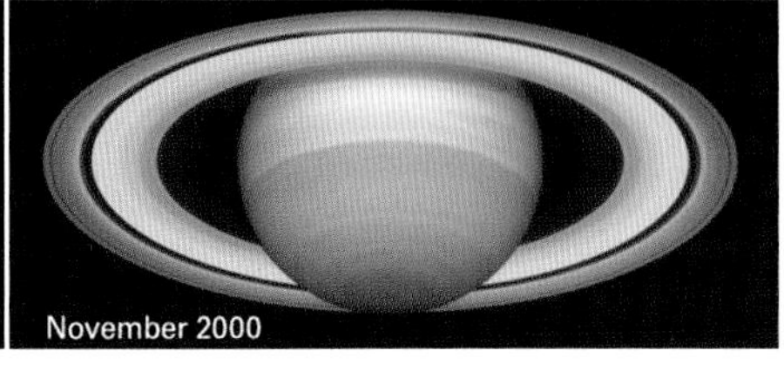
November 2000

Missions to Saturn

When missions to Saturn were first planned, it was – rather naturally – thought that stray ring particles might prove to be dangerous or even lethal. The initial foray, by Pioneer 11 in 1979, was something of an afterthought. The probe had been scheduled to survey Jupiter, and did so with success; at the end of its Jupiter programme, the NASA planners realized that an encounter with Saturn was practicable, and the opportunity was eagerly accepted. Pioneer passed within 21,000 kilometres (13,000 miles) of the cloud tops, and emerged unscathed. It sent back useful data as well as good images of the rings, before starting on its never-ending journey out of the Solar System. It was last heard from at the end of 1995, by which time it was about 6,300 million kilometres (4,000 million miles) from the Sun.

The two Voyagers, launched from Cape Canaveral in 1977, were highly successful. Both surveyed Jupiter, and then Saturn, but their subsequent careers were different. Titan, the main member of Saturn's satellite system, was of special interest, because it was known to have a substantial atmosphere, and nothing was known about the surface conditions; methane seas were thought to be quite possible, and certainly Titan was unlike any other world in the Solar System. It had been agreed that if for any reason Voyager 1 failed to survey Titan, its twin would do so. If Voyager 1 succeeded, Voyager 2 would not bother about Titan, but after completing its Saturn programme would be sent on a much longer journey, out to Uranus and Neptune. There was great relief when Voyager 1 did obtain close-range images of Titan, even though they showed nothing definite apart from the top of a layer of what was often referred to as 'orange smog'. The other satellites were not neglected; good pictures were obtained of the brighter members, and several new, very small satellites were discovered.

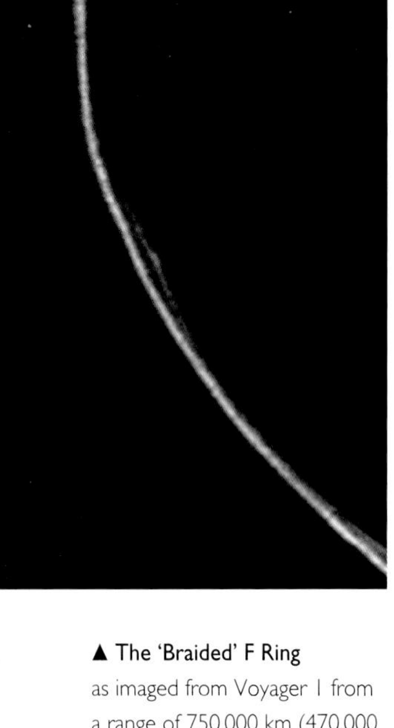

▲ The 'Braided' F Ring as imaged from Voyager 1 from a range of 750,000 km (470,000 mi). The complex structure of the ring was unexpected, and seems to be due to the gravitational effects of the small satellites Prometheus and Pandora.

The most surprising result from the Voyagers was that the rings were much more complex than anyone had suspected. They were made up of thousands of ringlets and minor divisions, and faint, narrow rings were even seen in the Cassini and Encke divisions. Strange radial 'spokes' were recorded crossing the B Ring; it is true that they had been glimpsed earlier by some Earth-based observers, notably Antoniadi, but the Voyagers were the first to show them really well. Logically they ought not to be there; following Kepler's Laws, the orbital speeds of the particles must decrease with increasing distance from the planet, and the difference in period between the inner and outer edges of the B Ring is over three hours – yet the spokes persisted for hours after emerging from the shadow of the globe. When I first saw the images of them, when I was at NASA headquarters during the pass of Voyager 1, I thought instinctively that they must be caused by particles elevated away from the ring-plane by gravitational or magnetic forces, and this does seem to be the correct explanation. The spokes are confined entirely to B Ring.

The Voyagers provided greatly improved images of the rings, all of which are very thin and are made up of water ice particles. The innermost, known generally as the D Ring, is not really a ring at all; there are no really sharp borders, and the icy particles may spread down almost to the cloud tops. In the C Ring the particles are around 2 metres (6 feet) across, and in the B Ring it is believed that the sizes may be as much as 1 metre (3 feet), with temperatures of –180°C in sunlight down to –200°C in shadow. In the A Ring the particles range between 'ice dust' and blocks the size of an office desk. The Encke Division contains the tiny satellite, Pan, discovered by M. Showalter in 1990 from a careful study of Voyager 2 photographs; it is a mere 35 kilometres (22 miles) across, but it keeps the division swept clear. A slightly larger satellite, Atlas, moves close to the outer edge of the A Ring, and is responsible for its well-defined border.

Outside the main system comes the F Ring, not visible in telescopes of the kind used by amateur observers. It is faint and complex, and is stabilized by two more small satellites, Prometheus and Pandora, which act as 'shepherds' and keep the ring particles in place. Prometheus, slightly closer to Saturn than the ring, moves faster than the ring particles, and will speed up a particle if it moves inwards, so returning it to the main ring zone. Pandora, on the far side, moves more slowly, and will drag any errant particle back.

The outer G and E Rings, discovered by the Voyagers, are very tenuous and very extensive; in fact the E Ring extends out almost to the orbit of Rhea, the fifth of Saturn's main satellites. The E Ring is brightest close to the satellite Enceladus, which is now known to be active and appears to be spraying ice particles into the ring.

The Voyagers did not neglect studies of Saturn itself. The disk is much blander than that of Jupiter. The cloud structure is of the same type, but the lower temperature means that ammonia crystals form at higher levels. The belts are obvious, and there are occasional bright white spots. Major outbreaks were seen in 1876, 1903 and 1933 (discovered by W.T. Hay, a skilled amateur astronomer better generally remembered today as Will Hay, the stage and screen comedian), 1960 and 1990. The 1990 spot was well imaged by the Hubble Space Telescope, and was obviously the result of an uprush of material from below. The time intervals between these white spots have been 27, 30, 27 and 30 years respectively. This is close to Saturn's orbital period of 29.5 years, which may or may not be significant; at any rate, observers will be looking out for a new white spot around 2020.

The Voyagers confirmed that Saturn, again like Jupiter, has a surface which is in constant turmoil (if not in the way that Proctor had supposed in 1882), and that wind speeds are very high. There is a wide equatorial jetstream, 80,000 kilometres (50,000 miles) broad and stretching from 35°S to

◀ **Rings of Saturn,** as seen from the Cassini spacecraft on 21 June 2004, nine days before it entered orbit about the planet. This natural-colour image was obtained from a distance of 6.4 million km (4 million mi). The B Ring is the bright, sandy coloured area stretching from lower left to upper right. Then comes the Cassini Division beyond the B Ring, followed by the outermost A Ring. The narrow Encke Division is visible at the lower right. Top left is the C Ring.

35°N, where the winds blow at up to 1800 kilometres an hour (1120 miles per hour), faster than any on Jupiter. A major surprise was that the wind zones do not follow the light and dark bands, but instead are symmetrical with the equator. One prominent 'ribbon' at 47°N was taken to be a wave pattern in a decidedly unstable jetstream.

The Voyagers had done more than their planners could have hoped, but there were many questions left unanswered. Even before the Voyagers had left the inner Solar System, bound for interstellar space, NASA was making ready a new Saturn probe: Cassini.

The Cassini mission was much more elaborate than any of its predecessors, and would be the first orbiter – which is why it took so long to plan. Not only was it to carry the Huygens probe to land on Titan but it was also to carry out an extended tour of the satellite system. It was launched on schedule, and proved to be a triumphant success in every respect.

◀ **Saturn's rings.** This image was obtained from Voyager 2, from a range of 4 million km (2.5 million miles). 'Spokes' can be seen in the B Ring. They are known to move due to the rotation of the ring system.

▶ **View of Saturn from Voyager 2,** showing the belts, the bright zones and the rings. The Cassini Division is obvious, as is the Encke Division in the A Ring. The other ring divisions are much less obvious.

Saturn: The Cassini mission

▲ **Saturn from Cassini.** This detailed mosaic of 126 images was obtained over two hours on 6 October 2004. The spacecraft was about 6.3 million km (3.9 million mi) from the planet, and the smallest features visible are 38 km (24 mi) across.

▼ **Panoramic view of Saturn's rings;** a mosaic of six Cassini images, covering a distance of 39,000 miles (62,000 km) along the ring plane (12 December, 2004).

Cassini, launched from Cape Canaveral on 15 October 1997, was a new kind of mission. Pioneer 11 and the Voyagers had been fly-by probes, paying one fleeting visit to Saturn and then departing for ever. Cassini was much more ambitious; it was an orbiter, capable of carrying out a research programme which would take years. Moreover it carried a smaller probe, Huygens, which would make a controlled landing on Titan and transmit direct from the surface. The names were well chosen. Giovanni Cassini was the best of the 17th-century observers of Saturn – the main division in the planet's ring system is named after him – while Christiaan Huygens discovered Titan, in 1655, as well as demonstrating that Saturn is surrounded by 'a thin, flat ring'.

Probably the most complicated probe built up to that time, Cassini measured 6.8 metres (22.3 feet) in height and over 4 metres (13 feet) in width; the various instruments required over 14 km (nearly 9 miles) of cabling. The probe was crammed with instruments, including a CCD imager, a cosmic dust analyser, an infra-red spectrometer and radar tracking equipment.

Cassini did not go straight to Saturn; it first made two passes of Venus (April 1998 and June 1999) as well as one of the Earth (August 1999). It also found time to take a picture of an asteroid, 1685 Masursky (23 January 2000). There was an important fly-by of Jupiter (December 2000). But this time Saturn was the main target, and by June 2004 Cassini was homing in on the ringed planet. On 11 June there was a close fly-by of the satellite Phoebe, which had not been well imaged by either of the Voyagers. Then, on 1 July, the spacecraft flew through the gap between the F and G Rings, and entered a closed orbit round Saturn. The ring particles gave no serious trouble, and Cassini was ready to begin its main programme.

There were several major investigations to be carried out. Wind speeds were known to be very high – the Voyagers had shown that – but there was still much to be learned about the

atmospheric circulation in general. In November 2006 a vast hurricane was tracked at the south pole; it was 8,000 kilometres (5,000 miles) across, 72 kilometres (45 miles) above the cloud deck, and had winds gusting up to 560 kilometres an hour (348 miles per hour). Unlike terrestrial cyclones it had two 'eyes', and seemed to hover over the south pole. Outwardly Saturn's surface may look placid, but appearances are deceptive.

The satellites provided several major surprises. The 'wispy' features on Dione proved to be towering ice cliffs; Iapetus had a long, very lofty ridge running along its equator; Enceladus was active, with geysers spouting water from its south pole – the last thing that would have been expected on such a tiny world.

It is fair to say that the most exciting part of the whole mission was the landing on Titan. Even if it all went well, nobody knew whether Huygens would land on solid rock, marshland or in a chemical ocean. Christmas 2004 was a tense time for Mission Control. On 25 December Huygens was separated from Cassini; then on 14 January it touched down, and data were transmitted almost at once. Contact was maintained for well over an hour.

The actual descent through Titan's thick atmosphere took over two hours; the spacecraft was – predictably – buffeted by winds, but the parachute arrangement worked perfectly, and Huygens made a gentle landing. The team back on Earth had an anxious wait; it took 84 minutes for a radio signal to travel from Titan to here! The surface of the satellite was firm, and Huygens settled comfortably down. It was always known that contact could not be maintained for long. Everything had to be relayed via the orbiting Cassini; before long the orbiter would go below Huygen's horizon, and there was no chance that the lander's power would last until Cassini came back. There could be no second chance. Mercifully there were no major mishaps; Huygens operated extremely well right up to the end of its brief but glorious career.

▲ **In Saturn's shadow.** On 15 September 2006 Cassini flew into Saturn's shadow, and obtained this view of the planet eclipsing the Sun. Two new faint rings were detected, one coincident with the shared orbit of the moons Janus and Epimetheus and the other coincident with Pallene's orbit. The narrowly confined G ring is visible outside the main rings. Encircling this is the more extended E ring. The icy plumes of Enceladus betray its position in the E ring's left-hand edge.

Saturn: Mimas and Enceladus

The system of Saturn is quite unlike that of Jupiter. here is one very large satellite (Titan), seven 'icy' satellites between 350 and 1550 kilometres (220 and 960 miles) in diameter, and many smaller bodies, some of which are probably asteroidal while one (Phoebe) is almost certainly from the Kuiper Belt. The innermost satellites are involved in the ring system, while Epimetheus and Janus have a curious relationship; they never come within 10,000 kilometres (6200 miles) of each other, and every four years actually exchange orbits!

Mimas, innermost of the satellites within easy range of most amateur-owned telescopes, was discovered by William Herschel in 1789, and the main crater on the surface is named after him. It is 130 km (80 miles) across; the impact which formed it must have almost broken Mimas entirely; it is well formed, with a prominent central peak. The low density of the globe indicates that it is composed mainly of water ice, with only a small amount of rock. The globe itself is not a perfect sphere; it measures 415 × 394 × 381 kilometres (258 × 249 × 237 miles). As with the other icy satellites apart from Hyperion and Phoebe, the rotation period is synchronous.

Enceladus, also discovered by Herschel in 1787, is very different. There are no giant craters, and there are smooth areas which look as though they have been recently resurfaced. Surprisingly, there is a thin but detectable atmosphere, even though the escape velocity is just 0.24 kilometres a second (0.15 miles an hour). In 2005 the Cassini spacecraft flew close to Enceladus, and discovered a water-rich plume venting from the south polar region, so Enceladus is active, probably with liquid water not far below the surface. Also in the south polar area are four roughly parallel linear depressions which have been nicknamed 'tiger stripes', each around 130 kilometres (80 miles) long, 2 km (1.2 miles) wide and 500 metres (1640 feet) deep . They are about 35 km (22 miles) apart. It is from these that the plumes of water are vented. Before the Cassini pass, there had been no suspicion of activity on Enceladus. It is highly likely that the E Ring is produced by the material sent out from the south polar geysers.

▲ **Mimas (Cassini).** The surface is dominated by one very large crater, now named Herschel, which has a diameter of 130 km (80 mi) – one-third that of Mimas itself – with walls which rise to 5 km (3 mi) above the floor, the lowest part of which is 10 km (6 mi) deep and includes a massive central mountain.

▲ **Fountains of Enceladus.** Cassini image showing Enceladus backlit by the Sun, and the fountain-like spray of icy material which is coming from the south polar region.

NAMED SATELLITES OF SATURN

SATELLITES OVER 200 KM IN DIAMETER

Name	Distance from Saturn, km	Period, days	Greatest diameter, km	Inclination, degrees	Eccentricity
Mimas	185,540	0.942	415	1.57	
Enceladus	238,040	1.370	513	0.01	
Tethys	294,760	1.887	1080	0.17	
Dione	377,420	2.737	1123	0.002	
Rhea	527,040	4.518	1535	0.33	
Titan	1,221,930	15.94	5151	1.63	
Hyperion	1,481,010	21.28	360	0.57	
Iapetus	3,560,820	79.32	1498	7.57	
Phoebe	12,869,700	545.1*	230	173.05	

SMALL INNER SATELLITES

Name	Greatest diameter, km	Position
Pan	35	In Encke Division
Daphnis	8	In Keeler Gap
Atlas	46	Outer Ring A shepherd
Prometheus	119	Inner Ring F shepherd
Pandora	103	Outer Ring F shepherd
Epimetheus	135	Co-orbital with Janus
Janus	193	Co-orbital with Epimetheus
Methone	3	Between orbits of Mimas and Enceladus
Pallene	4	Between orbits of Mimas and Enceladus

Tethys co-orbitals: Telesto (29 km) and Calypso (30).

Dione co-orbitals: Helene (36) and Polydeuces, (3.5).

OUTER SMALL SATELLITES

Inuit Group. Kiviuq, Ijiraq, Paaliaq and Siarnaq.

Norse Group. Phoebe is a member of this group. Others are Skathi, Narvi, Mundilfari, Suttungr, Theymr and Ymir. They have retrograde motion.

Gallic Group. Albiorix, Erriapo and Tarvos.

These make up the named satellites (by January 2007). Other unnamed small satellites bring the grand total up to 60.

(* = retrograde)

Saturn: Tethys and Dione

Both these icy moons were discovered in 1684 by G. D. Cassini; it is therefore appropriate that our best views of them have been obtained from the spacecraft named after him. The two are similar in size, but they are by no means identical twins.

Tethys, not far from spherical, at 1080 × 1062 × 1055 kilometres (671 × 660 × 656 miles) has a globe whose mean density is 0.97 of that of water; in other words it is made up of almost pure ice, with only a small admixture of rock. Craters are plentiful, with faults and valleys. The western hemisphere is dominated by a huge crater, 400 kilometres (250 miles) in diameter, which has been named Odysseus. Originally it must no doubt have had high walls and lofty central peak; its diameter is two-fifths of the satellite's – but over the ages the floor has relaxed to the spherical shape of Tethy's surface, and the central peak has collapsed through the slumping of the icy crust over geological time. The other main feature of Tethys is Ithaca Chasma, a huge valley 2000 kilometres (1240 miles) long – about three-quarters of the way round the globe – and up to 5 kilometres (3 miles) deep. It may have been formed at the time of the Odysseus impact, when a shock wave travelled through Tethys and fractured the brittle, icy crust on the opposite side. Tethys has two small, icy Trojans, Telesto and Calypso. Predictably, both have cratered surfaces.

Dione is only slightly larger than Tethys, but is much denser and more massive, so that the globe must contain a great deal of rock as well as ice. It is almost perfectly spherical, and is characterised by the fact that its two sides are not alike. The leading hemisphere is bright and heavily cratered; some craters, such as Dido and Aeneas, are well-formed and over 100 km across. On the darker grailing hemisphere the Voyager images showed 'wispy features', which Cassini has found to be towering ice cliffs. Another notable feature of the surface is Latium Chasma, a rimmed, flat-floored fracture over 300 kilometres (185 miles) long and up to 12 kilometres (7.5 miles) wide, but less than 1 kilometre (0.6 miles) deep. Ridges and troughs are common; Dione has a very varied landscape. Like Tethys, Dione has two Trojan co-orbitals, in this case Helene and Polydeuces.

▼ **Tethys,** from Cassini, September 2005; the globe is made up chiefly of water ice.

▼ **Dione,** from Cassini, showing the 'wispy features' now known to be ice cliffs.

▲ **Dione** (left) and Tethys (right) facing each other across Saturn's rings; Cassini, 2005.

Saturn: Rhea and Hyperion

▶ **Rhea** (Cassini). Rhea is heavily cratered, but with few really large formations. As with Dione, the trailing hemisphere is darkish, with wispy features which are not unlike those on Dione, though less prominent. Rhea seems to be made up of a mixture of rock and ice in almost equal amounts.

Rhea, largest of the icy satellites, was discovered by Cassini in 1672, not long after his discovery of Iapetus. Rhea is almost perfectly spherical; it is less dense than Dione, so about two-thirds of its material must be ice. In November 2005 the Cassini probe flew past it at 500 kilometres (310 miles), and sent back very detailed views of the surface.

Rhea is one of the most thickly cratered bodies in the Solar System, but there are few really large formations; as with Dione, the trailing hemisphere is the darker of the two, with wispy markings which are presumably ice cliffs. There seem to be two distinct types of terrain; the first contains craters over 40 kilometres (25 miles) across, while the second, in parts of the equatorial and polar regions, has many craters of smaller size. On the leading hemisphere there is one ray centre. No Rhean Trojans have been found.

Beyond the orbit of Rhea comes Titan, giant of Saturn's family, and then the small, irregular Hyperion, discovered in 1848 by W. C. Bond, in America. Cassini images have shown that the surface is saturated with deep, sharp-edged craters that make it look like a sponge! It is of low density, made up mainly of ice with a smaller amount of rock, and may indeed be little more than a rubble-pile. It is irregular in shape, measuring 360 × 280 × 225 kilometres (224 × 174 × 140 miles); the albedo (reflecting power) is less than 0.03, so that an icy surface must be coated with at least a thin layer of dark material. There are several craters with diameters up to 130 kilometres (80 miles), and one long ridge.

Hyperion has an orbital period of 21.3 days, but the rotation is not synchronous; in fact Hyperion's axial spin rate is variable, and has been described as 'chaotic'. The rotational axis wobbles so much that its orientation in space is unpredictable. The orbit is inclined by 0.43° to the Saturnian equator. There have been suggestions that Hyperion is a detached part of a larger body, but if so, where is the other part?

Hyperion is faint (below magnitude 14) and not easy to locate. With an adequate telescope, the best time to locate it is when it and Titan are at elongation simultaneously.

In 1904 W. H. Pickering reported a small satellite moving between the orbits of Titan and Hyperion. It was even given a name – Themis – but has not been seen since, and must be dismissed as a cosmic ghost.

▶ **Hyperion,** from Cassini, 20 September 2005. Hyperion is a cratered, irregularly shaped body. Its rotation is 'chaotic', and is not synchronous, as with the larger satellites.

Saturn: Iapetus and Phoebe

Iapetus, the outermost of Saturn's major satellites, is unique inasmuch as one of its hemispheres is bright and icy, while the other is almost as black as coal. This led to what I once called the 'Zebra Problem'. Was Iapetus a bright body with a dark stain, or a dark body overlaid in part by bright material? This question was answered when the mean density of the globe was found to be less than 1.3 times that of water, indicating that the main constituent is ice, so the first answer is the right one.

The rotation is synchronous. When Iapetus is west of Saturn it is an easy telescopic object, as the bright side is turned toward us; when east of Saturn, Iapetus shows us its dark side, and fades by several magnitudes. The demarcation line between the two regions is not sharp, and some craters on the icy areas are dark floored, so that in all probability the dark material has welled up from below, though there is also some evidence that suggests that it may have an external origin. We do not know just what it is, or why Iapetus alone presents such an appearance. Fittingly, the dark area is known as Cassini Regio.

The most extraordinary feature on Iapetus is the equatorial ridge, discovered by the Cassini orbiter in December 2004. It is 1300 kilometres (800 miles) long, extending through the centre of Cassini Regio and following the Iapetan equator almost perfectly; in places it is 20 kilometres (12.5 miles) wide and 13 kilometres (8 miles) high – far loftier than our Mount Everest. It forms a complex system, including isolated peaks and roughly parallel segments. The orbit of Iapetus is inclined by more than 15° to the plane of the planet's equator.

Phoebe, discovered by W. H. Pickering in 1899, has a retrograde motion, and is now classed as a member of the Norse group of satellites, though it is by far the largest of them. Cassini could make only one close pass – on 11 June 2004 – but obtained excellent images, showing a varied, cratered landscape with some relatively large, well-marked formations. It is certainly a captured body rather than a bona-fide satellite, and seems more likely to have come from the Kuiper Belt than the main asteroid zone. It never becomes brighter than magnitude 16.5, so that a fairly powerful telescope must be used to locate it.

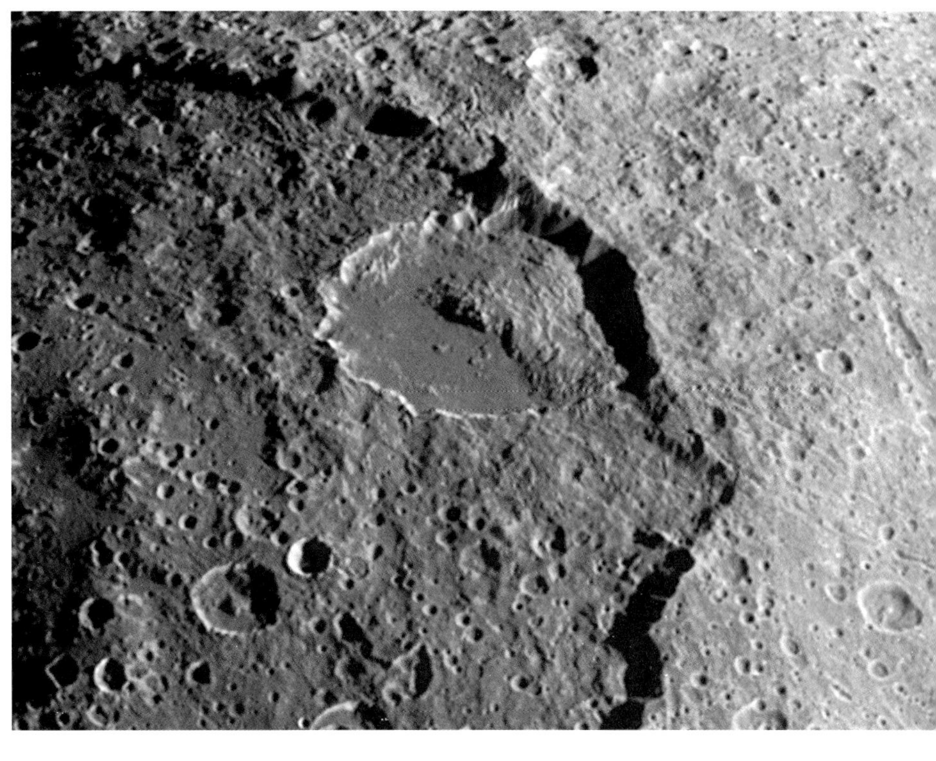

▲ **Giant Landslide on Iapetus** (Cassini, 31 December 2004 from 123,400 km (77,677 mi)). The landslide material seems to have collapsed from a scarp 15 km (9 miles) high forming the rim of an ancient 600-km (375-mi) crater basin. Unconsolidated rubble from the landslide extends half-way across a 129-km (75-mi) flat-floored impact crater lying just in the basin scarp.

▼ **Phoebe,** taken during the Cassini fly-by on 11 June 2004. Phoebe seems to have a dark layer over an icy globe. When meteoroids impacted Phoebe, they excavated icy material.

▼ **Iapetus** (Cassini). There are dark areas, high peaks and a curious equatorial ridge, the origins of which are unclear.

Saturn: Titan

▲ **The surface of Titan** as recorded by Huygens after its descent through the moon's dense atmosphere. The rocks in the foreground are only about 15 cm (6 in) across.

Apart from the Galileans, Titan was the first planetary satellite to be discovered – by Christiaan Huygens, in 1655. It is actually larger than the planet Mercury, though not so massive. In 1944 G. P. Kuiper showed spectroscopically that it has a dense atmosphere, which proved to consist mainly of nitrogen, with a good deal of methane. The Voyagers could do no more than image the upper clouds, and so Titan was a prime target for the Cassini-Huygens mission. After a 7-year journey Cassini, carrying Huygens, reached Saturn in July 2004.

On 14 January 2005 the Huygens spacecraft made a controlled landing on Titan; it had been released from the Cassini orbiter on 25 December 2004, and made an automatic landing, involving parachutes. The descent took 2¼ hours through Titan's dense atmosphere; during this period measurements were made. The touchdown speed was less than 20 kilometres per hour (13 miles an hour). After arrival, Huygens sent back data for 72 minutes – far longer than had been expected.

After Huygens fell silent, Cassini itself continued to make fly-bys of the satellite, and in July 2006 identified what seemed certainly to be lakes of liquid methane in the north polar region. More than 75 radar-dark patches were found, ranging from 3 to 70 kilometres across (1.8 miles to 43 miles). Some are completely filled with liquid, others partly so. Some have steep margins and very distinct edges, consistent with seepage or groundwater drainage lakes. Others have curved, channel-like extensions, similar in aspect to terrestrial flooded river valleys.

Huygens came down on a thin crust, and settled down some centimetres into it on spongy, hydrocarbon material with about the consistency of wet sand. The heat of Huygens' batteries caused some of the frozen surface to 'boil', causing puffs of methane. Titan is unique. There are icy pebble-sized objects near the landing site; the site was dry when Huygens arrived, but had been wet very recently. Methane rain lands on the icy uplands and washes the dark organic material off the hills; this is transported down to the plains in drainage channels, and eventually disappears. The hills themselves are of 'dirty water ice'. Huygens imaged snaking, branching river tracks; there are water-ice volcanoes, and liquid certainly flowed soon before the probe landed. There must be liquid a few centimetres below the surface.

The 'mud', as it was described, seems to be a mixture of sand, methane and complex organic molecules that form in the upper atmosphere. According to Martin Tomasko (University of Arizona), 'This smog falls out of the atmosphere and settles on everything. Then methane rain comes, washes it off the ice ridges and into rivers, then out into the broad plain where the rain settles into the ground and dries up. We are seeing evidence of Earth-like processes, but with very exotic materials.' Methane is constantly being destroyed and turned into complex chemical smog, so there must be some source inside Titan to replenish the atmosphere. Some gases, such as argon, are absent.

All the ingredients for life exist on Titan, but it seems virtually certain that the low temperature has prevented life from appearing there. There have been suggestions that in the future, when the Sun swells out and becomes much more luminous, Titan could become habitable. Unfortunately, there is a fatal objection to this idea. Titan has a low escape velocity – only 2.4 kilometres (1.5 miles) per second, almost exactly the same as that of the Moon – and it can hold on to its dense atmosphere only because it is so cold; low temperatures slow down the movements of atoms and molecules. Raise the temperature, and Titan's atmosphere will escape.

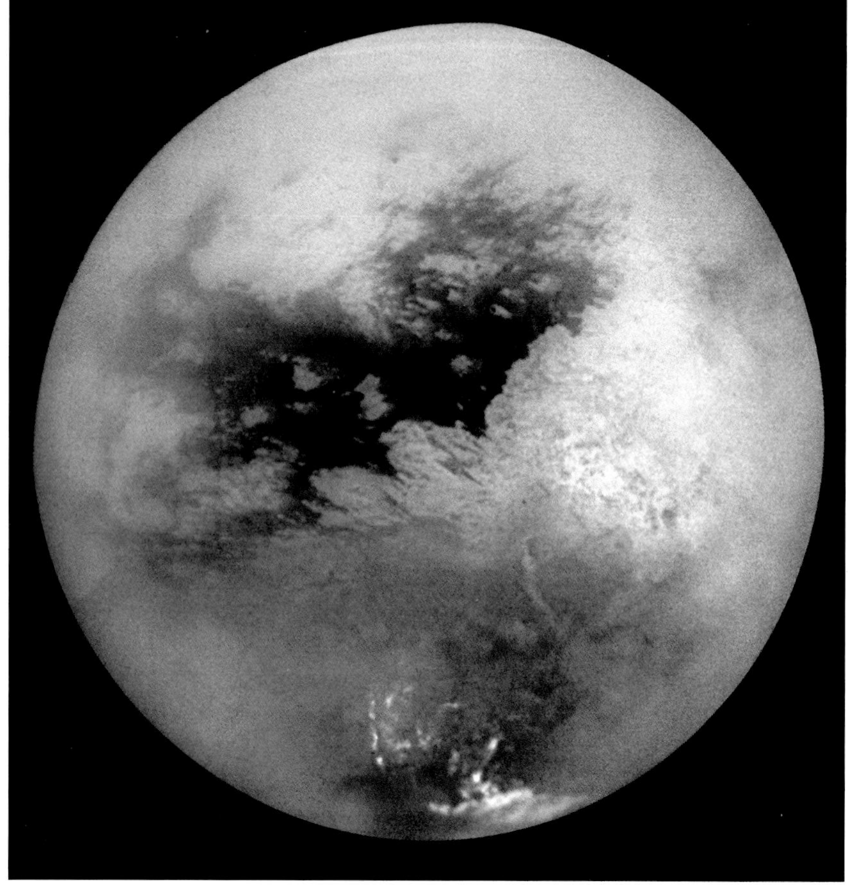

◀ **Titan from Cassini,** in a mosaic of nine images acquired on 26 October 2004, during the spacecraft's first close fly-by. Bright clouds can be seen, as well as some surface features.

▼ **Lakes in Titan;** Cassini orbiter, 23 September 2006; 73°N, 46°W. The lakes are each about 20 to 25km (12.5 to 16 mi) across, and they are joined by a relatively narrow channel. The lake on the right of the image shows lighter patches, indicating that it may be drying out with the approach of summer; as this is a radar picture, the lakes appear dark.

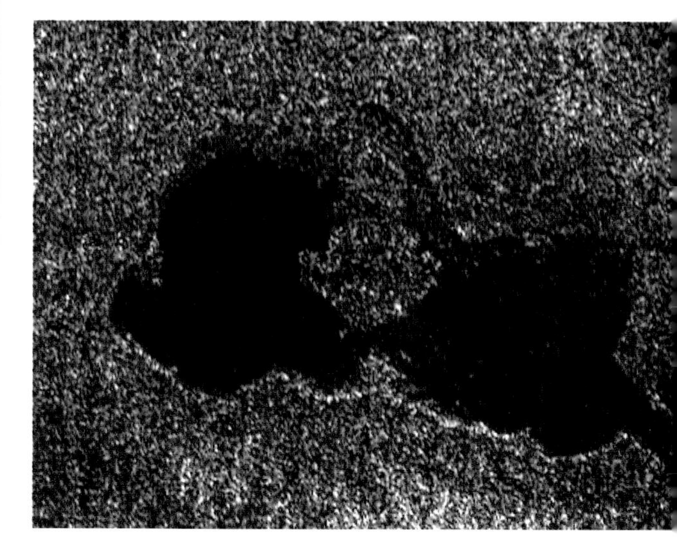

▼ **Mercator projection of Huygens' view** from different altitudes, taken on 14 January 2005. The instrument used was the descent imager/spectral radiometer carried on their probe. Huygens also carried a gas chromatograph and mass spectrometer.

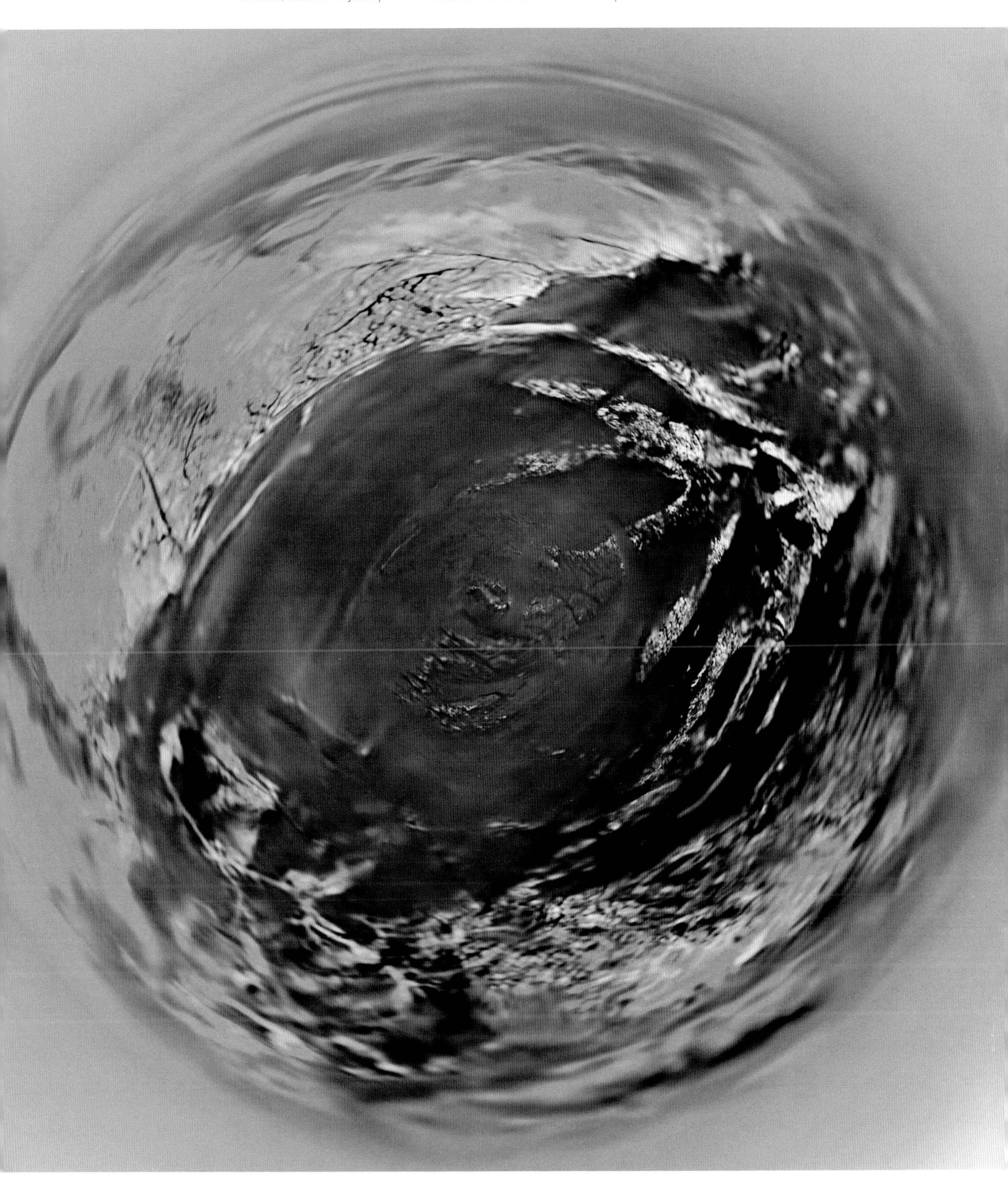

Uranus

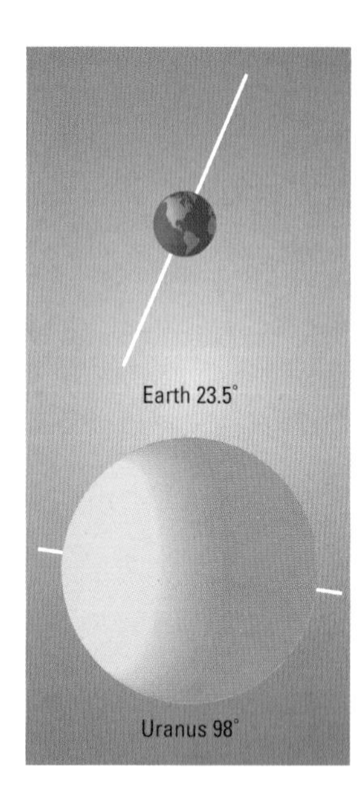

▲ **Axial inclination of Uranus.** The planets' inclinations have a wide range; 2° (Mercury), 178° (Venus), 24° (Mars), 3° (Jupiter), 26.5° (Saturn), 98° (Uranus), and 29° (Neptune). Uranus thus differs from all the other planets.

Uranus, third of the giant planets, was discovered by William Herschel in 1781. Herschel was not looking for a planet; he was engaged in a systematic 'review of the heavens' with a home-made reflecting telescope when he came across an object which was not a star. It showed a small disk, and it moved slowly from night to night. Herschel believed it to be a comet, but calculations soon showed it to be a planet, moving far beyond the orbit of Saturn. After some discussion it was named Uranus, after the mythological father of Saturn.

Uranus is just visible with the naked eye, and it had been seen on several occasions before. John Flamsteed, England's first Astronomer Royal, even included it in his star catalogue, and gave it a number: 34 Tauri. However, a small telescope will show its tiny, greenish disk. The equatorial diameter is 51,118 kilometres (31,770 miles), rather less than half that of Saturn; the mass is over 14 times that of the Earth, and the visible surface is made up of gas, mainly hydrogen together with a considerable amount of helium.

Irregularities in the movements of Uranus led to the tracking-down of the outermost giant, Neptune, in 1846. In size and mass the two are near twins, so that in many ways they may be considered together even though there are definite differences between them. As a pair they are very different from the Jupiter/Saturn pair, quite apart from being much smaller and less massive. If Jupiter and Saturn are, justifiably, called gas giants it may be more appropriate to class Uranus and Neptune as ice giants.

Unlike the other giants, Uranus has little or no source of internal heat, so the cloud-top temperatures of Uranus and Neptune are much the same even though Neptune is so much farther from the Sun (Uranus –214°C, Neptune –220°C). The outer atmosphere, which we see, is indeed made up of gas; approximately 83 per cent hydrogen, 15 per cent helium and 2 per cent methane, which does not leave much room for anything else. The main globe is composed of ices (mainly water, ammonia and methane) with some rock; the great pressure means that these materials behave as liquids. There may or may not be a definite rocky core, and even if a core exists it may not have a sharp upper boundary.

Uranus is a slow mover; it takes 84 years to orbit the Sun. The rotation period is 17 hours 14 minutes, though, as with the other giants, the planet does not spin in the way that a rigid body would do. The most extraordinary feature is the tilt of the axis, which amounts to 98°; this is more than a right angle, so that the rotation is technically retrograde. The Uranian calendar is very curious. Sometimes one of the poles is turned towards the Sun, and has a 'day' lasting for 21 Earth years, with a corresponding period of darkness at the opposite pole; sometimes the equator is presented. In total, the poles receive more heat from the Sun than does the equator. The reason for this exceptional tilt is not known. It is often thought that at an early stage in its evolution Uranus was hit by a massive body and literally knocked sideways. Alternatively, the tilt may have increased gradually because of gravitational perturbations by other planets in the early life of the Solar System. It may be significant that the rings and the orbits of the principal satellites lie virtually in the plane of Uranus' equator.

(*En passant*, which is the 'north' pole and which is the 'south'? The International Astronomical Union has decreed that all poles above the ecliptic, i.e. the plane of the Earth's orbit, are north poles, while all poles below the ecliptic are south poles. In this case it was the south pole which was in sunlight during the Voyager 2 pass of 1986. However, the Voyager team reversed this, and referred to the sunlit pole as the north pole. Take your pick!)

No Earth-based telescope will show definite markings on the disk of Uranus. Before the Voyager mission, five satellites were known – Miranda, Ariel, Umbriel, Titania and Oberon; Voyager added ten more, all close to the planet, and since then, 12 more have been added.

On 10 March 1977, Uranus passed in front of a star, and hid or occulted it. This gave astronomers an excellent chance to measure Uranus' apparent diameter – which is not easy by sheer visual observation, because the edge of the disk is not sharp, and the slightest error in measurement will make a tremendous difference to the final value. Therefore the phenomenon was carefully observed, with surprising results. Both before and after the actual occultation the star 'winked' several times, and this could be due only to a system of rings surrounding the planet. Subsequently D. A. Allen, at Siding Spring in Australia, managed to photograph the rings in infra-red light. However, our knowledge of Uranus and its system remained decidedly meagre, and a detailed survey had to await the fly-by of Voyager 2 in January 1986.

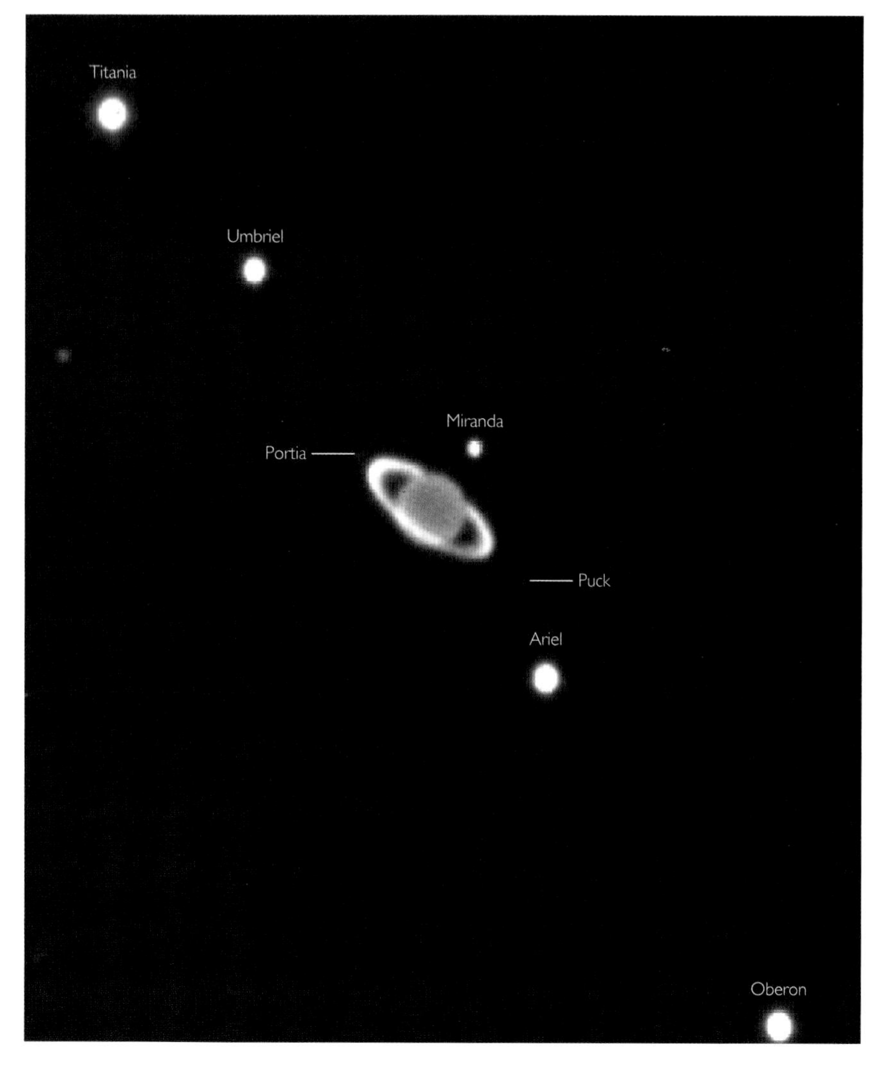

◀ **The best Earth-based views** have been obtained with the Hubble Space Telescope and the Very Large Telescope in Chile. This photograph of Uranus and some of its satellites was taken in near-infra-red light by the Antu Telescope, one unit of the VLT. The rings look unusually bright in comparison to the planet because the methane in the latter's atmosphere absorbs sunlight at this wavelength while the material in the rings reflects it strongly.

◀ **Uranus,** 24 August 1991 – a drawing I made with a magnification of 1000 on the Palomar 152-cm (60-inch) reflector. Even with this giant telescope, no surface details could be made out; all that could be seen of the planet was a greenish disk.

▼ **Uranus from Hubble** in a view obtained in 1994. The picture is a composite of three images taken six minutes apart. Because the moons move fairly rapidly, their positions change noticeably over a few minutes, so each appears in the image as three dots.

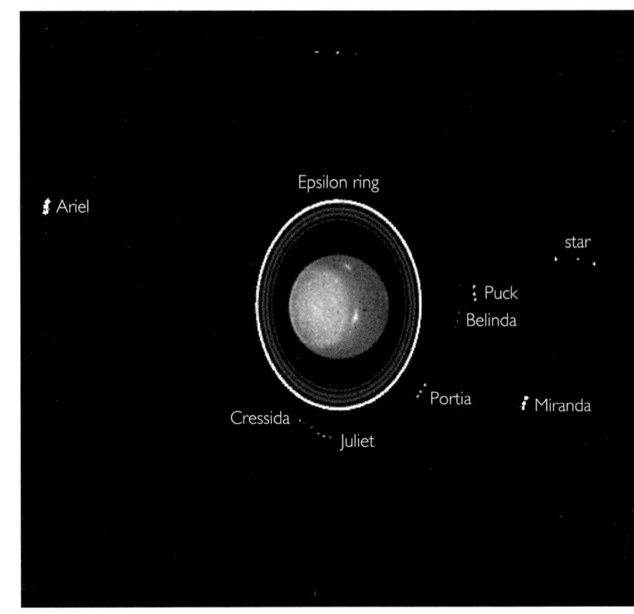

PLANETARY DATA – URANUS	
Sidereal period	30,684.9 days/84.01 years
Rotation period	17.2 hours
Mean orbital velocity	6.80 km/s (4.22 mi/s)
Orbital inclination	0.773°
Orbital eccentricity	0.047
Apparent diameter	max. 3.7", min. 3.1"
Reciprocal mass, Sun = 1	22,800
Density, water = 1	1.27
Mass, Earth = 1	14.6
Volume, Earth = 1	67
Escape velocity	22.5 km/s (14.0 mi/s)
Surface gravity, Earth = 1	1.17
Mean surface temperature	−214°C (at cloud tops)
Oblateness	0.24
Albedo	0.35
Maximum magnitude	+5.6
Diameter (equatorial)	51,118 km (31,770 mi)

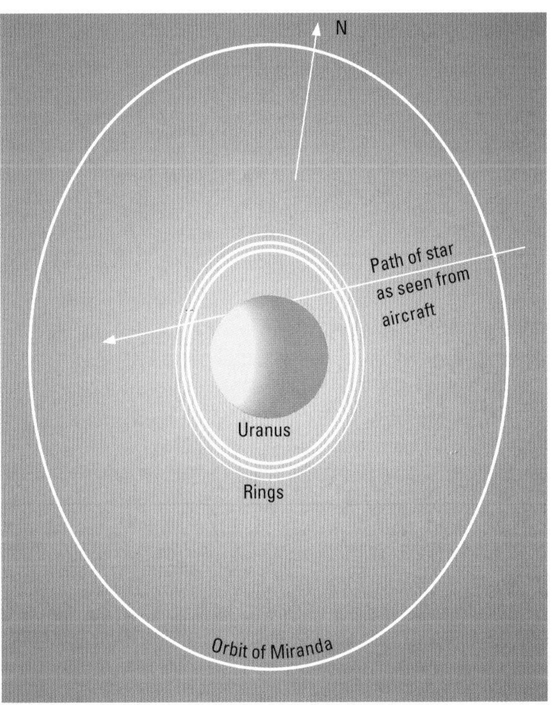

◀ **Discovery of the rings of Uranus.** On 10 March 1977 Uranus occulted the star SAO 158687 and observations from South Africa and from the Kuiper Airborne Observatory, flying over the Indian Ocean, established the existence of a ring system – confirmed by subsequent observations.

▶ **The changing presentation of Uranus.** Sometimes a pole appears in the middle of the disk as seen from Earth; sometimes the equator is presented. Adopting the International Astronomical Union definition, it was the south pole which was in sunlight during the Voyager 2 pass in 1986.

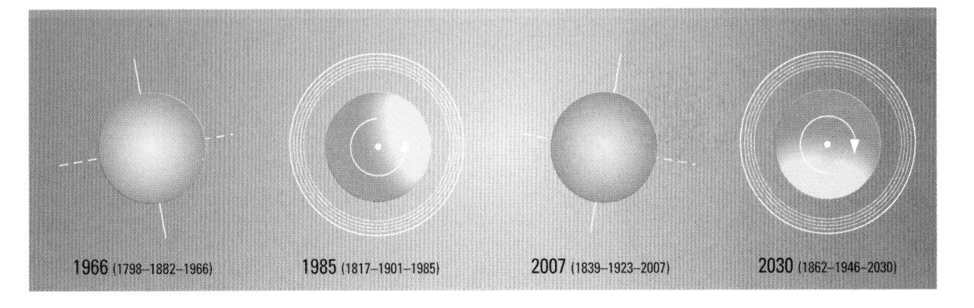

Satellites of Uranus

▶ **Umbriel.** The surface of Umbriel is much darker and more subdued than that of Ariel. The largest crater, Skynd, is 110 km (68 mi) in diameter, with a bright central peak. Wunda, diameter 140 km (87 mi), lies near Umbriel's equator; its nature is uncertain, but it is the most reflective feature on the satellite. (Remember that owing to the pole-on view, the equator lies round the limb in this picture.)

Uranus, like all the giant planets, has an extensive satellite family. The two outer members, Titania and Oberon, were discovered by William Herschel in 1787. Herschel also announced the discovery of four more satellites, but three of these are non-existent and must have been faint stars; the fourth may have been Umbriel, but there is considerable doubt. Umbriel and Ariel were found in 1851 by the English amateur William Lassell. The four first-discovered satellites are between 1100 and 1600 kilometres (680 to 1000 miles) in diameter, so that they are comparable with the medium-sized icy satellites in Saturn's system, but their greater distance makes them rather elusive telescopic objects.

During the 1890s W. H. Pickering (discoverer of Phoebe, the outermost satellite of Saturn) searched for further members of the system, but without success. The fifth moon, Miranda, was discovered by G. P. Kuiper in 1948; it is much fainter and closer in than the original four. Voyager 2 found another ten satellites, all moving inside Miranda's orbit. The

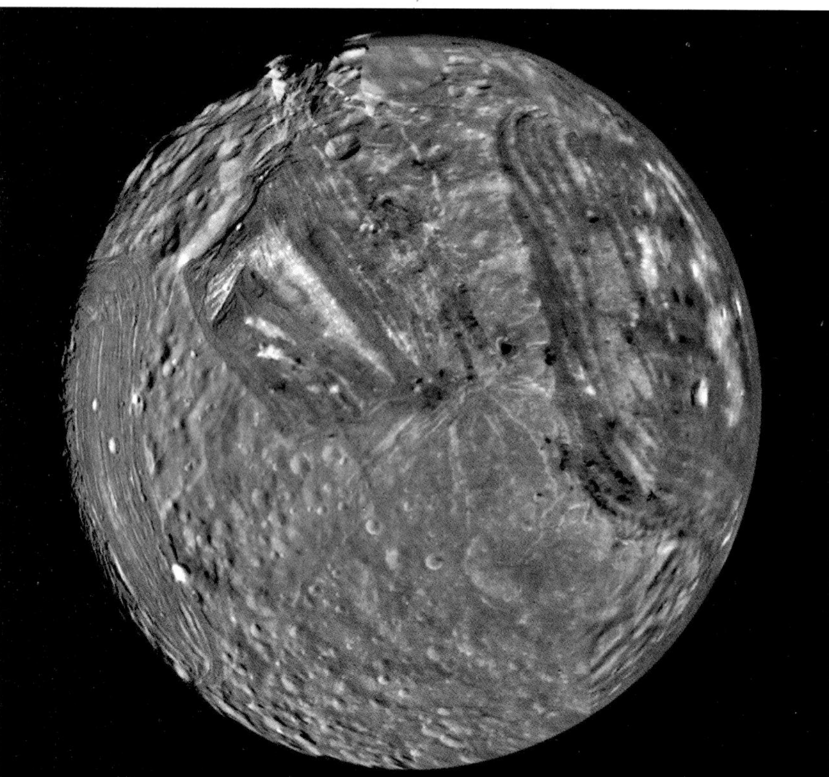

◀ **Miranda.** The innermost of Uranus' large satellites is seen at close range in this image from Voyager 2, taken from a distance of 35,000 km (22,000 mi). Scarps, ice cliffs and craters are visible.

▼ **Oberon.** Voyager 2 took this picture from around 660,000 km (410,105 mi), with a resolution of 11 km (6.8 mi). Note the high peak, which is about 6 km (3.7 mi) high, projecting from the lower left limb.

SATELLITES OF URANUS

Name	Distance from Uranus, km	Orbital period, days	Orbital inclination, °	Orbital eccentricity	Diameter, km	density, water = 1	Escape velocity, km/s	Magnitude
Cordelia	49,471	0.330	0.14	0.0005	26	?	very low	24.2
Ophelia	53,796	0.372	0.09	0.0101	32	?	very low	23.9
Bianca	59,173	0.433	0.16	0.0009	42	?	very low	23.1
Cressida	51,777	0.463	0.04	0.0001	62	?	very low	22.3
Desdemona	62,676	0.475	0.16	0.0002	54	?	very low	22.5
Juliet	64,352	0.493	0.04	0.0002	84	?	very low	21.7
Portia	66,085	0.513	0.09	0.0002	106	?	very low	21.1
Rosalind	69,941	0.558	0.08	0.0006	54	?	very low	22.5
Belinda	75.258	0.622	0.03	0.0001	66	?	very low	22.1
Perdita	76.420	0.638	0.0	0.0012	30	?	very low	23
Puck	86,000	0.762	0.31	0.0001	154	?	very low	20.4
Miranda	129,400	1.414	4.22	0.0027	481 × 466 × 466	1.3	0.5	16.3
Ariel	191,000	2.520	0.31	0.0034	1158	1.6	1.2	14.2
Umbriel	256,300	4.144	0.36	0.0050	1169	1.4	1.2	14.8
Titania	435,000	8.706	0.014	0.0022	1578	1.6	1.6	13.7
Oberon	583,500	13.463	0.10	0.0008	1523	1.5	1.5	13.9
Caliban	7,170,000	579	140	0.082	60	?	very low	22.3
Stephano	7,940,000	676	141.5	0.146	30	?	very low	24
Sycorax	12,214,000	1203	153	0.51	120	?	very low	20.7
Prospero	16,110,000	1993	146.3	0.327	40	?	very low	23
Setebos	18,200,000	2202	148.8	0.494	40	?	very low	23

only newcomer to exceed 100 kilometres (60 miles) in diameter is Puck, which was imaged from a range of 500,000 kilometres (310,000 miles) and found to be dark and roughly spherical; three craters were seen, and given the rather bizarre names of Bogle, Lob and Butz.

Incidentally, it may be asked why the names of the Uranian moons come from literature, not mythology. The names Titania and Oberon were suggested by Sir John Herschel, and the later satellites were also given names coming either from Shakespeare or from Pope's poem *The Rape of the Lock*. This is certainly a departure from the norm, and in my opinion an undesirable one, but the names are now well established, and have all been ratified by the International Astronomical Union.

The nine innermost satellites are presumably icy, but nothing is known about their physical make-up. Cordelia and Ophelia act as shepherds to the Epsilon ring. A careful search was made for similar shepherds inside the main part of the ring system, but without success; if any such shepherds exist, they must be very small indeed.

The four largest members of the family are unalike. In general they are denser than the icy satellites of Saturn, and so must contain more rock and less ice; the proportion of rocky material is probably between 50 and 55 per cent. All have icy surfaces, but there are marked differences between the four of them.

Umbriel is the darkest of the four, with a rather subdued surface and one bright feature, called Wunda, which lies almost on the equator, so in the pole-on view obtained by Voyager 2 it appears near the edge of the disk; it may be a crater, but its nature is uncertain. Umbriel is fainter than the other major satellites, and in pre-Voyager days was assumed to be the smallest, though in fact it is marginally larger than Ariel. Oberon is heavily cratered, and some of the craters such as Hamlet, Othello and Falstaff have dark floors, perhaps because of a mixture of ice and carbonaceous material erupted from the interior; on the limb, near the crater Macbeth, there is a high mountain. Titania is distinguished by high ice cliffs, and there are broad, branching and interconnected valleys, so that there seems to have been more past internal activity than on Oberon. Ariel also has very wide, branching valleys which look as though they have been cut by liquid – though, needless to say, all the satellites are far too lightweight to retain any trace of atmosphere.

Miranda has an amazingly varied surface. There are regions of totally different types – some cratered, some relatively smooth; there are ice-cliffs up to 20 kilometres (12.5 miles) high and large trapezoidal areas or 'coronae' which were initially nicknamed 'race tracks'. The three main coronae (Arden, Elsinore and Inverness) cover much of the hemisphere which was imaged by Voyager 2. It has been suggested that during its evolution Miranda has been broken up by collision, perhaps several times, and that the fragments have subsequently reformed. This may or may not be true, but certainly it would go some way to explaining the jumble of surface features now seen.

Five small outer satellites, all with retrograde motion, have been found: Caliban, Stephano, Sycorax, Prospero and Setebos. All are reddish, and presumably asteroidal in origin.

Other small outer satellites have been found; those so far (2007) named are Francisco, Trinculo, Margaret, Ferdinand, Cupid and Mab. All are reddish, below 150 km (93 miles) in diameter, and presumably asteroidal. Margaret, Cupid and Mab have direct motion; the others are retrograde.

▲ **Titania, the largest moon of Uranus.** The image shows a 1,600 km (1,000 mi) long trench, as well as other features.

▼ **Ariel.** This Voyager 2 image was taken from 169,000 km (105,000 mi); the resolution is 3.2 km (2 mi). The surface is cratered, with fault scarps and graben, suggesting considerable past tectonic activity, and there is evidence of erosion.

Neptune

Some years after Uranus was discovered, it became clear that it was not moving as had been expected. The logical cause was perturbation by an unknown planet at a greater distance from the Sun. Two mathematicians, U. J. J. Le Verrier in France and J. C. Adams in England, independently calculated the position of the new planet, and in 1846 Johann Galle and Heinrich D'Arrest, at the Berlin Observatory, identified it. After some discussion it was named Neptune, after the mythological sea-god.

Neptune is too faint to be seen with the naked eye; its magnitude is 7.7, so it is within binocular range. Telescopes show it as a small, bluish disk. In size it is almost identical with Uranus, but it is appreciably more massive. The orbital period is almost 165 years. Like all of the giants, it is a quick spinner, with an axial rotation period of 16 hours 7 minutes. Neptune does not share Uranus' unusual inclination; the axis is tilted by only 28° 48' to the perpendicular.

Though Uranus and Neptune are near twins, they are not identical. Unlike Uranus, Neptune has a strong source of internal heat, so that the temperature at the cloud tops is almost the same as that of Uranus, even though Neptune is over 1600 million kilometres (1000 million miles) farther from the Sun. In composition Neptune is presumably dominated by planetary 'ices', such as water ice; there may be a silicate core surrounded by the mantle, but it is quite likely that the core is not sharply differentiated from the ice components.

▲ **Neptune's blue-green atmosphere** seen by Voyager at a distance of 16 million km (10 million mi). The Great Dark Spot at the centre is about 13,000 × 6600 km (8000 × 4100 mi). The 'Cirrus-type' clouds are higher.

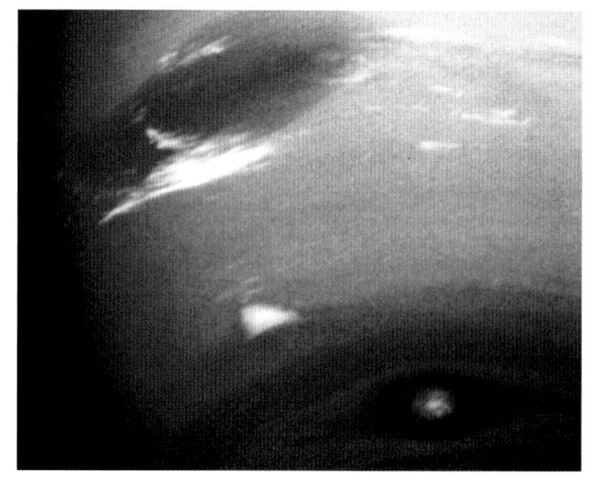

► **Three prominent features** reconstructed from two Voyager images. Towards the north (top) is the Great Dark Spot. To the south is the 'Scooter' which rotates around the globe faster than other features. Still farther south is the feature called 'Dark Spot 2'. Each moves at a different velocity.

PLANETARY DATA – NEPTUNE	
Sidereal period	60,190.3 days/164.8 years
Rotation period	16h 7m
Mean orbital velocity	5.43 km/s (3.37 mi/s)
Orbital inclination	1° 45' 19.8"
Orbital eccentricity	0.009
Apparent diameter	max. 2.2", min. 2.0"
Reciprocal mass, Sun = 1	19,300
Density, water = 1	1.77
Mass, Earth = 1	17.2
Volume, Earth = 1	57
Escape velocity	23.9 km/s (14.8 mi/s)
Surface gravity, Earth = 1	1.2
Mean surface temperature	−220°C (at cloud tops)
Oblateness	0.02
Albedo	0.35
Maximum magnitude	+7.7
Diameter	50,538 km (31,410 mi)

▼ **Neptune's clouds** two hours before Voyager 2's closest approach. In this view, reminiscent of Earth from an airliner, fluffy white clouds are seen high above Neptune, with their shadows on the cloud deck below. Cloud shadows have not been seen on any other planet.

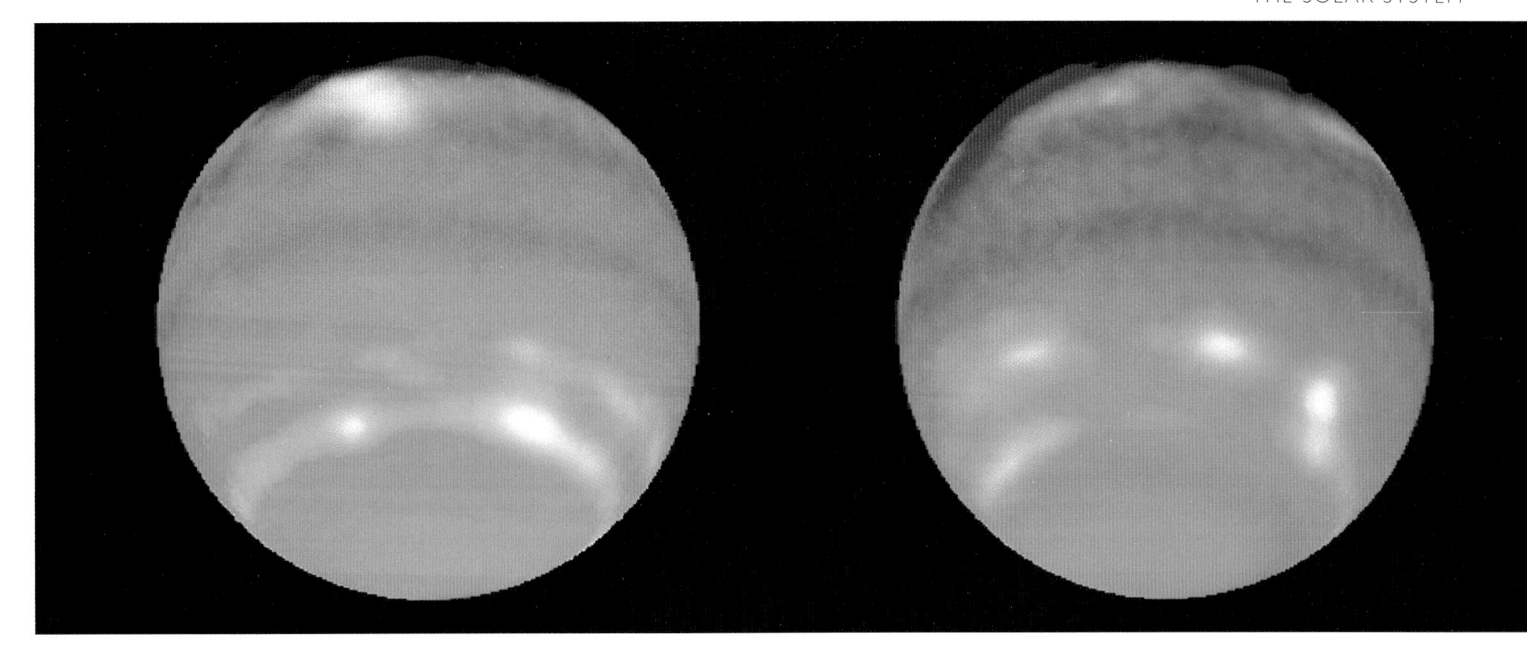

Much of our detailed knowledge of Neptune has been provided by just one spacecraft, Voyager 2, which flew past the planet on 25 August 1989 – at 4425 million kilometres (2750 million miles) from the Earth. Voyager passed over the darkened north pole at a relative velocity of just over 17 kilometres (10 miles) per second; at that time the southern hemisphere was having its long 'summer'.

Well before Voyager closed in, the images showed that Neptune is a far more dynamic world than Uranus. The most conspicuous feature on the blue surface was a huge oval, the Great Dark Spot, at latitude 8° 28'S; it had a rotation period of over 18 hours, so that it drifted westwards relative to the nearby clouds at 30 metres (100 feet) per second. It rotated in an anticlockwise direction, and showed more or less predictable changes in shape and orientation. Above it lay wispy clouds made of methane crystals ('methane cirrus') and between these and the main cloud deck there was a 50-kilometre (31-mile) clear zone. Farther south (42°S) was a smaller, very variable feature with a bright centre, which had a shorter rotation period and was nicknamed the 'Scooter'; still farther south (55°S) was a second dark spot. Details on Neptune are now within the range of the Hubble Space Telescope – and, surprisingly, the images taken in August 1996 (and since) showed no trace of the Great Dark Spot. Smaller features were seen, and there seems no escape from the conclusion that the Great Dark Spot has disappeared. Other spots have been recorded, and it is becoming clear that Neptune is much more variable than had been expected.

Neptune is a windy place. At the equator the winds blow westwards (retrograde) at up to 450 metres (1500 feet) per second; farther south the winds slacken, and beyond latitude 50° they become eastwards, reaching 300 metres (1000 feet) per second but decreasing once more near the south pole. Temperature measurements show that there are cold mid-latitude regions with a warmer equator and pole.

The upper atmosphere is made up chiefly of hydrogen (85 per cent), with a considerable amount of helium and a little methane. There are various cloud layers, above which lies the general methane haze.

Neptune is a source of radio waves, which was expected, but the magnetic field proved to be very surprising. The magnetic axis makes an angle of 47° with the axis of rotation, so that in this respect Neptune resembles Uranus more than Jupiter or Saturn; here also the magnetic axis does not pass through the centre of the globe, but is displaced by 10,000 kilometres (6200 miles). The magnetic field itself is weaker than those of the other giants. Aurorae were confirmed, though they are of course brightest near the magnetic poles.

Voyager confirmed that Neptune has a ring system, though it is much less evident than those of the other giants. Altogether there seem to be five separate rings, plus the so-called 'plateau', a diffuse band of material made up of very small particles. There may also be 'dust' extending down almost to the cloud-tops.

The rings have been named in honour of astronomers who were involved in Neptune's discovery. The Adams ring is the most pronounced and is 'clumpy' with three brighter arcs which may be due to the gravitational pull of Galatea, one of the newly discovered small satellites. The ring which is at 62,000 kilometres (38,750 miles) is close to Galatea's orbit. The rings are dark and ghostly, and the fainter sections were only just above the threshold of visibility from Voyager.

THE RINGS OF NEPTUNE

Name	Distance from centre of Neptune, km	Width, km
Galle	41,900	50
Le Verrier	53,200	50
'Plateau'	53,200–59,100	4000
Arago	57,100	outer edge of the 'Plateau'
Unnamed	62,000	30
Adams	62,900	50

▲ **Views of Neptune's weather,** on opposite hemispheres. Taken on 11 August 1998 with Hubble's Wide Field Planetary Camera 2, these composite images show Neptune's blustery weather. The predominant blue colour is a result of the absorption of red and infra-red light by the planet's methane atmosphere. Clouds elevated above most of the methane appear white, while the very highest clouds tend to be yellow-red. Neptune's powerful equatorial jet – where winds blow at nearly 1500 km/h (900 mph) – is centred on the dark blue belt just south of Neptune's equator. Farther south, the green belt indicates a region where the atmosphere absorbs blue light.

◀ Neptune's two main rings, about 53,000 km (33,000 mi) and 63,000 km (39,000 mi) from the centre of the planet, were backlit by the Sun as Voyager 2 swept past. The rings appear bright as microscopic ring particles scatter sunlight towards the camera. Particle-size distribution in Neptune's rings is quite different from that in Uranus' rings. The outer Adams ring is clumpy with three bright arcs of material, as shown here at the top of the image.

Satellites of Neptune

▲ **Leaving the solar system.** As it begins its neverending journey into space, Voyager 2 sends back a final picture of Neptune and Triton, both showing up as crescents.

Two satellites of Neptune were known before the Voyager fly-by: Triton and Nereid. Each was exceptional in its own way. Triton, discovered by Lassell a few weeks after Neptune itself had been found, is large by satellite standards but has retrograde motion; that is to say, it moves round Neptune in the opposite direction to that in which Neptune rotates. This makes it unique among major satellites, since all other attendants with retrograde motion are asteroidal in origin. Nereid is only 240 kilometres (140 miles) across; and though it moves in the direct sense its eccentric orbit is more like that of a comet than a satellite; its distance from Neptune varies by over 8 million kilometres (5 million miles), and its revolution period is only one week short of an Earth year, so obviously the axial rotation is not synchronous.

Voyager discovered six new inner satellites, one of which (Proteus) is actually larger than Nereid, but is virtually unobservable from Earth because of its closeness to Neptune. Proteus and one of the other new discoveries, Larissa, were imaged by Voyager, and both turned out to be dark and cratered; Proteus shows a major depression, Pharos, in its southern hemisphere, with a rugged floor. No doubt the other inner satellites are of the same type. Nereid was in the wrong part of its orbit during the Voyager pass, and only one very poor image was obtained, but Triton more than made up for this omission.

Three new satellites – Halimede, Sao and Laomedeia – were discovered in 2002 by a team of astronomers led by M. Holman and J. J. Kavelaars, using the 4-metre (157-inch) Blanco Telescope at Cerro Tololo and the 6-metre (236-inch) CFH Telescope in Hawaii. They were missed by Voyager 2 because they are very faint (mag 25) and very distant from Neptune, with highly inclined orbits. Halimede is in a retrograde orbit; the other two are prograde. Their diameters are probably between 30 and 40 km (19 and 25 miles). Two small retrograde satellites have since been found, Psamanthe and Neso. Psamanthe orbits at a distance of 46,695,000 kilometres (91,152,000 miles); in 9115.9 days, Neso at 48,387,000 kilometres (30,066,000 miles) in 9374 days.

Estimates of the diameter of Triton had been discordant, and at one time it was thought to be larger than Mercury, with an atmosphere dense enough to support clouds similar to those of Titan. Voyager proved otherwise. Triton is smaller than the Moon, and is well over twice as dense as water, so that its globe is made up of more rock than ice. The surface temperature is around −236°C, so that Triton is the chilliest world so far encountered by a spacecraft.

The escape velocity is 1.4 kilometres (0.9 miles) per second, and this is enough for Triton to retain a very tenuous atmosphere, made up chiefly of nitrogen with an appreciable amount of methane. Pressure is 14 microbars, which is 1⁄70,000 of the surface pressure on Earth. There is considerable haze, seen by Voyager above the limb and which extends to at least 6 kilometres (3.7 miles) above the surface; it is probably composed of tiny particles of methane or nitrogen ice. Winds in the atmosphere average about 5 metres (16 feet) per second in a westwards direction.

The surface of Triton is very varied, but there is a general coating of water ice, overlaid by nitrogen and methane ices. There are very few craters, but many flows which are probably due to ammonia-water fluids; surface relief is very muted, and certainly there are no mountains. The most striking feature is the southern polar cap, which is pink and makes Triton look quite different from any other planet or satellite. The pink colour is thought to be the result of the methane ice

SELECTED FEATURES ON TRITON

Name	Lat. °	Long. E °
Abatos Planum	35–8S	35–81
Akupara Maculae	24–31S	61–65
Bin Sulci	28–48S	351–14
Boyenne Sulci	18–4S	351–14
Bubembe Regio	25–43S	285–25
Medamothi Planitia	16S–17N	50–90
Monad Regio	30S–45N	330–90
Namazu Macula	24–28S	12–16
Ob Sulci	19–14N	325–37
Ruach Planitia	24–31S	20–28
Ryugu Planitia	3–7S	25–29
Tuonela Planitia	36–42N	7–19
Uhlanga Regio	60–0S	285–0
Viviane Macula	30–32S	34–38
Zin Maculae	21–27S	65–72

SATELLITES OF NEPTUNE

Name	Dist. from Neptune, km	Orbital period, days	Orbital incl.,°	Orbital eccentricity	Diameter, km	Mag.
Naiad	48,000	0.296	4.5	0	54	26
Thalassa	50,000	0.312	0	0	80	24
Despina	52,500	0.333	0	0	180	23
Galatea	62,000	0.429	0	0	150	23
Larissa	73,500	0.544	0	0	192	21
Proteus	117,600	1.121	0	0	416	20
Triton	354,800	5.877	159.9	0.0002	2705	13.6
Nereid	1,345,500–9,688,500	360.15	27.2	0.749	240	18.7

Three more outer satellites were discovered in 2002 and two more in 2003. All are over 20 million km (12.4 million mi) from Neptune, and are less than 50 km (32 mi) in diameter.

reacting under sunlight to form pink or red compounds. The long Tritonian season means that the south pole has been in constant sunlight for over a century now, and along the borders of the cap there are signs of evaporation. North of the cap there is an 'edge' which looks darker and redder, perhaps because of the action of ultra-violet light upon methane, and running across this region is a slightly bluish layer, caused by the scattering of incoming light by tiny crystals of methane.

The surface imaged by Voyager 2 is divided into three main regions: Uhlanga Regio (polar), Monad Regio (eastern equatorial) and Bubembe Regio (western equatorial). It is in Uhlanga that we find the nitrogen geysers. The most plausible explanation is that there is a layer of liquid nitrogen 20 or 30 metres (65 to 100 feet) below the surface. If for any reason this liquid migrates towards the upper part of the crust, the pressure will be eased, and the nitrogen will explode in a shower of ice and gas, travelling up the nozzle of the geyser-like vent at a rate of up to 150 metres (500 feet) per second – fast enough to send the material up for many kilometres before it falls back. The outrush sweeps dark debris along with it, and this debris is wafted downwind, producing plumes of dark material such as Viviane Macula and Namazu Macula. Some of these plumes are over 70 kilometres (40 miles) long.

Monad Regio is part smooth and in part hummocky, with walled plains or 'lakes' such as Tuonela and Ruach; these have flat floors, and water must be the main material from which they were formed, because nitrogen ice and methane ice are not rigid enough to maintain surface relief over long periods. Bubembe Regio is characterized by the 'cantaloupe terrain', a name given because of the superficial resemblance to a melon skin. Fissures cross the surface, meeting in huge X or Y junctions, and there are subdued circular pits with diameters of around 30 kilometres (19 miles).

It may well be that there will be marked changes in Triton's surface over the coming decades, because the seasons there are very long indeed, and the pink snow may migrate across to the opposite pole – which was in darkness during the Voyager pass.

▼ **Triton photomosaic** made up of Voyager 2 images – the pinkish south polar cap lies at the bottom.

The Kuiper Belt

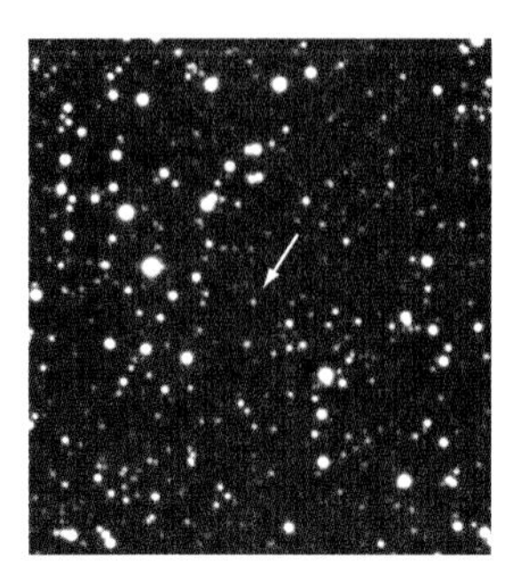

▲ **Quaoar.** This large TNO was discovered in June 2002 by Chad Trujillo and Mike Brown at Palomar, California. It is likely to be made of rock and ice.

▼ **The Kuiper Belt,** (upper) with the orbits of the planets out from the asteroid belt. Inclination of the orbit of Pluto (below) against the main plane of the Solar System; it amounts to 17°.

The outermost part of the Solar System is more populous than was thought a few decades ago. For example, there are several Neptune Trojans, which share the orbit of the giant planet; five are known, and no doubt others exist. More importantly, we have the swarm of asteroid-sized bodies moving beyond Neptune's orbit. These make up what is called the Kuiper Belt, after the Dutch astronomer Gerard Kuiper, who discussed the possibility in 1951. Pluto is the brightest of the Kuiper Belt objects (KBOs), though not the largest.

A few small bodies known as Twotinos move round the Sun well beyond the orbit of Neptune and have 1:2 resonance, i.e. a Twotino will orbit the Sun in the same time that Neptune takes to do so twice. But the most numerous members of the Kuiper Belt are the Cubewanos, sometimes referred to as classical KBOs, including 20000 Varuna, 50000 Quaoar and 58534 Logos, as well as 1992 QB1 itself. In general their orbits are of low eccentricity, and not highly inclined to the main plane of the Solar System. They move in the region between 40 and 50 astronomical units from the Sun – i.e. at distances between about 6000 million and 8000 million kilometres (380 million and 500 million miles), far enough out to avoid being markedly perturbed by Neptune. The swarm has a fairly sharp outer boundary; there are very few Cubewanos beyond a range of 50 astronomical units.

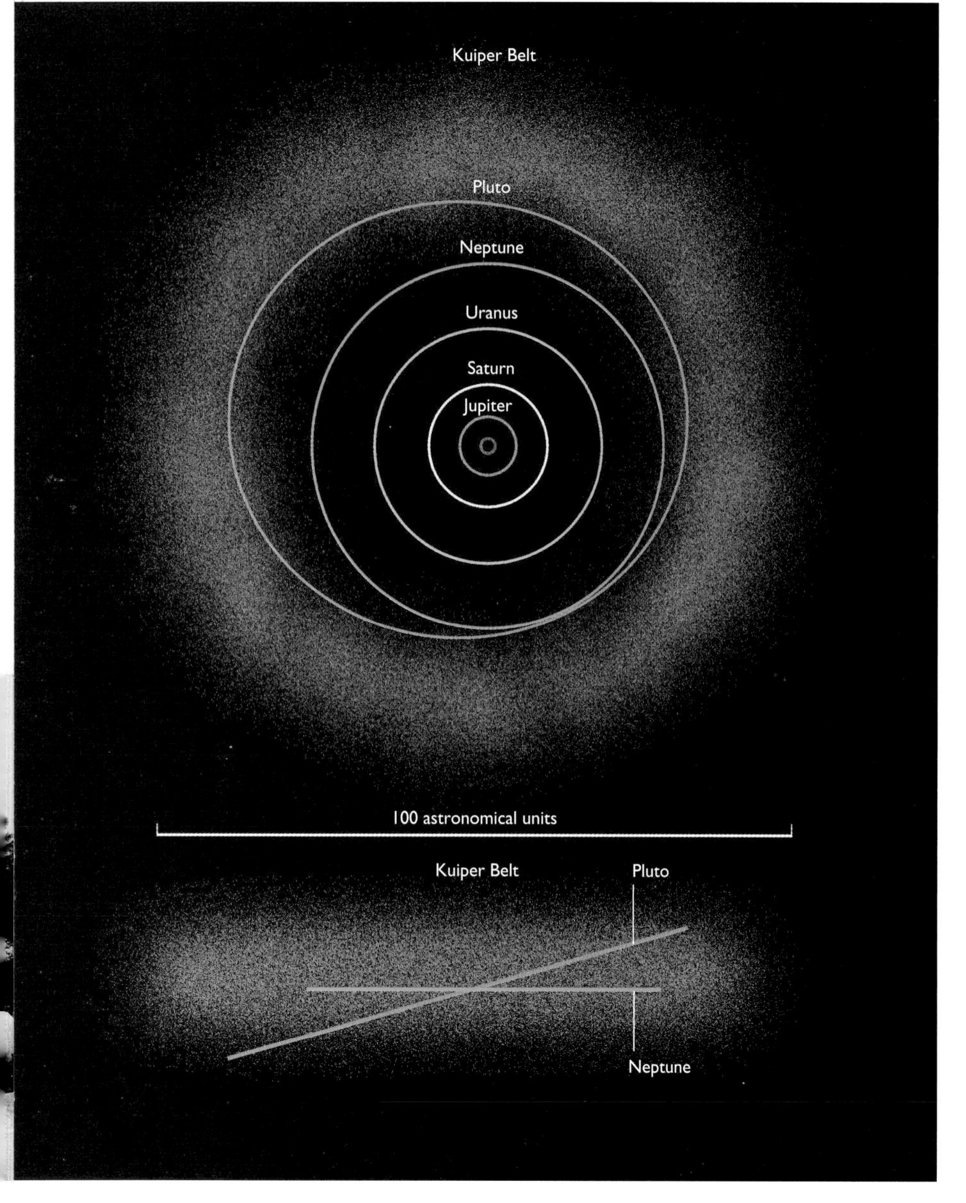

Plutinos do have definite resonance with Neptune (3:2), and their orbits are more eccentric and inclined than those of the Cubewanos; some of them, including Pluto itself, actually cross Neptune's orbit, though the resonance and inclination make them safe from collision. Notable Plutinos are 90482 Orcus, 28978 Ixion and 38083 Rhadamanthus. There was some sentimental opposition when Pluto was given the indignity of a KBO number, 134340, but at least it was officially classed as a dwarf planet, and it is the only KBO upon which definite surface markings have been recorded. The Hubble Space Telescope has shown bright areas and darker patches; there seems to be a dusky equatorial band and bright polar caps. Pluto's companion Charon, discovered in 1977, has a diameter of 1210 kilometres (750 miles) against 2320 kilometres (1440 miles) for Pluto, and the two seem to be better regarded as a binary KBO than as a primary and a satellite. Pluto has a rotation period of 6.3 days, and this is also the orbital period of Charon, so the two bodies are tidally locked. To an observer on Pluto, Charon would remain motionless in the sky.

Pluto has a very tenuous but surprisingly extensive atmosphere, and a surface coated with methane and nitrogen ices; it is likely that in the outer part of the orbit, the cold will become so intense that the atmosphere will freeze out until well after aphelion. The axial tilt is 122°, more similar to that of Uranus than that of the Earth or, for that matter, Neptune. There are two more satellites, Nix and Hydra, but both are less than 100 kilometres (60 miles) across.

It was once thought that Pluto and Charon were parts of a larger body which was broken up by collision, but this theory does not explain why the surfaces are so different. Also, Pluto is much the denser of the two; both seem to be made up of rock and ice, but Pluto has a greater percentage of rock (76 per cent as against 45 per cent for Charon), and Charon shows no sign of any atmosphere. It is dark in colour, while Pluto is reddish.

Farther out come the Scattered Disk objects, all of which have perihelia beyond 35 astronomical units from the Sun (roughly 5.3 thousand million kilometres or 3.3 thousand million miles) and are not therefore strongly influenced by Neptune. They have erratic, unstable paths, usually very eccentric and inclined; they may recede to great distances. 136199 Eris, the largest of them so far discovered, goes out to almost 100 astronomical units, so that it is within our range for only part of its 557-year period; we happened to catch it near its perihelion. It is quite definitely larger than Pluto, and has one substantial satellite, Dysnomia. (Most of these names are mythological; Eris and Dysnomia are deities of strife.) 903775 Sedna, named after an Inuit sea goddess, is even more extreme. It has a period of over 12,000 years, and at aphelion moves out to 975 astronomical units. At a distance of about a light year it is thought that there is a vast spherical cloud of icy bodies; this was suggested by the Dutch astronomer Jan Oort, and probably does exist, though we cannot see it. Some astronomers believe Sedna is a refugee from the Oort Cloud.

The Kuiper Belt itself may well be a 'left-over' portion of the solar nebula from which the planets were formed, but we cannot yet claim to have a really good knowledge of the early history of the Solar System. It does however seem certain that Lowell's predictions for his Planet X were based upon an incorrect value for the mass of Neptune, so it was sheer chance that Pluto turned up in the right place at the right time.

▲ **Comparative Sizes** of the Earth, Moon and some of the larger Trans Neptunian Objects. Note that Pluto, long classed as a planet but now designated a dwarf planet, is much smaller than the Moon, and also smaller than Eris.

▼ **Map of Pluto's surface,** assembled by computer image-processing software from four separate images of Pluto's disk taken with the European Space Agency's (ESA) Faint Object Camera (FOC) aboard the Hubble Space Telescope. Hubble imaged nearly the entire surface, as Pluto rotated on its axis in late June and early July 1994. The map, which covers 85 per cent of the planet's surface, confirms that Pluto has a dark equatorial belt and bright polar caps, as inferred from ground-based light curves obtained during the mutual eclipses that occurred between Pluto and its satellite Charon in the late 1980s. The brightness variations in this composite image may be the result of topographic features such as basins and fresh impact craters. The black strip across the bottom corresponds to the region surrounding Pluto's south pole, which was pointed away from Earth at the time when the observations were made, and so could not be imaged by the FOC.

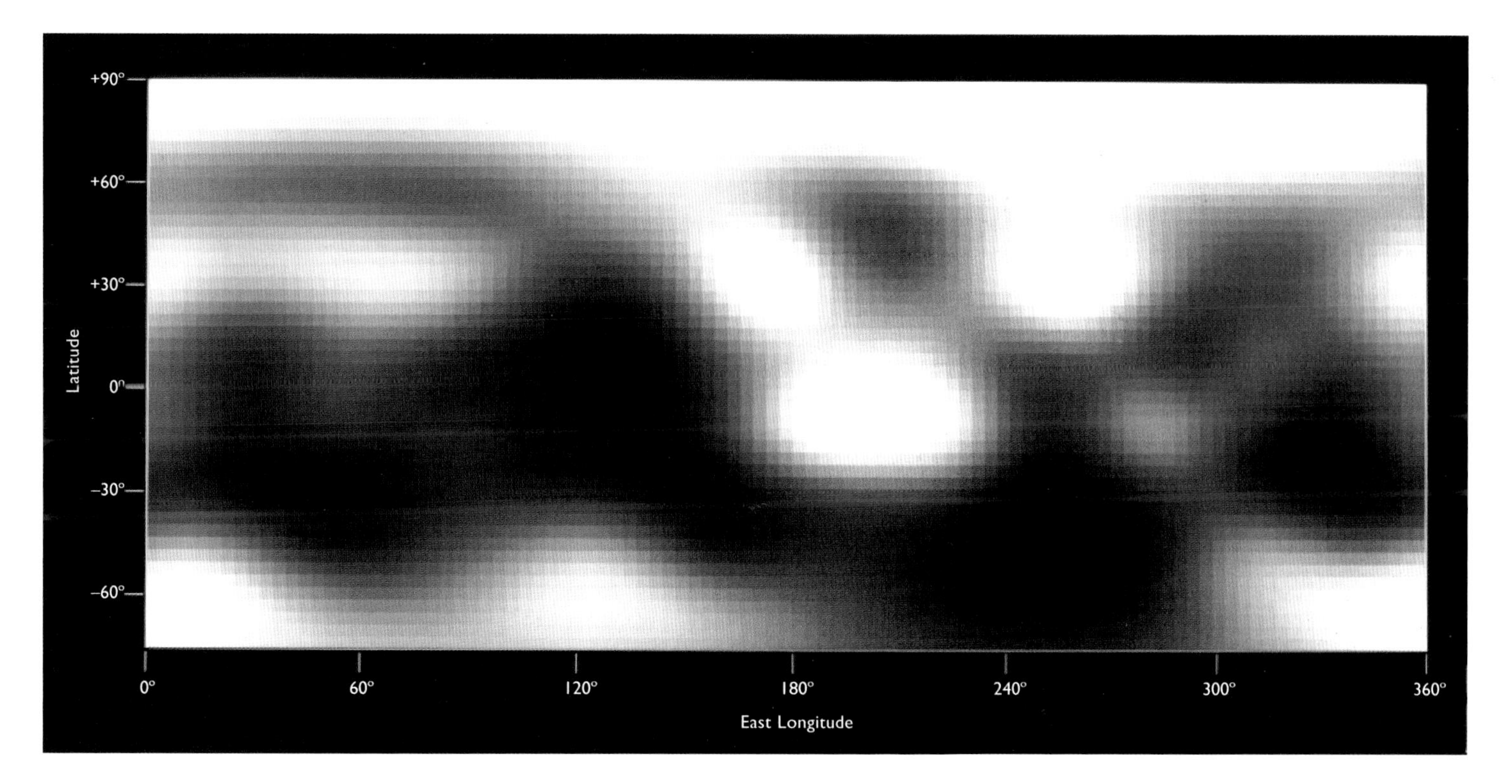

Comets

▼ **Comet West (C/1975 V1),** below left, and Comet Neat (C/2002 V1), below right. Comet West was a bright naked-eye object for several mornings in March 1976, when this photograph was taken by Akira Fujii. As the comet receded from the Sun it showed signs of disruption in its nucleus, so that it may not be so bright when it next returns to perihelion in approximately 550,000 years! Comet Neat is seen here in a photograph taken on 1 February 2003 by Gordon Rogers, with his 41-cm (16-in) reflector. It has a very eccentric path, and an orbital period of about 37,000 years.

Comets are the least predictable members of the Solar System. A comet is a wraith-like object. The only substantial part is the nucleus, which has been described as a dirty ice-ball and is usually no more than a few kilometres across. When the comet is heated, as it nears the Sun, ices in the nucleus start to evaporate, and the comet develops a head, or coma, which may be huge; the coma of the Great Comet of 1811 was larger than the Sun. There may be one or more tails, though many small comets never produce tails.

A cometary nucleus is composed of rocky fragments held together with ices such as frozen ammonia, methane and water. Tails are of two kinds. The gas or ion tail is produced as the solar wind interacts with the gas in the coma, ionizing it and sweeping it back in a direction directly opposite to the Sun. The dust tail consists of tiny dust grains pushed back from the comet's head for solar radiation pressure; in general an ion tail is straight, while a dust tail is curved. Tails always point more or less away from the Sun, so that when a comet is travelling outwards it moves tail-first.

Each time a comet passes through perihelion it loses material to produce a coma and (in some cases) tails; for example Halley's (1P/Halley), with a period of 76 years, loses about 250 million tons of material at each return to the Sun. This means that by cosmical standards, comets must be short-lived. Some short-period comets which used to be seen regularly have now disappeared; such as 3D/Biela, 5D/Brorsen and 20D/Westphal. (Comets are usually named after their discoverers, though occasionally after the mathematician who first computed the orbit – as with Halley's Comet.) A prefix P/ indicates that the comet is periodical, while long-period comets are designated C/ and defunct ones D/. A comet leaves a 'dusty' trail behind it as it moves along, and if the Earth plunges through one of these trails the result is a meteor shower. Most meteor showers have known parent comets; for instance the annual Perseid shower of early August is associated with 109P/Swift-Tuttle, which has a period of 130 years and last returned to perihelion in 1992.

Because a comet is so flimsy and of such low mass, it is at the mercy of planetary perturbations, and orbits may be drastically altered from one cycle to another. The classic case is that of Comet Lexell (D1770/L1), which became a bright naked-eye object. A few years later it made a close approach to Jupiter, and its orbit was completely changed, so we have no idea where the comet is now.

Short-period Comets

Short-period comets come from the Kuiper Belt, but long-period comets come from the so-called Oort Cloud, a swarm of these icy bodies orbiting the Sun at a distance of more than a light year. If one of the members of the cloud is perturbed for any reason, it may start to fall inwards towards the Sun, and eventually it will invade the inner part of the Solar System. One of several things may happen. The comet may simply swing round the Sun and return to the Oort Cloud, not to be back for many thousands or millions of years. It may fall into the Sun, and be destroyed. It may be perturbed by a planet (usually Jupiter) and either thrown out of the Solar System altogether, or else forced into a short-period orbit which brings it back to perihelion after a few years. Or it may collide with a planet, as D/1993 F2 (Shoemaker–Levy 9) did in July 1994, when it impacted on Jupiter. But really brilliant comets have periods so long that we cannot predict the date of their appearance, and they are always apt to arrive without warning and take us by surprise.

It is a pity that all the comets with periods of less than half a century are faint. No doubt they were much more imposing when they first plunged sunwards, but by now they are mere ghosts of their former selves. Comet 2P/Encke, the first to be identified, is a case in point. It was originally found in 1786 by the French astronomer Pierre Méchain, when it was magnitude 5, and had a short tail. It was seen again in 1795 by Caroline Herschel, William Herschel's sister, and yet again in 1805 by Thulis, from Marseilles. In 1818 it turned up once more, and was detected by Jean Louis Pons, whose grand total of comet discoveries amounted to 37. (Pons' story is unusual, because he began his career as an observatory door-keeper and ended it as an observatory director.) The orbit was calculated by J. F. Encke, of Berlin, who concluded that the comets of 1786, 1795, 1805 and 1818 were one and the same; he gave the period as 3.3 years, and predicted a return for 1822. The comet duly appeared just where Encke had expected, and, very appropriately, was named after him. Since then it has been seen at every return except that of 1945, when it was badly placed and when most astronomers had other things on their minds.

At some returns during the last century 2P/Encke was quite prominent; in 1829 it reached magnitude 3.5, with a tail 18 minutes of arc in length. Nowadays it does not achieve such eminence, and although it is hard to be sure – estimating comet magnitudes is far from easy – it does seem to have faded. Whether it will survive into the twenty-second century remains to be seen.

Comet 2P/Encke has a small orbit; at perihelion it ventures just inside the orbit of Mercury, while at its farthest from the Sun it moves out into the asteroid zone. Modern instruments can follow it all around its path; its period is the shortest known. In 1949 a new comet, 107P/Wilson–Harrington, was believed to have a period of only 2.4 years, but it was not seen again until 1979, when it was recovered – this time as an asteroid, designated No 4015! There is little doubt that it has changed its status, and it may well be that many of the small close-approach asteroids, such as 3200 Phaethon, are ex-comets which have lost all their volatiles.

Comet 3D/Biela met with a sad fate. It was discovered in 1772 by Montaigne from Limoges, recovered by Pons in 1805 and again by an Austrian amateur, Wilhelm von Biela, in 1826. The period was given as between six and seven years, and it returned on schedule in 1832, when it was first sighted by John Herschel. (It was unwittingly responsible for a major panic in Europe. The French astronomer Charles Damoiseau had predicted that the orbit of the comet would cut that of the Earth; he was quite right, but at that time the comet was nowhere near the point of intersection.) Comet 3D/Biela was missed in 1839 because of its unfavourable position, but it came back again in 1846, when it astonished astronomers by splitting into two. The pair returned in 1852, were missed in 1859 again because they were badly placed, and failed to appear at the expected return of 1866 – in fact they have never been seen again. When they ought to have returned, in 1872, a brilliant meteor shower was seen coming from that part of the sky where the comet had been expected, and there is no doubt that the meteors represented the funeral pyre of the comet. The shower was repeated in 1885, 1892 and 1899, but no more brilliant displays have been seen since then; to all intents and purposes the shower has ceased, so that we really have, regretfully, seen the last of 3D/Biela.

Other periodical comets have been 'mislaid', only to be found again after many years; thus Comet 17P/Holmes, which reached naked-eye visibility in 1892 and had a period of nearly seven years, was lost between 1908 and 1965; it has since been seen at several returns, but is excessively faint.

Comet 16P/Brooks (Brooks 2) made a close approach to Jupiter in 1886, when it moved inside the orbit of Io and was partially disrupted, spawning four minor companion comets which soon faded away. During the Jupiter encounter, the orbit was changed from 29 years to its present seven years.

Comet 29P/Schwassmann–Wachmann (Schwassmann–Wachmann 1) is of unusual interest. Its orbit lies wholly between those of Jupiter and Saturn, and normally it is very faint, but sometimes it shows sudden outbursts which bring it within the range of small telescopes. Large instruments can follow 29P/Schwassmann–Wachmann all round its orbit, as is also the case with a few other comets with near-circular paths, such as 74P/Smirnova–Chernykh and 5P/Gunn. Comet 39P/Oterma used to have a period of 7.9 years, but an encounter with Jupiter in 1973 altered this to 19.3 years, and the comet now comes nowhere near the Earth, so that its future recovery is very doubtful. It cannot be said that these short-period comets are spectacular objects, but they are always worth locating, and amateurs find them easy to photograph. Of comets which return regularly, only Halley's can ever become really bright.

▲ **Comet 2P/Encke,** as photographed by Jim Scotti on 5 January 1994, using the 0.91-m (36-in) Spacewatch Telescope on Kitt Peak.

PERIODICAL COMETS WHICH HAVE BEEN OBSERVED AT TEN OR MORE RETURNS

Name	Year of discovery	Period, years	Eccentricity	Incl.	Dist. from Sun, astr. units min.	Dist. from Sun, astr. units max.
2P/Encke*	1786	3.3	0.85	12.0	0.34	4.10
26P/Grigg–Skjellerup	1902	5.1	0.66	21.1	0.99	4.93
10P/Tempel (Tempel 2)	1873	5.3	0.55	12.5	1.38	4.70
7P/Pons–Winnecke	1819	6.4	0.64	22.3	1.25	5.61
6P/D'Arrest	1851	6.4	0.66	16.7	1.29	5.59
22P/Kopff	1906	6.4	0.55	4.7	1.58	5.34
31P/Schwassmann–Wachmann (2)	1929	6.5	0.39	3.7	2.14	4.83
21P/Giacobini–Zinner	1900	6.6	0.71	13.7	1.01	6.00
19P/Borrelly	1905	6.8	0.63	30.2	1.32	5.83
16P/Brooks (Brooks 2)	1889	6.9	0.49	5.6	1.85	5.41
15P/Finlay	1886	7.0	0.70	3.6	1.10	6.19
4P/Faye	1843	7.4	0.58	9.1	1.59	5.96
14P/Wolf 1	1884	8.4	0.40	27.3	2.42	5.73
8P/Tuttle	1790	13.7	0.82	54.4	1.01	10.45
29P/Schwassmann–Wachmann* (1)	1908	15.0	0.11	9.7	5.45	6.73
1P/Halley	240 BC	76.0	0.97	162.2	0.59	34.99

(* = comets which can be followed throughout their orbits)

Halley's Comet

Of all the comets in the sky
There's none like Comet Halley.
We see it with the naked eye,
And periodically.

Nobody seems to know who wrote this piece of doggerel, but certainly Halley's Comet (formally called Comet 1P/Halley) is in a class of its own. It has been seen at every return since that of 240 BC, and the earliest record of it, from Chinese sources, may date back as far as 1059 BC. Note that the interval between successive perihelion passages is not always 76 years; like all of its kind, Halley's Comet is strongly affected by the gravitational pulls of the planets.

Edmond Halley, later to become Astronomer Royal, observed the return of 1682. He calculated the orbit, and realized that it was strikingly similar to those of comets previously seen in 1607 and 1531, so that he felt confident in predicting a return for 1758. On Christmas night of that year – long after Halley's death – the comet was recovered by the German amateur astronomer Palitzsch, and it came to perihelion in March 1759, within the limits of error given by Halley. This was the first predicted return of any comet; previously it had been thought by most astronomers that comets travelled in straight lines.

Halley's Comet has a very elliptical orbit. At its closest it is about 88 million kilometres (55 million miles) from the Sun, within the orbit of Venus; at aphelion it recedes to 5250 million kilometres (3260 million miles), beyond the orbit of Neptune and the Kuiper Belt. At its brightest recorded return, that of AD 837, it passed by the Earth at only 6 million kilometres (3.75 million miles), and contemporary reports tell us that its head was as brilliant as Venus, with a tail stretching 90 degrees across the sky. Another bright return was that of April 1066, before the Battle of Hastings; the comet caused great alarm among the Saxons, and it is shown on the Bayeux Tapestry, with King Harold toppling on his throne and the courtiers looking on aghast. In 1301 it was seen by the Florentine artist Giotto di Bondone, who used it as a model for the Star of Bethlehem in his picture *The Adoration of the Magi* – even though Halley's Comet was certainly not the Star of Bethlehem; it returned in 12 BC, years before the birth of Christ. At the return of 1456 the comet was condemned by Pope Calixtus III as an agent of the Devil. It was prominent in 1835 and in 1910, but unfortunately not in 1986, when it was badly placed and never came within 39 million kilometres (24 million miles) of the Earth, and though it became an easy naked-eye object it was by no means spectacular. The next return, that of 2061, will be no better. We must wait until 2137, when it will again be a magnificent sight.

The fact that Halley's Comet can still become brilliant shows that it came in fairly recently from the Oort Cloud. It loses about 250 million tons of material at each perihelion passage, but it should survive in more or less its present form for at least 150,000 years to come.

The comet was first photographed at the 1910 return, after which it remained out of range until 16 October 1982, when it was recovered by D. Jewitt and E. Danielson, at the Palomar Observatory, only six minutes of arc away from its predicted position; it was moving between the orbits of Saturn and Uranus. As it drew in towards perihelion, six spacecraft were sent towards it: one American, two Japanese, two Russian and one European. The European probe Giotto, named after the painter, invaded the comet's head, and on the night of 13–14 March 1986 passed within 605 kilometres (370 miles) of the nucleus. Giotto's camera functioned until about 14 seconds before closest approach, when the spacecraft was struck by a particle about the size of a rice-grain and contact was temporarily broken; in fact the camera never worked again, and the closest image of the nucleus was obtained from a range of 1675 kilometres (about 1000 miles).

The nucleus was found to be shaped rather like a peanut, measuring 15 × 8 × 8 kilometres (9 × 5 × 5 miles), with a total volume of over 500 cubic kilometres (120 cubic miles) and a mass of from 50,000 million to 100,000 million tonnes (it would need 60,000 million comets of this mass to equal the mass of the Earth). The main constituent is water ice, insulated by an upper layer of black material which cracks in places when heated by the Sun, exposing the ice below and resulting in jets of dust and gas. Jet activity was very marked during

► **Halley's Comet** taken during the 1910 return. When far from the Sun, a comet has no tail; the tail starts to develop when the comet draws inward and is heated, so that the ices in its nucleus begin to evaporate. This series shows the tail increasing to a maximum.

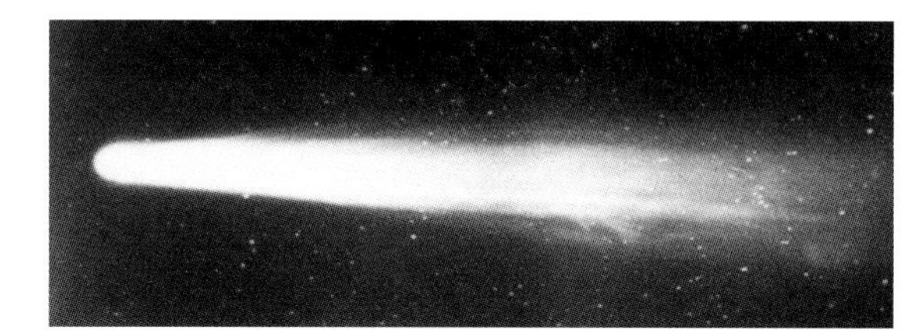

OBSERVED RETURNS OF HALLEY'S COMET

Year	Date of perihelion
1059 BC	3 Dec
240 BC	25 May
164 BC	12 Nov
87 BC	6 Aug
12 BC	10 Oct
AD 66	25 Jan
141	22 Mar
218	17 May
295	20 Apr
374	16 Feb
451	28 June
530	27 Sept
607	15 Mar
684	2 Oct
760	20 May
837	28 Feb
912	18 July
989	5 Sept
1066	20 Mar
1145	18 Apr
1222	28 Sept
1301	25 Oct
1378	10 Nov
1456	9 June
1531	26 Aug
1607	27 Oct
1682	15 Sept
1759	13 Mar
1835	16 Nov
1910	10 Apr
1986	9 Feb

▲ **Halley's Comet,** 12 April 1986. I took this photograph with an ordinary camera to show what the comet really looked like in the sky. At this return it was by no means spectacular. It will be no better next time round, but the return the following century it should be bright enough to cast shadows.

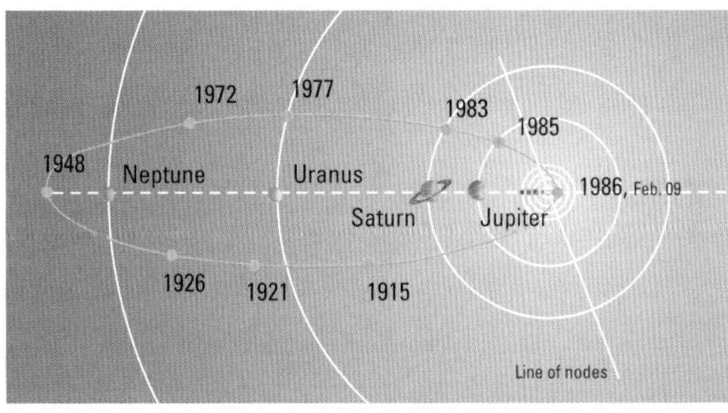

◀ **Orbit of Halley's Comet.** Aphelion was reached in 1948; the comet passed perihelion in 1986, and has now receded once more as far as the orbit of Uranus. The next perihelion will be in 2061.

▼ **Nucleus of Halley's Comet,** from the Halley Multi-colour Camera carried in Giotto. The range was 20,000 km (12,500 mi).

the Giotto pass, though the jets themselves were confined to a small area on the sunward side of the nucleus. The central region of the nucleus was smoother than the ends; a bright 1.5-kilometre (1-mile) patch was presumably a hill, and there were features which appeared to be craters with diameters of around a kilometre (3280 feet). The comet was rotating in a period of 55 hours with respect to the long axis of the nucleus.

Tails of both types were formed, and showed marked changes even over short periods. By the end of April the comet had faded below naked-eye visibility, but it provided a major surprise in February 1991, when observers using the Danish 154-cm (60-inch) reflector at the European Southern Observatory's site at La Silla in Chile found that it had flared up by several magnitudes. There had been some sort of outburst, though the reason for this increase in activity is unclear.

Giotto survived the Halley encounter, and was then sent on to rendezvous with a much smaller and less active comet, 26P/Grigg–Skjellerup, in July 1992. Despite the loss of the camera, a great deal of valuable information was obtained. Unfortunately Giotto did not have enough propellant remaining for a third cometary encounter.

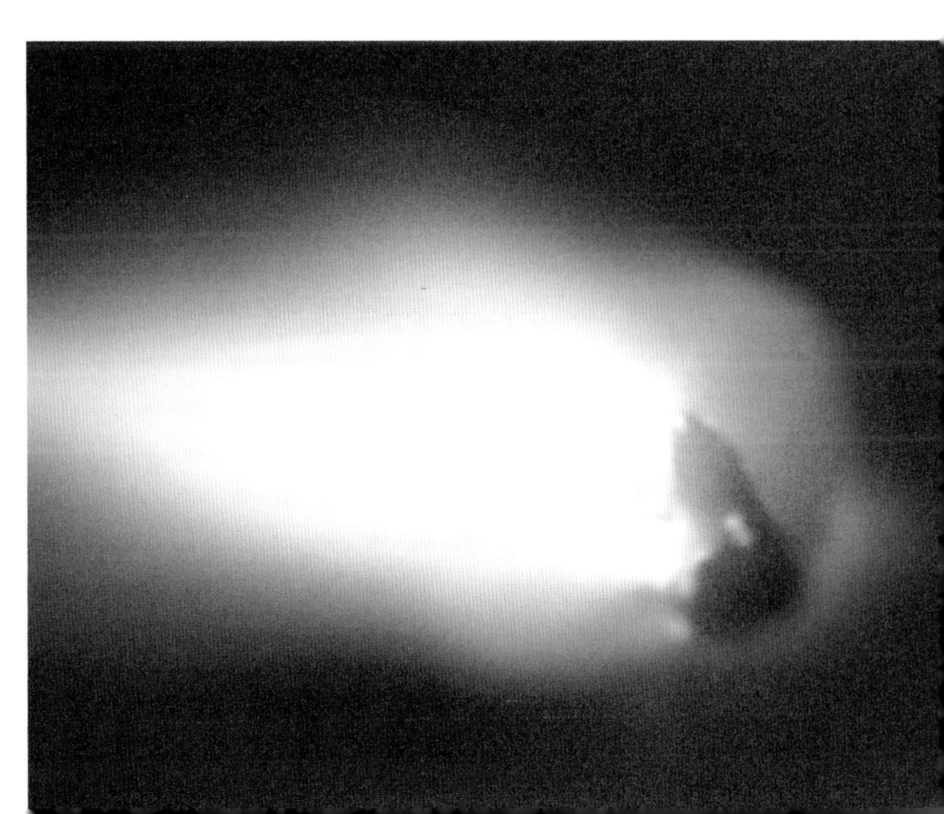

Great Comets

▲ **Comet C/1811 F1 (The Great Comet of 1811),** discovered by Honoré Flaugergues. This impression shows the comet on 15 October, from Otterbourne Hill, near Winchester in England.

▼ **Comet C/1956 R1 (Arend–Roland)** in 1957, photographed by E. M. Lindsay from Armagh, Northern Ireland. It will never return; it has been perturbed into an open hyperbolic orbit.

It is not surprising that ancient peoples were alarmed whenever a brilliant comet appeared. These so-called 'hairy stars' were regarded as unlucky; remember Shakespeare's lines in Julius Caesar – 'When beggars die, there are no comets seen; the heavens themselves blaze forth the death of princes.' There have been various comet panics, one of which was sparked off in 1736 by no less a person than the Rev. William Whiston, who succeeded Newton as Lucasian Professor of Mathematics at Cambridge. Mainly on religious grounds, Whiston predicted that the world would be brought to an end by a collision with a comet on 16 October of that year, and the alarm in London was so great that the Archbishop of Canterbury felt bound to issue a public disclaimer!

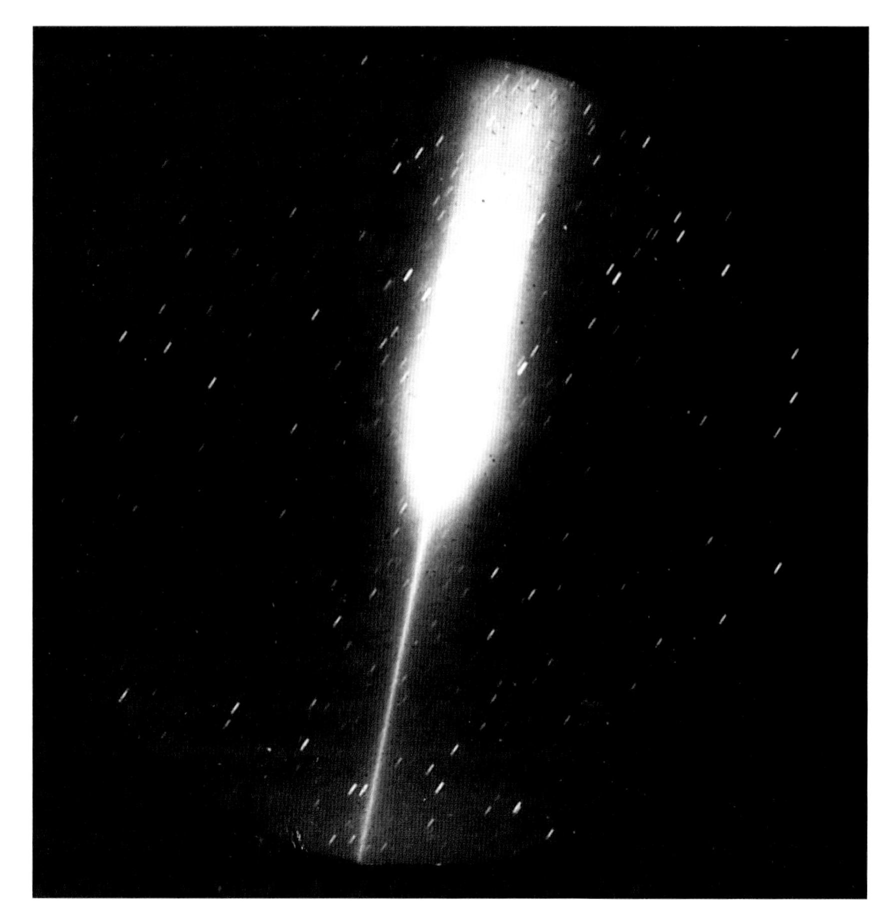

If the Earth were struck by a cometary nucleus a few kilometres in diameter there would undoubtedly be widespread damage, but the chances are very slight. A theory of a different type has been proposed in recent years by Sir Fred Hoyle and Chandra Wickramasinghe, who believe that comets can deposit viruses in the upper air and cause epidemics such as smallpox. It must be said, however, that these ideas have met with practically no support either from astronomers or from medical experts.

Great comets were rare during the twentieth century, but many have been seen in the past. For example, Comet C/1743 X1 (Klinkenberg–de Chéseaux) developed multiple tails, and a contemporary drawing of it has been likened to a Japanese fan. Even more impressive was the Great Comet of 1811 (C/1811 F1), discovered by the French astronomer Honoré Flaugergues. The coma was 2 million kilometres (1.2 million miles) across, and the 16-million-kilometre (10-million-mile) tail stretched out to over 90°, while the tail of the Great Comet of 1843 (C/1843 D1) extended to 330 million kilometres (205 million miles), considerably greater than the distance between the Sun and Mars. It is not easy to remember that these huge bodies are so flimsy, and that their masses are absolutely negligible by planetary standards.

Comet C/1858 L1 (Donati) of 1858 is said to have been the most beautiful ever seen, with its brilliant head, two straight ion tails and dust-tail curved like a scimitar. Three years later came Comet C/1861 J1 (Tebbutt), discovered by an Australian amateur, which came within 2 million kilometres (1.2 million miles) of the Earth; we may even have passed through the tip of the tail, though nothing unusual was reported apart from a slight, unconfirmed yellowish tinge over the sky.

The Great September Comet of 1882 (C/1882 R1) was bright enough to cast shadows, and to remain visible even when the Sun was above the horizon. This was the first comet to be properly photographed. Sir David Gill, at the Cape of Good Hope, obtained an excellent picture of it, and this led to an important development. Gill's picture showed so many stars that he realized that the best way to map the stellar sky was by photographic methods rather than by laborious visual measurement. The 1882 comet was a member of the Kreutz Sun-grazing group, distinguished by very small perihelion distance.

The Daylight Comet of 1910 (C/1910 A1) appeared a few weeks before Halley, and was decidedly the brighter of the two. It also was visible at the same time as the Sun, and it had a long, imposing tail. The orbit is elliptical, but we will not be seeing the comet again yet awhile, because the estimated period is of the order of 4 million years. Obviously we cannot be precise; we can measure only a very small segment of the orbit, and it is very difficult to distinguish between a very eccentric ellipse and a parabola.

Comet C/1927 X1 (Skjellerup–Maristany) of 1927 was also very brilliant, but its glory was brief, and it remained inconveniently close to the Sun in the sky. This was also true, though not to so great an extent, of Comet C/1965 S1 (Ikeya–Seki), discovered independently by two Japanese observers. This was also a Kreutz sungrazer. From some parts of the world it was brilliant for a while, but it soon faded, and will not be back for at least 880 years. Comet C/1973 E1 (Kohoutek) was expected to become extremely brilliant, but was a disappointment and was none too conspicuous with the naked eye.

Of lesser comets, special mention should be made of 1957's C/1956 R1 (Arend–Roland), 1970's C/1969 Y1 (Bennett) and 1976's C/1975 V1 (West). C/1956 R1 (Arend–Roland) was quite conspicuous in the evening sky for a week or two in April 1957, and showed a curious sunward spike which was not a reverse tail, but was merely thinly spread material in the comet's orbit catching the sunlight at a favourable angle. C/1969 Y1 (Bennett) was rather brighter, with a long tail; its period is about 1700 years. C/1975 V1 (West) was also bright, but suffered badly as it passed through perihelion, and the nucleus was broken up.

The only really bright comets of very recent years came in 1996, 1997 and 2007 – C/1995 O1 (Hyakutake), C/1996 B2 (Hale–Bopp), which was a naked-eye object for 18 months, and P/2006 P1 (McNaught). At least the appearance of three bright comets in just over a decade is encouraging. Comet McNaught was discovered by the Anglo-Australian astronomer Rob McNaught at Siding Spring Observatory in August 2006 and by early 2007 it was a magnificent sight, for observers in the southern hemisphere. The brightest comet since Ikeya–Seki, at its peak its tail reached 35° across the sky. When the next will appear we do not know, but we hope it will not be too long delayed.

SELECTED LIST OF GREAT COMETS

Year	Name	Date of discovery	Greatest brightness	Mag.	Min. dist. from Earth, 10^6 km
1577	C/1577 V1	1 Nov	10 Nov	−4	94
1618	C/1618 V1	16 Nov	6 Dec	−4	54
1665	C/1665 F1	27 Mar	20 Apr	−4	85
1743	C/1743 X1 (Klinkenberg–de Chéseaux)	29 Nov	20 Feb 1744	−7	125
1811	C/1811 F1 (Flaugergues)	25 Mar	20 Oct	0	180
1843	C/1843 D1	5 Feb	3 Jul	−7	125
1858	C/1858 L1 (Donati)	2 June	7 Oct	−1	80
1861	C/1861 J1 (Tebbutt)	13 May	27 June	0	20
1874	C/1874 H1 (Coggia)	17 Apr	13 July	0	44
1882	C/1882 R1 (Great September Comet)	1 Sep	9 Sept	−10	148
1910	C/1910 A1 (Daylight Comet)	13 Jan	30 Jan	−4	130
1927	C/1927 X1 (Skjellerup–Maristany)	27 Nov	6 Dec	−6	110
1975	C/1975 V1 (West)	10 Aug	25 Feb 1976	−3	120
1965	C/1965 S1 (Ikeya–Seki)	18 Sept	14 Oct	−10	135
1996	C/1996 B2 (Hyakutake)	30 Jan	1 May	−1	21
1997	C/1995 O1 (Hale–Bopp)	22 July 1995	30 Apr	−1.5	193
2007	C/2006 P1 (McNaught)	7 Aug 2006	14 Jan	−6.0	123

▲ **C/1843 D1** (the Great Comet of 1843), from the Cape of Good Hope on 3 March. This may have been the brightest comet for many centuries.

▲ **C/1858 L1 (Donati).** The frontispiece to Agnes Giberne's *Sun, Moon and Stars* (1880) the comet over St Paul's Cathedral in London.

Millennium Comets

The closing years of the old century were graced by two bright comets. The first C/1996 B2 (Hyakutake) was discovered on 30 January 1996 by the Japanese amateur Yuji Hyakutake, using 25 × 150 binoculars; it was then at magnitude 11. It brightened steadily, and moved north; it reached perihelion on 1 May 1996, at 34 million kilometres (21 million miles) from the Sun. On 24 March it passed Earth at 15 million kilometres (9.3 million miles) – 40 times as far away as the Moon. At this time it was near Polaris in the sky; the magnitude was –1, and there was a long, gossamer-like tail extending for 100°. Its main feature was its beautiful green colour. It faded quickly during April. It will not return for at least another 29,500 years. It was a small comet, with a nucleus estimated at just 3.2 kilometres (2 miles) in diameter.

The second bright comet C/1995 O1 (Hale–Bopp) was discovered on 22 July 1995 independently by two American observers, Alan Hale and Thomas Bopp. It was by no means a faint telescopic object, but was 900 million kilometres (560 million miles) from the Sun, beyond the orbit of Jupiter. It brightened steadily; by the autumn of 1996 it had reached naked-eye visibility, and became brilliant in March and April 1997, at about magnitude –1.5. It passed Earth on 22 March

▼ **C/1995 01 (Hale–Bopp),** photographed by Akira Fujii on 10 March 1997. Note the clear separation of the long blue ion tail and the dust tail.

1997, at over 190 million kilometres (120 million miles); had it come as close as Hyakutake had done, it would have cast shadows. Perihelion was reached on 1 April, at over 125 million kilometres (over 80 million miles) from the Sun. It will return in several centuries time – sometime around AD 4380.

Hale–Bopp was not only very large, by cometary standards, but also very active. In addition it developed a third tail, consisting of sodium, extending for some 50 million kilometres (30 million miles). Spectroscopic analysis showed that the comet contained organic chemicals, some of which had never before been seen in comets.

C/2006 M4 (Swan) was not 'great', only just attaining naked-eye visibility; it reached magnitude 4 in late October 2006. In contrast, C/2006 P1 (McNaught) was, for a few days, visible in daylight during January 2007. It was notable for its complex, striated dust tail and was far better seen in the southern hemisphere.

▼ **C/1995 O1 (Hale–Bopp),** photographed by the author on 1 April 1997, at Selsey, West Sussex (Nikon F3, 50 mm, exposure 40 seconds, Fuji ISO 800).

► **C/2006 R1 (McNaught) at dusk,** from Selsey by Pete Lawrence. Unfortunately the comet was never well seen in the northern hemisphere as it was too far south.

▼ **C/1996 B2 (Hyakutake),** photographed in 1996 by Akira Fujii. Note the lovely green colour of the comet – particularly the coma – in this photograph.

Meteors

▲ **The Leonid Meteor Storm of 1833,** when it was said that meteors 'rained down like snowflakes'. Other major Leonid storms were those of 1799, 1866, 1966, 1999 and 2001.

Meteors are cometary debris. They are very small, and we see them only during the last moments of their lives as they enter the upper atmosphere at speeds of up to 72 kilometres (45 miles) per second. What we actually observe, of course, are not the tiny particles themselves (known more properly as meteoroids) but the luminous effects which they produce as they plunge through the air. On average a 'shooting star' will become visible at a height of about 115 kilometres (70 miles) above ground level, and the meteoroid will burn out by the time it has penetrated to 70 kilometres (45 miles), finishing its journey in the form of fine 'dust'. Still smaller particles, no more than a tenth of a millimetre (0.004 inches) across, cannot produce luminous effects, and are known as micrometeorites; these drift down through the atmosphere.

When the Earth moves through a trail of cometary debris we see a shower of meteors, but there are also sporadic meteors, not connected with known comets, which may appear from any direction at any moment. The total number of meteors of magnitude 5 or brighter entering the Earth's atmosphere is around 75 million per day, so an observer under clear, dark skies may expect to see something of the order of ten naked-eye meteors per hour, though during a shower the number will naturally be higher.

It is worth noting that more meteors may be expected after midnight than before. During evenings, the observer will be on the trailing side of the Earth as it moves round the Sun, so incoming meteors will have to catch up with it; after midnight the observer will be on the leading side, so meteors meet the Earth head-on and the relative velocities are higher.

The meteors of a shower will seem to issue from one particular point in the sky, which is known as the radiant. The particles are travelling through space in parallel paths, so this is an effect of perspective – just as the parallel lanes of a motorway appear to 'radiate' from a point on the horizon.

The richness of a shower is measured by its Zenithal Hourly Rate (ZHR). This is the number of naked-eye meteors which could be seen by an observer under ideal conditions, in an hour with the radiant at the zenith. These conditions are never met, so the observed rate is always appreciably lower than the theoretical ZHR.

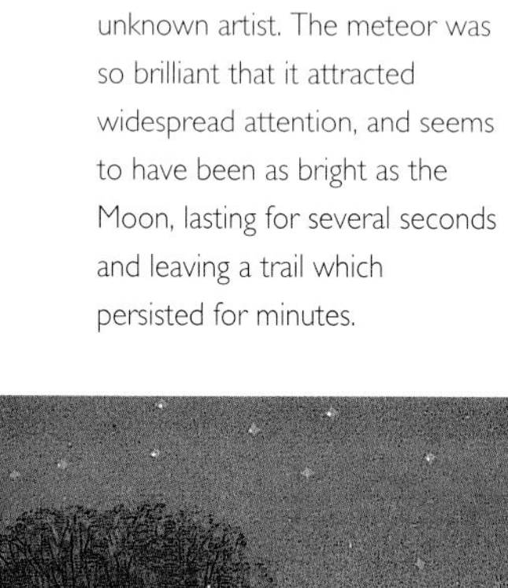

▼ **Great Meteor of 7 October 1868.** Old painting by an unknown artist. The meteor was so brilliant that it attracted widespread attention, and seems to have been as bright as the Moon, lasting for several seconds and leaving a trail which persisted for minutes.

Each shower has its own particular characteristics. The Quadrantids of early January are possibly associated with the comets 2003 EH1 or C/1490 Y1; the radiant lies in the constellation of Boötes (the Herdsman), the site of a former constellation, the Quadrant, which was rejected by the International Astronomical Union and has now disappeared from the maps. The ZHR can be over 100 meteors an hour, but the maximum is very brief. The April Lyrids are associated with Comet C/1861 G1 (Thatcher), which has an estimated period of 415 years; the ZHR is not usually very high, but there can be occasional rich displays, as last happened in 1982. Two showers, the Eta Aquarids of April–May and the Orionids of October, come from Halley's Comet. Though they were not particularly rich around the time of the comet's last return in 1986, they had an unexpected outburst in October 2006. The October Draconids are associated with 21P/Giacobini–Zinner, and are sometimes referred to as the Giacobinids. Usually they are sparse, but produced major storms in 1933 and 1946, when for a brief period the ZHR reached 4000–5000 meteors an hour. Ever since then, unfortunately, the Draconids have been rather disappointing.

Two important showers occur in December: the Geminids and the Ursids. The Geminids have an unusual parent – the asteroid 3200 Phaethon, which is very probably a dead comet. The Ursids, with the radiant in the Little Bear, are associated with Comet 8P/Tuttle and can sometimes be rich, as in 1945 and again in 1986.

Some showers appear to have decreased over the years. The Andromedids are now almost extinct. The Taurids, associated with Comet 2P/Encke, are not usually striking, though they last for over a month; reports seem to indicate that in past centuries they were decidedly richer than they are now.

Probably the most interesting showers are the Perseids and the Leonids. The Perseids are very reliable, and last for several weeks with a sharp maximum on 12/13 August each year; if you look up into a clear, dark sky for a few minutes during the first fortnight in August, you will be very unlucky if you don't see several Perseids. The fact that the display never fails us shows that the particles have had time to spread all round the orbit of the parent comet, 109P/Swift–

◀ **Perseid meteor.** Against the backdrop of a glorious aurora in Colorado, by Jimmy Westlake.

▼ **A meteor shower.** Of course not all these meteors would be seen at the same moment.

Tuttle, which has a period of 130 years and was last at perihelion in 1992. The comet was not then conspicuous, but at its next return it will come very near the Earth – certainly within a couple of million kilometres, perhaps even closer – and there have been suggestions that it might hit us. The chances of a collision are many hundreds to one against, but 109P/Swift–Tuttle will be a magnificent spectacle. It is a pity that nobody born before the early 21st century will see it.

The Leonids are quite different. The parent comet, 55P/Tempel–Tuttle, has a period of 33 years, and it is when the comet returns to perihelion that we see major Leonid displays; the particles are not yet spread out all round the comet's orbit. Superb meteor storms were seen in 1799, 1833 and 1866. The expected displays of 1899 and 1933 were missed, because the swarm had been perturbed by Jupiter and Saturn, but in 1966 the Leonids were back with a vengeance, reaching a peak rate of over 140,000 per hour. Sadly, this lasted for only about 40 minutes, and it occurred during daylight in Europe, so observers in the Americas had the best view. The Leonids were rich in 1999 and 2001, though there was no display comparable with that of 1833. Leonid showers have been traced back for many centuries, and 902 was known as 'the Year of the Stars'.

▼ **The 'radiant' principle.** I took this picture in Alaska in 1992; the parallel tracks seem to radiate from a point near the horizon.

SELECTED ANNUAL METEOR SHOWERS

Shower	Begins	Max.	Ends	Max. ZHR	Parent comet	Notes
Quadrantids	1 Jan	4 Jan	6 Jan	100?	2003 EH1 or C/1490 Y1	Radiant in Boötes. Short, sharp max.
Lyrids	19 Apr	22 Apr	25 Apr	10	C/1861 G1 (Thatcher)	Occasionally rich, as in 1922 and 1982.
Eta Aquarids	24 Apr	5 May	20 May	35	1P/Halley	Broad maximum.
Delta Aquarids	15 July	29 July 6 Aug	20 Aug	10–20	–	Double radiant. Faint meteors.
Perseids	23 Jul	12 Aug	20 Aug	75	109P/Swift–Tuttle	Rich; consistent.
Orionids	16 Oct	21 Oct	27 Oct	25	1P/Halley	Swift; fine trails.
Draconids	10 Oct	10 Oct	10 Oct	var.	21P/Giacobini–Zinner	Usually weak, but occasional great displays, as in 1933 and 1946.
Taurids	20 Oct	3 Nov	30 Nov	10	2P/Encke	Slow meteors. Fine display in 1988.
Leonids	15 Nov	17 Nov	20 Nov	var.	55P/Tempel–Tuttle	Usually sparse, but occasional storms at intervals of 33 years: good displays from 1999 to 2001. No more Leonid storms expected in the near future.
Geminids	7 Dec	13 Dec	16 Dec	100	3200 Phaethon (asteroid)	Rich, consistent.
Ursids	17 Dec	22 Dec	25 Dec	5–10	8P/Tuttle	Can be rich, as in 1945 and 1986.

Meteorite Craters

Go to Arizona, not far from the town of Winslow, and you will come to what has been described as 'the most interesting place on Earth'. It is a huge crater, 1265 metres (4150 feet) in diameter and 175 metres (575 feet) deep; it is well preserved, and has become a well known tourist attraction, particularly as there is easy access from Highway 99. There is no doubt about its origin; it was formed by the impact of a meteorite which hit the Arizonan desert in prehistoric times. The date of its origin is not known with certainty, but current estimates give its age as about 50,000 years. Settlers first came across it in 1871.

The crater is circular, even though the impactor came in at an angle. When the meteorite struck, its kinetic energy was converted into heat, and it became what was to all intents and purposes a very powerful bomb. What is left of the meteorite is probably buried beneath the crater's south wall. Incidentally, the popular name is wrong. It is called Meteor Crater, but this should really be 'Meteorite' Crater.

A smaller but basically similar impact crater is Wolfe Creek in a remote part in the north of Western Australia. There are various local legends about it. The Kjaru Aborigines call it Kandimalal, and describe how two rainbow snakes made sinuous tracks across the desert, forming Wolfe Creek and the adjacent Sturt Creek, while the crater marks the spot where one of the snakes emerged from below the ground. It is much older than the Arizona crater; the age is around 300,000 years. Wolfe Creek is more difficult to reach than Meteor Crater, and the road from the nearest settlement, Halls Creek, is usually open for only part of the year, but it has now been well studied since aerial surveys first identified it in 1947. The wall rises at an angle of 15 to 35°, and the floor is flat, 55 metres (180 feet) below the rim and 25 metres (80 feet) below the level of the surrounding plain. The diameter is 675 metres (2200 feet). Meteoritic fragments found in the area leave no doubt that it really is of cosmic origin.

Also in Australia there are other impact craters; one at Boxhole and a whole group at Henbury, both in the Northern Territory. Equally intriguing is Gosse's Bluff, which is at least 140 million years old and very eroded, though there is the remnant of a central structure and indications of the old walls.

Lists of impact craters include structures in America, Arabia, Argentina, Estonia and elsewhere, but one must be wary of jumping to conclusions; for example, there is no crater associated with the giant Hoba West Meteorite.

It has often been suggested that the Earth was struck by a large missile 65 million years ago, and that this caused such a change in the Earth's climate that many forms of life became extinct, including the dinosaurs. It has been claimed

▼ **Site of the Siberian impact of 1908,** Tunguska; photographed by Don Trombino in 1991. No crater was produced, so presumably the projectile broke up before landing, but the results of the impact are still very evident.

▼ **Wolfe Creek Crater** in Western Australia; an aerial photograph which I took in 1993. This is a very well-formed crater and possibly the most perfect example of an impact structure on the Earth, apart from the famous Meteor Crater in Arizona, USA.

▶ **'Saltpan'** near Pretoria, South Africa, was identified as an impact crater recently. It is larger than the Arizona crater and the associated breccia are clearly seen. The water in the lake is salty. The surrounding wall is uniform in height. Photograph by Dr Kelvin Kemm, 1994.

that the buried Chicxulub impact crater in the Yucatan Peninsula, Mexico, was the site of this impact.

Most impact craters found on Earth are on ancient parts of the crust that have changed little or in dry areas where there is little erosion. Elsewhere, old craters hve been eroded, covered up or deformed beyond recognition by geological processes. However, younger impact craters may still lurk: The town of Nördlingen in Bavaria sits in the centre of a crater 25 kilometres (15 miles) across, which was recognized as being meteoritic in origin only in the 1960s.

No doubt further craters will be formed in the future; there are plenty of potential impactors moving in the closer part of the Solar System. Although the chances of a major collision are slight, they are not nil, which is partly why constant watch is now being kept to identify wandering bodies. It is even possible that if one of these bodies could be seen during approach, we might be able to divert it by nuclear warheads carried on ballistic missiles – though whether we would be given enough advance warning to enable us to do anything about it is problematical.

In January 2000, the British government set up a special committee to look into the whole question of danger from asteroidal or cometary impact. If there is such an impact, let us hope that we cope with the situation better than the dinosaurs did.

▲ **Meteor Crater in Arizona, USA,** photographed from the air. This is the most famous of all impact structures, though not now the largest one to be found. It is also known as Barringer Crater.

▼ **Gosse's Bluff,** Northern Territory of Australia; photograph by Gerry Gerrard. Its impact origin is not in doubt, but it is very ancient, and has been greatly eroded.

SOME IMPORTANT METEORITIC CRATERS

Name	Diameter, m	Date of discovery
Meteor Crater, Arizona	1265	1871
Wolfe Creek, Australia	675	1947
Henbury, Australia	200 × 110	1931 (13 craters)
Boxhole, Australia	175	1937
Odessa, Texas, USA	170	1921
Waqar, Arabia	100	1932
Oesel, Estonia	100	1927

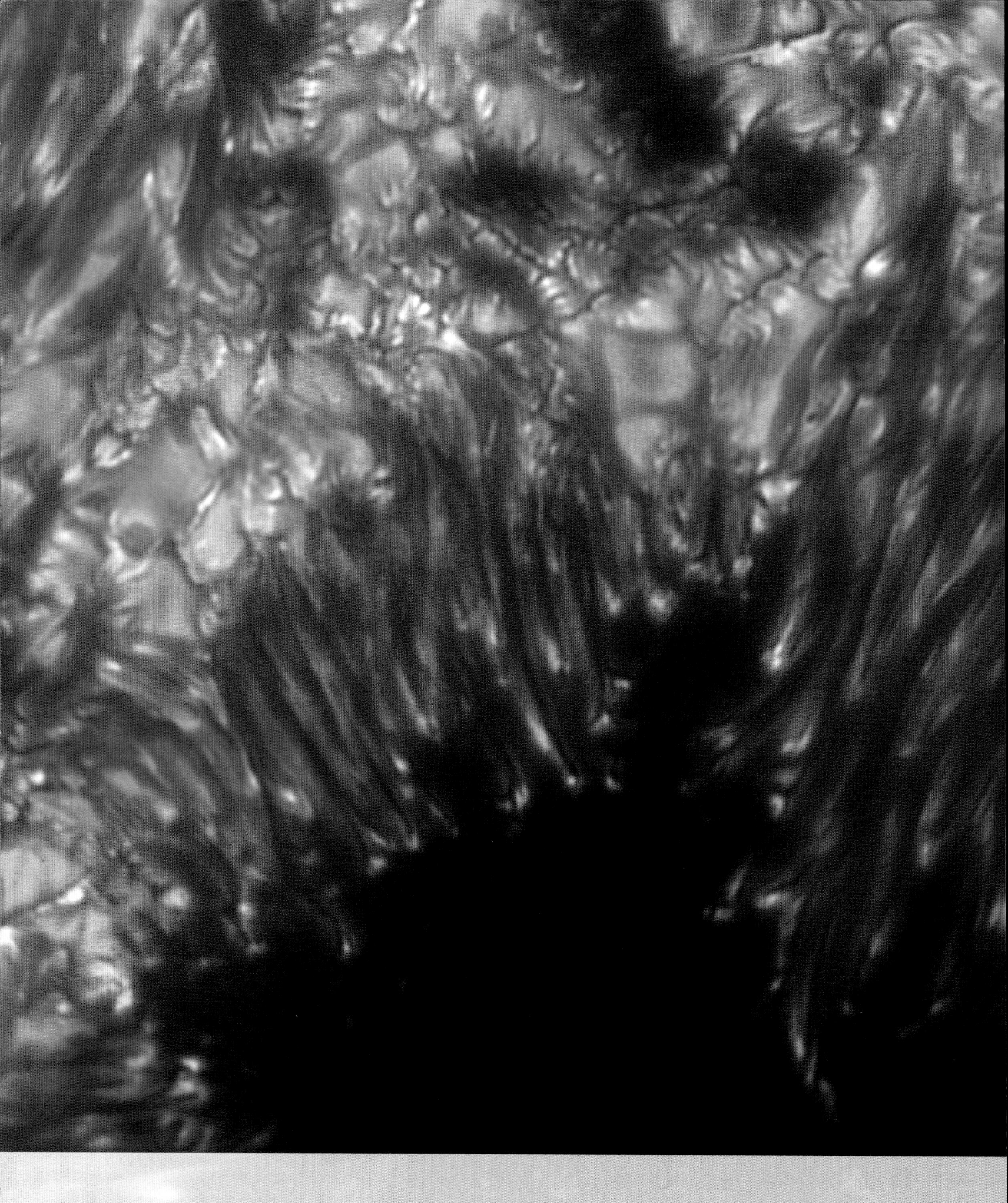

Part of a large sunspot, recorded on 15 July 2002 with the Swedish 1-m (39-in) Solar Telescope on La Palma. The centre of the sunspot, known as the umbra, appears dark because strong magnetic fields there stop hot gas upwelling from the solar interior.

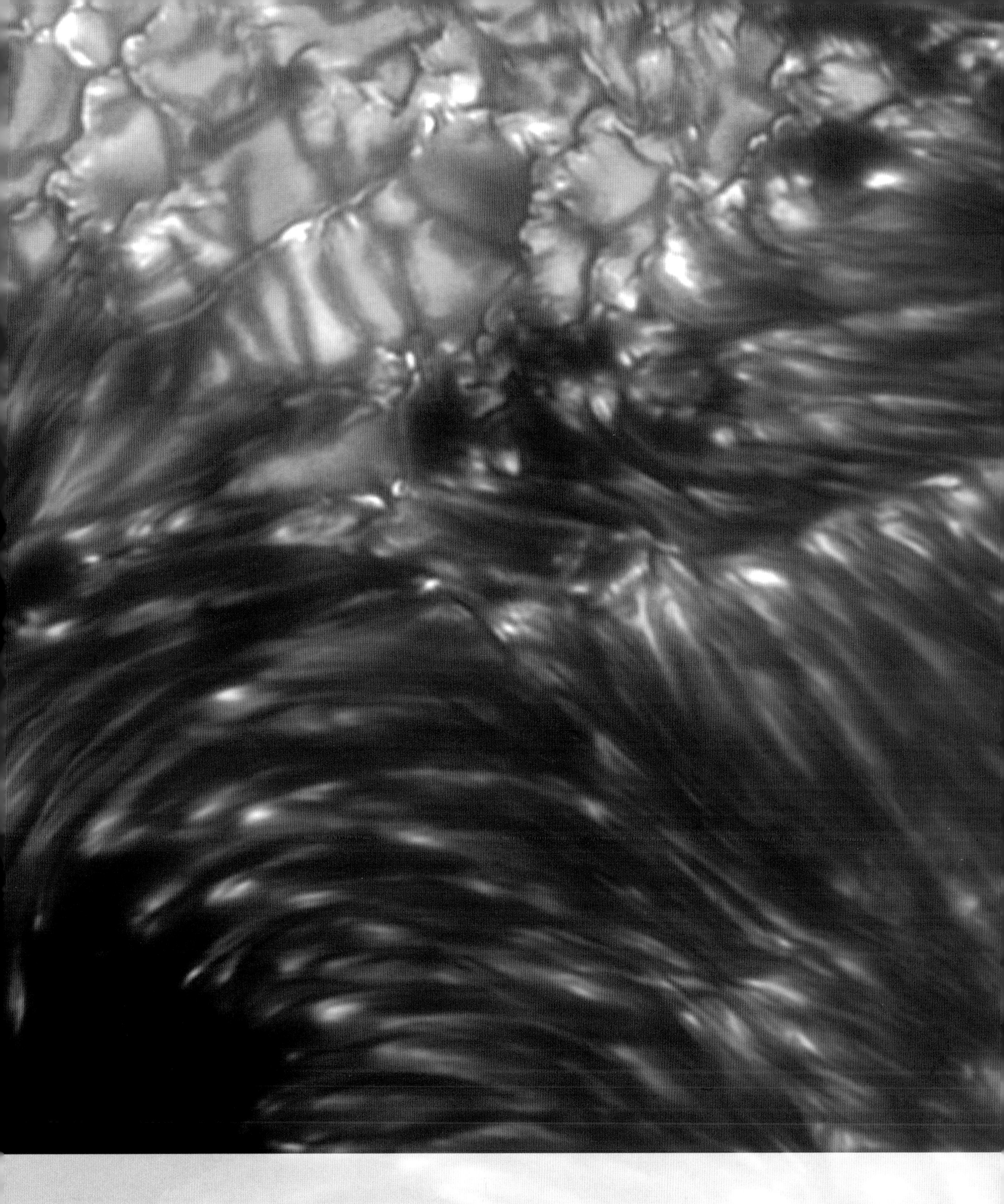

THE SUN

Our Star: The Sun

The Sun's diameter is 1,392,000 kilometres (865,000 miles), and it could engulf over a million globes the volume of the Earth, but it is very much less dense, because it is made up of incandescent gas. At the core, where the energy is being produced, the temperature may be as high as 15,000,000°C; even the bright surface which we can see – the photosphere – is at a temperature of 5500°C. It is here that we see sunspots and the bright regions known as faculae. Above the photosphere comes the chromosphere, a layer of much more rarefied gas, and finally the corona, which may be regarded as the Sun's outer atmosphere.

The Sun is nowhere near the centre of the Galaxy; it is around 25,000 light years from the nucleus. It shares in the general rotation of the Galaxy, moving at 220 kilometres (140 miles) per second and taking 225 million years to make one complete circuit – a period often called the cosmic year; one cosmic year ago, even the age of the dinosaurs lay in the future!

The Sun is rotating on its axis, but it does not spin in the way that a solid body would do. The rotation period at the equator is 25.4 days, but near the poles it is about 34 days. The rotation is easy to observe by the drift of the sunspots across the disk; it takes about a fortnight for a group to cross the disk from one limb to the other.

The greatest care must be taken when observing the Sun. Looking directly at it with any telescope, or even binoculars, means focusing all the light and (worse) the heat on to the observer's eye, and total and permanent blindness will result. Even using a dark filter is unsafe; filters are apt to shatter without warning, and in any case cannot give full protection. The only sensible method is to use the telescope as a projector, and observe the Sun's disk on a screen held or fastened behind the telescope eyepiece.

We know that the Earth is approximately 4600 million years old, and the Sun is certainly older than this. A Sun made up entirely of coal, and burning furiously enough to emit as much energy as the real Sun actually does, would be reduced to ashes in only 5000 years. In fact, the Sun's energy is drawn from nuclear reactions near its core, where the temperatures and pressures are colossal. Not surprisingly, the Sun consists largely of hydrogen (over 70 per cent), and near the core the nuclei of hydrogen atoms are combining to form nuclei of the next lightest element, helium. It takes four hydrogen nuclei to make one helium nucleus; each time this happens, a little energy is released and a little mass is lost. It is this energy which keeps the Sun shining, and the mass-loss amounts to 4 million tons per second. Fortunately there is no cause for immediate alarm; the Sun will not change dramatically for at least a thousand million years yet.

The photosphere extends down to about 300 kilometres (190 miles), and below this comes the convection zone, which has a depth of about 200,000 kilometres (125,000 miles); here, energy is carried upwards from below by moving streams and masses of gas. Next comes the radiative zone, and finally the energy-producing core, which seems to have a diameter of around 450,000 kilometres (280,000 miles). The theoretical models seem satisfactory enough, and a major problem has recently been solved. The Sun sends out vast numbers of strange particles called neutrinos, which are difficult to detect because they have no electrical charge and negligible mass. The Sun appeared to emit far fewer neutrinos than predicted.

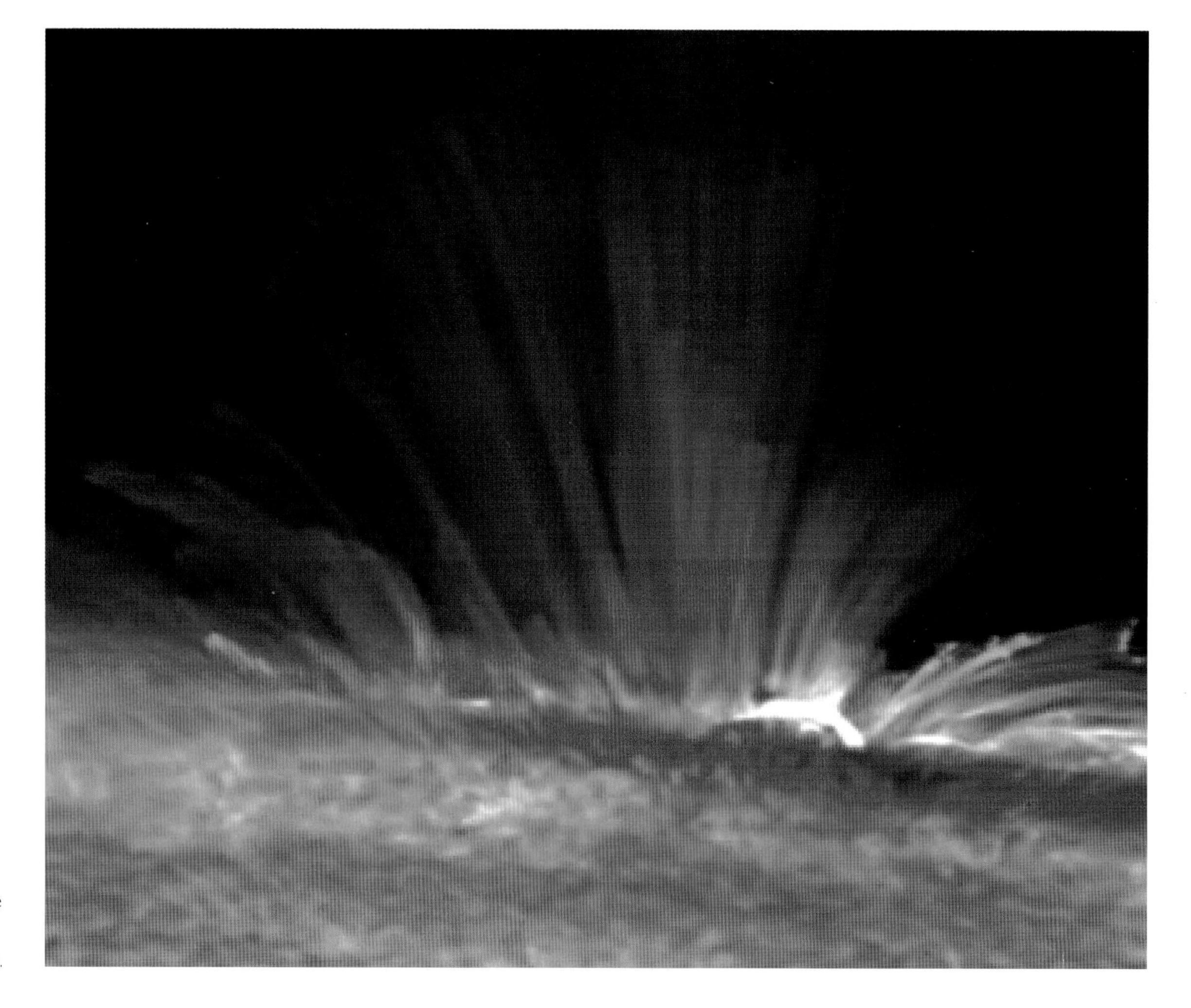

▶ **Magnetic field around a sunspot** The image shows the structure of the solar magnetic field rising vertically fom a sunspot. Taken by the telescope on the Japanese solar probe Hinode, on 20 November 2006.

If a neutrino hits an atom of chlorine, the latter may be changed into a form of radioactive argon. Deep in Homestake Gold Mine in South Dakota, Ray Davis and his colleagues filled a tank with over 450,000 litres (12,000 gallons) of cleaning fluid, which is rich in chlorine; every few weeks they flushed out the tank to see how much argon had been produced by neutrino hits. The numbers were far lower than expected, and similar experiments elsewhere confirmed this. (It was essential to install the tank deep below the ground to avoid the results being affected by cosmic ray particles which, unlike neutrinos, cannot penetrate far below the Earth's surface.) The problem was solved when it was found that there are several kinds of neutrinos (which can switch between types during their journey from the Sun's core to the detectors) and the Homestake detector cannot register all of them. Japan's Super-Kamiokande detector, 1 kilometre (0.6 miles) down in the Mozumi mine, uses 50,000 tons of pure water; it too detects fewer neutrinos than expected.

▼ Spectacular solar prominence Image of a huge, handle-shaped prominence taken on 14 September 1999 by SOHO. Prominences are huge clouds of relatively cool dense plasma suspended in the Sun's hot, thin corona. At times, they can erupt, escaping the Sun's atmosphere.

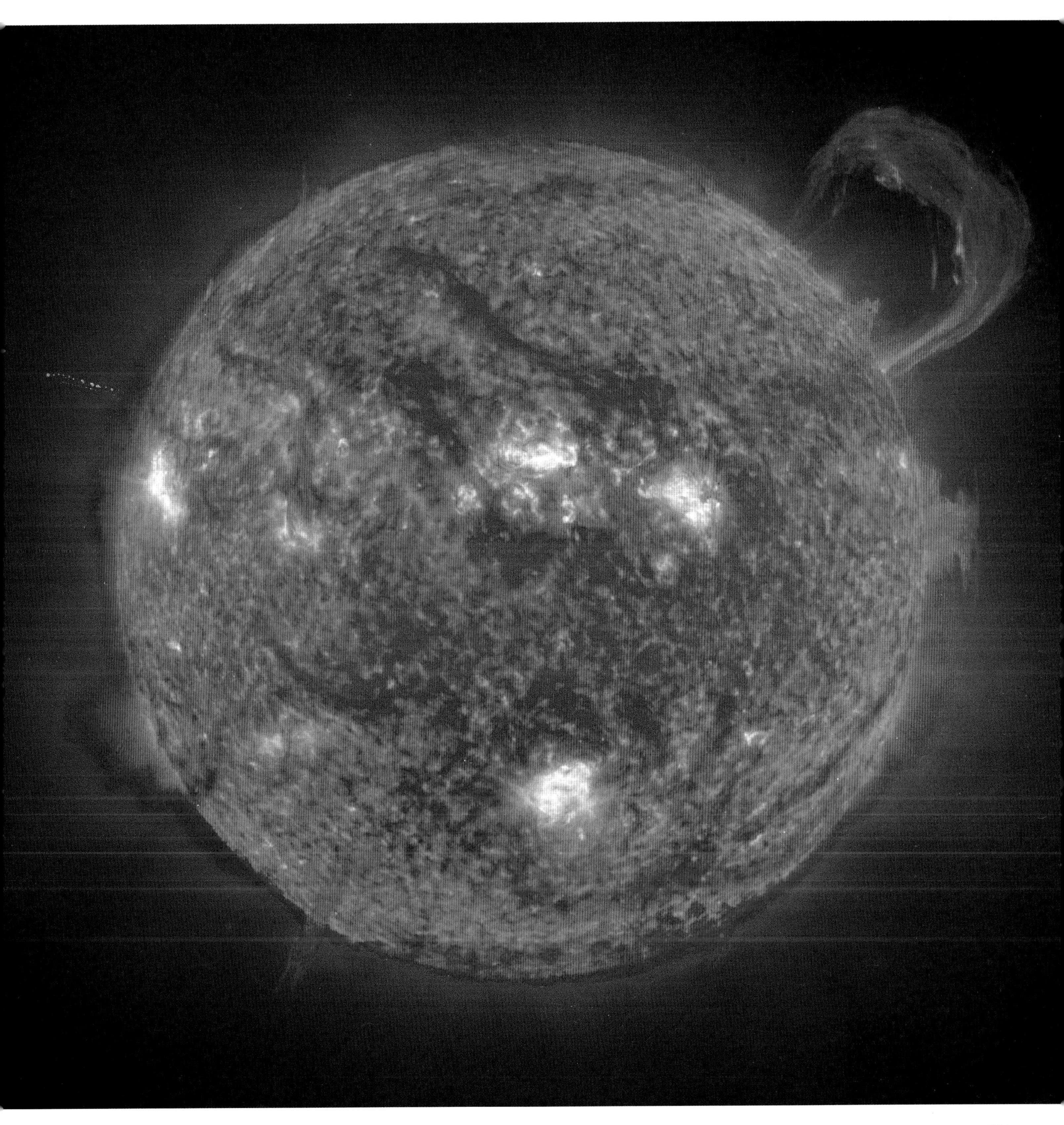

The Surface of the Sun

► **The Sun in hydrogen light (H-Alpha)** Image taken by Pete Lawrence from Selsey on 15 August 2006 using a Coronado telescope. The structure round the sunspot group is very evident.

A major spot is made up of a dark central portion or umbra, surrounded by a lighter penumbra. Sometimes the shapes are regular; sometimes they are very complex, with many umbrae contained in a single mass of penumbra. The temperature of the umbra is about 4500°C, and of the penumbra 5000°C (as opposed to 5500°C for the surrounding unaffected photosphere), so that if a spot could be seen shining on its own, the surface brilliance would be greater than that of an arc lamp.

Spots generally appear in groups. An 'average' two-spot group begins as a pair of tiny pores at the limit of visibility. The pores develop into proper spots, growing and separating in longitude; within two weeks the group has reached its maximum length, with a fairly regular leading spot and a less regular follower, together with many smaller spots spread around in the area. A slow decline then sets in, usually leaving the leader as the last survivor. Around 75 per cent of groups fit into this pattern, but there are many variations, and single spots are also common.

Sunspots may be huge; the largest on record, that of April 1947, covered an area of over 18,000 million square kilometres (7000 million square miles) when at its maximum. Obviously they are not permanent. A major group may persist for anything up to six months, though very small spots often have lifetimes of less than a couple of hours.

Spots are essentially magnetic phenomena, and there is a fairly predictable cycle of events. Maxima, with many groups on view simultaneously, occur every 11 years or so; activity then dies down, until at minimum the disk may be free of spots for many consecutive days or even weeks, after which activity starts to build up once more towards the next maximum. The cycle is not perfectly regular, but 11 years is a good average length, so that there were maxima in 1957–58, 1968–69, 1979–80, 1990–91 and 2000–2001.

The maxima are not equally energetic, and there seems to have been a long spell, between 1645 and 1715, when there were almost no spots at all, so that the cycle was suspended. This is termed the Maunder Minimum, after the British astronomer E. W. Maunder, who was one of the first to draw attention to it. Obviously the records at that time are not complete, but certainly there was a dearth of spots for reasons which are not understood. There is also evidence of earlier periods when spots were either rare or absent, and it may well be that other prolonged minima will occur in the future. During the Maunder Minimum the Earth went through a cool spell; during the 1680s the Thames froze every year, and frost fairs were held on it. Earlier spot-minima are also coincident with cool spells.

There is a further peculiarity, first noted by the German amateur F. W. Spörer. At the start of a new cycle, the spots break out at latitudes between 30° and 45° north or south of the solar equator. As the cycle progresses, new spots appear closer and closer to the equator, until at maximum the average latitude is only 15° north or south of the equator. After solar maximum new spots become less common, but may break out at latitudes down to 7° north or south of the equator. They never appear on the equator itself, and before the last spots of the old cycle die away the first spots of the new cycle begin to appear at higher latitudes.

According to the generally accepted theory, proposed by H. Babcock in 1961, spots are caused by the effects of the Sun's magnetic field lines, which run from one pole to the other just below the bright surface. The rotation period at the equator is shorter than that at higher latitudes, so the field lines are dragged along more quickly, and magnetic 'tunnels' or flux tubes, each about 500 kilometres (320 miles) in diameter, are formed below the surface. These float upwards and break through the surface, producing pairs of spots with opposite polarities. At maximum the magnetic field lines are looped and tangled, but then rejoin to make a more stable configuration, so at the end of the cycle activity fades away and the field lines revert to their original state.

The polarities of leader and follower are reversed in the two hemispheres, and at the end of two cycles there is a complete reversal, so that there are grounds for suggesting that the true length of a sunspot cycle is 22 years rather than 11.

Tracking sunspots is a fascinating pastime. A group takes slightly less than two weeks to cross the disk from one limb to the other, and after an equivalent period it will reappear at the following limb if, of course, it still exists. A spot is foreshortened when near the limb, and the penumbra of a regular spot appears broadened to the limbward side. This 'Wilson effect' indicates that the spot is a depression rather than a hump, but not all spots show it.

Many spots are associated with faculae (Latin for 'torches') which may be described as bright, cloudlike features at higher levels; they are often seen in regions where spots are about to appear, and persist for some time after the spots have died out. And even in non-spot zones, the surface is not calm. The photosphere has a granular structure; each granule is about 1000 kilometres (600 miles) in diameter with a lifetime of about eight minutes. They represent convection currents, and it is estimated that the surface includes about 4 million granules at any one time.

It would be idle to pretend that we have anything like a complete understanding of the Sun. Many problems have been solved, but we still have much to learn about our 'daytime star'.

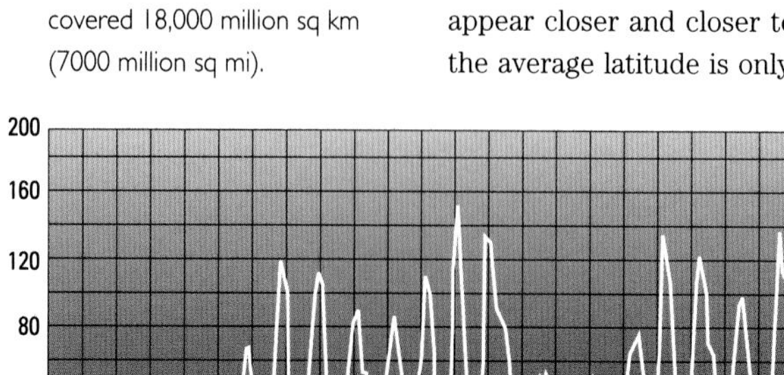

▼ **The solar cycle,** 1650 to present. Not all maxima are equally energetic, and during the 'Maunder Minimum', 1645–1715, it seems that the cycle was suspended, though the records are incomplete. The vertical scale is the Zürich number, calculated from the number of groups and the number of spots.

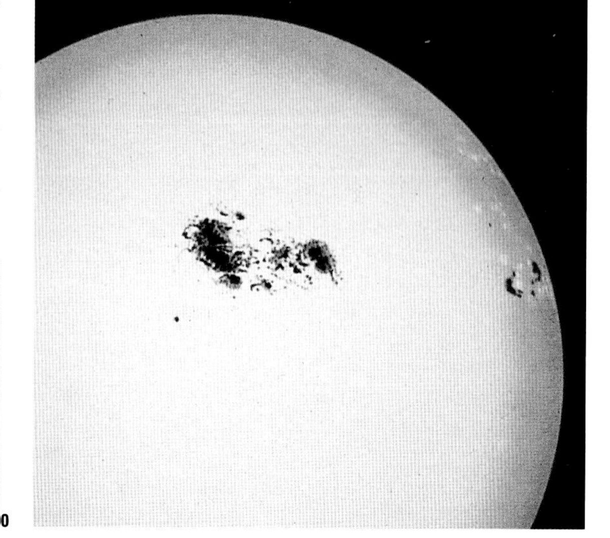

► **The great sunspot of 1947** – the largest known. On 8 April, it covered 18,000 million sq km (7000 million sq mi).

The Sun in Action

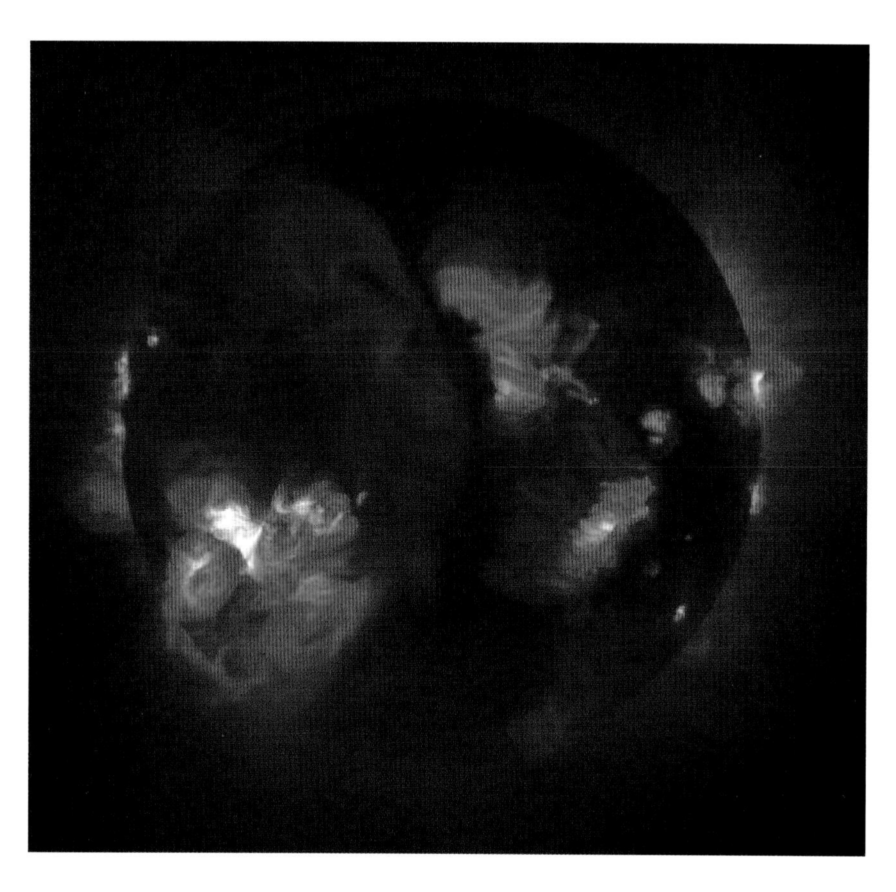

The Sun is never calm. Even more dramatic are the flares, which usually, though not always, occur above active spot-groups; they are seldom seen in ordinary light, so spectroscopic equipment has to be used to study them. They are short-lived, and generally last for no more than 20 minutes or so, though a few have been known to persist for several hours. They produce shock waves in the chromosphere and the corona, and considerable quantities of material may be blown away from the Sun altogether in a coronal mass ejection (CME); the temperatures may rocket to many millions of degrees. Flares are essentially magnetic phenomena, and it seems that the rapid rearrangement of magnetic fields in active regions of the corona results in a sudden release of energy which accelerates and heats matter in the Sun's atmosphere. Radiation is emitted at all wavelengths, and is particularly strong in the X-ray and ultra-violet regions of the electromagnetic spectrum.

The solar wind is made up of charged particles sent out from the Sun at all times. It is made up of a plasma (that is to say, an ionized gas, made up of a mixture of electrons and the nuclei of atoms), and is responsible for repelling the ion tails of comets, making them point away from the Sun. When these charged particles reach the Earth they are responsible for the lovely displays of aurorae or polar lights – aurora borealis in the

◄ **The X-Ray Sun,** 28 September 1991; imaged by the Japanese X-ray satellite Yohkoh. This picture shows regions of different X-ray emission; there is clear evidence of coronal holes.

northern hemisphere, aurora australis in the southern. The average velocity of the solar wind as it passes the Earth is 300 to 400 kilometres per second (190 to 250 miles per second), but during a major solar storm the solar wind's velocity can increase to 1000 kilometres (650 miles) a second.

The solar wind escapes most easily through coronal holes, where the magnetic field lines are open instead of looped. In 1990 a special spacecraft, Ulysses, was launched to study the polar regions of the Sun, which have never been well known simply because from Earth, and from all previous probes, we have always seen the Sun more or less broadside-on. Ulysses had to move well out of the ecliptic plane, which it did by going first out to Jupiter in 1992 and using the powerful gravitational pull of the Giant Planet to put it into the correct path.

Many solar probes have been launched, and have provided an immense amount of information, particularly in the X-ray and ultra-violet regions of the electromagnetic spectrum. The Solar and Heliospheric Observatory (SOHO) was launched in 1995, and stationed at a vantage point 1.5 million kilometres (900,000 miles) sunward of the Earth, keeping the Sun continuously in view. It has obtained superb pictures of solar flares, as well as CMEs in which thousands of millions of charged particles are hurled into space at speeds of around 3.5 million kilometres (2.2 million miles) per hour. CMEs are often (not always) associated with flares.

Another recent solar probe is Hinode, launched on 22 September 2006; it carries optical and X-ray telescopes, and an Extreme Ultra-violet Imaging Spectrograph. Like SOHO, it has been an immediate and lasting success. We have come a long way since William Herschel believed that the Sun might be inhabited, and we are learning more all the time.

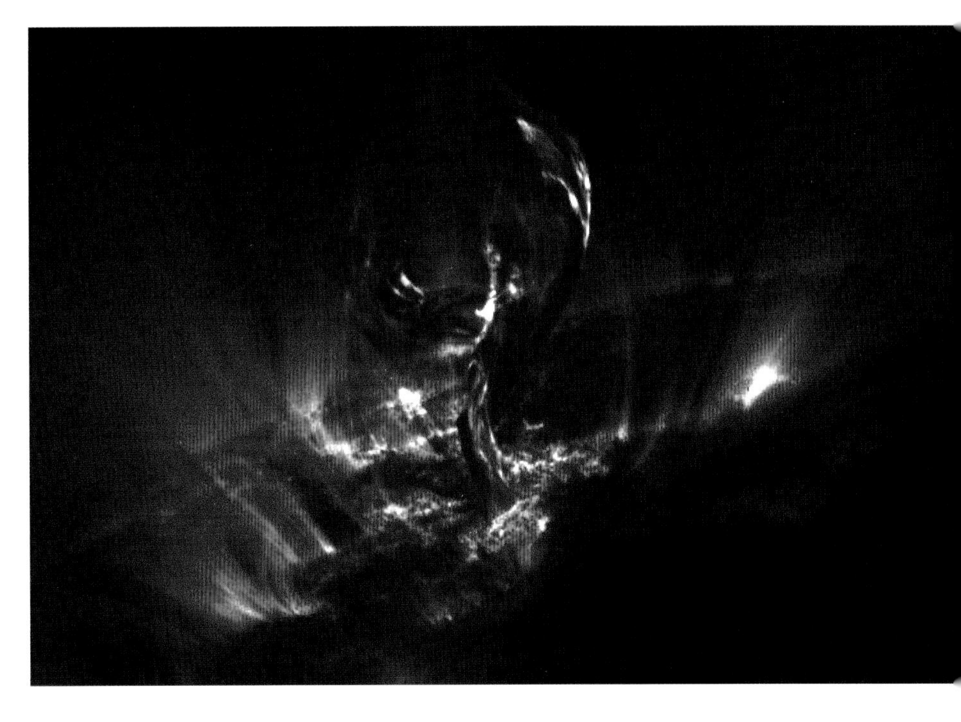

▲ **Active region of the Sun,** imaged by the TRACE satellite on 19 July 2000. The filament seen erupting from the Sun is about 121,000 km (75,000 mi) long. It is shaped by the Sun's magnetic field.

◄ **Coronal Mass Ejection,** imaged by the SOHO's LASCO C3 coronagraph on 29 March 1988. The central disk blocks out the light from the Sun's photosphere.

◄ **Solar prominences.** This image, taken by Pete Lawrence in Selsey with a Coronado H-alpha telescope, shows the granular structure of the solar surface and the spectacular prominences. Taken in April 2006 when the Sun was approaching the minimum of the sunspot cycle.

Eclipses of the Sun

The Moon moves round the Earth; the Earth moves round the Sun. Therefore, there are times when the three bodies line up, with the Moon in the mid position. The result is what is termed a solar eclipse, though it should more properly be called an occultation of the Sun by the Moon.

Eclipses are of three types: total, partial and annular. At a total eclipse the photosphere is completely hidden, and the sight is probably the most magnificent in all Nature. As soon as the last segment of the bright disk is covered, the Sun's atmosphere flashes into view, and the chromosphere and corona shine out, together with any prominences which happen to be present. The sky darkens sufficiently for planets and bright stars to be seen; the temperature falls sharply, and the effect is dramatic by any standards. Unfortunately, total eclipses are rare as seen from any particular locality. The Moon's shadow can only just touch the Earth, and the track of totality can never be more than 272 kilometres (169 miles) wide; moreover, the total phase cannot last more than 7 minutes 31 seconds, and is generally much shorter.

To either side of the main cone of shadow the eclipse is partial, and the glorious phenomena of totality cannot be seen; many eclipses are only partial and not total anywhere. Finally there are annular eclipses, when the alignment is perfect but the Moon is near its greatest distance from Earth; its disk is not then large enough to cover the photosphere completely, and a ring of sunlight is left showing round the dark mass of the Moon (from Latin *annulus*, a ring).

For obvious reasons, a solar eclipse can happen only when the Moon lies on the Sun side of the Earth and is thus new. If the lunar orbit lay in the same plane as that of the Earth, there would be an eclipse every month, but in fact the Moon's orbit is tilted at an angle of just over 5°, so in general the New Moon passes unseen either above or below the Sun in the sky.

The points at which the Moon's orbit cuts the ecliptic are known as the nodes, so to produce an eclipse the Moon must be at or very near a node. Because of the gravitational pull of the Sun, the nodes shift slowly but regularly. After a period of 18 years, 11.3 days, the Earth, Sun and Moon return to almost the same relative positions, so that a solar eclipse is likely to be followed by another eclipse 18 years and 11.3 days later – a period known as the Saros. It is not exact, but it was good enough for ancient peoples to predict eclipses with fair certainty. For example, the Greek philosopher Thales is said to have forecast the eclipse of 25 May 585 BC, which put an abrupt end to a battle being fought between the armies of King Alyattes of the Lydians and King Cyraxes of the Medes; the combatants were so alarmed by the sudden darkness that they made haste to conclude peace.

From any particular point on the Earth's surface, solar eclipses are less common than lunar eclipses. This is because to see a solar eclipse, the observer has to be in just the right place at just the right time, whereas a lunar eclipse is visible from any location where the Moon is above the horizon. Britain had two total eclipses during the twentieth century, those of 29 June 1927 and 11 August 1999. The track of the 1927 eclipse crossed northern Wales and northern England, but at the 'return' at the end of the Saros (9 July 1945) the track missed Britain altogether, though it crossed Canada, Greenland and North Europe. The 11 August 1999 total eclipse crossed the Scilly Isles, Cornwall, South Devon and

▼ Total eclipse of the Sun Photograph by Chris Doherty from a ship in the South China Seas, the corona was beautifully displayed.

Alderney, and then crossed Europe. It was a pity that most of mainland Britain was clouded out.

The main features seen during totality are the chromosphere, the prominences and the corona. The chromosphere has an overall depth of about 10,000 kilometres (6250 miles). The temperature is about 5000°C at the boundary with the photosphere, first decreasing to a temperature minimum around 3,500°C about 600 kilometres (370 miles) into the chromosphere, and then increasing very rapidly to several hundred thousand degrees at the upper boundary of the chromosphere. Prominences – once, misleadingly, called Red Flames – are masses of red, glowing hydrogen. Quiescent prominences may hang in the chromosphere for many weeks, but eruptive prominences show violent motion, often rising to tens of thousands of kilometres. They can be seen with the naked eye only during totality, but spectroscopic equipment now makes it possible for them to be studied at any time. By observing in hydrogen light, prominences may also be seen against the bright disk as dark filaments, sometimes termed flocculi. (Bright flocculi are caused by calcium.)

Shadow bands are wavy lines seen moving across the Earth's surface just before and just after totality. They are caused by atmospheric effects, and are remarkably difficult to photograph well; neither are they seen at every total eclipse.

During totality, the scene is dominated by the glorious pearly corona, which stretches out from the Sun; at times of spot-maxima it is reasonably symmetrical, but near spot-minimum there are long streamers. It is extremely rarefied, with a density less than one million-millionth of that of the Earth's atmosphere at sea level. Its temperature is well over a million degrees, but this does not indicate that it sends out much heat. Scientifically, temperature is measured by the average speeds at which the various atoms and molecules move around; the greater the speeds, the higher the temperature. In the corona the speeds are very high, but there are so few particles that the heat is negligible. The cause of the high temperature seems to be linked with releases of magnetic energy in the layers below, though it is not yet fully understood.

Eclipse photography is fascinating, but there is one point to be borne in mind. Though it is quite safe to look directly at the totally eclipsed Sun, the slightest trace of the photosphere means that the danger returns, and it is essential to remember that pointing an SLR camera at the Sun is tantamount to using a telescope. As always, the greatest care must be taken – but nobody should ever pass up the chance of seeing the splendour of a total solar eclipse.

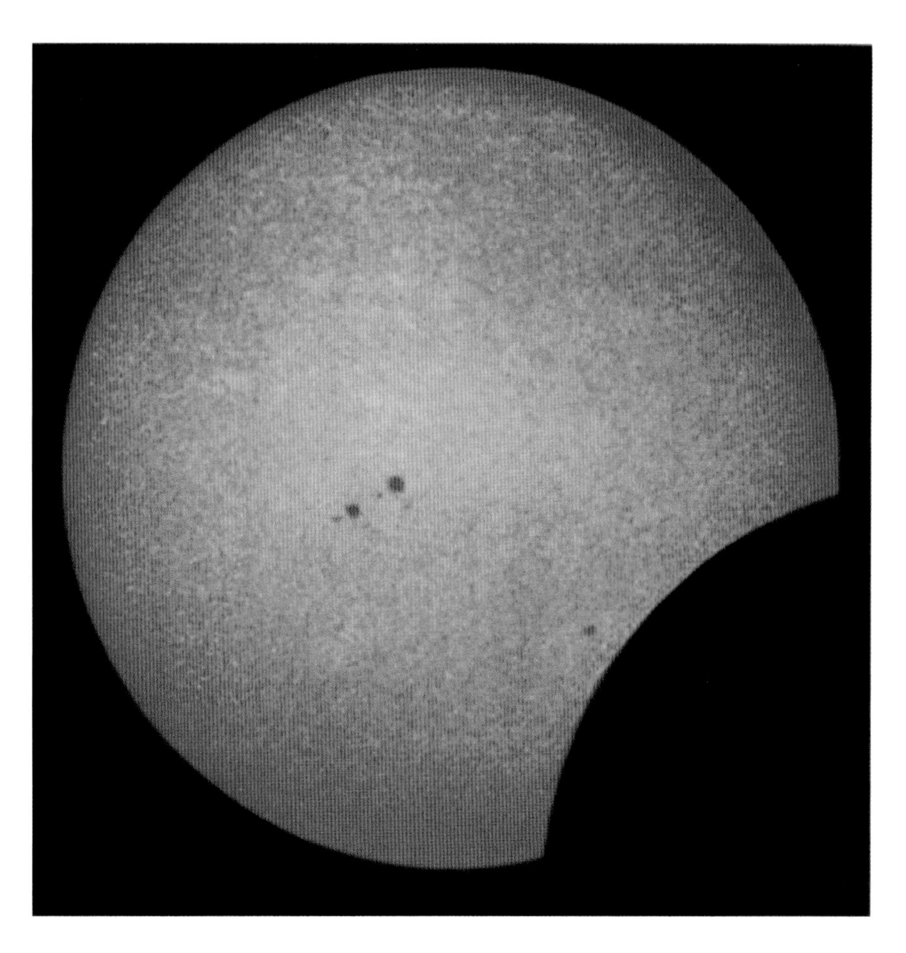

▲ **The partial eclipse** of 21 November 1966 photographed from Sussex by Henry Brinton with a 10-cm (4-in) reflector.

▼ **The lovely Diamond Ring effect,** seen just before and just after totality. This photograph was taken from Java on 11 June 1983 by Dr Bill Livingston.

► **The annular eclipse** of 10 May 1994, photographed by the author from Mexico. The ring of bright photosphere drowns out the corona.

The Milky Way in Cygnus. Image composed of six two-minute exposures. Canon 20D DSLR and 28-mm lens from La Palma. The rich star fields of the Milky Way are bifurcated by the Cygnus Rift, a huge dark nebula. Nik Szymanek.

THE STARS

Introduction to the Stars

THE GREEK ALPHABET

α	Alpha
β	Beta
γ	Gamma
δ	Delta
ε	Epsilon
ζ	Zeta
η	Eta
θ	Theta
ι	Iota
ϰ	Kappa
λ	Lambda
μ	Mu
ν	Nu
ξ	Xi
ο	Omicron
π	Pi
ϱ	Rho
σ	Sigma
τ	Tau
υ	Upsilon
φ	Phi
χ	Chi
ψ	Psi
ω	Omega

How many stars can you see with the naked eye on a clear, dark night? Many people will say, 'Millions', but this is quite wrong. There are roughly 5800 stars within naked-eye range. Only about half of these will be above the horizon at any one time, and faint stars which are low down will probably not be seen. This means that if you can see a grand total of 2500 stars, you are doing very well.

The ancients divided up the stars into groups of constellations, which were named in various ways. The Egyptians had one method, the Chinese another, and so on; the constellations we use today are those of the Greeks (admittedly with Latin names) and if we had used one of the other systems our sky maps would look very different, though the stars themselves would be exactly the same. In fact, a constellation pattern has no real significance, because the stars are at very different distances from us, and we are dealing with nothing more than line of sight effects.

Ptolemy, last of the great astronomers of Classical times, listed 48 constellations, all of which are still given on modern maps even though they have been modified in places. Some of the groups were named after mythological characters, such as Orion and Perseus; others after animals or birds, such as Cygnus (the Swan) and Ursa Major (the Great Bear), and there are a few inanimate objects, such as Triangulum (the Triangle). Other constellations have been added since, notably those in the far south of the sky which never rose above the horizon in Egypt, where Ptolemy seems to have spent the whole of his life.

Some of the new groups have modern-sounding names, such as Telescopium (the Telescope) and Octans (the Octant). During the seventeenth century various astronomers compiled star catalogues, usually by appropriating stars from older groups. Some of the additions have survived (including Crux, the Southern Cross), while others have mercifully been deleted; our maps no longer show constellations such as Globus Aerostaticus (the Balloon), Officina Typographica (the Printing Press) and Sceptrum Brandenburgicum (the Sceptre of Brandenburg). Today we recognize a total of 88 constellations. They are very unequal in size and importance, and some of them are so obscure that they seem to have little claim to a separate identity. One can sympathize with Sir John Herschel, who once commented that the constellation patterns seemed to have been drawn up so as to cause the maximum possible inconvenience and confusion.

Very bright stars such as Sirius, Canopus, Betelgeux and Rigel have individual names, most of which are Arabic, but in other cases a different system is used. In 1603 Johann Bayer, a German amateur astronomer, drew up a star catalogue in which he took each constellation and gave its stars Greek letters, starting with alpha (α) for the brightest star and working through to omega (ω). This proved to be very satisfactory, and Bayer's letters are still in use, though in many cases the proper alphabetical sequence has not been followed; thus in Sagittarius (the Archer), the brightest stars are Epsilon (ε), Sigma (σ) and Zeta (ζ), with Alpha (α) and Beta (β) Sagittarii very much 'also rans'. Later in the century John Flamsteed, the first Astronomer Royal, gave numbers to the stars, and these too are still in use; thus Sirius, in Canis Major (the Great Dog) is not only α Canis Majoris but also 19 Canis Majoris.

The stars are divided into classes or magnitudes depending upon their apparent brilliance. The scheme works rather like a golfer's handicap, with the more brilliant performers having the lower values; thus magnitude 1 is brighter than 2, 2 brighter than 3, and so on. The faintest stars normally visible with the naked eye are of magnitude 6, though modern telescopes using electronic equipment can reach down to at least 30. On the other end of the scale, there are few stars with zero or even negative magnitudes; Sirius is −1.46, while on the same scale the Sun would be −26.8. The scale is logarithmic, and a star of magnitude 1.0 is exactly a hundred times as bright as a star of 6.0.

Note that the apparent magnitude of a star has nothing directly to do with its real luminosity. A star may look bright either because it is very close on the cosmic scale, or because it is genuinely very large and powerful, or a combination of both. The two brightest stars are Sirius, at −1.46, and Canopus, at −0.73, so Sirius is over half a magnitude the more brilliant of the two; yet Sirius is 'only' 26 times as luminous as the Sun, while the much more remote Canopus could match 15,000 Suns. Appearances can often be deceptive.

There is one anomaly. It is customary to refer to the 21 brightest stars as being of the first magnitude; they range from Sirius down to Regulus (α Leonis), whose magnitude is 1.35. Next in order comes Adhara (η Canis Majoris); its magnitude is 1.50, but even so it is not included among the élite.

The stars are not genuinely fixed in space. They are moving about in all sorts of directions at all sorts of speeds, but they are so far away that their individual or proper motions are very slight; the result is that the constellation patterns do not change appreciably over periods of many lifetimes and

they look virtually the same today as they must have done in the time of Julius Caesar or even the builders of the pyramids. It is only our nearer neighbours, the members of the Solar System, which move about from one constellation into another. The nearest star beyond the Sun lies at a distance of 4.2 light years, a light year being the distance travelled by a ray of light in one year – over 9.5 million million kilometres (around 6 million million miles).

This, of course, is why the stars appear relatively small and dim; no normal telescope will show a star as anything but a point of light. Yet some stars are huge; Betelgeux (α Orionis) is so large that it could contain the entire orbit of Mars round the Sun. Other stars are much smaller than the Sun, or even than the Earth, but the differences in mass are not so great as might be expected, because small stars are denser than large ones. It is rather like balancing the mass of a meringue against a lead pellet.

There is a tremendous range in luminosity. We know of stars which are more than a million times as powerful as the Sun, while others have only a tiny fraction of the Sun's power. The colours, too, are not the same; our Sun is yellow, while other stars may be bluish, white, orange or red. These differences are due to real differences in surface temperature. The hottest known stars have temperatures of up to 80,000°C, while the coolest are so dim that they barely shine at all.

Many stars, such as the Sun, are single. Others are double or members of multiple systems. There are stars which are variable in light; there are clusters of stars, and there are vast clouds of dust and gas which are termed nebulae. Our star system or Galaxy contains about 200,000 million stars, and beyond we come to other galaxies, so remote that their light takes millions, hundreds of millions or even thousands of millions of years to reach us. Look at these distant systems today, and you see them not as they are now, but as they used to be when the universe was young – long before the Earth or even the Sun came into existence.

Our Sun is an ordinary star, and is attended by a system of planets. Other stars have planetary systems of their own, and by 2007 over two hundred 'extra-solar planets' had been located – though not seen directly: a planet, with no light of its own, is drowned by the glare of its parent star, just as the bulb of a pocket torch would be hard to see close to a distant searchlight. Fortunately, there are ways round this problem. A massive planet moving round a normal star will make the star 'wobble' very slightly and these tiny shifts can be measured. Also, a large planet passing in front of the star will block some of the light and the star's brightness will drop slightly until the planet moves out of the way. The trouble is that all the planets so far detected are gas-giants, more like Jupiter than like the Earth, and many are often referred to as 'hot Jupiters'. Planets like our Earth must surely exist, and in the fairly near future we hope to find them.

Other Earths? Other life-forms – and other astronomers? There seems no reason why not. It is a sobering thought.

▼ **Ursa Major.** The Great Bear photographed by Nik Szymanek. The seven famous stars that make up the pattern nicknamed the Plough or Big Dipper. Over Britain they never set.

The Celestial Sphere

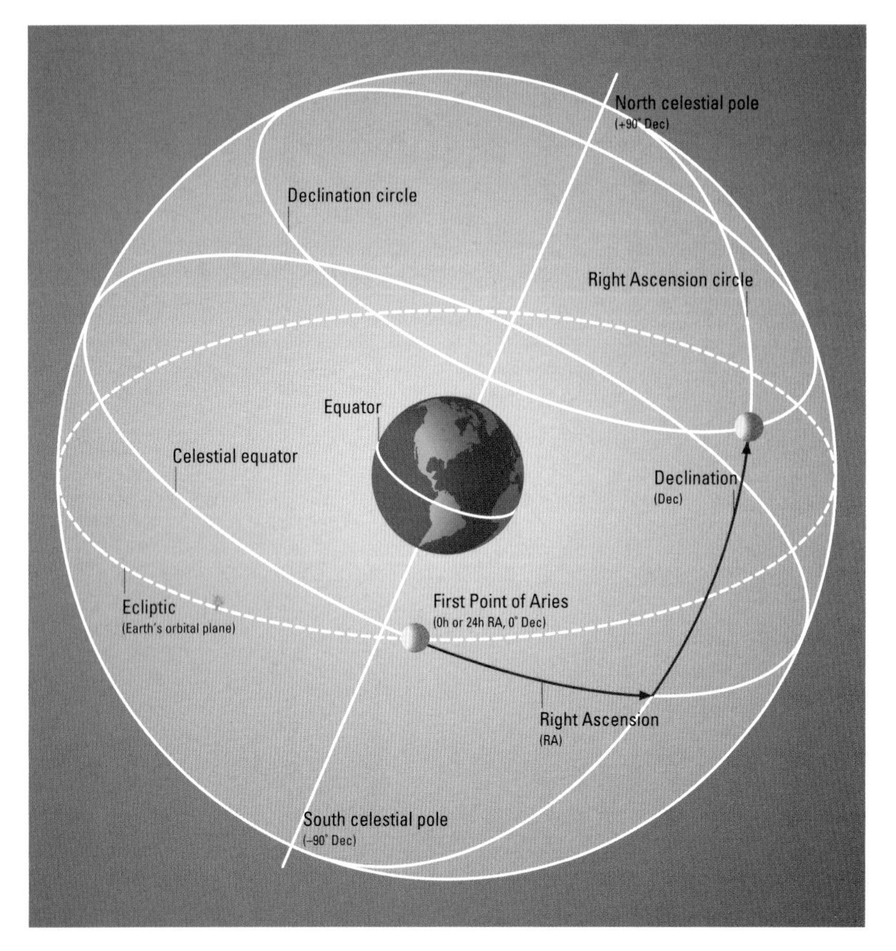

▲ **The celestial sphere.** For some purposes it is still convenient to assume that the sky really is solid, and that the celestial sphere is concentric with the surface of the Earth. We can then mark out the celestial poles, which are defined by the projection of the Earth's axis on to the celestial sphere; the north pole is marked closely by the bright star Polaris (α Ursae Minoris), but there is no bright south pole star, and the nearest candidate is the dim σ Octantis. The celestial equator is the projection of the Earth's equator on to the celestial sphere; it divides the sky into two hemispheres. Declination is the angular distance of a body north or south of the celestial equator, reckoned from the centre of the Earth (or the centre of the celestial sphere, which is the same thing); it therefore corresponds to terrestrial latitude. Right ascension is measured eastwards from the First Point of Aries; it is measured in hours and corresponds roughly to terrestrial longitude.

Ancient peoples believed the sky to be solid, with the stars fixed on to an invisible crystal sphere. This is a convenient fiction, so let's assume that the celestial sphere really exists, making one revolution round the Earth in a few minutes under 24 hours and carrying all the celestial bodies around with it.

The north pole of the sky is simply the point on the celestial sphere which lies in the direction of the Earth's axis; it is marked within one degree by the second-magnitude star Polaris (α Ursae Minoris). Of course there is also a south celestial pole, but unfortunately there is no bright star anywhere near it, and we have to make do with the obscure σ Octantis, which is none too easy to see with the naked eye even under good conditions.

Just as the Earth's equator divides the world into two hemispheres, so the celestial equator divides the sky into two hemispheres – north and south. The celestial equator is defined as the projection of the Earth's equator on to the celestial sphere, as shown in the diagram on the left.

To define a position on Earth, we need to know its latitude, and its longitude. Latitude is the angular distance north or south of the equator, as measured from the centre of the globe; for example, the latitude of London is approximately 52°N and that of Sydney 34°S. The latitude of the north pole is 90°N, that of the south pole 90°S. The celestial equivalent of latitude is known as declination, and is reckoned in precisely the same way; thus the declination of Betelgeux (α Orionis) is 7° 24′N (or +7° 24′), and that of Sirius (α Canis Majoris) 16° 43′S (or +16° 43′). Northern values are also given as + or positive and southern values as – or negative.

All this is quite straightforward, but when we consider the celestial equivalent of longitude, matters are less simple. On Earth, longitude is defined as the angular distance of the site east or west of a particular scientific instrument, the Airy Transit Circle, in the Royal Observatory, Greenwich. Greenwich was selected as the zero for longitude over a century ago, when international agreement was much easier to obtain; there were very few dissenters apart from France.

We need a celestial equivalent of Greenwich, and there is only one obvious candidate: the northern vernal equinox, or First Point of Aries. To explain this, it is necessary to say something about the way in which the Sun seems to move across the sky.

Because the Earth goes round the Sun in a period of one year (just over 365 days), the Sun appears to travel right round against the background of the sky in the same period. The apparent yearly path of the Sun against the stars is known as the ecliptic, and passes through the twelve constellations of the Zodiac (plus a small part of a thirteenth constellation, Ophiuchus, the Serpent-bearer). The Earth's equator is tilted to the orbital plane by 23½°, and so the angle between the ecliptic and the celestial equator is also 23½°. Each year, the Sun crosses the equator twice. On or about 22 March – the date is not quite constant, owing to the vagaries of our calendar – the Sun reaches the equator, travelling from south to north; its declination is then 0°, and it has reached the northern vernal equinox or First Point of Aries, which is again unmarked by any bright star. The Sun then spends six months in the northern hemisphere of the sky. On

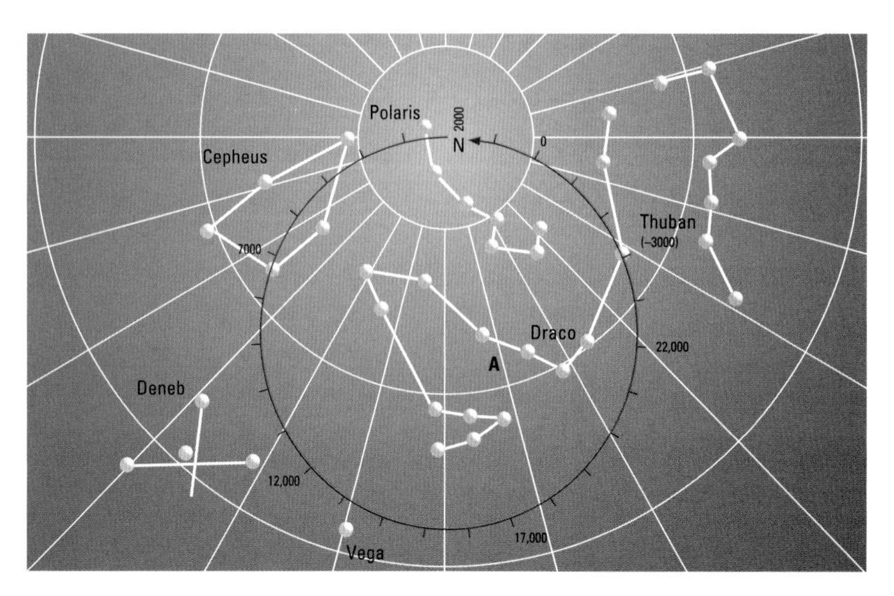

◀ **Precession.** The precession circle, 47° in diameter, showing the shift in position of the north celestial pole around the pole of the ecliptic (A). In Egyptian times (c. 3000 BC) the polar point lay near Thuban in Draco; it is now near Polaris in Ursa Minor (declination +89° 15′); in AD 12,000 it will be near Vega. The south celestial pole describes an analogous precession circle.

or about 22 September it again reaches the equator, this time moving from north to south; it has reached the northern autumnal equinox or First Point of Libra, and spends the next six months in the southern hemisphere. (Of course, for the southern hemisphere the Sun crossing the First Point of Aries in March heralds autumn and its crossing the First Point of Libra in September heralds spring.)

The celestial equivalent of longitude is termed right ascension. Rather confusingly, it is measured not in degrees, but in units of time. As the Earth spins, each point in the sky must reach its highest point above the horizon once every 24 hours; this is termed culmination. The right ascension of a star is simply the time which elapses between the culmination of the First Point of Aries, and the culmination of the star. Betelgeux culminates 5 hours 53 minutes after the First Point has done so; therefore its right ascension is 5h 53m.

Oddly enough, the First Point is no longer in the constellation of Aries (the Ram); it has moved into the adjacent constellation of Pisces (the Fish). This is because of the phenomenon of precession. The Earth is not a perfect sphere; the equator bulges out slightly. The Sun and Moon pull on this bulge, and the effect is to make the Earth's axis wobble slightly in the manner of a gyroscope which is running down and has started to topple. But whereas a gyroscope swings round in a few seconds, the Earth's axis takes 25,800 years to describe a small circle in the sky. Thousands of years ago, when the Egyptians were building their pyramids, the north celestial pole lay not close to Polaris, but near a much fainter star, Thuban (α Draconis); in 12,000 years' time we will have a really brilliant pole star, Vega (α Lyrae).

For the moment, let us suppose that Polaris, our present pole star, lies exactly at the pole instead of being rather less than one degree away (its exact declination is +89° 15′ 51″). To an observer standing at the North Pole of the Earth, Polaris will have an altitude of 90°; in other words it will lie at the zenith or overhead point. From the Earth's equator the altitude of Polaris will be 0°; it will be on the horizon, while from southern latitudes it will never be seen at all. When observed from the northern hemisphere, the altitude of Polaris is always the same as the latitude of the observer. Thus from London, latitude 52°N, Polaris will be 51° above the horizon. From Sydney, the altitude of σ Octantis will be 34°.

A star which never sets, but merely goes round and round the pole without dipping below the horizon, is said to be circumpolar. To decide which stars are circumpolar and which are not, simply subtract the latitude of the observing site from 90. In the case of London, 90 − 51 = 39; it follows that any star north of declination +39° will never set, and any star south of declination −39° will never rise. Thus constellations such as Ursa Major (the Great Bear) and Cassiopeia are circumpolar from anywhere in the British Isles, but not from the southern Mediterranean.

Another useful example concerns Crux (the Southern Cross), which is as familiar to Australians and New Zealanders as Ursa Major is to Britons. The declination of Acrux (α Crucis), the brightest star in the constellation, is −63°. 90 − 63 = 27, so Acrux can never be seen from any part of Europe, though it does rise in Hawaii, where the latitude is 20°N.

Incidentally, it was this sort of calculation which gave an early proof that the Earth is round. Canopus (α Carinae), the second brightest star in the sky, has a declination of −53°; therefore it can be seen from Alexandria (at 31°N) but not from Athens (at 38°N), where it grazes the horizon. The Greeks knew this, and realized that such a situation could arise only if the Earth is a globe rather than a flat plane. From Wellington, in New Zealand, Canopus is circumpolar, so it can always be seen whenever the sky there is sufficiently dark and clear. Canopus, though it is not nearly so bright as Sirius, is in fact a far more powerful star than Sirius. Sirius appears brighter because is is only 8.5 light years away – making it one of our nearest stellar neighbours.

▼ **Star trails over the William Herschel Telescope.** Imaged by Nik Szymanek.

The Lives of the Stars

When we try to work out the life history of a star, we are faced with the difficulty that we cannot – usually – see a star change its condition; just as on a city street we cannot see a boy change into a man. All we can do is to estimate stars are young and which are old, after which we can do our best to trace the sequence of events. Earlier theorists, through no fault of their own, picked wrongly. The mistake lay in faulty interpretation of the HR Diagram.

We now understand the sequence of events better and according to current theory, a star begins by condensing out of the tenuous material in a nebula; the Orion Nebula is a typical stellar nursery. In an area of above average density, gravity causes clumps, which start to shrink; this means that the temperature at the middle of the clump will rise. The temperature becomes higher and higher; the mass of material becomes what is called a protostar. What happens after this depends upon the initial mass.

Remember, a star such as the Sun shines by nuclear reactions – in normal stars, the conversion of hydrogen into helium. To trigger this off requires a temperature of about 10,000,000°C. A very low-mass protostar cannot become as hot as this, and so nuclear reactions never begin; the star simply shines feebly for an immensely long period until its energy is used up. A star of this kind is known as a brown dwarf – a misleading name because the actual colour is purple. It remains dimly luminous until it becomes a cold, dead black dwarf.

If the mass is between 0.1 and 8 times that of the Sun, the story is very different. Nuclear reactions begin, and the star

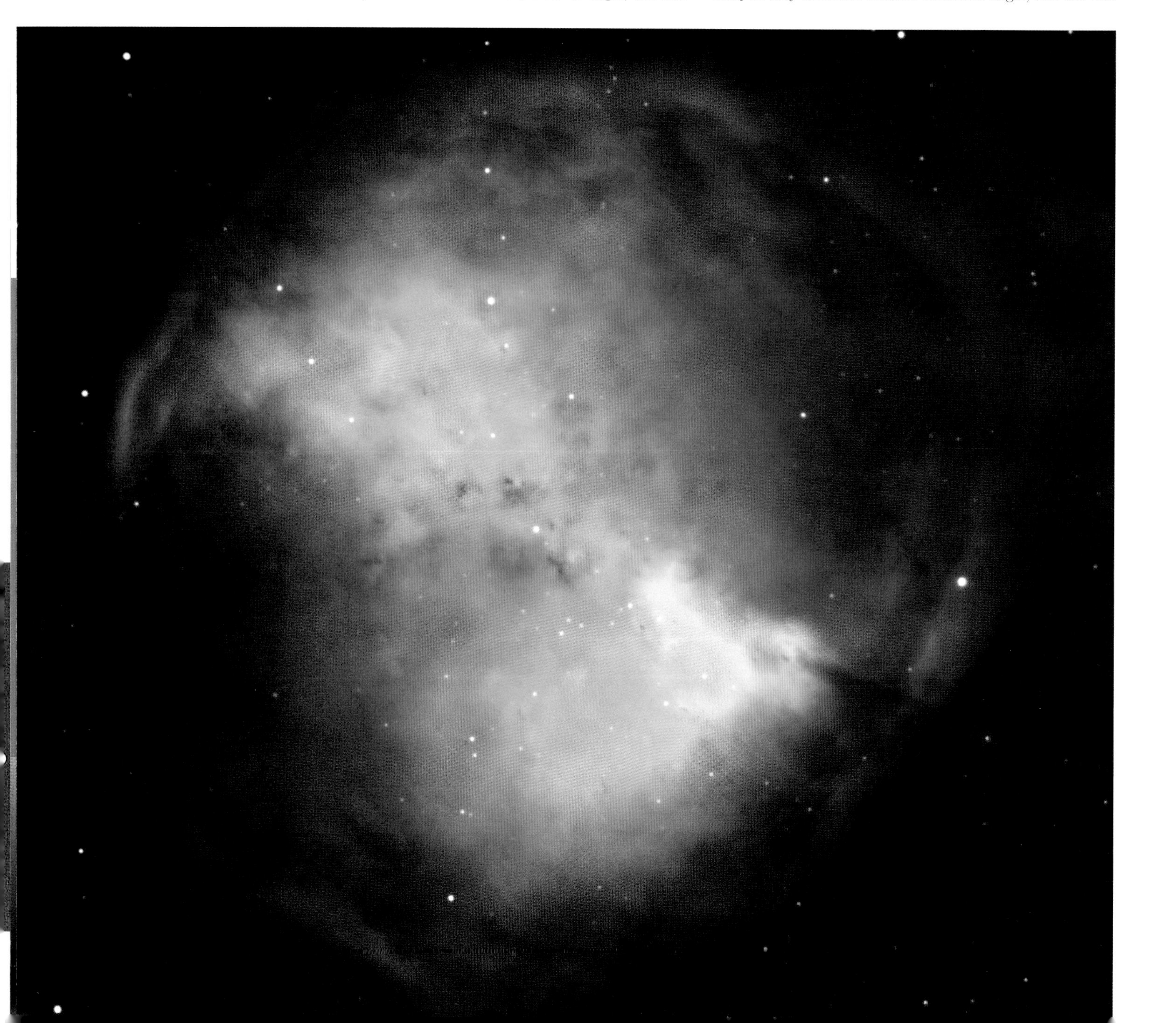

▼ M27 (NGC 6853). A planetary nebula in Vulpecula always known as the Dumbbell Nebula. It is just under 1000 light years away. This image was taken in October 1998 with Antu, the first unit of the VLT (Very Large Telescope) at Cerro Paranal in Chile. The Antu mirror is 8.2 m (323 in) in diameter.

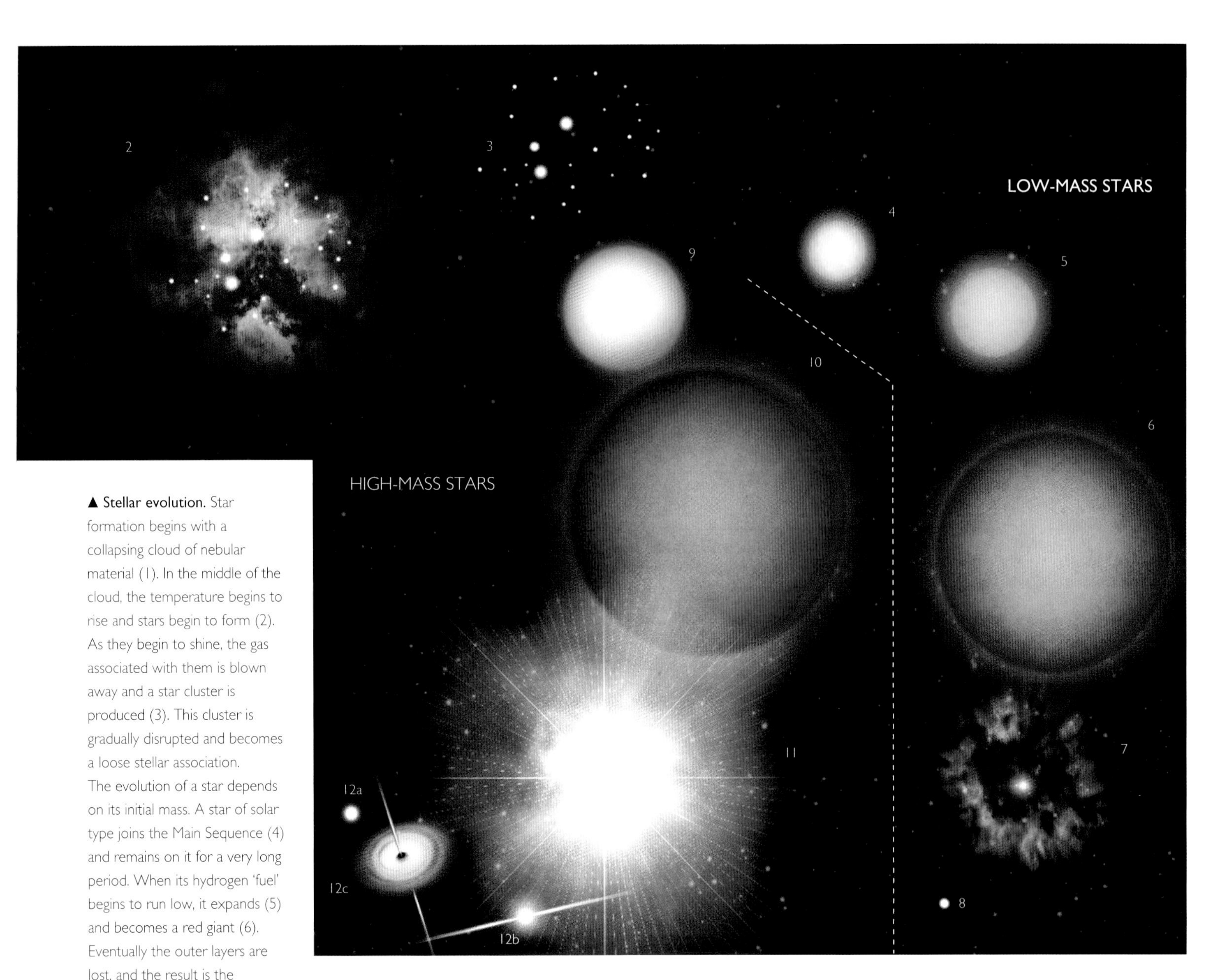

▲ Stellar evolution. Star formation begins with a collapsing cloud of nebular material (1). In the middle of the cloud, the temperature begins to rise and stars begin to form (2). As they begin to shine, the gas associated with them is blown away and a star cluster is produced (3). This cluster is gradually disrupted and becomes a loose stellar association. The evolution of a star depends on its initial mass. A star of solar type joins the Main Sequence (4) and remains on it for a very long period. When its hydrogen 'fuel' begins to run low, it expands (5) and becomes a red giant (6). Eventually the outer layers are lost, and the result is the formation of a planetary nebula (7). The 'shell' of gas expands and is finally dissipated, leaving the core of the old star as a white dwarf (8). The white dwarf continues to shine feebly for an immense period before losing the last of its heat and becoming a cold, dead black dwarf. With a more massive star (more than 8 times the mass of the Sun) the sequence of events is much more rapid. After its main sequence period (9) the star becomes a red supergiant (10) and may explode as a supernova (11). It may end as a neutron star (12a), a particular class of neutron star, a pulsar (12b) or if its mass is even greater, a black hole (12c).

goes on shrinking, sending out a strong stellar wind – made up of streams of particles – and fluctuating irregularly. Eventually it blows away its original cocoon of dust. This is the so-called T Tauri stage, which with the Sun may have lasted for 30,000,000 years. Once nuclear reactions begin, the star settles down to a long period of stability on the Main sequence; it will stay there for around 10,000 million years, so that our Sun is now about half-way through its Main Sequence stage. At first it was only about 70 per cent as luminous as it is now, but it heated up until the core temperature had reached 15,000,000°C. This is the Sun we know today.

When the supply of available hydrogen runs low the star is forced to change its structure. Helium in the core starts to react, producing carbon; around this core is a shell where hydrogen is still producing energy. The star becomes unstable, and the outer layers expand, cooling as they do so. The star becomes a red giant.

From this point onward the story becomes more complicated, but the end result is definite enough. The star's outer layers are puffed off altogether, producing what is called a planetary nebula – not a good name, because the object is not a true nebula and has absolutely nothing to do with a planet. Finally, when the outer layers have been lost, we are left with a white dwarf star; thus simply the original core, but now 'degenerate', so the atoms are broken up and packed closely together with almost no waste of space. The density is amazingly high; if a spoonful of white dwarf material could be brought to Earth, it would weigh as much as a steam-roller. The best-known white dwarf is the dim companion of Sirius. It is no larger than the Earth, but is as massive as the Sun.

Bankrupt though it is, a white dwarf still has a high surface temperature when it is first formed – up to 100,000°C in some cases, and it continues to radiate. Gradually it fades, and must end up as a cold, dead black dwarf, but at the moment no white dwarf with a temperature much below 3000°C has been found, and it may be that the universe is not yet old enough for any black dwarfs to have evolved.

With stars of a greater initial mass, i.e. more than 8 tmes the mass our our Sun, everything happens at an accelerated rate. The core temperatures become so high that new reactions occur, producing heavier and heavier elements. Finally the core is made up principally of iron, which cannot 'burn' in the same way. There is a sudden collapse, followed by an explosion in which most of the star's material is blown away in what is called a supernova outburst, leaving only a very small super-dense core made up of neutrons – so dense that a pin's head of its material would outweigh an ocean liner. A teaspoonful of it would weigh a thousand million tons. If the star's mass is even greater, the core of the original star is compressed by the supernova outburst beyond that of a helium star and eventually the gravitational pull will be so strong that not even light can escape from the wreck of the star. The outcome – a black hole.

I will have more to say about these bizarre objects shortly. At least we know that our Sun can never become a supernova or a black hole, even though it will eventually destroy us.

Variable Stars

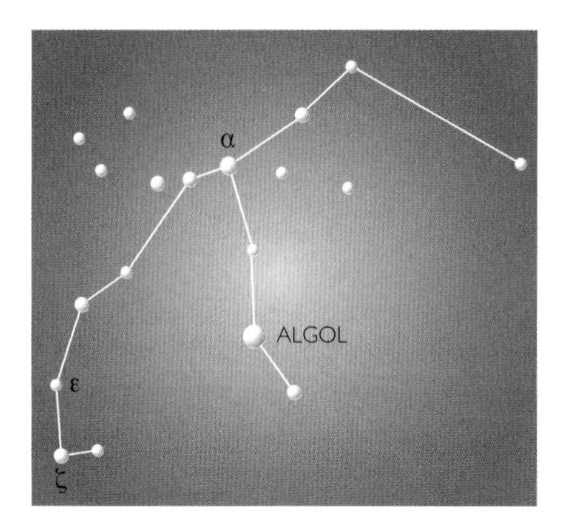

▲ **Perseus** Prominent northern constellation, brightest star α (Mirphak) (mag. 1.8). At maximum Algol is 2.1. Comparison stars are ε and ζ (each 2.9).

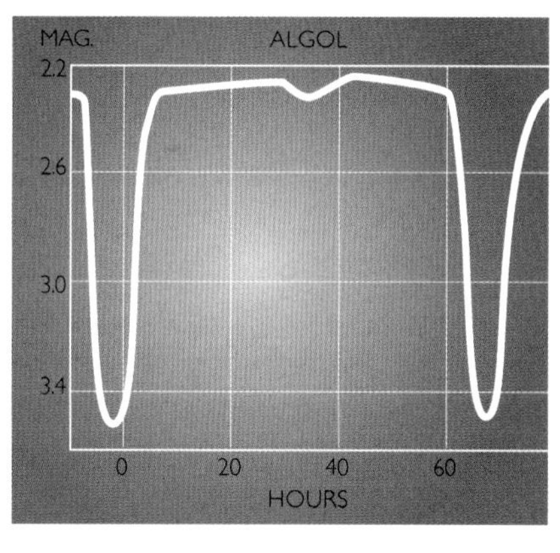

▲ **Algol** (β Persei). Range 2.1–3.4, period 2.87 days; period absolutely regular. The secondary minimum is too slight to be noticed with the naked eye.

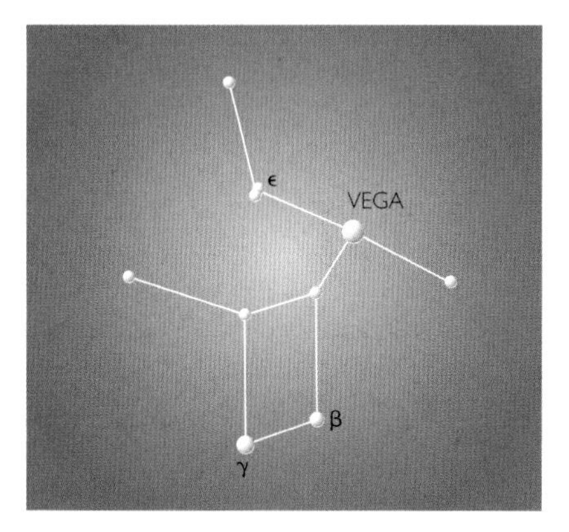

▲ **Lyra** Dominated by the brilliant blue star Vega (mag. 0.0). γ (3.2) is a convenient comparison star for the variable eclipsing binary β.

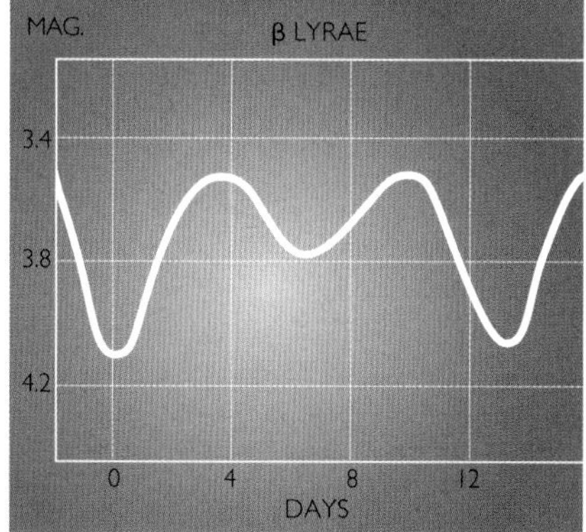

▲ **Sheliak** Prototype of β Lyrae type of eclipsing binary; two minima, unequal in depth. The period is absolutely regular at 12.9 days.

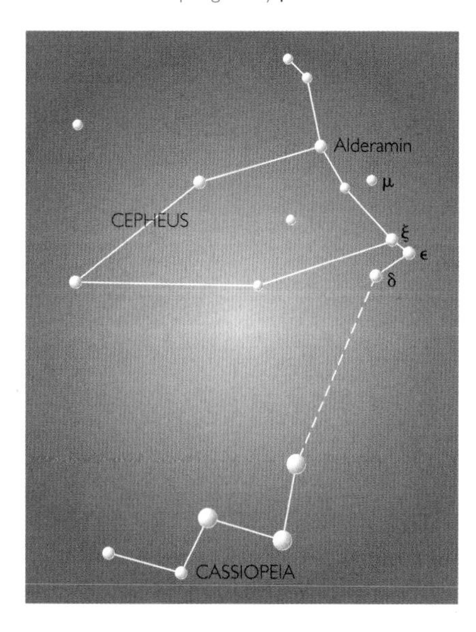

▲ **Cepheus** The brightest star is Alderamin (α, mag. 2.4). Good comparisons for the variable δ are ζ (3.3) and ε (4.2). μ is a very red variable, range 3.4–5.1.

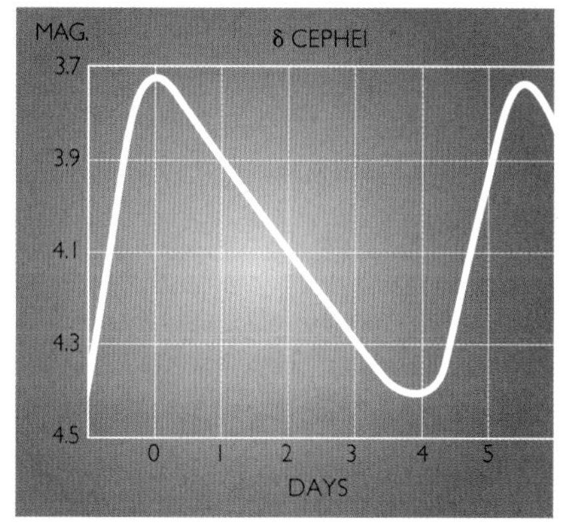

▲ **Delta Cephei** (δ) makes a small triangle with ζ and ε; period absolutely regular, at 5.37 days range 3.5–4.5. The rise to maximum is always steeper than the subsequent fall to minimum.

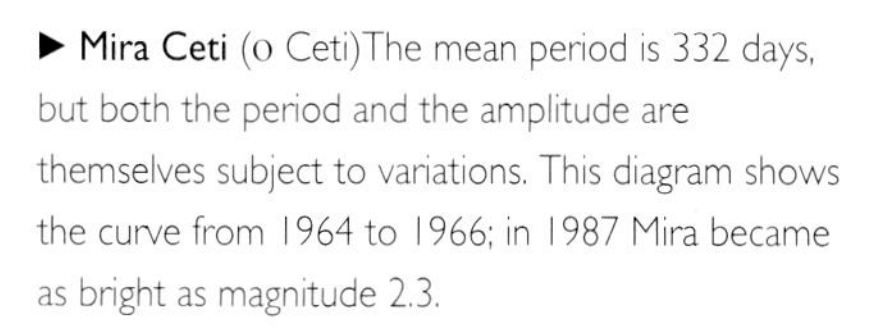

► **Mira Ceti** (o Ceti)The mean period is 332 days, but both the period and the amplitude are themselves subject to variations. This diagram shows the curve from 1964 to 1966; in 1987 Mira became as bright as magnitude 2.3.

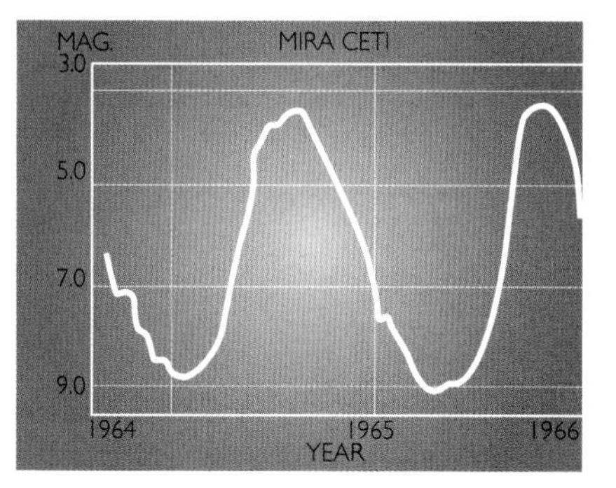

Most stars shine more or less steadily for year after year, century after century. True, there are usually small fluctuations, and our Sun is no exception – hence the alternate spells of warming and cooling – but the true variable stars change much more obviously, and are common enough in the Galaxy.

They are of various types. First there are eclipsing binaries, which are not actually variable at all even though they appear to brighten and fade. The prototype is Algol (β Persei), which normally shines just below second magnitude, but every 2.9 days it begins to fade, dropping to magnitude 3.4 in just over four hours. It remains at minimum for a mere twenty minutes, after which it brightens up again. The changes are very easy to follow with the naked eye.

Algol is a binary, whose components move round their common centre of mass. Algol A is a white star, 100 times as luminous as the Sun; the secondary, Algol B, is a K-type subgiant, larger than A but dimmer and less massive. When B passes in front of A part of A's light is blocked, and the magnitude falls; when A passes in front of B there is a much shallower minimum. The eclipses are not total, and if Algol could be observed from a different vantage point there would be no variation at all.

Incidentally, we have here a good example of what is called mass transfer. The K-type component was originally the more massive of the two, so that it left the Main Sequence earlier and swelled out. As it did so, its gravitational grip on its outer layers was weakened, and material was 'captured' by the companion, which is now the senior partner. The process is still going on, and material continues to stream from B to A.

Other naked-eye eclipsing pairs are λ Tauri (Map 17) and δ Librae (Map 6). Sheliak (β Lyrae), near the brilliant Vega (Map 8) is different, because the two components are less unequal, and there are alternate deep and shallow minima; a full cycle lasts for almost thirteen days. The two components are almost intact, and each must be drawn out into the shape of an egg, though of course no optical telescope will separate them. Some eclipsing pairs have much longer periods – 27 years in the case of Almaaz (ε Aurigae), one of a triangle of naked-eye stars close beside Capella.

Pulsating stars are intrinsically variable. To astronomers, the most important are the Cepheids, named after the prototype star, δ Cephei (Map 3), in the far north of the sky. They are yellow supergiants well advanced in their evolution, and they have become unstable, swelling and shrinking. The period of pulsation is the time needed for a vibration to travel from the star's surface to the deep interior and back again, so that large, very luminous stars have longer periods than stars which are smaller and less powerful. The light-curves are regular, and the cycles repeat themselves, so we always know how bright a Cepheid will be at any particular moment. This is evident from the light-curve given here.

There is a link between a Cepheid's period and its luminosity. This means that as soon as we measure the period, we know how powerful the star is, and this in turn gives us the distance, so that Cepheids are invaluable as 'standard candles'. They can be seen over a vast range, and can be detected in other galaxies. In fact, it was the discovery of Cepheids in the so-called 'starry nebulae', such as the Andromeda Spiral, which proved that we were dealing with separate galaxies rather than minor features of the Milky Way. Other short-period variables are the W Virginis stars, which are similar to Cepheids but less luminous, and the RR Lyrae stars, all of which are about 50 times as luminous as the Sun and are therefore also useful as 'standard candles'.

Next we find the longer-period Mira stars, named after Mira (o Ceti), the best-known example. Unlike Cepheids, the Mira stars – all of which are red giants or supergiants – are not perfectly regular either in period or amplitude. Mira itself (Map 15) has a mean period of 332 days, but this may vary by a few days either way. At some maxima the magnitude may rise to 2 while at other maxima the star never exceeds magnitude 4; at minimum Mira descends to tenth magnitude, and is a naked-eye object for only a few weeks in each year. In general the amplitudes are large, over ten magnitudes in the case of χ Cygni in the Swan (Map 8), and there is no Cepheid-type period-luminosity law. Quite a number of Mira stars reach naked-eye visibility at maximum, but at minimum most of them fall below binocular range.

Semi-regular variables are not unlike Mira stars, but have smaller amplitudes, and their periods are very rough. The brightest of them is Betelgeux (α Orionis, Map 10), which has been known to equal Rigel (β Orionis), but is usually no brighter than Procyon (α Canis Minoris). The variations are slow, and the official period of around 5 years is by no means well marked. Betelgeux is huge, with a diameter of over 1,500 million kilometres (950 million miles) – its outer very rarefied layers would extend out to the orbit of Jupiter.

Telescopic variables abound, and again there are many different types. R Coronae Borealis, in the Northern Crown (Map 4) is usually on the fringe of the naked-eye visibility, but at unpredictable intervals accumulates clouds of soot in its atmosphere, and becomes very dim until the soot is blown away. Use binoculars to look at the bowl of constellation, and you will probably see two stars of around magnitude 6. If only one is on view, you may be sure that R Coronae Borealis is going through minimum.

There are so many variable stars that professional astronomers cannot hope to keep track of them all, and amateurs can do very valuable work. The classic method has been to estimate the brightness of the variable by comparing it with nearby stars which do not change. This method can be surprisingly accurate, but amateurs of today in general use photoelectric photometers, which are more precise. In any case, variable star observation is one of the most popular and most important branches of amateur astronomy in the twenty-first century.

BASIC CLASSIFICATION OF VARIABLE STARS

Symbol	Type	Example	Notes
EA	Algol	Algol	Periods 0.2d–27y. Maximum for most of the time.
EB	Beta Lyrae	β Lyrae	Periods over 1 day. Less unequal components. Continuous variation.
EW	W Ursae Majoris	W UMa	Dwarfs; periods usually less than 1 day.
PULSATING			
M	Mira	Mira	Long-period red giants. Periods 80–1000 days. Periods and amplitudes vary from cycle to cycle.
SR	Semi-regular	η Geminorum	Red giants. Periods and amplitudes very rough.
RV	RV Tauri	R Scuti	Red supergiants; alternate deep and shallow minima. Marked irregularities.
CEP	Cepheids	δ Cephei	Regular; periods 1–135 days; spectra F to K.
CW	W Virginis	κ Pavonis	Population II Cepheids.
RR	RR Lyrae	RR Lyrae	Regular; short periods, 0.2–1.2 days; all of equal luminosity.
ERUPTIVE			
GCAS	γ Cassiopeiae	γ Cassiopeiae	Shell stars: rapidly rotating; small amplitudes.
IT	T Tauri	T Tauri	Very young, irregularly varying stars.
RCB	R Coronae Borealis	RCrB	Unpredictable deep minima. Large amplitude. Highly luminous.
SDOR	S Doradûs	S Doradûs	Very luminous supergiants with expanding shells.
CATACLYSMIC			
UG	U Geminorum	SS Cygni	Dwarf novae
UG2	Z Camelopardalis	Z Cam.	Dwarf novae with occasional standstill.
N	Novae	DQ Herculis	Violent outburst.
SN	Supernovae	B Cassiopeiae	Extremely violent outburst. Type Ia; destruction of the white dwarf component of a binary system. Type II; collapse of a supergiant star.

◀ **Cassiopeia and Cepheus** This area of the sky contains many variable stars, including one of the most famous of all, δ Cephei; also shown is the red variable μ Cephei, known as the 'Garnet Star'. In Cassiopeia, the central star of the 'W' pattern, γ Cassiopeiae, is an irregular variable which suffers occasional outbursts. There are many fainter variables in this picture.

Black Holes

In many ways, stellar evolution is now reasonably well understood. We know how stars are born, and how they create their energy; we know how they die – some with a whimper, others with a very pronounced bang. But when we come to consider stars of really enormous mass, that there are still some details about which we are far from clear.

Consider a very massive star (at least 40 solar masses) at the end of its life. When the star's energy runs out, gravity will take over, and it will start to collapse. The process is remarkably rapid and whether there is an outburst or not – the star simply goes on becoming smaller and smaller, denser and denser. As it does so, the escape velocity rises, and there comes a time when the escape velocity reaches 300,000 kilometres (186,000 miles) per second. This is the speed of light, so that not even light can escape from the shrunken star – and if light cannot do so, then certainly nothing else can, because light is as fast as anything in the universe. The old star has surrounded itself with a 'forbidden area' from which absolutely nothing can escape. It has created a black hole. If the collapse is very rapid, it might swallow the entire star before the supernova blast can reach the surface.

For obvious reasons, we cannot see a black hole – it emits no radiation at all. Therefore, our only hope of locating such an object is by detecting its effects upon something which we can see. A typical example is Cygnus X-1, so called because it is a powerful X-ray source; it lies near the star η Cygni (Map 8). The system consists of a ninth-magnitude B-type supergiant, HDE 226868. It seems to have about thirty times the mass of the Sun, with a diameter of perhaps 18 million kilometres (11.25 million miles); it is associated with an invisible secondary with 14 times the Sun's mass. The orbital period is 5.6 days as we can tell from the behaviour of the supergiant; the distance from us is about 8000 light years. What seems to be happening is that the black hole is pulling material away from the supergiant, and swallowing it up. Before this material disappears, it is whirled around the supergiant, and is so intensely heated that it gives off the X-ray radiation which we can pick up.

▶ **The Vela supernova remnant** from the 3.9-m (153-inch) Anglo-Australian telescope. The red, glowing filaments are caused by hydrogen. The Vela pulsar was the second to be identified optically; its magnitude is 24, one of the faintest objects ever observed. Its rate of slowing down indicates that the supernova outburst occurred about 11,000 years ago. The green line across the photograph is the path of a satellite that traversed the field of view during the exposure of the green-sensitive plate.

The size of a black hole depends upon the mass of the collapsed star. The critical radius of a non-rotating black hole is called the Schwarzschild radius, after the German astronomer – Karl Schwarzschild – who investigated the problem mathematically as long ago as 1916; the boundary around the collapsed star having this radius is termed the 'event horizon'. Once any material passes over the event horizon, it is forever cut off from the rest of the universe. For a body of the mass of the Sun, the Schwarzschild radius would be about 3 kilometres (1.9 miles), while for a body of the mass of the Earth the value would be less than a single centimetre (less than half an inch).

There seems no doubt that black holes exist in the cores of many galaxies (including ours) and, for example, the movements of stars near the core of the giant elliptical galaxy M87 show that a supermassive black hole must lurk there. Whether all galaxies have central black holes is still being debated.

What are conditions like inside a black hole? Frankly we are reduced to guess-work, because beyond the event horizon all the laws of science seem to break down. Does the old, collapsed star crush itself out of existence altogether – if not, what happens to it? All manner of exotic theories have been proposed, but we have to admit that our present-day knowledge is painfully scanty. Stephen Hawking has suggested that a black hole may not even be permanent, but will gradually lose energy and finally evaporate, but this too is little more than speculation.

At least we have the satisfaction of knowing that our Sun will never produce a black hole. It is not nearly massive enough, and it will end its career much more gently, passing through the white dwarf stage and coming to its final state as a cold, dead globe.

▲ **NGC7742,** a spiral galaxy imaged in 1998 by the Hubble Space Telescope. This is an active Seyfert galaxy, probably powered by a black hole at its core. The core of NGC7742 is the large yellow 'yolk' in the centre of the image. The thick, lumpy ring round this core is an area of active star birth. The ring is about 3000 light years from the core. Tightly wound spiral arms are visible. Surrounding the inner ring is a wispy band of material, probably the remains of a once very active stellar breeding area.

◄ **Impression of a black hole** (Paul Doherty). Material can be drawn into the black hole, but nothing – absolutely nothing – can escape. Therefore, a black hole can only be detected by its effects upon objects which we can record.

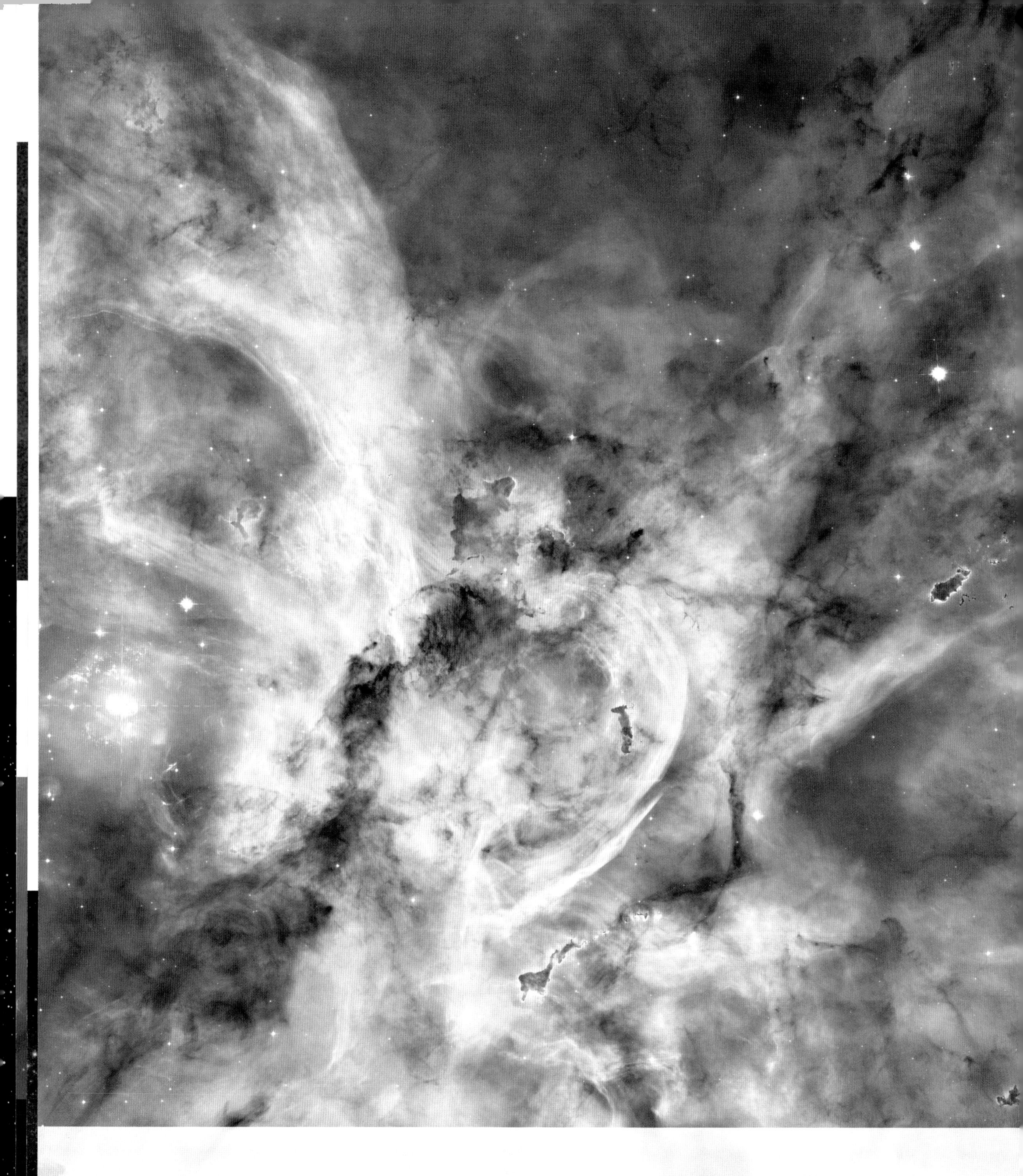

Star formation in the region of the Eta Carinae Nebula This immense nebula, where star formation is in progress at a furious rate, contains stars at least 100 times the mass of the Sun; Eta Carinae itself is at far left. The image, taken with the Hubble Space Telescope to celebrate 17 productive years, shows a region 50 light years across.

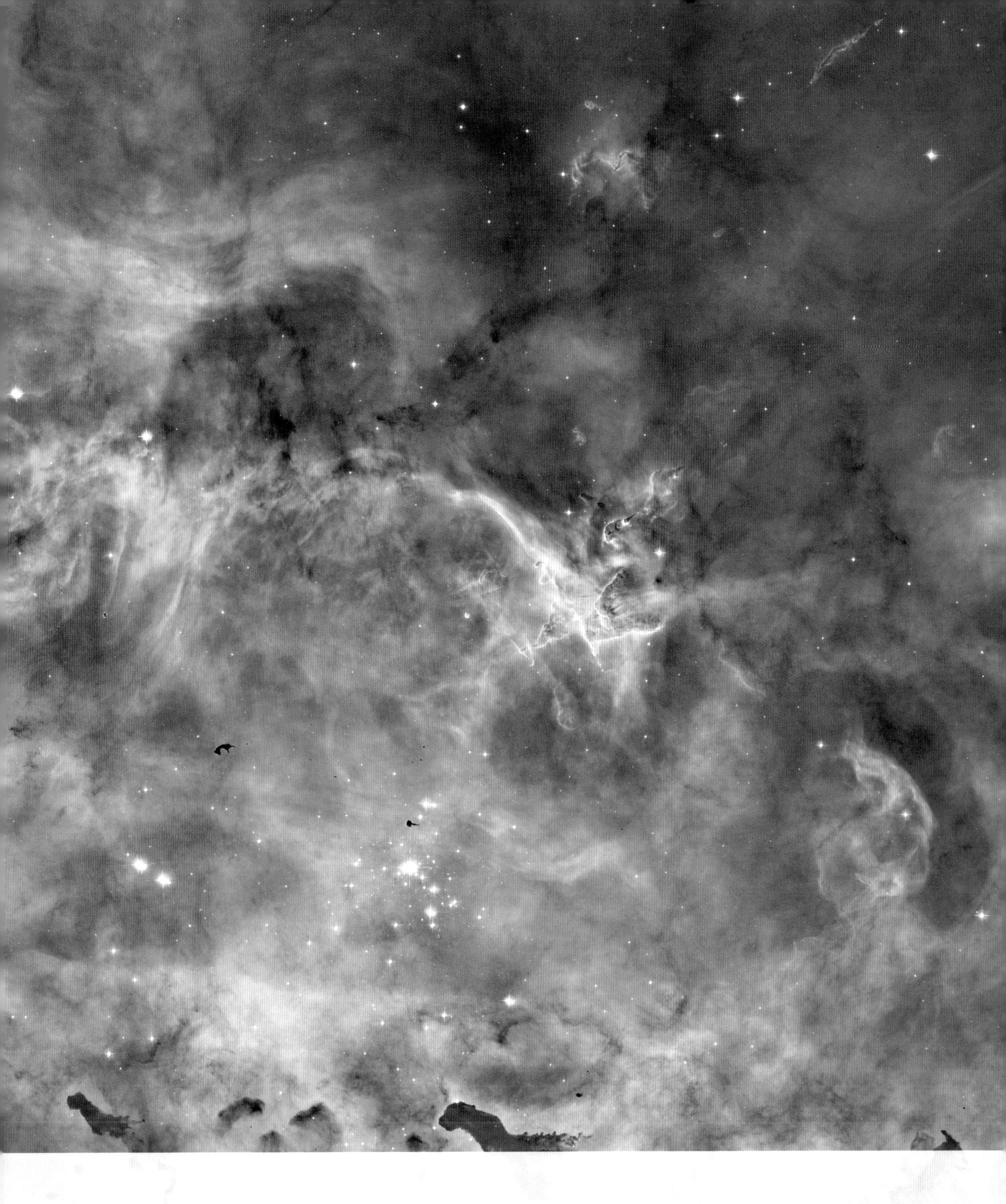

INTO THE UNIVERSE

The Realm of Galaxies

▶ **The Tadpole Galaxy (ARP 188) in Draco;** Hubble Space Telescope image, 2006. The galaxy is 420 million light years away; the 280,000 light-year-long 'tail' has been drawn out by a passing galaxy, which can be seen through the foreground spiral arms at the upper left.

Less than a hundred years ago, our views about the make-up of the universe were very different from those of today. Everything was contained in one huge system, the Milky Way Galaxy; our Sun was a normal star, unusual only in being attended by a system of planets. The universe was certainly very old, but nobody had any real idea of its precise age, or how it had been created.

Astronomically, the most important development of the twentieth century – perhaps the most important since the Copernican Revolution – was the realization that the Milky Way is not unique, but is only one of innumerable galaxies, most of them many millions of light years away. The main credit for this must go to Edwin Hubble, but the story goes back well before his time.

Charles Messier's 1781 catalogue of nebular objects contained not only open and globular clusters, but also 'nebulae' of two distinct types. Some, such as M42 in Orion's Sword, looked like (and were) clouds of dust and gas, but others, notably M31 in Andromeda, gave every impression of being made up of stars; from 1845 Lord Rosse and his assistants showed that some of these, though not all were spiral in form. One vital question had to be answered. Were the starry nebulae minor features of the Milky Way, or were they independent galaxies in their own right? Of America's two leading astronomers, Harlow Shapley believed in the first of these ideas, Hubble in the second.

It was Shapley who, in 1917, became the first man to make a reasonable measurement of the size of the Milky Way system. Of course he knew that the globular clusters lie around the boundary of the main part of the Galaxy, and he located Cepheid variables in them; once had found the periods of these useful stars, he was able to calculate their distances. He also established that the Sun lies nowhere near the centre of the Galaxy; we now know that we are about 25,000 light years away from the galactic nucleus.

The obvious next step was to locate Cepheid variables in the starry nebulae, but this was easier said than done. Fortunately the Hooker 2.5-metre (100-inch) reflector at Mount Wilson had just been completed (in 1917), and was far more powerful than any other telescope in the world at that time. Hubble was able to use it, and in 1923 he detected short-period variables in the Andromeda Galaxy and a few other systems. As soon as he measured their periods, he realized that they were far too remote to belong to the Milky Way. He gave the distance of M31 as 750,000 light years. This was a considerable underestimate, but at least the status of the spirals was no longer in doubt.

Working together with his colleague Milton Humason (who began his career as a mule driver, and ended it as one of the world's most celebrated astronomers), Hubble made another major discovery. It was already known that almost all the 'starry nebulae', now re-christened galaxies, showed red shifts in their spectra, and if these were Doppler effects, as seemed overwhelmingly probable, it followed that the galaxies were receding from us. Moreover, the farther away they were, the faster were their speeds of recession. There was a definite relationship between distance and velocity, now known, fittingly, as Hubble's Law. A slight qualification is needed here. Galaxies collect into clusters or groups; each group is racing away from each other group so the whole universe is expanding. The Milky Way system is a member of what we call the Local Group, of which the largest member is the Andromeda Galaxy, which is approaching our Galaxy, and will eventually collide with it, but by then the Earth will no longer exist anyway.

At least we now appreciate the scale of the universe, and we have what we believe to be a reasonable idea of its past history, but there is still much that we do not know. In particular, we cannot give a firm answer to the most fundamental question of all: How did the universe come into existence?

▼ **The Tarantula Nebula** (NGC2070) 30 Doradûs, in the Large Magellanic Cloud, imaged with the Kueyen unit telescope of the Very Large Telescope (VLT).

▼ **M31, the Andromeda Galaxy.** Until astronomers could use the behaviour of variable stars to calculate distances to such objects, no one knew whether they were part of our Galaxy or not. The largest member of our Local Group of galaxies, this spiral is, in fact, 2.5 million light years away.

The Big Bang

▼ **The observable universe.** (left to right) As instruments are used to peer farther out into space, they are also looking back farther in time. We see the edge of the Solar System as it was a few hours ago; the light from the nearest stars reach us within a few years; that from the nearest galaxies hundreds of thousands of years and that from other clusters takes millions of years. Astronomers think that the oldest objects we will observe are distant, very early quasars from about 13,200 million years ago, only 500 million years younger than the universe itself.

It used to be thought that the Earth was only a few thousand years old. Today, all lines of investigation show that the Earth is around 4600 million years old and the Sun, 5000 to 6000 million years old. This seems fairly definite, but measuring the age of the universe itself is much more of a problem.

According to most authorities, the universe came into existence at one definite moment in time. This event is always referred to as the Big Bang – a term introduced by Sir Fred Hoyle, who never believed in it! At once the infant universe began to expand, and the fact that this expansion is still going on makes it possible for us to estimate how long ago the Big Bang happened. We depend upon the red shifts in the spectra of the galaxies. A galaxy is made up of stars, and its spectrum is bound to be rather jumbled, but the main dark absorption lines can be identified, and their positions measured. As we have noted, if a line is shifted towards the red or long-wave end of the electromagnetic spectrum, it shows that the light source is receding; a blue shift indicates that the source is approaching. Beyond our Local Group, all galaxies show red shifts. The greater the distance, the greater the amount of red shift.

The red shifts give us the clues we need, because we can work backwards and judge when the explosion occurred; the method may be likened to a film played in reverse. It seems that the Big Bang happened 13,700 million years ago. But... if the universe sprang into existence 13,700 million years ago, what happened before that?

We have two possible answers, neither of which is particularly palatable. To say that nothing happened before the creation of the universe is merely begging the question and solves nothing, but otherwise we have to visualize a period of time which has no beginning, a concept quite beyond our mental capabilities. We are well and truly caught between Scylla and Charybdis. There must have been a beginning – otherwise, you would not now be reading this book – but just how this happened we do not know. Neither do we know whether the beginning was abrupt or gradual.

Of course every culture has its own creation myths. My personal favourite comes from Polynesia: 'A heavenly woman was tossed out of heaven and fell upon a turtle, which became the Earth'. At least this is refreshingly different. In the so-called 'Steady State' theory, the universe has always

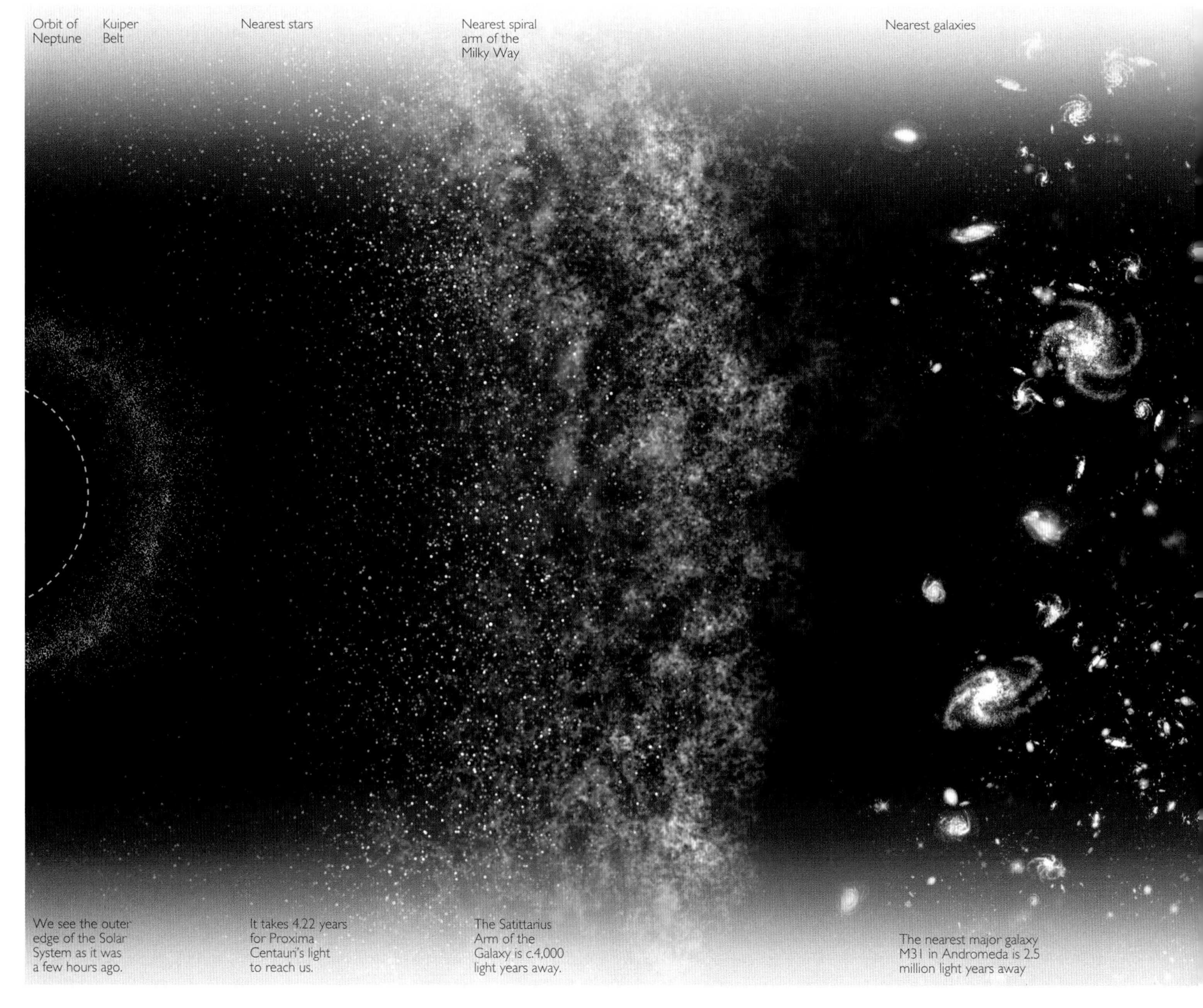

existed; there was no beginning and there will be no end. As old galaxies die, they will be replaced by new ones, produced by matter created spontaneously in space in the form of hydrogen atoms. If we could come back and look at the universe in, say, a million million years' time, we would find the same numbers of galaxies as we see today. The overall aspect would be unchanged.

This can be checked. When looking out into space, we are also looking back in time. For exampled, the Andromeda Galaxy is 2.5 million light years away, so that we see it not as it is now, but as it used to be 2.5 million years ago. If we peer into the depths of the universe, we see objects well over 10,000 million light years away – and 10,000 million years ago the universe was young. In fact, we can have a reliable picture of the youthful universe. According to the Steady-State theory, the distribution and numbers of galaxies should be the same as in the regions closer to us – but this is not so. The young universe is not the same the older universe. The universe is not in a steady state; the evidence cannot be challenged.

We are back to the Big Bang.

Blueshifted, approaching

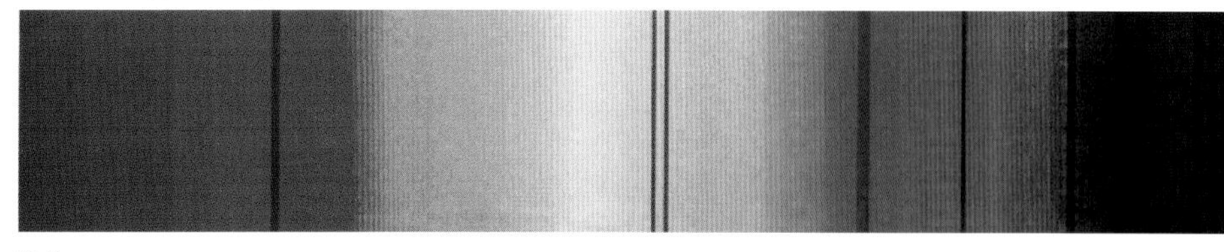

Stationary

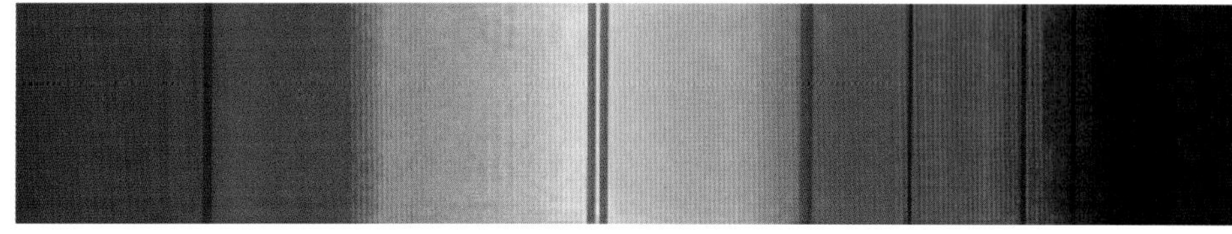

Redshifted, receding

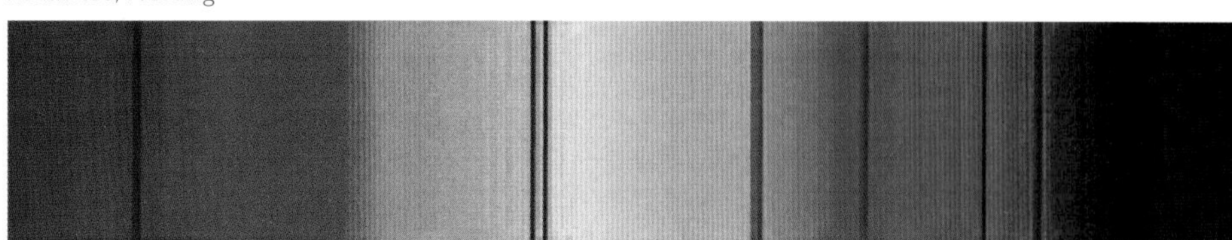

▲ The Doppler Effect
Stellar spectra; the double D line, the signature of sodium, is prominent in the yellow to green part of the spectrum. The central spectrum is the laboratory view; top, the lines are shifted towards the blue end of the spectrum (velocity of approach); bottom, the shift is to the red (velocity of recession).

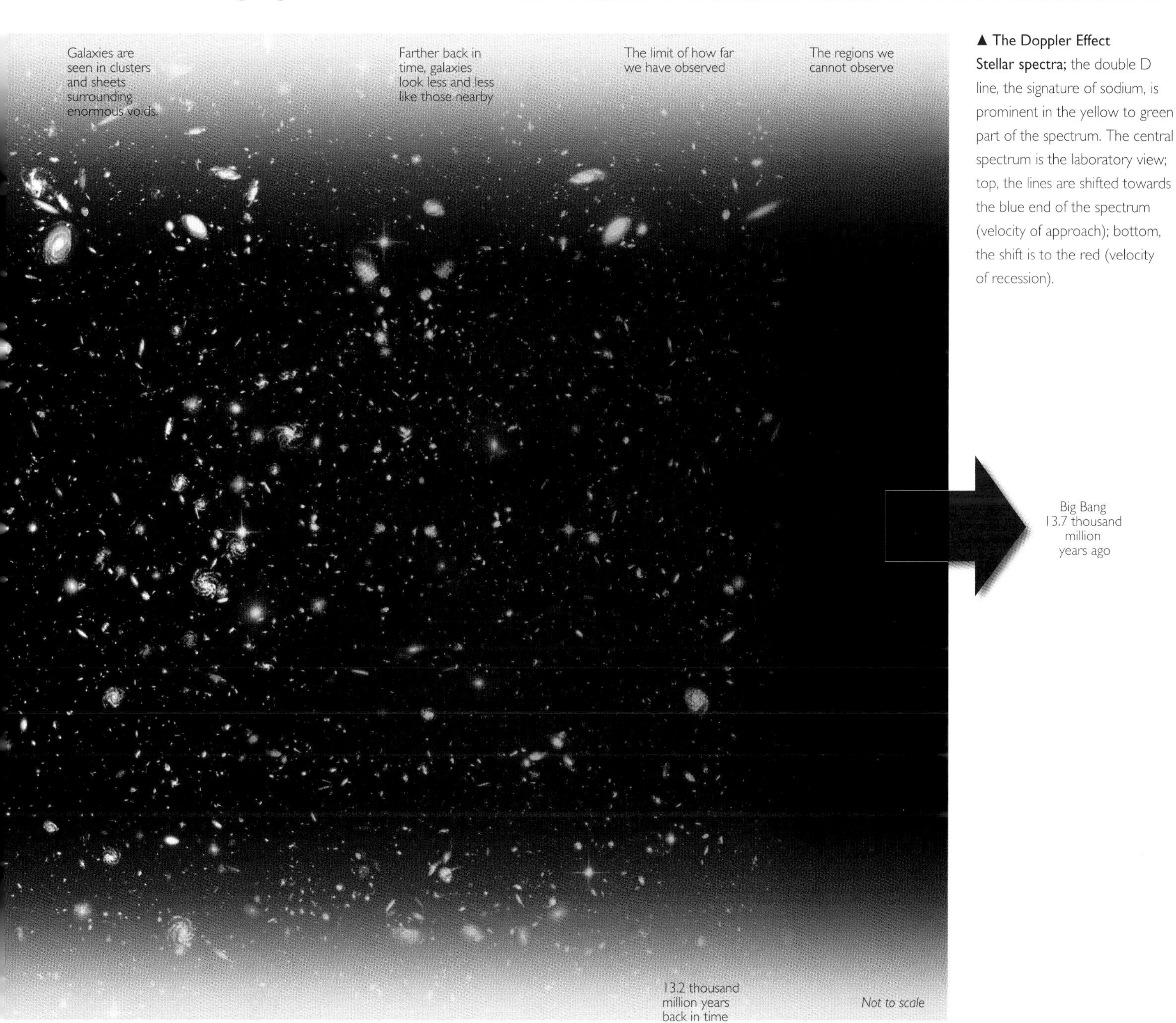

The Local Group

▶ **Barnard's Galaxy,** NGC6822 Named after its discoverer, E. E. Barnard. It is a dwarf irregular system near the star 54 Sagittarii; it is well worth finding, but is not a particularly easy telescopic object.

Galaxies tend to congregate in groups or clusters. Our Milky Way is a member of such cluster, which we call the Local Group. This is made up of three large spirals, in order of size M31 in Andromeda, the Milky Way Galaxy, and M33 in Triangulum; several galaxies of moderate size, notably the two Magellanic Clouds, and more than two dozen dwarf galaxies, some of which are not much more populous than globular clusters. The members of the Local Group are the only galaxies which are not racing away from us; indeed, the Andromeda Galaxy is approaching at a speed of over 300 kilometres (180 miles) per second, and must eventually collide with the Milky Way, though not for several thousands of millions of years yet.

Only a few Local Group systems are visible with the naked eye. Much the brightest are the two Magellanic Clouds, which resemble detached portions of the Milky Way; northern-hemisphere observers never cease to regret that they lie so far south in the sky that they can never be seen from Europe or most parts of the United States. (They were named in honour of Ferdinand Magellan, but they were known long before the Portuguese explorer's voyage round the world; and were recorded by the Persian astronomer Al-Sufi in AD 964.) The Large Magellanic Cloud, straddling the border between the constellations Mensa and Dorado, is 169,000 light years from us, and is about 1/20 the size of our Galaxy. It contains objects of all kinds – stars of every type, open and globular clusters and galactic nebulae; the magnificent Tarantula Nebula (NGC2070, 30 Doradûs), 1000 light years in diameter, is so rich that if it were as close to us as the Orion Nebula, it would cast shadows. The Small Magellanic Cloud, about 190,000 light years away, is not so huge, but even so it contains several hundred million stars. The clouds are 75,000 light years apart, and are generally believed to be satellite galaxies of the Milky Way. This has recently been disputed, but certainly the clouds are linked with each other, and with our Galaxy, by a stream of hydrogen gas.

▼ **M33, also known as the Triangulum Galaxy.** This image was obtained by the National Science Foundation's 0.9-m (35.4-in) telescope on Kitt Peak. The reddish areas are regions of star formation. In 2005, astronomers using the VLBA equipment succeeded in measuring the proper motion of M33. It is not exactly racing along against its background. The annual motion amounts to 30 micro-arc seconds, which is 1/100 the apparent speed of a snail, crawling on the surface of Mars, as observed from the Earth!

Far beyond the Milky Way, the clouds and various dwarf systems, we come to M31, the Andromeda Galaxy. It is dimly seen with the naked eye, and binoculars show it clearly, but even with a telescope it is, frankly, unspectacular. It is a well-formed spiral, with an inconspicuous central 'bar', but it lies at an angle of 77° to us, and its full beauty is lost; if it were face-on, it would be a wonderful sight. Its distance has been disputed; the Hipparcos astrometric satellite, expected to be of pinpoint accuracy, gave a value of 2.9 million light years, but it has now been established that the Hipparcos data are somewhat suspect, and 2.5 million light years may be closer to the truth. The Andromeda Galaxy contains around a million million stars – many more than the Milky Way and objects of all kinds. One supernova has flared up there in what we may call 'telescopic' times, S Andromedae in 1885. It was not well studied, because nobody knew quite what it was. One independent discoverer was a Hungarian baroness who was giving a garden party and had set up a small telescope on her front lawn! (In February 1987, a supernova was observed in the Large Magellanic Cloud, which at 169,000 light years is far closer than the Andromeda Galaxy. The 1987 supernova, SN1987A, reached second magnitude and temporarily altered the aspect of the whole of that part of the sky. The progenitor star could be idenitified; unexpectedly, it proved to be a blue star rather than a red supergiant.)

SELECTED MEMBERS OF THE LOCAL GROUP

Name	Type	Absolute Mag.	Distance 1000 l.y.	Diameter
The Milky Way Galaxy	Barred spiral	−20.5	–	100
Large Cloud of Magellan	Barred spiral	−18.5	169	30
Small Cloud of Magellan	Barred spiral	−16.8	190	16
Ursa Minor dwarf	Dwarf elliptical	−8.8	250	2
Draco dwarf	Dwarf elliptical	−8.6	250	3
Sculptor dwarf	Dwarf elliptical	−11.7	280	5
Fornax dwarf	Dwarf elliptical	−13.6	420	7
Leo I dwarf	Dwarf elliptical	−11.0	750	2
Leo II dwarf	Dwarf elliptical	−9.4	750	3
NGC6822 (Barnard's Galaxy)	Irregular	−15.7	1700	5
IC1613	Irregular	−14.8	2400	8
M31 (Andromeda Galaxy)	Spiral	−21.1	2500	130
M32	Elliptical	−16.4	2500	12
NGC205	Elliptical	−16.4	2500	8
NGC147	Dwarf elliptical	−14.9	2500	2
M33 (Triangulum Galaxy)	Spiral	−18.9	2900	52

The third important member of the Local Group, M33 in Triangulum, is on the fringe of naked-eye visibility; it is easy to pick up with binoculars, but tends to be elusive for a small telescope, because of its low surface brightness; it is a normal spiral, just under 3 million light years away according to the latest estimates. It used to be thought that two other substantial galaxies, Maffei 1 and Maffei 2 (named after their discoverer) belonged to the Local Group, but it has now been established that they lie well beyond.

The Local Group is well defined, but it belongs to a much larger unit, the Virgo Supercluster – often called the Local Supercluster – which includes at least a hundred groups of galaxies, of which the most important member is the exceptionally massive M87 in Virgo, which is around 60 million light years away and is known to have a supermassive black hole in its centre. The more we learn, the more we realize that the distribution of the galaxies in space is more orderly than might be expected.

▼ **The Small Magellanic Cloud** is a system about one-sixth the size of our Galaxy. It is comparatively close, at about 190,000 light years, and is a prominent naked-eye object in the far south of the sky.

The Outer Galaxies

▲ M81, Spiral galaxy in Ursa Major. It and the smaller irregular galaxy M82 lie near the mag. 4.5 star 24 Ursae Majoris. M81 is 8.5 million light years away; it is mag. 7.9. The two galaxies are in the same low-power telescopic field.

Beyond the Local Group we come to galaxies which are millions of light years away, and are racing away from us at tremendous speeds. They are of many kinds. Some are much smaller than our Galaxy, others much larger; some are violently active, others much more quiescent. There are spirals, ellipticals and systems which are quite irregular; we have even located galaxies which appear to have no stars at all, There is an endless variety. Our Galaxy is normal enough, though its size and mass do seem to be slightly above average.

Not surprisingly, the first man to work out a useful classification system was Edwin Hubble, who produced a diagram which, for obvious reasons, is always known as the Tuning Fork. The main types are:

Spirals, from S0 (large nucleus, tightly wound arms) through to Sd (small nucleus, loosely wound arms). M51, the Whirlpool Galaxy, is a perfect example of an S0 system; the Pinwheel Galaxy in Triangulum belongs to Type Sc. Spiral galaxies are spinning round, and in almost every case they rotate clockwise, so the arms trail Yet the situation is not nearly as straightforward as Hubble originally thought. The spiral arms are caused by 'density waves' which sweep round the galaxy's nucleus more slowly than the stars and gas; when gas enters a density wave it is compressed and star formation is triggered. Many of these new stars are blue and highly luminous, so the spiral is very prominent, but on the cosmic timescale they do not last for long. The end result is that although the system will continue to show spiral arms, the individual arms themselves will not be permanent.

Barred spirals. Here the arms issue from the ends of a 'bar' of bright stars running across the centre of the galaxy. Barred spirals range from SBa through to SBd in order of increasing looseness; it is thought that the centre of a bar structure may favour conditions for star formation. In many cases the bars are very obscure, and this is so with our own Galaxy, which is now officially placed in type SBb.

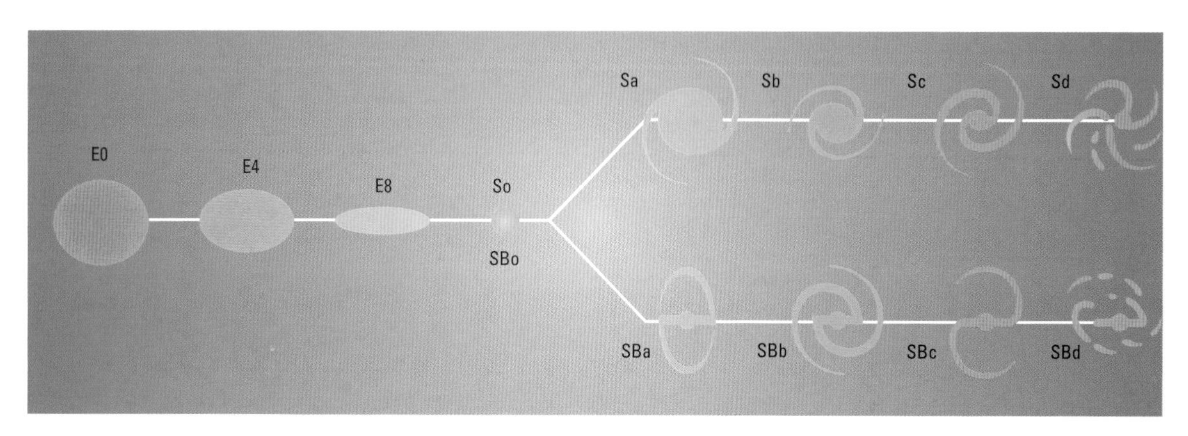

◄ Hubble classification of galaxies. There are elliptical galaxies (E0 to E8), spirals (Sa to Sd) and barred spirals (SBa to SBd), as well as irregular systems which are not shown here. There are many refinements; for instance, Seyfert galaxies (many of which are radio sources) have very bright, condensed nuclei.

Ellipticals, from E0 (virtually spherical) to E8 (very elongated). Unlike the spirals, they have little interstellar gas and dust left, so star formation in them has practically ceased and the brightest stars are old red giants and supergiants which have evolved off the Main Sequence. Giant ellipticals, such as M87 in Virgo, are much more massive than any spiral, and many of them have central black holes – though of course this is also true of many spirals and barred spirals, including our own. There are also many dwarf ellipticals, several of which are satellites of the Milky Way.

Irregulars, which have no definite shape. Many of them are minor and one of these is the Canis Major Dwarf, which was discovered in 2003 and is actually the nearest of all the galaxies – it is 25,000 light years from the Sun, and 42,000 light years from the centre of our Galaxy and seems to be in the process of being gravitationally disrupted. One particularly famous irregular is the 'starburst galaxy' M82 in Ursa Major, which is being strongly affected by the pull of its more massive neighbour, the spiral M81.

It was formerly believed that the Tuning Fork represented a true evolutionary sequence, so that an elliptical would evolve into a spiral or vice versa. This is not now accepted, and it is more likely that much depends upon the initial mass of the system, but we have to admit that our knowledge of the evolution of galaxies is very far from complete. We must take particularly careful note of collisions and mergers; the favoured theory is that large galaxies are built up when groups of smaller systems come together, but though we know much more than we did only a few years ago there still many points which are unclear.

At least there does seen every likelihood that when M31 collides with the Milky Way, as will eventually happen, the beautiful spirals of both systems will be destroyed, and the end product will be a single giant elliptical. We will not be there to see; by that time the Sun will have passed through its red giant stage, and the Galaxy will be a very different place.

▲ **Barred spiral galaxy NGC1300 in Eridanus** is 70 million light years away; this HST picture shows striking detail in the dominant central bar and the majestic arms. The nucleus is about 3000 light years in diameter.

◀ **The Witch's Cauldron Galaxy in Circinus** A Seyfert galaxy 13 million light years away. The intense activity is powered by a central black hole. Seyfert galaxies are part of a larger class of objects known as Active Galactic Nuclei or AGNs.

Active Galactic Nuclei: Quasers

By cosmic standards our Milky Way Galaxy is placid enough, and does not undergo violent outbursts. Other systems are much more energetic – for example, the so-called radio galaxies, which emit very strongly at radio wavelengths. Some of these are systems with extremely bright, variable nuclei; they are known as Seyfert galaxies (after Carl Seyfert, who drew attention to them in 1942). Some starburst galaxies are also very powerful in the radio range. Of course, galaxies are not the only radio sources; supernova remnants are also prominent, including of course the Crab Nebula. At Cambridge, catalogues of these sources were drawn up. Among the objects in their third catalogue was one, No 273 (3C273), which led to a remarkable discovery.

Radio sources did not always correspond to prominent visible objects; brilliant stars such as Sirius and Canopus remained obstinately silent at radio wavelengths. There was also the problem that in those days no radio telescope was capable of pinpointing a source really accurately. For once Nature was helpful. Radio source 3C273 lay in a part of the sky where it could be occulted by the Moon, and this happened on 5 August 1962. Using the powerful radio telescopes at Parkes, in Australia, astronomers timed the exact moment when the radio emissions were cut off. Since the position of the Moon was known, the position of 3C273 could also be found. It coincided with an ordinary-looking bluish star.

▼ **Diagram of an Active Galactic Nucleus.** Material spirals through the accretion disk on to a central black hole, and excess material is shot out in jets along the axis of the accretion disk by the shockwave around the central vortex.

The results were sent to the Palomar Observatory in California, where Maarten Schmidt used the Hale telescope to examine the optical spectrum of the source. The results were astonishing; 3C273 was not a star at all, but something much more dramatic. The spectral red shift indicated a distance of about 2000 million light years, and it followed that the luminosity was much greater than that of an average galaxy, even though the object looked stellar. Other discoveries followed, and it became clear that we were dealing with objects of an entirely new type. At first they were called QSOs (Quasi-Stellar Radio sources), but then it emerged that by no means all QSOs were strong radio emitters, and the name was shortened to 'quasar'.

More than 100,000 quasars are now known. All are very remote; the closest, so far as we know at present (2007) is about 700 million light years away, while the farthest are around 13,000 million light years from us, not far from what we believe to be the boundary of the observable universe. We are seeing them as they used to be when the universe was young, and since none have been found in our part of the cosmos it seems that no quasars have been formed for a very long time indeed.

What exactly were quasars? For a while astronomers did not know, and all kinds of theories were proposed. Some astronomers, such as Sir Fred Hoyle, maintained that the spectral red shifts were misleading, and that quasars were minor features shot out from relatively nearby galaxies; a few cosmologists still believe this, but it has to be said that mod-

▶ **Jet from the centre of the giant elliptical galaxy M87** The jet made up of electrons and other sub-atomic particles travelling at nearly the speed of light, is shown in this Hubble Space Telescope image. The blue jet contrasts with the yellow glow from the combined light of many millions of stars.

ern evidence is against it. Quasars are the cores of very active galaxies, where supermassive black holes are powered by the infall of material. In many cases it has been possible to image the host galaxies in visible light. It may well be that a quasar remains active for only a limited period, and it is possible that many large galaxies go through a quasar period which does not last for very long.

In our sky, the brightest quasar is the first to be identified, 3C273 in Virgo. Its apparent magnitude is 12.8, so a telescope will show it easily. If it could be seen from the 'standard distance' of 10 parsecs (32.6 light years), it would be as brilliant as the Sun.

The remoteness of the quasars means that they can be used to study interstellar and intergalactic material. A quasar's light will have to pass through this material before reaching us, and the material will leave its imprint upon the quasar's spectrum; we can tell which of the lines are caused by the quasar and which are not, because the non-quasar lines will not share in the overall red shift. There is also the 'gravitational lensing' effect. If the light from a remote source passes near a massive object *en route*, the rays of light will be 'bent', and the result will be that several images will be formed of the object in the background. If the alignment is not precise, there will still be obvious effects. A good example is G2237+0305. Here we have a galaxy 400 million light years away, behind which is a quasar lying at 8000 million light years. The light from the quasar is split, producing four images surrounding that of the lensing galaxy. This is often termed the Einstein Cross, because it was Albert Einstein who first predicted that such effects could occur.

If the alignment is less perfect, incomplete 'arcs' of the background object will appear. A precise alignment produces an Einstein Ring, of which several are now known. Finally, an intervening object may focus and magnify the image of a much more remote source which we would otherwise be unable to see.

Obviously things have changed since 3C273 was identified more than forty years ago. Modern radio methods make it possible to pinpoint discrete sources very precisely; there is no longer any need to rely upon the help of the Moon. Quasars abound, and there are also the BL Lacertae objects (BL Lacs), which are similar to quasars, and may indeed simply be quasars seen from a narrow angle. They take their name from the first-discovered member of the type, which was taken for an ordinary variable star and given a conventional variable star designation.

So far we have been discussing the visible universe. But what about those parts of the universe which we cannot see even with our most powerful telescopes?

◀ **Gravitational lens.** The massive galaxy cluster Abell 2218 in the constellation of Draco acts as a gravitational lens, producing distorted arc-shaped images of more distant objects that happen to lie in the same line of sight.

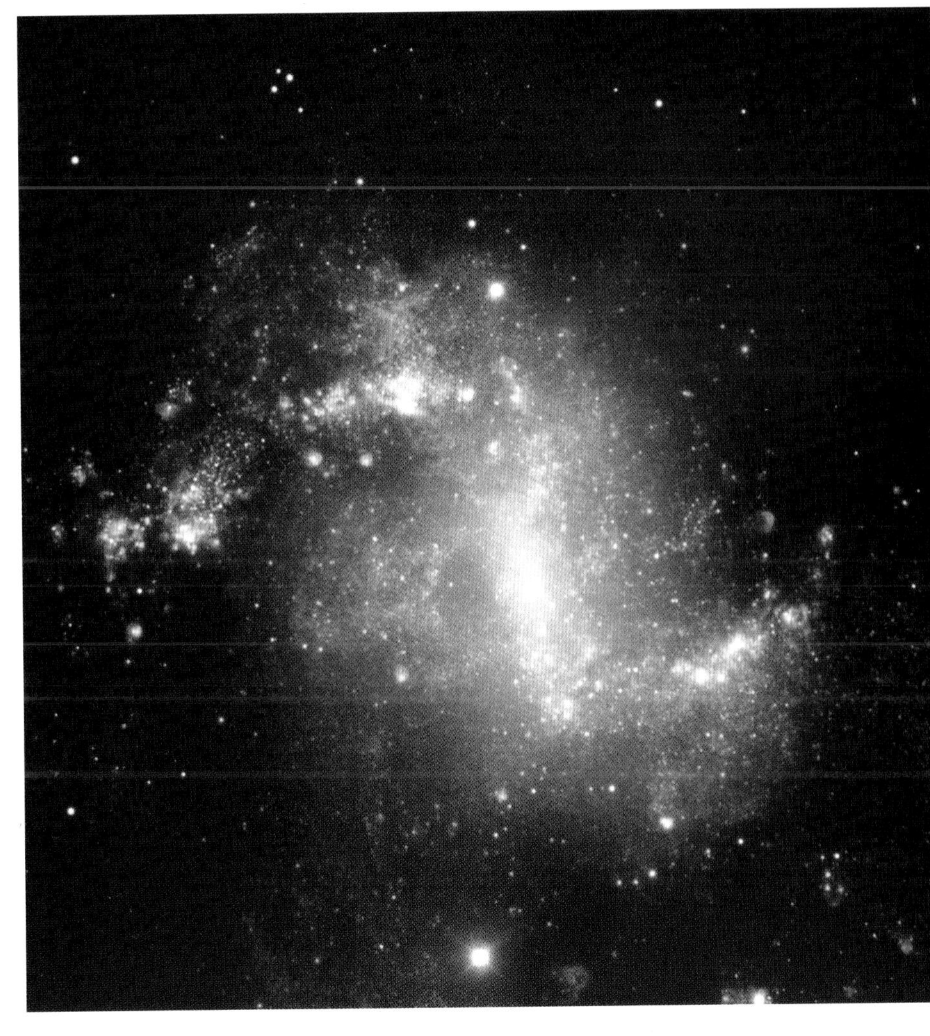

▶ **The Topsy-Turvy Galaxy NGC1313.** This composite image, taken with the FORS1 instrument on the VLT (Very Large Telescope) at Cerro Paranal, Chile, shows the many star-forming regions in this very active starburst galaxy.

Astrobiology

Since the start of the BBC 'Sky At Night' series, more than fifty years ago, I have been asked thousands of questions. Of these the most common, by far, is 'Is there life on other worlds?' followed by 'If life does exist, where will we find it – and what will it be like?'

At the moment I can give no definite answer, because we have no firm evidence of life anywhere beyond the Earth. But everyone can speculate, and this introduces us to the new science of astrobiology, which combines astronomy and astrophysics with chemistry, biology and geology. It is in effect the study of life in all its aspects, both here and elsewhere.

There is one point worth making at the outset. All life of the kind we can understand is based upon one type of atom: carbon. No other element has the ability to produce the large, complex molecules needed for life, and this rules out entirely alien beings of the kind known to science-fiction writers as BEMs (Bug-Eyed Monsters). If we are wrong about this, then the whole of our modern science is wrong too, so that speculation becomes both endless and pointless. We have to confine ourselves to life as we know it'.

Some people, including eminent scientists, maintain that because appearance of life involves so many complex processes that it is unlikely to have happened more than once, in which case life on Earth is unique. Yet there are a 400,000 million stars in our Galaxy alone, and many of these stars are attended by planets; we can see 50,000 million galaxies at least, and the total number of planets in the universe must be staggeringly great. Many of these planets must be similar to Earth, and to me it seems inconceivable that we are alone.

Unfortunately, we have no real idea of how living things are created, and there are suggestions that life on our own world might have been brought here from space. This idea – 'panspermia' – was proposed a century ago by Svante Arrhenius, who believed that life was brought to Earth by a comet. He never received a great deal of support, because theory seemed to raise more difficulties than it solved, and in any case came no nearer to explaining how life originated in the first place. In a modified form, panspermia has been revived much more recently by Sir Fred Hoyle and Chandra Wickramasinghe, but again has been met with little enthusiasm. Also, we have no idea of how to create life, though unfortunately we can destroy it with surprising ease.

There is one fundamental problem which we must solve if we are to make any real progress. If conditions on a planet are suitable for life to appear – will it? And if so, will it develop as far as its environment allows? On Earth, life has developed from tiny single-celled creatures to you and me, and has also shown that life-forms can survive in unexpected places – such as hydrothermal vents on the ocean floor! All that the astrobiologist of 2007 can really do is to weigh what evidence we have, and from it draw the most logical conclusions.

It has been suggested that we might not recognize extraterrestrial life even if we found it. Again we have few positive facts to guide us. Encountering intelligent life is quite another matter; we can rule out any advanced beings in the Solar System, and so any extra-terrestrial civilization must be light years away. Radio contact with another race cannot be ruled out, and there have been – and still are – organizations devoted to SETI, the Search for Extraterrestrial Intelligence. The chances of making contact are low, but if we make no attempts they will be nil. If we do pick up an intelligible message, our whole way of thinking will be dramatically changed.

Realistically, the only world we may hope to reach in the foreseeable future is Mars (discounting the Moon, of course). We are now sure that there is no life on the Martian surface, but not necessarily beneath the surface, but when we set up bases there, as we may hope to do well before the end of the present century, it would be helpful to be able to grow plants which could be used in everyday life. Astrobiologists are already at work, hoping to be able to produce plants which can survive under conditions of low atmospheric pressure, bitterly cold nights and bombardment by damaging radiation from the Sun and space.

So – is there life on other worlds? In my view, Mars may give us the answer. If our spacecraft prove that there is any life at all, even of the most lowly kind, we will know. The conditions below the Martian surface will be no more unfriendly than our hydrothermal vents, and if life survives there we may be fairly confident that living things will appear wherever they can. In this case, astrobiology will have a fascinating future.

▼ **Hydrothermal vent.** Also known as black smokers, such vents are found on the ocean floor far below the surface. Superheated gases escape from fissures in the Earth's crust; the high temperatures and acidity here make it seem an unlikely place to foster life – and yet life-forms seem to thrive in this environment, feeding off the chemical nutrients.

▶ **Evidence of water flowing on Mars.** Mars Global Surveyor imaged this long streak which had been absent in previous images of the same terrain. It is direct evidence of a short-lived gush of water flowing down a slope on Mars.

Into the future

▼ The Hubble Ultra Deep Field
This infra-red view of a tiny patch of sky (just one-tenth the diameter of the Full Moon) in Fornax contains an estimated 10,000 galaxies. The million-second exposure, obtained by combingin images from the HST's ACS and NICMOS, reveals galaxies that are too faint to be seen by ground-based telescopes. It shows galaxies that existed when the universe was less than a billion years old – a 'zoo' of oddball galaxies that bear little similarity to the classic spirals and ellipticals of today.

We can look back into the past with a certain amount of confidence. We can say that the Earth is some 4,600 million years old, that the Sun dates back around 5,000 million years, and that the universe as we know it began about 13,700 million years ago. But when we try to look into the future, things are not so clear cut. The Earth will probably be destroyed when the Sun enters its red giant stage, and even if the planet survives life upon it certainly cannot do so. Stars have finite lives, and the Sun is no exception: after its spell of glory as a red giant it will subside to a dim white dwarf and eventually become a cold, dead black dwarf. But what about the universe itself? Will it ever come to an end – and if so, how? There are several possibilities, so let us consider them one by one and see which, if any, is the most plausible.

(1) The universe has always existed, and will exist forever; as old galaxies die, new ones replace them, which means that matter is being continuously created in the form of Hydrogen atoms. This, of course, is the Steady-State theory championed by Sir Fred Hoyle, but all the observational evidence is against it, and by now it has been abandoned by almost everyone.

(2) The Oscillating Universe, a theory which I irreverently call the Concertina Universe. The present phase of expansion will be followed by a period of contraction. The galaxies will start to come together again, and after perhaps another 80,000 million years there will be another Big Bang. Quite possibly the whole cycle will be repeated over again. This means that there may have been an infinite number of Big Bangs in the past, and there may be an infinite number in the future, in which case the universe will never die. The trouble here is that it is hard to see what mechanism could cause the expansion to be reversed.

(3) The universe will expand forever. The groups of galaxies will lose contact with each other; all the stars in them will die, and eventually even protons will disintegrate, leaving only radiation. Nothing more will ever happen. Even time will come to an end. It is a dismal outlook, but it is accepted by many modern cosmologists, and the fact that the rate of expansion is accelerating does seem to support it.

(4) The Big Rip. The rate of expansion will go on increasing until it becomes overwhelming; the galaxies are ripped apart, then stars, planets, molecules, atoms – in fact, everything. Nothing will remain but shattered fragments, with no hope that the universe can be re-born.

All of these forecasts seem very pessimistic, but we must remember that we are still literally, as well as metaphorically, in the dark with regard to the main fundamentals. If time ends, what happens after that? To say 'Nothing' is no answer at all. The plain truth is that our brains are not advanced enough to understand infinity, either in space or in time. Until we can, we must accept the fact that we are limited. One day we may break through this wall of ignorance, but for the moment we must wait. Perhaps the great enlightenment will never come; perhaps a super-Newton or a super-Einstein has already been born Moreover, despite all that modern science can tell us, there is always an inner feeling that we may, after all, be missing something – a 'something' that could overturn all our ideas about the universe. Here there is a danger of wandering into the realm of science fiction, and speculation is endless; it may also be futile. If life survives on Earth for, say, another 500,000 years, we may well be in a far better position to judge.

Look up at the Sun and Moon, the planets, the stars and the galaxies – the whole of the observable universe is there for our inspection. It is ever more wonderful than we can appreciate, and it is always changing. Let us enjoy it.

Spitzer Space Telescope

▼ **Herbig-Haro 49/50.** A shock front created as a cosmic jet from an object out of picture lances through dust and gas.

The Spitzer Space Telescope is the final mission of NASA's Great Observatories programme – four orbiting observatories each looking at the universe at different wavelengths, equating to different temperatures. Spitzer works in infra-red light, which allows it to penetrate clouds of dust and gas that block visible light and to observe regions where stars – and perhaps planets – are forming, as well as deep into the centre of galaxies.

▲ **NGC2207 and IC2163.** In this false-colour image, Spitzer's infra-red vision (red) sees regions of star formation unseen at visible wavelengths (HST, blue-green and inset).

▼ **The Helix (NGC7293).** The glow around the central white dwarf is the heat from a disk of dust and gas – the remnants of comets – which is not seen in visible light images.

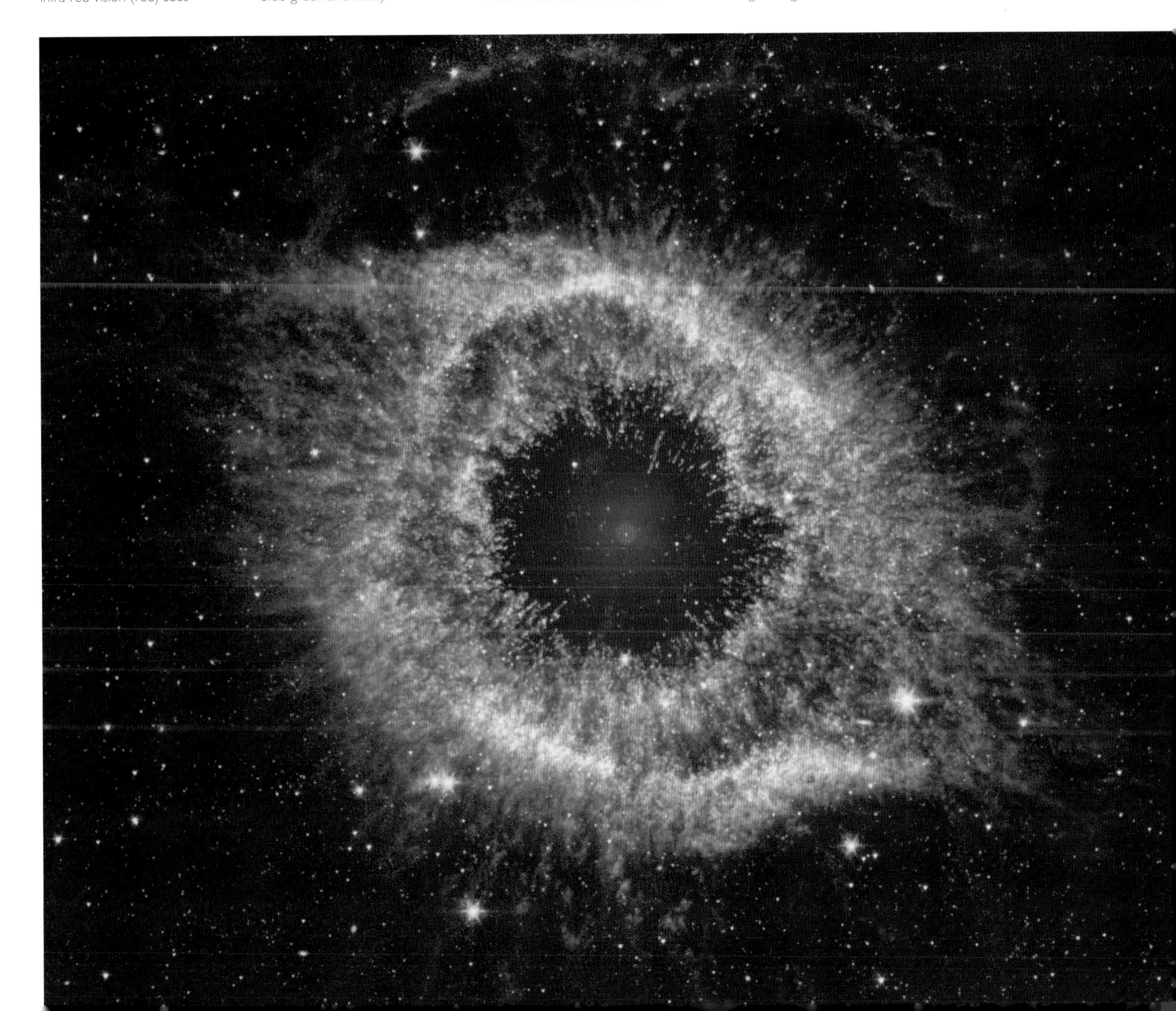

The constellations, as depicted by Giovanni de Vecchi and Raffaellino da Reggio on the ceiling of the Sala del Mappamondo of the Palazzo Farnese.

STAR MAPS

Whole Sky Maps

▼ **Turn the map** for your hemisphere so that the current month is at the bottom. The map will then show the constellations on view at approximately 11 pm GMT (facing south in the northern hemisphere and north in the southern hemisphere). Rotate the map clockwise 15° for each hour before 11 pm; anticlockwise for each hour after 11 pm.

The origin of the constellation patterns is not known with any certainty. The ancient Chinese and Egyptians drew up fanciful sky maps (two of the Egyptian constellations, for example, were the Cat and the Hippopotamus) and so, probably, did the Cretans. The pattern followed today is based on that of the ancient Greeks and all of the 48 constellations given by Ptolemy in his book the *Almagest*, written about AD 150, are still in use.

Ptolemy's list contains most of the important constellations visible from the latitude of Alexandria. Among them are the Ursa Major and Ursa Minor, Cygnus, Hercules, Hydra and Aquila, as well as the 12 Zodiacal constellations. There are also some small, obscure constellations, such as Equuleus (the Foal) and Sagitta (the Arrow), which are surprisingly faint and, one would have thought, too ill defined to be included in the original 48.

It has been said that the sky is a mythological picture book, and certainly most of the famous old stories are commemorated there. All the characters of the Perseus tale are to be seen – including the sea monster, although nowadays it is better known as Cetus, a harmless whale! Orion, the Hunter, sinks below the horizon as his killer, the Scorpion, rises; Hercules lies in the north, together with his victim the Nemean lion (Leo). The largest of the constellations, Argo Navis – the ship which carried Jason and his companions in quest of the Golden Fleece – has been unceremo-

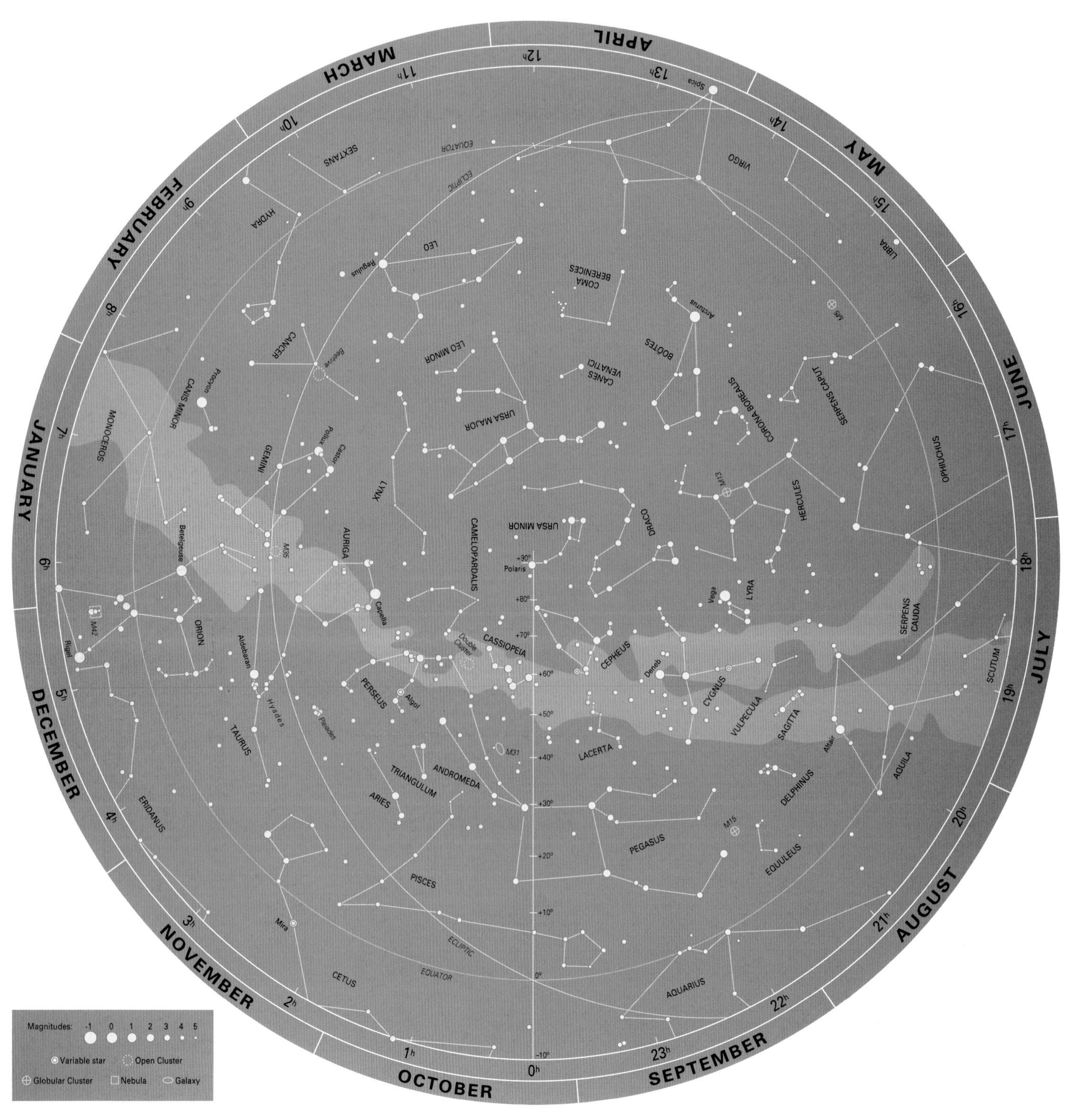

niously cut up into its keel (Carina), poop (Puppis) and sails (Vela), because the original constellation was thought to be too large and unwieldy.

Ptolemy's constellations did not cover the entire sky. There were gaps between them, and these were filled. Later astronomers added new constellations, sometimes modifying the original boundaries. Later still, the stars of the far south were divided into constellations, and some of the names have a modern flavour. The Telescope (Telescopium), the Microscope (Microscopium) and the Air-pump (Antlia) are three of the more recent groups. Even the Southern Cross, Crux, is a seventeenth-century constellation. It was formed by Royer in 1679, and so has no great claim to antiquity.

Many additional constellations have been proposed from time to time, but these have not been adopted, although one of the rejected groups – Quadrans Muralis (the Mural Quadrant) is remembered in the name of the annual Quadrantid meteor shower, which peaks in early January.

The nineteenth-century astronomer Sir John Herschel said that the patterns of the constellations had been drawn up to be as inconvenient as possible. In 1933, modified constellation boundaries were laid down by the International Astronomical Union. There have been occasional attempts to revise the entire nomenclature, but it is unlikely any radical change will now be made. The present-day constellations have been accepted for too long to be altered.

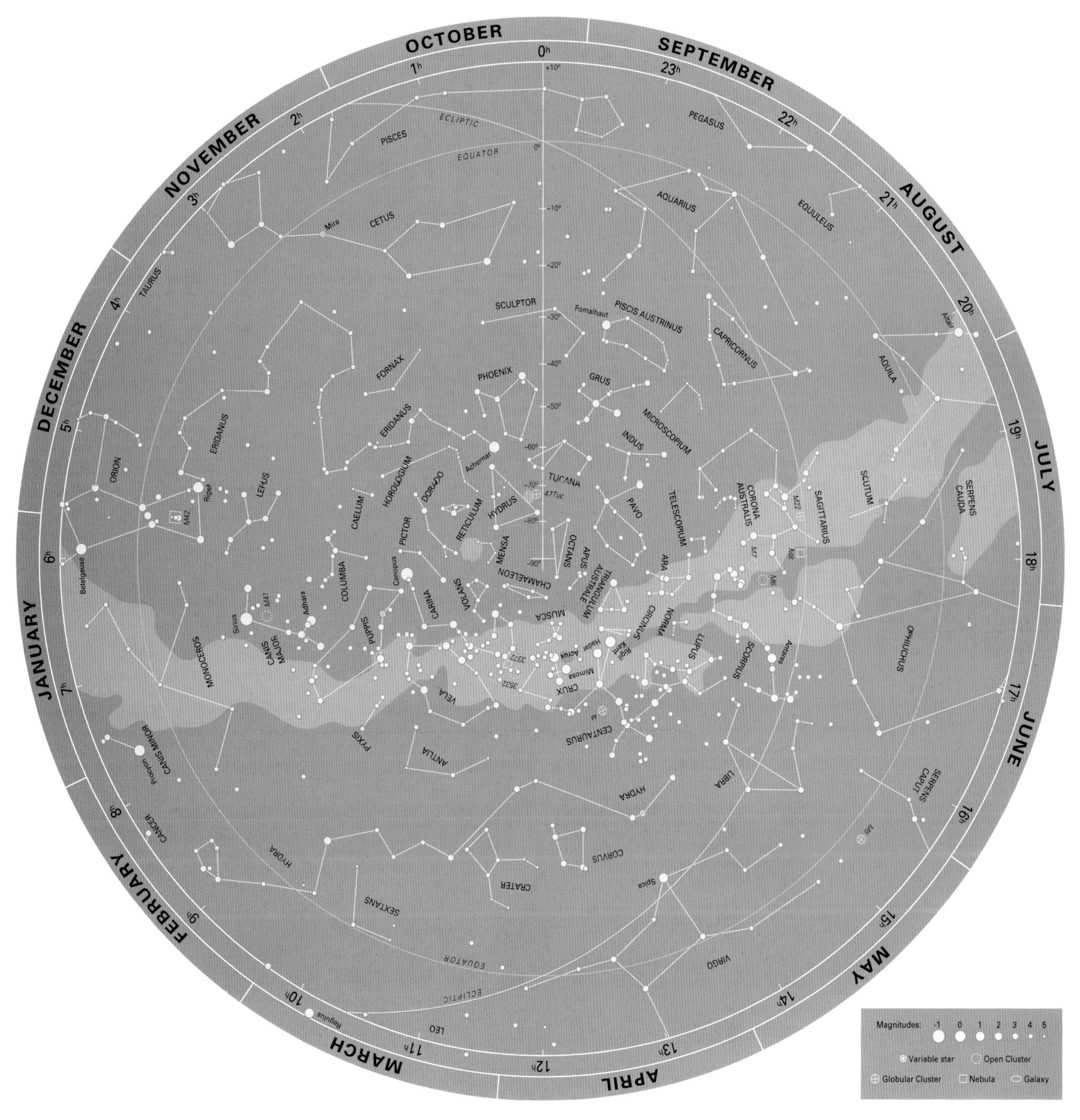

Seasonal Charts: North

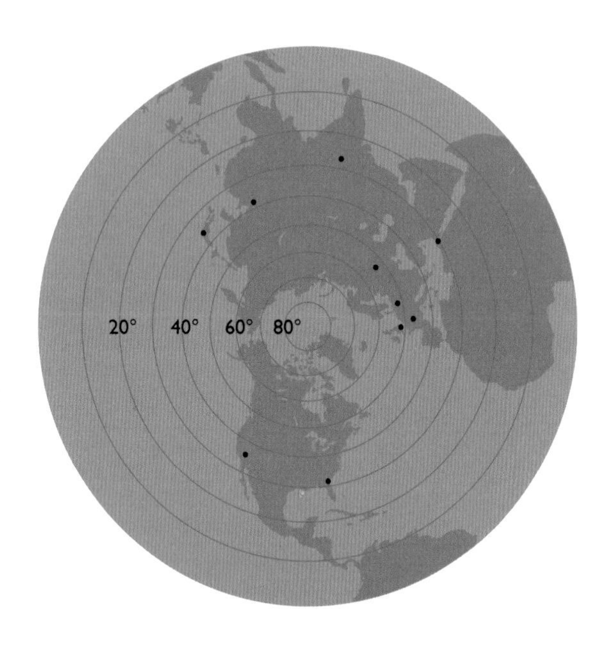

► **Latitudes** of the major cities of the northern hemisphere. For the observer, all the stars of the northern sky are visible in the course of a year, but he or she can see only a limited distance south of the equator. At a latitude of x°N, the most southerly point that can be seen in the sky is 90 minus x°S. Thus, for example, to an observer at latitude 50°N, only the sky north of 90 minus 50 (or 40°S) is ever visible.

The charts given on this page are suitable for observers who live in the northern hemisphere, between 60°N and 30°N. The horizon is given by the latitude marks near the bottom of the charts. Thus, for an observer who lives at 30°N, the northern horizon in the first map will pass just above Deneb, which will be visible.

A star rises earlier, on average, by two hours a month; thus the chart for 8 pm on 1 January will be valid for 6 pm on 1 February and 10 pm on 1 December.

The limiting visibility of a star for an observer at any latitude can be worked out from its declination. To an observer in the northern hemisphere, a star is at its lowest point in the sky when it is due north; a star which is 'below' the pole by the amount of one's latitude will touch the horizon when at its lowest point. If it is closer to the pole than that it will be circumpolar. From latitude 51°N, for example, a star is circumpolar if its declination is 90 minus 51 (or 39°N), or greater. Thus Capella, at dec. +45° 57′, is circumpolar to an observer in London, Cologne or Calgary. A minor allowance must be made for atmospheric refraction.

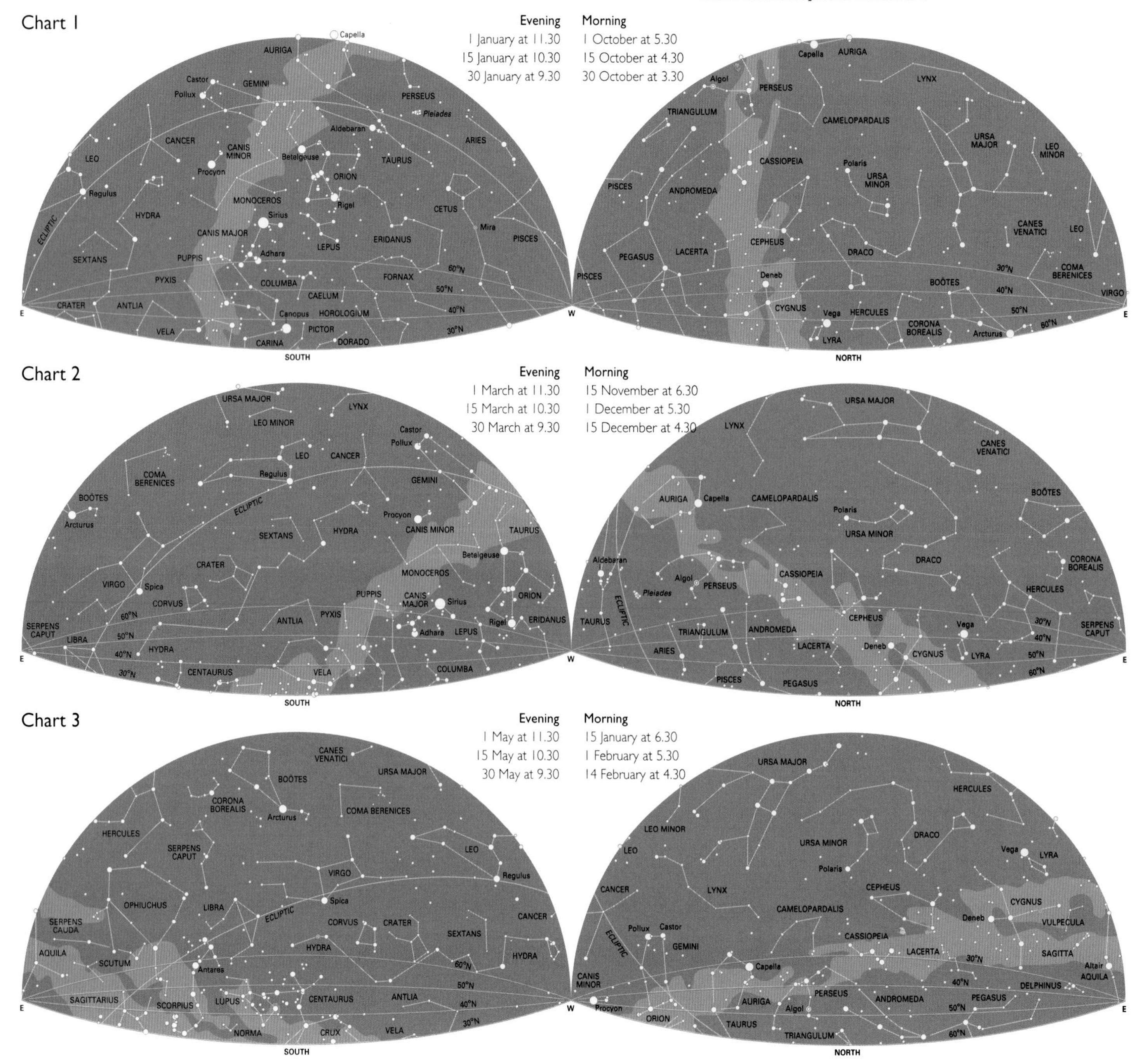

Similarly, to an observer who is at latitude 51°N, a star that lies at a declination farther south than −39° will never rise. Canopus lies at declination −52° 42′; therefore it is invisible from London, but can be seen from any latitude south of 37° 20′ north, again ignoring the effects of atmospheric refraction.

The charts given here show the northern (right) and the southern (left) aspects of the sky from the viewpoint of an observer in northern latitudes. They are self explanatory; the descriptions given below apply in each case to the late evening, but more accurate calculations can be made by consulting the notes at the side of each chart.

Chart 1. In winter, the southern aspect is dominated by Orion and its retinue. Capella is almost at the zenith or overhead point, and Sirius is at its best. Observers in Britain can see part of Puppis, but Canopus is too far south to be seen from any part of Europe. The sickle of Leo is very prominent in the east; Ursa Major is to the north-east, while Vega is at its lowest in the north. It is circumpolar from London but not from New York, and is not on the first chart.

Chart 2. In spring, Orion is still above the horizon until past midnight; Leo is high up, with Virgo to the east. Capella is descending in the north-west, Vega is rising in the north-east; these two stars are so nearly equal in apparent magnitude (0.1 and 0.0) that, in general, whichever is the higher in the sky will also seem the brighter. In the west, Aldebaran and the Pleiades are still visible.

Charts 3–6. In early summer (Chart 3), Orion has set and, to British observers, the southern aspect is relatively barren, but observers in more southerly latitudes can see Centaurus and its neighbours. During summer evenings (Chart 4), Vega is at the zenith and Capella low in the north; Antares is at its highest in the south. By early autumn (Chart 5), Aldebaran and the Pleiades have reappeared, and the Square of Pegasus is conspicuous in the south, with Fomalhaut well placed. By early winter (Chart 6), Orion is back in view, with Ursa Major lying low in the northern sky.

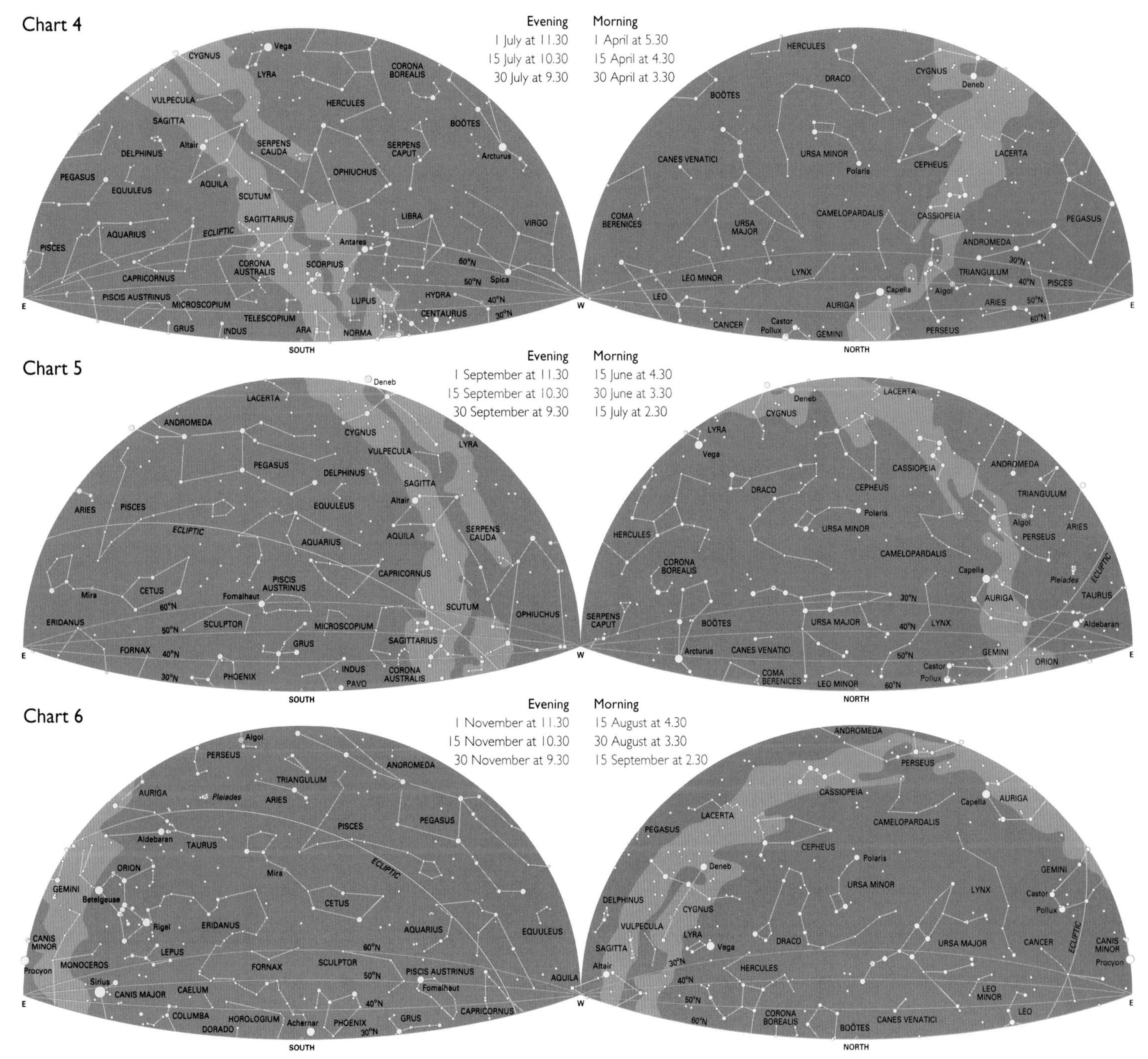

Seasonal Charts: South

Generally, the stars in the South Polar area of the sky are brighter than those of the far north, even though the actual pole lies in a barren region, and there is no pattern of stars so distinctive as Ursa Major – apart from Crux (the Southern Cross), which covers a much smaller area. Canopus, the brightest star in the sky apart from Sirius, has a declination of some −53°, and is not visible from Europe, but rises well above the horizon from Mexico, and from Australia and New Zealand it is visible for much of the year. In the far south, too, there are the Magellanic Clouds. They are prominent naked-eye objects, and the Large Magellanic Cloud can be seen without optical aid even under conditions of full moonlight.

An observer at one of the Earth's poles would see one hemisphere of the sky only, and all the visible stars would be circumpolar. It is not even strictly correct to say that Orion is visible from the entire surface of the Earth. An observer at the South Pole would never see Betelgeux, whose declination is +7°. From latitudes farther south than 83°S (90 minus 7) Betelgeux would never rise.

These charts may be used for almost all the densely populated regions of the southern hemisphere which lie between 5°S and 35°S. The northern view is given in the left chart, the southern in the right.

Chart 1. In January, the two most brilliant stars, Sirius and Canopus, are high up. Sirius seems appreciably the brighter of the two (magnitude −1.5 as against −0.8), but its eminence is due to its closeness rather than its real luminosity. It is an A-type Main Sequence star, only 26 times as luminous as the Sun; Canopus is an F-type supergiant, whose luminosity may be 15,000 times that of the Sun, according to one estimate, though both its distance and its luminosity are uncertain and estimates vary widely. Lower down, the Southern Cross (Crux) is a prominent feature, and the brilliant pair of stars Alpha and Beta Centauri (Agena) are also found in the same area. In the north, Capella is well above the horizon; Orion is not far from the zenith, and if the sky is clear a few stars of Ursa Major may be seen low over the northern horizon.

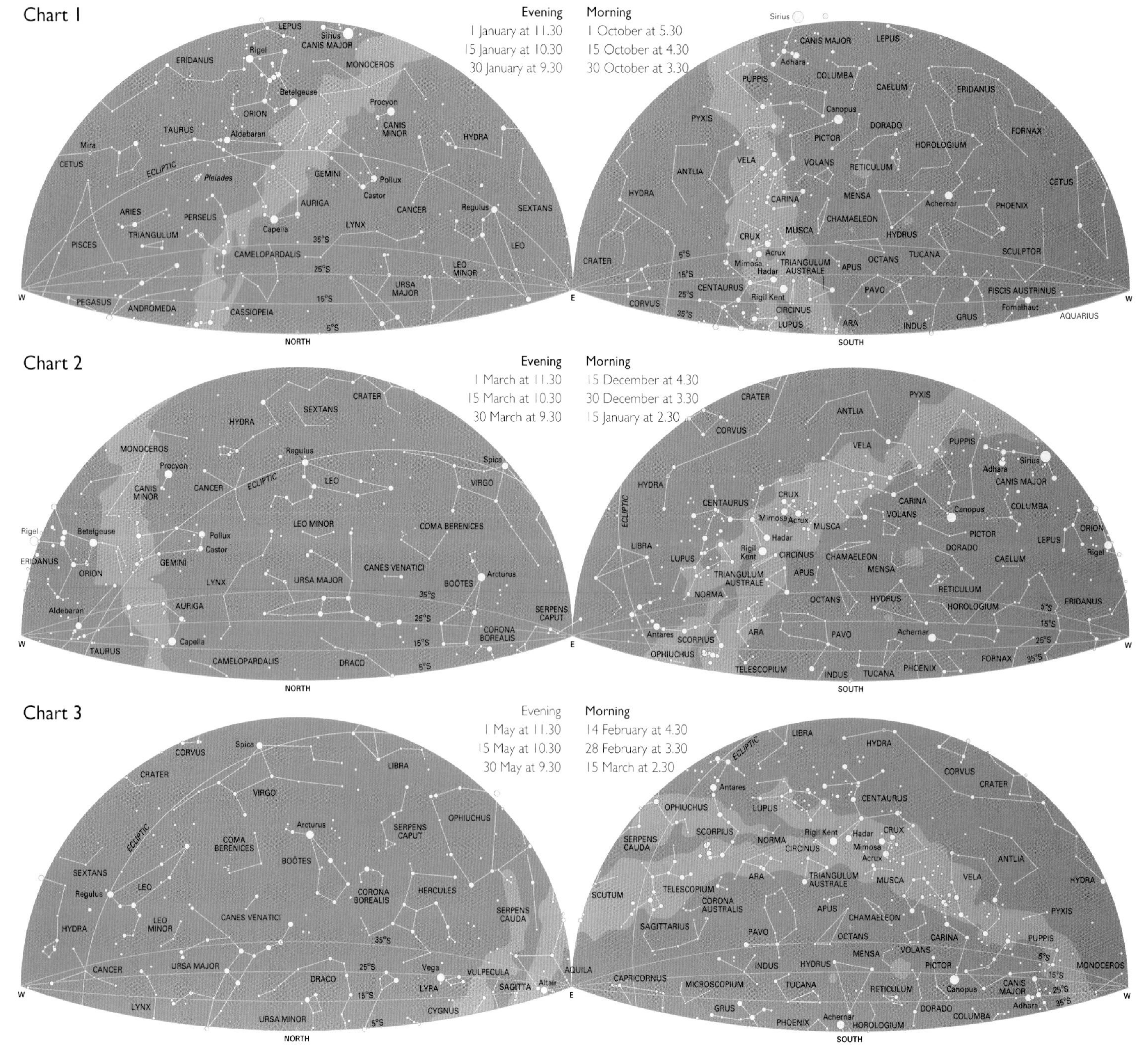

Chart 2. In March, Canopus is descending in the south-west, and Crux is rising to its greatest altitude; the south-east is dominated by the brilliant constellations of Scorpius and Centaurus. (Scorpius is a magnificent constellation. Its leading star, Antares, is quite visible from Europe, but the 'tail' and 'sting' are too far south to be seen properly.) To the north, Ursa Major is seen; Orion is descending in the west.

Charts 3–4. The May aspect (Chart 3) shows Alpha and Beta Centauri very high up, and Canopus in the south-west; Sirius and Orion have set, but Scorpius is brilliant in the south-east. In the north, Arcturus is prominent, with Spica in Virgo near the zenith. By July (Chart 4) Vega, Altair and Deneb are all conspicuous in the north. Arcturus is still high above the north-west horizon. Antares is not far from its zenith.

Charts 5–6. The September view (Chart 5) shows Pegasus in the north, and the 'W' of Cassiopeia is above the horizon. Crux is almost at its lowest. By November (Chart 6) Sirius and Canopus are back in view; Alpha and Beta Centauri (Agena) graze the horizon, and the region of the zenith is occupied by large, comparatively barren groups such as Cetus and Eridanus.

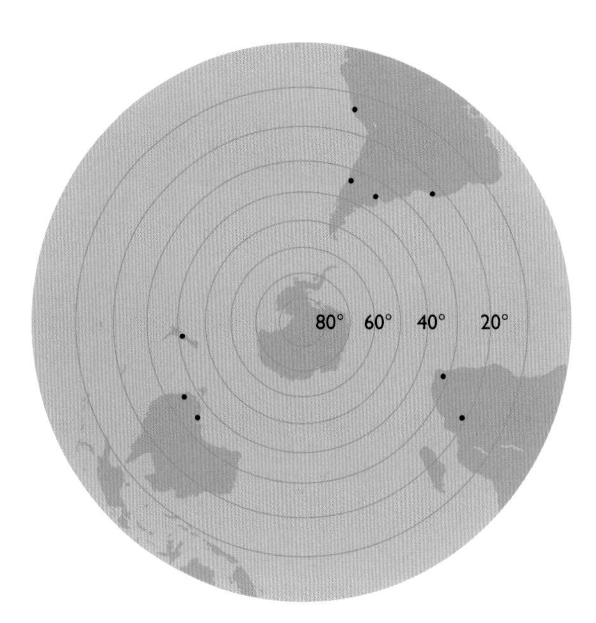

◀ **For the observer** in the southern hemisphere all the stars of the southern sky are visible in the course of a year, but he or she can only see a limited distance north of the celestial equator. At a latitude of x°S, the most northerly point that can be seen is 90 minus x°N. Thus, for example, to an observer at latitude 50°S only the sky south of 90 minus 50 (or 40°N) is ever visible.

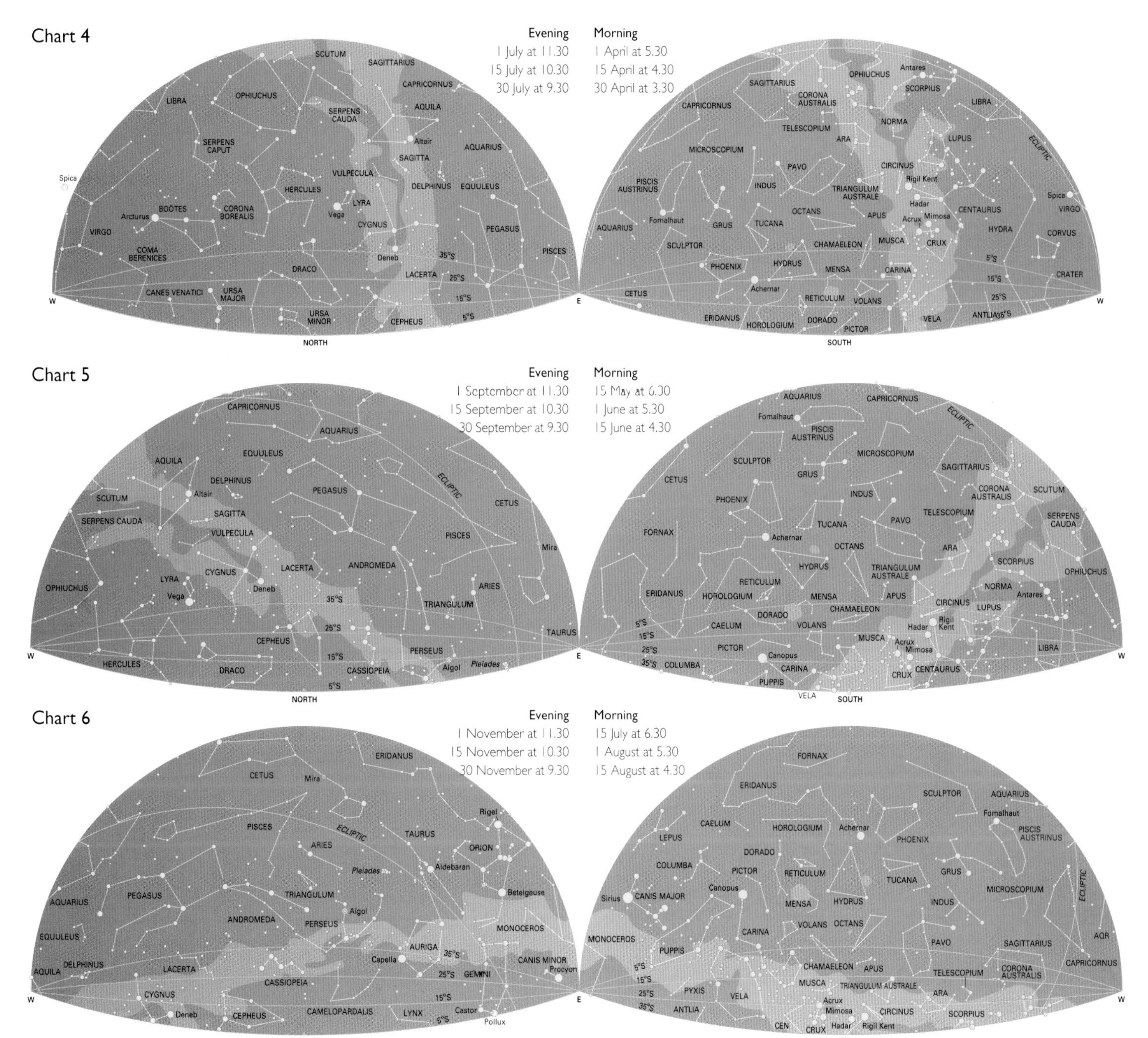

Messier Catalogue

M	NGC/IC	Constellation	Type	RA (2000.00) h	m	Dec. °	′	Magnitude	Size (′)	Comment
1	1952	Taurus	SN remnant	05	34.5	+22	01	10	6 × 4	CRAB NEBULA
2	7089	Aquarius	globular cluster	21	33.5	−00	49	6.5	12.9	
3	5272	Canes Venatici	globular cluster	13	42.2	+28	23	6.4	16	
4	6121	Scorpius	globular cluster	16	23.6	−26	32	5.9	26.3	
5	5904	Serpens	globular cluster	15	18.6	+02	05	5.8	17.4	
6	6405	Scorpius	open cluster	17	40.1	−32	13	4.2	50	BUTTERFLY CLUSTER
7	6475	Scorpius	open cluster	17	53.9	−34	49	3.3	80	
8	6523	Sagittarius	bright nebula	18	03.8	−24	23	6.0	90 × 40	LAGOON NEBULA
9	6333	Ophiuchus	globular cluster	17	19.2	−18	31	7.9	9.3	
10	6254	Ophiuchus	globular cluster	16	57.1	−04	06	6.6	15.1	
11	6705	Scutum	open cluster	18	51.1	−06	16	5.8	14	WILD DUCK CLUSTER
12	6218	Ophiuchus	globular cluster	16	47.2	−01	57	6.6	14.5	
13	6205	Hercules	globular cluster	16	41.7	+36	28	5.9	16.6	GREAT GLOBULAR IN HERCULES
14	6402	Ophiuchus	globular cluster	17	37.6	−03	15	7.6	11.7	
15	7078	Pegasus	globular cluster	21	30.0	+12	10	6.3	12.3	
16	6611	Serpens	nebula and cluster	18	18.8	−13	47	6.4	35 × 28	EAGLE NEBULA
17	6618	Sagittarius	bright nebula	18	20.8	−16	11	7.0	46 × 37	OMEGA NEBULA
18	6613	Sagittarius	open cluster	18	19.0	−17	08	6.9	9	
19	6273	Ophiuchus	globular cluster	17	02.6	−26	16	7.1	13.5	
20	6514	Sagittarius	bright nebula	18	02.6	−23	02	7.5	29 × 27	TRIFID NEBULA
21	6531	Sagittarius	open cluster	18	04.6	−22	30	5.9	13	
22	6656	Sagittarius	globular cluster	18	36.4	−23	54	5.1	24	
23	6494	Sagittarius	open cluster	17	56.8	−19	01	5.5	27	
24	6603	Sagittarius	star cloud	18	16.9	−18	29	4.5	90	
25	IC4725	Sagittarius	open cluster	18	31.6	−19	15	4.6	31	
26	6694	Scutum	open cluster	18	45.2	−09	24	8.0	15	
27	6853	Vulpecula	planetary nebula	19	59.6	+22	43	7.6	8 × 4	DUMBBELL NEBULA
28	6626	Sagittarius	globular cluster	18	24.5	−24	52	6.9	11	
29	6913	Cygnus	open cluster	20	23.9	+38	32	6.6	7	
30	7099	Capricornus	globular cluster	21	40.4	−23	11	7.5	11	
31	224	Andromeda	spiral galaxy	00	42.7	+41	16	3.5	178 × 63	ANDROMEDA GALAXY
32	221	Andromeda	elliptical galaxy	00	42.7	+40	52	8.2	7.6 × 6.8	COMPANION TO M31
33	598	Triangulum	spiral galaxy	01	33.9	+30	39	5.7	67 × 39	TRIANGULUM GALAXY
34	1039	Perseus	open cluster	02	42.0	+42	47	5.2	35	
35	2168	Gemini	open cluster	06	08.9	+24	20	5.0	28	
36	1960	Auriga	open cluster	05	36.1	+34	08	6.0	12	
37	2099	Auriga	open cluster	05	52.4	+32	33	5.6	24	
38	1912	Auriga	open cluster	05	28.4	+35	50	6.4	21	
39	7092	Cygnus	open cluster	21	32.2	+48	26	4.6	32	
40*		Ursa Major	double star	12	22.4	+58	05	9.0, 9.3	–	
41	2287	Canis Major	open cluster	06	47.0	−20	44	4.5	38	
42	1976	Orion	bright nebula	05	35.4	−05	27	5.0	66 × 60	ORION NEBULA
43	1982	Orion	bright nebula	05	35.6	−05	16	7	20 × 15	EXTENSION OF M42
44	2632	Cancer	open cluster	08	40.1	+19	59	3.1	95	PRAESEPE
45	–	Taurus	open cluster	03	47.0	+24	07	1.2	110	PLEIADES
46	2437	Puppis	open cluster	07	41.8	−14	49	6.1	27	
47*	2422	Puppis	open cluster	07	36.6	−14	30	4.4	30	
48*	2548	Hydra	open cluster	08	13.8	−05	48	5.8	54	
49	4472	Virgo	elliptical galaxy	12	29.8	+08	00	8.4	8.9 × 7.4	
50	2323	Monoceros	open cluster	07	03.2	−08	20	5.9	16	
51	5194–5	Canes Venatici	spiral galaxy & companion	13	29.9	+47	12	8.4	11.0 × 7.8	WHIRLPOOL GALAXY
52	7654	Cassiopeia	open cluster	23	24.2	+61	35	7.3	13	
53	5024	Coma Berenices	globular cluster	13	12.9	+18	10	7.7	12.6	
54	6715	Sagittarius	globular cluster	18	55.1	−30	29	7.7	9.1	
55	6809	Sagittarius	globular cluster	19	40.0	−30	58	6.9	19	

* identification uncertain

MESSIER CATALOGUE (CONTINUED)

M	NGC	Constellation	Type	RA (2000.00) h	m	Dec. °	′	Magnitude	Size (′)	Comment
56	6779	Lyra	globular cluster	19	16.6	+30	11	8.2	7.1	
57	6720	Lyra	planetary nebula	18	53.6	+33	02	9.7	1.1 × 2.5	RING NEBULA
58	4579	Virgo	spiral galaxy	12	37.7	+11	49	9.8	5.4 × 4.4	
59	4621	Virgo	elliptical galaxy	12	42.0	+11	39	9.8	5.1 × 3.4	
60	4649	Virgo	elliptical galaxy	12	43.7	+11	33	8.8	7.2 × 6.2	
61	4303	Virgo	spiral galaxy	12	21.9	+04	28	9.7	6.0 × 5.5	
62	6266	Ophiuchus	globular cluster	17	01.2	−30	07	6.6	14.1	
63	5055	Canes Venatici	spiral galaxy	13	15.8	+42	02	8.6	12.3 × 7.6	SUNFLOWER GALAXY
64	4826	Coma Berenices	spiral galaxy	12	56.7	+21	41	8.5	9.3 × 5.4	BLACK EYE GALAXY
65	3623	Leo	spiral galaxy	11	18.9	+13	05	9.3	10 × 3.3	
66	3627	Leo	spiral galaxy	11	20.2	+12	59	9.0	8.7 × 4.4	
67	2682	Cancer	open cluster	08	50.4	+11	49	6.9	30	
68	4590	Hydra	globular cluster	12	39.5	−26	45	8.2	12	
69	6637	Sagittarius	globular cluster	18	31.4	−32	21	7.7	7.1	
70	6681	Sagittarius	globular cluster	18	43.2	−32	18	8.1	7.8	
71	6838	Sagittarius	globular cluster	19	53.8	+18	47	8.3	7.2	
72	6981	Aquarius	globular cluster	20	53.5	−12	32	9.3	5.9	
73	6994	Aquarius	asterism	20	58.9	−12	38	9.0	3	
74	628	Pisces	spiral galaxy	01	36.7	+15	47	9.2	10.2 × 9.5	
75	6864	Sagittarius	globular cluster	20	06.1	−21	55	8.6	6.0	
76	650 + 651	Perseus	planetary nebula	01	42.4	+51	34	12.2	2.7 × 1.8	LITTLE DUMBBELL
77	1068	Cetus	spiral galaxy	02	42.7	−00	01	8.8	6.9 × 5.9	
78	2068	Orion	gaseous nebula	05	46.7	+00	03	8.0	8 × 8	
79	1904	Lepus	globular cluster	05	24.5	−24	33	9.9	8.7	
80	6093	Scorpius	globular cluster	16	17.0	−22	59	7.2	8.9	
81	3031	Ursa Major	spiral galaxy	09	55.6	+69	04	6.9	25.7 × 14.1	BODE'S NEBULA
82	3034	Ursa Major	irregular galaxy	09	55.8	+69	41	8.4	11.3 × 10.2	
83	5236	Hydra	spiral galaxy	13	37.0	−29	52	8.2	10 × 8	SOUTHERN PINWHEEL
84	4374	Virgo	spiral galaxy	12	25.1	+12	53	9.3	5.0 × 4.4	
85	4382	Coma Berenices	spiral galaxy	12	25.4	+18	11	9.2	7 × 5	
86	4406	Virgo	elliptical galaxy	12	26.2	+12	57	9.2	7.4 × 5.5	
87	4486	Virgo	elliptical galaxy	12	30.8	+12	24	8.6	7.2 × 6.8	RADIO SOURCE VIRGO A
88	4501	Coma Berenices	spiral galaxy	12	32.0	+14	25	9.5	6.9 × 3.9	
89	4552	Virgo	elliptical galaxy	12	35.7	+12	33	9.8	4	
90	4569	Virgo	spiral galaxy	12	36.8	+13	10	9.5	9.5 × 4.7	
91*	4548	Coma Berenices	barred spiral galaxy	12	35.4	+14	30	10.2	5 × 4	
92	6341	Hercules	globular cluster	17	17.1	+43	08	5.5	11.2	
93	2447	Puppis	open cluster	07	44.6	−23	52	6.2	22	
94	4736	Canes Venatici	spiral galaxy	12	50.9	+41	07	8.2	11.0 × 9.1	
95	3351	Leo	barred spiral galaxy	10	44.0	+11	42	9.7	7.4 × 5.1	
96	3368	Leo	spiral galaxy	10	46.8	+11	49	9.2	7.1 × 5.1	
97	3587	Ursa Major	planetary nebula	11	14.8	+55	01	12.0	3 × 3	OWL NEBULA
98	4192	Coma Berenices	spiral galaxy	12	13.8	+14	54	10.1	9.5 × 3.2	
99	4254	Coma Berenices	spiral galaxy	12	18.8	+14	25	9.8	5.4 × 4.8	
100	4321	Coma Berenices	spiral galaxy	12	22.9	+15	49	9.4	6.9 × 6.2	
101	5457	Ursa Major	spiral galaxy	14	03.2	+54	21	7.7	27 × 26	PINWHEEL GALAXY
102*	5586	Draco	lenticular galaxy	15	06.5	+55	46	9.2	5 × 2	
103	581	Cassiopeia	open cluster	01	33.2	+60	42	7.4	6	
104	4594	Virgo	spiral galaxy	12	40.0	−11	37	8.3	8.9 × 4.1	SOMBRERO GALAXY
105	3379	Leo	elliptical galaxy	10	47.8	+12	35	9.3	4.5 × 4.0	
106	4258	Ursa Major	spiral galaxy	12	19.0	+47	18	8.3	18.2 × 7.9	
107	6171	Ophiuchus	globular cluster	16	32.5	+13	03	8.1	10	
108	3556	Ursa Major	spiral galaxy	11	11.5	+55	40	10.0	8 × 2	
109	3992	Ursa Major	spiral galaxy	11	57.6	+53	23	9.8	7 × 4	
110	205	Andromeda	elliptical galaxy	00	40.4	+41	41	8.0	17 × 10	COMPANION TO M31

* identification uncertain

Caldwell Catalogue

CALDWELL CATALOGUE										
				RA (2000.0)		Dec.				
C	NGC/IC	Constellation	Type	h	min	°	′	Magnitude	Size (′)	Comment
1	188	Cepheus	open cluster	00	44.4	+85	20	8.1	14	
2	40	Cepheus	planetary nebula	00	13.0	+72	32	10.7	1.0 × 0.7	
3	4236	Draco	Sb galaxy	12	16.7	+69	27	9.6	23 × 8	
4	7023	Cepheus	reflection nebula	21	01.8	+68	10	—	18 × 18	
5	IC342	Camelopardalis	SBc galaxy	03	46.8	+68	06	9.2	18 × 17	
6	6543	Draco	planetary nebula	17	58.6	+66	38	8.8	0.3 × 0.3	CAT'S EYE NEBULA
7	2403	Camelopardalis	Sc galaxy	07	36.9	+65	36	8.4	18 × 10	
8	559	Cassiopeia	open cluster	01	29.5	+63	18	9.5	4	
9	Sh2-155	Cepheus	bright nebula	22	56.8	+62	37	—	50 × 10	CAVE NEBULA
10	663	Cassiopeia	open cluster	01	46.0	+61	15	7.1	16	
11	7635	Cassiopeia	bright nebula	23	20.7	+61	12	8	15 × 8	BUBBLE NEBULA
12	6946	Cepheus	SAB galaxy	20	34.8	+60	09	8.9	11 × 10	
13	457	Cassiopeia	open cluster	01	19.1	+58	20	6.4	13	PHI CASSIOPEIAE CLUSTER
14	869/884	Perseus	open clusters	02	20.0	+57	08	5.3 and 6.1	29 and 29	DOUBLE CLUSTER
15	6826	Cygnus	planetary nebula	19	44.8	+50	31	8.8	27 × 24	BLINKING PLANETARY
16	7243	Lacerta	open cluster	22	15.3	+49	53	6.4	21	
17	147	Cassiopeia	dE4 galaxy	00	33.2	+48	30	9.3	13 × 8	
18	185	Cassiopeia	dE0 galaxy	00	39.0	+48	20	9.2	12 × 10	
19	IC5146	Cygnus	bright nebula	21	53.4	+47	16	—	10	COCOON NEBULA
20	7000	Cygnus	bright nebula	20	58.8	+44	20	6.0	120 × 100	NORTH AMERICA NEBULA
21	4449	Canes Venatici	Irregular galaxy	12	28.2	+44	06	9.4	6 × 5	
22	7662	Andromeda	planetary nebula	23	25.9	+42	33	9.2	0.3 × 0.3	BLUE SNOWBALL NEBULA
23	891	Andromeda	Sb galaxy	02	22.6	+42	21	9.9	14 × 3	
24	1275	Perseus	Seyfert galaxy	03	19.8	+41	31	11.6	3.5 × 2.5	PERSEUS A
25	2419	Lynx	globular cluster	07	38.2	+38	53	10.4	4.1	
26	4244	Canes Venatici	Scd galaxy	12	17.5	+37	49	10.2	18 × 2	
27	6888	Cygnus	bright nebula	20	12.0	+38	20	—	20 × 10	CRESCENT NEBULA
28	752	Andromeda	open cluster	01	57.8	+37	41	5.7	50	
29	5005	Canes Venatici	Sb galaxy	13	10.9	+37	03	9.5	6 × 3	
30	7331	Pegasus	Sb galaxy	22	37.1	+34	25	9.5	11 × 4	
31	IC405	Auriga	bright nebula	05	16.2	+34	16	—	30 × 19	FLAMING STAR NEBULA
32	4631	Canes Venatici	Sc galaxy	12	42.1	+32	32	9.3	17 × 3	WHALE GALAXY
33	6992/5	Cygnus	SN remnant	20	56.8	+31	28	—	60 × 8	Eastern VEIL NEBULA
34	6960	Cygnus	SN remnant	20	45.7	+30	43	—	70 × 6	Western VEIL NEBULA
35	4889	Coma Berenices	E4 galaxy	13	00.1	+27	59	11.4	3 × 2	
36	4559	Coma Berenices	Sc galaxy	12	36.0	+27	58	9.8	13 × 5	
37	6885	Vulpecula	open cluster	20	12.0	+26	29	5.9	7	
38	4565	Coma Berenices	Sb galaxy	12	36.3	+25	59	9.6	16 × 2	NEEDLE GALAXY
39	2392	Gemini	planetary nebula	07	29.2	+20	55	9.2	0.25	ESKIMO NEBULA
40	3626	Leo	Sb galaxy	11	20.1	+18	21	10.9	3 × 2	
41	Melotte 25	Taurus	open cluster	04	27	+16		0.5	330	HYADES
42	7006	Delphinus	globular cluster	21	01.5	+16	11	10.6	2.8	
43	7814	Pegasus	Sb galaxy	00	03.3	+16	09	10.3	6 × 3	
44	7479	Pegasus	SBb galaxy	23	04.9	+12	19	10.9	4.4 × 3.4	
45	5248	Boötes	Sc galaxy	13	37.5	+08	53	10.2	7 × 5	
46	2261	Monoceros	bright nebula	06	39.2	+08	44	—	3.5 × 1.5	HUBBLE'S VARIABLE NEBULA
47	6934	Delphinus	Globular cluster	20	34.2	+07	24	8.9	5.9	
48	2775	Cancer	Sa galaxy	09	10.3	+07	02	10.1	5 × 4	
49	2237-9	Monoceros	bright nebula	06	32.3	+05	03	—	80 × 60	ROSETTE NEBULA
50	2244	Monoceros	open cluster	06	32.4	+04	52	4.8	24	
51	IC1613	Cetus	irregular galaxy	01	04.8	+02	07	9.2	12 × 11	
52	4697	Virgo	E4 galaxy	12	48.6	−05	48	9.3	6 × 4	
53	3115	Sextans	SO galaxy	10	05.2	−07	43	8.9	8 × 3	SPINDLE GALAXY
54	2506	Monoceros	open cluster	08	00.2	−10	47	7.6	7	
55	7009	Aquarius	planetary nebula	21	04.2	−11	22	8.3	0.3 × 1.7	SATURN NEBULA

CALDWELL CATALOGUE (CONTINUED)

C	NGC/IC	Constellation	Type	RA (2000.0) h	min	Dec. °	′	Magnitude	Size (′)	Comment
56	246	Cetus	planetary nebula	00	47.0	−11	53	8.6	4 × 3	
57	6822	Sagittarius	irregular galaxy	19	44.9	−14	48	8.8	20 × 10	
58	2360	Canis Major	open cluster	07	17.8	−15	37	7.2	13	
59	3242	Hydra	planetary nebula	10	24.8	−18	38	8.6	0.2 × 0.3	GHOST OF JUPITER
60	4038	Corvus	Sc galaxy	12	01.9	−18	52	10.5	2.6 × 2	ANTENNAE
61	4039	Corvus	Sp galaxy	12	01.9	−18	53	10.5	2.6 × 2	ANTENNAE
62	247	Cetus	SAB galaxy	00	47.1	−20	46	9.1	20 × 7	
63	7293	Aquarius	planetary nebula	22	29.6	−20	48	6.5	12.8	HELIX NEBULA
64	2362	Canis Major	open cluster	07	18.8	−24	57	4.1	8	TAU CANIS MAJORIS CLUSTER
65	253	Sculptor	Scp galaxy	00	47.6	−25	17	7.1	25.1 × 7.4	SILVER COIN GALAXY
66	5694	Hydra	globular cluster	14	39.6	−26	32	10.2	3.6	
67	1097	Fornax	SBb galaxy	02	46.3	−30	16	9.2	9 × 7	
68	6729	Corona Australis	bright nebula	19	01.9	−36	58	—	1.0	R CORONAE AUSTRALIS NEBULA
69	6302	Scorpius	planetary nebula	17	13.7	−37	06	9.6	2 × 1	BUG NEBULA
70	300	Sculptor	Sd galaxy	00	54.9	−37	41	8.7	20.0 × 14.8	
71	2477	Puppis	open cluster	07	52.3	−38	33	5.8	27	
72	55	Sculptor	SB galaxy	00	14.9	−39	11	8.2	32.4 × 6.5	
73	1851	Columba	globular cluster	05	14.1	−40	03	7.3	11	
74	3132	Vela	planetary nebula	10	07.7	−40	26	8.2	1.4 × 0.9	SOUTHERN RING NEBULA
75	6124	Scorpius	open cluster	16	25.6	−40	40	5.8	29	
76	6231	Scorpius	open cluster	16	54.0	−41	48	2.6	15	
77	5128	Centaurus	radio galaxy	13	25.5	−43	01	7.7	18.2 × 14.3	CENTAURUS A
78	6541	Corona Australis	globular cluster	18	08.0	−43	42	6.6	13	
79	3201	Vela	globular cluster	10	17.6	−46	25	6.7	18	
80	5139	Centaurus	globular cluster	13	25.8	−47	29	3.6	36	OMEGA CENTAURI
81	6352	Ara	globular cluster	17	25.5	−48	25	8.1	7.1	
82	6193	Ara	open cluster	16	41.3	−48	46	5.2	15	
83	4945	Centaurus	SBc galaxy	13	05.4	−49	28	8.7	20 × 4	
84	5286	Centaurus	globular cluster	13	46.4	−51	22	7.6	9	
85	IC2391	Vela	open cluster	08	40.2	−53	04	2.5	50	O VELORUM CLUSTER
86	6397	Ara	globular cluster	17	40.7	−53	40	5.6	25.7	
87	1261	Horologium	globular cluster	03	12.3	−55	13	8.4	7	
88	5823	Circinus	open cluster	15	05.7	−55	36	7.9	10	
89	6087	Norma	open cluster	16	18.9	−57	54	5.4	12	S NORMAE CLUSTER
90	2867	Carina	planetary nebula	09	21.4	−58	19	9.7	12	
91	3532	Carina	open cluster	11	06.4	−58	40	3.0	55	
92	3372	Carina	bright nebula	10	43.8	−59	52	—	120 × 120	ETA CARINAE NEBULA
93	6752	Pavo	globular cluster	19	10.9	−59	59	5.4	20.4	
94	4755	Crux	open cluster	12	53.6	−60	20	4.2	10	KAPPA CRUCIS CLUSTER
95	6025	Triangulum Australe	open cluster	16	03.7	−60	30	4.0	12	
96	2516	Carina	open cluster	07	58.3	−60	52	3.8	30	
97	3766	Centaurus	open cluster	11	36.1	−61	37	5.3	12	
98	4609	Crux	open cluster	12	42.3	−62	58	6.9	5	
99	—	Crux	dark nebula	12	53	−63	—	—	420 × 300	COALSACK
100	IC2944	Centaurus	cluster & nebula	11	36.6	−63	02	4.5	60 × 40	LAMBDA CENTAURI CLUSTER
101	6744	Pavo	SBb galaxy	19	09.8	−63	51	8.3	16 × 10	
102	IC2602	Carina	open cluster	10	43.2	−64	24	2.0	50	SOUTHERN PLEIADES
103	2070	Dorado	bright nebula	05	38.7	−69	06	3	40 × 25	TARANTULA NEBULA (30 DORADUS)
104	362	Tucana	globular cluster	01	03.2	−70	51	6.6	12.9	
105	4833	Musca	globular cluster	13	00.0	−70	53	7.3	13.5	
106	104	Tucana	globular cluster	00	24.1	−72	05	4.0	30.9	47 TUCANAE
107	6101	Apus	globular cluster	16	25.8	−72	12	9.3	11	
108	4372	Musca	globular cluster	12	25.8	−72	40	7.8	18.6	
109	3195	Chamaeleon	planetary nebula	10	09.5	−80	52	8.4	40 × 30	

Constellations

CONSTELLATIONS						
Constellation	**Name**	**Genitive**	**Abbr.**	**Area (sq deg)**	**First-magnitude stars**	**Star Map**
Andromeda*	Andromeda	Andromedae	And	722		12
Antlia	The Air Pump	Antliae	Ant	239		19
Apus	The Bird of Paradise	Apodis	Aps	206		22
Aquarius*	The Water Carrier	Aquarii	Aqr	980 Z		14
Aquila*	The Eagle	Aquilae	Aql	652	Altair	8
Ara*	The Altar	Arae	Ara	237		20
Aries*	The Ram	Arietis	Ari	441 Z		12
Auriga*	The Charioteer	Aurigae	Aur	657	Capella	18
Boötes*	The Herdsman	Boötis	Boo	907	Arcturus	4
Caelum	The Chisel	Caeli	Cae	125		22
Camelopardalis	The Giraffe	Camelopardalis	Cam	757		3
Cancer*	The Crab	Cancri	Cnc	506 Z		5
Canes Venatici	The Hunting Dogs	Canum Venaticorum	CVn	465		1
Canis Major*	The Great Dog	Canis Majoris	CMa	380	Sirius	16
Canis Minor*	The Little Dog	Canis Minoris	CMi	183	Procyon	16
Capricornus*	The Sea Goat	Capricorni	Cap	414 Z		14
Carina	The Keel	Carinae	Car	494	Canopus	19
Cassiopeia*	Cassiopeia	Cassiopeiae	Cas	598		3
Centaurus*	The Centaur	Centauri	Cen	1060	Alpha, Beta Centauri	20
Cepheus*	Cepheus	Cephei	Cep	588		3
Cetus*	The Whale	Ceti	Cet	1232		15
Chamaeleon	The Chameleon	Chamaeleontis	Cha	132		22
Circinus	The Compasses	Circini	Cir	93		20
Columba	The Dove	Columbae	Col	270		16
Coma Berenices	Berenice's Hair	Comae Berenices	Com	386		4
Corona Australis*	The Southern Crown	Coronae Australis	CrA	128		11
Corona Borealis*	The Northern Crown	Coronae Borealis	CrB	179		4
Corvus*	The Crow	Corvi	Crv	184		7
Crater*	The Cup	Crateris	Crt	282		7
Crux	The Southern Cross	Crucis	Cru	68	Alpha, Beta Crucis	20
Cygnus*	The Swan	Cygni	Cyg	804	Deneb	8
Delphinus*	The Dolphin	Delphini	Del	189		8
Dorado	The Goldfish	Doradus	Dor	179		22
Draco*	The Dragon	Draconis	Dra	1083		2
Equuleus*	The Foal	Equulei	Equ	72		8
Eridanus*	The River	Eridani	Eri	1138	Achernar	15; 22
Fornax	The Furnace	Fornacis	For	398		15
*Gemini	The Twins	Geminorum	Gem	514 Z	Pollux	17
Grus	The Crane	Gruis	Gru	366		21
Hercules*	Hercules	Herculis	Her	1225		9
Horologium	The Pendulum Clock	Horologii	Hor	249		22
Hydra*	The Water Snake	Hydrae	Hya	1303		7
Hydrus	The Little Water Snake	Hydri	Hyi	243		22
Indus	The Indian	Indi	Ind	294		21

** Ptolemy's original 48 constellations (excluding Argo, since divided into Carina, Puppis and Vela)   Z = Zodiacal constellations*

CONSTELLATIONS (CONTINUED)						
Constellation	Name	Genitive	Abbr.	Area (sq deg)	First-magnitude stars	Star Map
Lacerta	The Lizard	Lacertae	Lac	201		3
Leo*	The Lion	Leonis	Leo	947 Z	Regulus	5
Leo Minor	The Little Lion	Leonis Minoris	LMi	232		1
Lepus*	The Hare	Leporis	Lep	290		16
Libra*	The Balance	Librae	Lib	538 Z		6
Lupus*	The Wolf	Lupi	Lup	334		20
Lynx	The Lynx	Lyncis	Lyn	545		18
Lyra*	The Lyre	Lyrae	Lyr	286	Vega	8
Mensa	The Table	Mensae	Men	153		22
Microscopium	The Microscope	Microscopii	Mic	210		21
Monoceros	The Unicorn	Monocerotis	Mon	482		16
Musca	The Fly	Muscae	Mus	138		22
Norma	The Set Square	Normae	Nor	165		20
Octans	The Octant	Octantis	Oct	291		20
Ophiuchus*	The Serpent Bearer	Ophiuchi	Oph	948		10
Orion*	Orion	Orionis	Ori	594	Rigel, Betelgeux	16
Pavo	The Peacock	Pavonis	Pav	378		21
Pegasus*	The Winged Horse	Pegasi	Peg	1121		13
Perseus*	Perseus	Persei	Per	615		12
Phoenix	The Phoenix	Phoenicis	Phe	469		21
Pictor	The Painter	Pictoris	Pic	247		19
Pisces*	The Fish	Piscium	Psc	889 Z		13
Piscis Austrinus*	The Southern Fish	Piscis Austrini	PsA	245	Fomalhaut	14
Puppis	The Stern	Puppis	Pup	673		19
Pyxis	The Compass	Pyxidis	Pyx	221		19
Reticulum	The Net	Reticuli	Ret	114		22
Sagitta*	The Arrow	Sagittae	Sge	80		8
Sagittarius*	The Archer	Sagittarii	Sgr	867 Z		11
Scorpius*	The Scorpion	Scorpii	Sco	497 Z	Antares	11
Sculptor	The Sculptor	Sculptoris	Scl	475		21
Scutum	The Shield	Scuti	Sct	109		8
Serpens*	The Serpent	Serpentis	Ser	637		10
Sextans	The Sextant	Sextantis	Sex	314		5
Taurus*	The Bull	Tauri	Tau	797 Z	Aldebaran	17
Telescopium	The Telescope	Telescopii	Tel	252		20
Triangulum*	The Triangle	Trianguli	Tri	132		12
Triangulum Australe	The Southern Triangle	Trianguli Australis	TrA	110		20
Tucana	The Toucan	Tucanae	Tuc	295		21
Ursa Major*	The Great Bear	Ursae Majoris	UMa	1280		1
Ursa Minor*	The Little Bear	Ursae Minoris	UMi	256		2
Vela	The Sails	Velorum	Vel	500		19
Virgo*	The Virgin	Virginis	Vir	1294 Z	Spica	6
Volans	The Flying Fish	Volantis	Vol	141		19
Vulpecula	The Fox	Vulpeculae	Vul	268		8

** Ptolemy's original 48 constellations (excluding Argo, since divided into Carina, Puppis and Vela)   Z = Zodiacal constellations*

Ursa Major, Canes Venatici, Leo Minor

Magnitudes

−1

0

1

2

3

4

5

Variable star

Galaxy

Planetary nebula

Gaseous nebula

Globular cluster

Open cluster

Ursa Major. There can be few people who cannot recognize the group of stars known as the Plough – alternatively nicknamed King Charles' Wain or, in America, the Big Dipper. The seven main stars make up an unmistakable pattern, but in fact only five of them share a common motion in space and presumably have a common origin; the remaining two – Dubhe (α) and Alkaid (η) – are moving through space in the opposite direction, so that after a sufficient length of time the Plough shape will become distorted. Of the seven, six are hot and white, but Dubhe is obviously orange; the colour is detectable with the naked eye, and binoculars bring it out well.

It is interesting that Megrez (δ) is about a magnitude fainter than the rest. In 1603 Bayer, who drew up a famous star catalogue and gave the stars their Greek letters, gave its magnitude as 2; but earlier cataloguers ranked it as 3, and there has probably been no real change. It is 65 light years away, and 17 times as luminous as the Sun.

Of course the most celebrated star in the constellation is ζ (Mizar) with its naked-eye companion Alcor (80 Ursae Majoris). Strangely, the Arabs of a thousand years ago regarded Alcor as a test of keen eyesight, but today anyone with average eyes can see it when the sky is reasonably dark and clear. A small telescope will show that Mizar itself is double, but the separation (14.4 seconds of arc) is too small for the two stars to be seen separately with the naked eye, or even binoculars. Between Alcor and the two Mizars is a fainter star which was named Sidus Ludovicianum in 1723 by courtiers of Emperor Ludwig V, who believed that it had appeared suddenly. Ludwig's Star can be seen with powerful binoculars, and it has been suggested that it might have been the 'test' referred to by the Arabs, but certainly it would have been a very severe one – even if the star is slightly variable.

Outside the Plough pattern is a triangle of fainter stars: ψ, λ and μ. The two latter stars are in the same binocular field, and make a good colour contrast. λ is white, while μ, with its M-type spectrum, is very red.

ξ, close to ν, was one of the first binary stars to have its orbit computed. The components are of magnitudes 4.3 and 4.8, and the period is 59.8 years; but the separation is currently only about 1.7 seconds of arc, so a very small telescope will not split the pair. (Generally speaking, a 7.6-centimetre (3-inch) refractor will be able to divide pairs down to a separation of about 1.8 seconds of arc, assuming that the components are more or less equal and are not too faint.) There are not many notable variables in the constellation, but the red semi-regular Z Ursae Majoris, easy to find because of its closeness to Megrez, is a favourite test subject for newcomers to variable-star work.

There are four Messier objects in the constellation. One of these, M97, is the famous planetary nebula known as the Owl Nebula. It was discovered by Pierre Méchain in 1781, who recorded it as being 'difficult to see', and certainly it can be elusive; the two embedded stars which give it its owlish appearance are no brighter than magnitude 14, and the whole nebula is faint. It lies not far from β (Merak), and it can be seen with a 7.6-centimetre (3-inch) telescope when the sky is dark and clear. M81 and M82 are within binocular range, not far from 24 Ursae Majoris; M81 is a spiral, while M82 is a peculiar system which is a strong radio source. Each is about 8.5 million light years away, and they are associated with each other. The other Messier object, M101, was also discovered by Méchain; it forms an equilateral triangle with Mizar and Alkaid, and is a loose spiral whose surface brightness is rather low. It is face on to us, and photographs can often show it beautifully.

Though all the main stars of Ursa Major are well below the first magnitude, their proper names are often used. There are, incidentally, two alternatives; η may be called Benetnasch as well as Alkaid, while γ is also known as Phekda or Phecda.

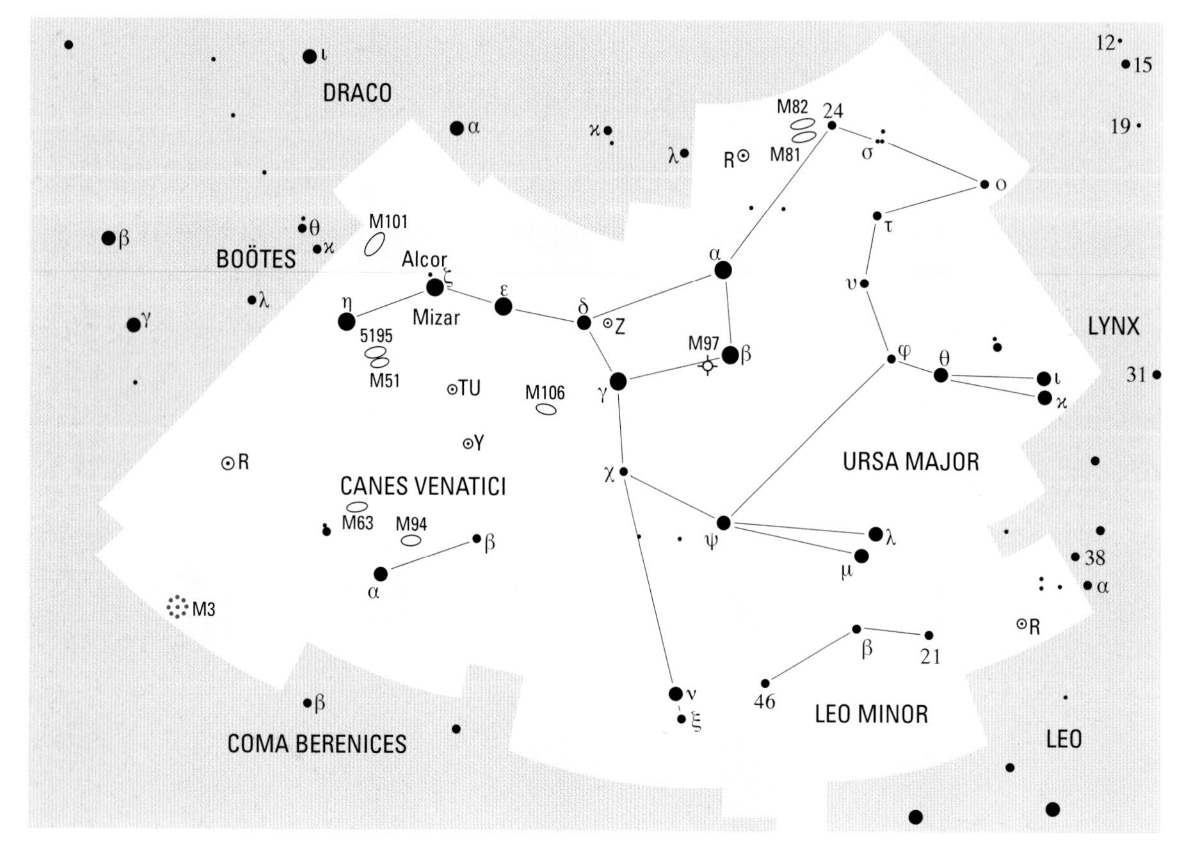

◀ **Ursa Major** is the most famous of all the northern constellations, and can be used as a guide to find many of the less prominent groups. The 'Plough' is only part of the entire constellation, but its seven main stars cannot be mistaken: they are circumpolar over the British Isles and parts of Europe and North America. They are always low over South Africa and Australia, but only from parts of New Zealand and South America and from Antarctica is the 'Plough' completely lost. Canes Venatici and Leo Minor adjoin Ursa Major.

◀ **The Whirlpool Galaxy, M51, in Canes Venatici,** imaged by the NOAO Mosaic CCD camera at the Kitt Peak National Observatory. The Whirlpool Galaxy consists of NGC5194, a large spiral galaxy, and a smaller companion, NGC5195. The red areas are nebulae within the galaxy. M51 is about 37 million light years away.

Canes Venatici, the Hunting Dogs – Asterion and Chara – were added to the sky by Hevelius in 1690; they are held by the herdsman Boötes – possibly to stop them from chasing the bears round the celestial pole. The only bright star, α^2, was named Cor Caroli (Charles' Heart) by the second Astronomer Royal, Edmond Halley, in honour of King Charles I. The star shows interesting periodical changes in its spectrum, due probably to variations in its magnetic field. It is 65 light years away, and 80 times as luminous as the Sun. Its companion, of magnitude 5.5, lies at a separation of over 19 seconds of arc, so that this is a very easy pair.

The semi-regular variable Y Canum Venaticorum lies about midway between β (Chara) and Mizar. It is one of the reddest stars known, and has been named La Superba; at maximum it is visible with the naked eye, but binoculars are needed to bring out its vivid colour.

M51, the Whirlpool Galaxy, lies near the border of Canes Venatici, less than 4° from Alkaid. It was discovered by Messier himself in 1773, and is the perfect example of a face-on spiral; its distance is around 37 million light years. It was the first spiral to be seen as such, by Lord Rosse in 1845. Though it is a difficult binocular object, a modest telescope – say a 30-centimetre (12-inch) – is adequate to show its form; it is linked with its companion, NGC5195. M94, not far from Cor Caroli, is also a face-on spiral, and though it is small it is not difficult to find, because its nucleus is bright and distinct.

The other Messier spirals in Canes Venatici, M63 and M106, are less striking. M63 is also a spiral, but the arms are much less obvious. M106 – added later to the Messier catalogue – has one arm which is within range of a 25-centimetre (10-inch) reflector.

M3 is one of the most splendid globular clusters in the sky. It lies almost midway between Cor Caroli and Arcturus (α Boötis), near the fainter star β Comae (magnitude 4.6) and is easy to find with binoculars, while it can be partly resolved into stars using a telescope with an aperture of more than 7.6 centimetres (3 inches). Like all globular clusters, it is a long way away – more than 48,000 light years – and is particularly rich in RR Lyrae variables. The total mass has been given as around 245,000 times that of the Sun. Not surprisingly, it is a favourite target for amateur astrophotographers. The integrated magnitude is about 6.4, so it is not very far below naked-eye visibility; Messier discovered it in 1764.

Leo Minor is a small constellation with very dubious claims to a separate identity; it was first shown by Hevelius, on his maps of 1690. The system of allotting Greek letters has gone badly wrong here, and the only star so honoured is β, which is not even the brightest star in the group. The leader is 46 Leonis Minoris, which has been given a separate name: Praecipua. It is in the same binocular field as ν and ξ Ursae Majoris, and can be identified by its decidedly orange colour. The only object of any interest is the Mira-type variable R Leonis Minoris, which can reach magnitude 6.3 at maximum, but sinks to below 13 when at its faintest.

Hevelius had a habit of creating new constellations. Some of these have survived; as well as Leo Minor, there are Camelopardalis, Canes Venatici, Lacerta, Lynx, Scutum, Monoceros, Sextans and Vulpecula, while others, such as Triangulum Minor (the Little Triangle) and Cerberus (Pluto's three-headed dog), have now been rejected. The constellation Leo Minor – which has no mythological significance – was formerly included in Ursa Major, and logically should probably have remained there.

URSA MAJOR

BRIGHTEST STARS

No.	Star	R.A.			Dec.			Mag.	Spectrum	Proper name
		h	m	s	°	'	"			
77	ε	12	54	02	+55	57	35	1.77	A0	Alioth
50	α	11	03	44	+61	45	03	1.79	K0	Dubhe
85	η	13	47	32	+49	18	48	1.86	B3	Alkaid
79	ζ	13	23	56	+54	55	31	2.09	A0	Mizar
48	β	11	01	50	+56	22	56	2.37	A1	Merak
64	γ	11	53	50	+53	41	41	2.44	A0	Phad
52	ψ	11	09	40	+44	29	54	3.01	K1	
34	μ	10	22	20	+41	29	58	3.05	M0	Tania Australis
9	ι	08	59	12	+48	02	29	3.14	A7	Talita
25	θ	09	32	51	+51	40	38	3.17	F6	
69	δ	12	15	25	+57	01	57	3.31	A3	Megrez
1	ο	08	30	16	+60	43	05	3.36	G4	Muscida
33	λ	10	17	06	+42	54	52	3.45	A2	Tania Borealis
54	ν	11	18	29	+33	05	39	3.48	K3	Alula Borealis

Also above mag. 4.3: ϰ (Al Kaprah) (3.60), h (3.67), χ (Alkafzah) (3.71), ξ (3.79), 10 (4.01).

VARIABLES

Star	R.A.		Dec.		Range	Type	Period	Spectrum
	h	m	°	'	(mags)		(d)	
R	10	44.6	+68	47	6.7–13.4	Mira	302	M
Z	11	56.5	+57	52	6.8–9.1	Semi-reg.	196	M

DOUBLES

Star	R.A.		Dec.		P.A.	Sep.	Mags	
	h	m	°	'	°	"		
ν	11	18.5	+33	06	147	7.2	3.5, 9.9	
ζ	13	23.9	+54	56	AB 152	14.4	2.3, 4.0	Mizar/Alcor
					AC 071	708.7	2.1, 4.0	

GALAXIES AND NEBULAE

M	NGC	R.A.		Dec.		Mag.	Dimensions	Type
		h	m	°	'		'	
81	3031	09	55.6	+69	04	6.9	25.7 × 14.1	Sb galaxy
82	3034	09	55.8	+69	41	8.4	11.2 × 4.6	irregular galaxy
97	3587	11	14.8	+55	01	12	194"	planetary nebula (Owl)
101	5457	14	03.2	+54	21	7.7	26.9 × 26.3	Sc galaxy

CANES VENATICI

BRIGHTEST STARS

No.	Star	R.A.			Dec.			Mag.	Spectrum	Proper name
		h	m	s	°	'	"			
12	α^2	12	56	02	+38	19	06	2.90	A0p	Cor Caroli

Also above mag. 4.3: β (Chara) (4.26).

VARIABLES

Star	R.A.		Dec.		Range	Type	Period	Spectrum
	h	m	°	'	(mags)		(d)	
R	13	49.0	+39	33	6.5–12.9	Mira	329	M
TU	12	54.9	+47	12	5.6–6.6	Semi-reg.	50	M
Y	12	45.1	+45	26	4.8–6.6	Semi-reg.	157	N

DOUBLE

Star	R.A.		Dec.		P.A.	Sep.	Mags
	h	m	°	'	°	"	
α^2	12	56.0	+38	19	22.9	19.4	2.9, 5.5

CLUSTERS AND GALAXIES

M	NGC	R.A.		Dec.		Mag.	Dimensions	Type
		h	m	°	'		'	
3	5272	13	42.2	+28	23	6.4	16.2	globular cluster
51	5194–5	13	29.9	+47	12	8.4	11.0 × 7.8	Sc galaxy (Whirlpool)
63	5055	13	15.8	+42	02	8.6	12.3 × 7.6	Sb galaxy
94	4736	12	50.9	+41	07	8.2	11.0 × 9.1	Sb galaxy
106	4258	12	19.0	+47	18	8.3	18.2 × 7.9	Sb galaxy
	5195	13	30.0	+47	16	9.6	5.4 × 4.3	companion to M51

LEO MINOR

The brightest star is 46 (Praecipua), R.A. 10h 53m, dec. +34° 13', mag. 3.83. Also above mag. 4.3: β (4.21).

VARIABLE

Star	R.A.		Dec.		Range	Type	Period	Spectrum
	h	m	°	'	(mags)		(d)	
R	09	45.6	+34	31	6.3–13.2	Mira	372	M

Cassiopeia, Cepheus, Camelopardalis,

Cassiopeia. The W shape of the constellation Cassiopeia is unmistakable, and is of special interest because one member of the pattern is definitely variable, while another probably is.

The confirmed variable is γ, with a peculiar spectrum which shows marked variations. No changes in light seem to have been recorded until about 1910, and the magnitude had been given as 2.25. The star then slowly brightened, and there was a rapid increase during late 1936 and early 1937, when the magnitude rose to 1.6. A decline to below magnitude 3 followed by 1940, and then came a slow brightening; ever since the mid-1950s the magnitude has hovered at around 2.2, slightly fainter than Polaris and slightly brighter than β (Chaph). There is certainly no period; what apparently happens is that the star throws off shells of material and brightens during the process. A few other stars of the same type are known – Pleione in the Pleiades (28 Tauri) and Dschubba (δ Scorpii) are good examples – but what are now known as 'GCAS' or Gamma Cassiopeiae variables, are rare. All of them seem to be rapid rotators. There may be a new brightening of γ at any time, so the luminosity rises to about 6000 times that of our own Sun.

α (Shedir) is decidedly orange, with a K-type spectrum. It is 230 light years away, and over 800 times as luminous as the Sun. Previously it was accepted as being variable, with a probable range of between magnitudes 2.2 and 2.8; it was even suggested that there might be a rough period of about 80 days. Later observers failed to confirm the changes, and in modern catalogues α is often listed as 'constant', though my own observations between 1933 and the present time indicate that there are slight, random fluctuations between magnitudes 2.1 and 2.4, with a mean of 2.3. Generally speaking, the order of brilliance of the three main members of the W is γ, α, β, but this is not always the case, and watching the slight variations is a good exercise for the naked-eye observer. β itself fluctuates very slightly, but the range is less than 0.04 of a magnitude, so that in estimating γ and α it is safe to take the magnitude of β as 2.27. β is the nearest and brightest of the δ Scuti class of variables

ϱ, which lies close to β, is one of the rare class of 'hypergiant' stars – 10,000 light years away, and 500,000 times as luminous as the Sun. Normally it is of around magnitude 4.8, comparable with σ (4.88) and τ (4.72), but occasionally it drops by two magnitudes, as it did in 1946 and 2000. It is an unstable star, and may well suffer a supernova outburst at any moment. It is an excellent target for the binocular observer.

R Cassiopeiae, a normal Mira star, can reach naked-eye visibility at maximum. The supernova of 1572 flared up near ϰ; the site is now identified by its radio emissions. η (Archid) is a wide, easy double. ι is also easy, and there is another seventh-magnitude companion at a separation of just over 8 seconds of arc.

There are two Messier open clusters in Cassiopeia, neither of which is of special note; indeed M103 is less prominent than its neighbour NGC663 (C10), and it is not easy to see why Messier gave it preference. NGC457 (C13) is of more interest. It contains several hundred stars, and is an easy binocular object. φ Cassiopeiae, at magnitude 4.98, lies in its south-eastern edge, but it is thought to be a foreground object and not a member of the cluster. NGC457 is superior to the Messier clusters in Cassiopeia.

The Milky Way crosses Cassiopeia, and the whole constellation is very rich. Here too we find the galaxies Maffei 1 and 2, which are so heavily obscured that they are difficult to see; Maffei 1 is the closest giant elliptical galaxy to our Galaxy, but is no longer considered part of the Local Group.

Cepheus, the King, is much less prominent than his Queen. α (Alderamin) is of magnitude 2.4, and is 45 light years away, with a luminosity 14 times that of the Sun. The quadrilateral made up of α, β (Alphirk), ι and ζ is not particularly hard to identify.

Magnitudes
–1
0
1
2
3
4
5
Variable star
Galaxy
Planetary nebula
Gaseous nebula
Globular cluster
Open cluster

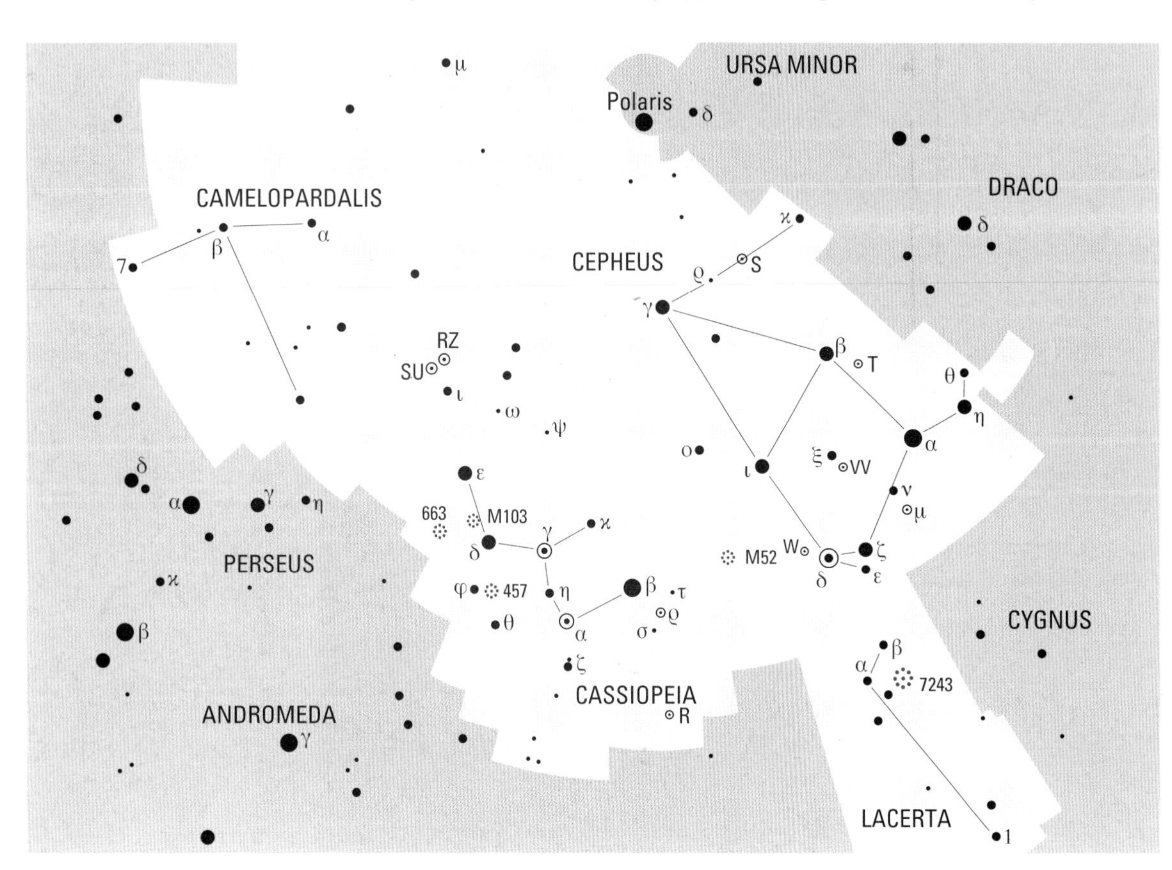

◄ **Apart from Ursa Major,** Cassiopeia is much the most conspicuous of the far northern constellations. It and Ursa Major lie on opposite sides of the celestial pole, so when Ursa Major is high up, Cassiopeia is low down, and vice versa – though neither actually sets over any part of the British Isles or the northern United States. Cepheus is much less prominent, and is all but lost from southern countries; Lacerta and Camelopardalis are very obscure.

Lacerta

The main interest in Cepheus is centred upon three variable stars: δ, μ and VV. δ is the prototype Cepheid, with a range from 3.5 to 4, and has given its name to the whole class; its behaviour was explained in the 18th century by the young deaf-mute astronomer John Goodricke. It forms a small triangle with ζ (3.35) and ε (4.19) which make good comparison stars, though δ never becomes as bright as ζ. The magnitude-7.5 companion is an easy telescopic object, and seems to be genuinely associated, since it and the variable share a common motion in space.

μ Cephei is so red that William Herschel nicknamed it the Garnet Star; although the light level is too low for the colour to be evident with the naked eye, binoculars bring it out beautifully. The range of μ Cephei is between magnitudes 3.4 and 5.1, but the usual value is about 4.3, so that the nearby ν (at magnitude 4.29) makes a convenient comparison. It has been suggested that μ may be of the semi-regular type, but it is difficult to find any real periodicity. The distance has been given as 1500 light years, in which case the luminosity is more than 50,000 times that of the Sun – making it much more powerful than Betelgeux (α Orionis), which is of the same type. The Garnet Star is so luminous that if it were as close as, say, Pollux in Gemini, the apparent magnitude would be −7, and it would be conspicuous in the sky even in broad daylight.

VV Cephei, close to ξ (4.29), is a huge eclipsing binary of the Zeta Aurigae type. The system consists of a red supergiant together with a smaller hot blue companion; the range is small – magnitude 4.7 to 5.4 – and the orbital period is 7430 days, or 20.3 years. It is thought that the diameter of the supergiant may be as much as 1800 times that of the Sun, in which case it is one of the largest stars known. The last eclipse occurred in 1996.

Two other variables in the constellation are worth mentioning. W, close to δ, is a red semi-regular with a long but uncertain period. Telescope users may care to pick out the Mira variable S, which is one of the reddest of all stars. There are no Messier objects in Cepheus.

Camelopardalis (alternatively known as Camelopardus). This is a very barren far-northern constellation, and was introduced to the sky by Hevelius in the year 1690. There is little really of much interest to observers here, but it is worth noting that the three brightest stars, α, β and 7, are all very remote and luminous; 7 is well over 50,000 times as powerful as the Sun.

Lacerta. Although Lacerta – the Lizard – was one of the original constellations listed by Ptolemy, it is very small and obscure. There is a small 'diamond' of dim stars, of which α is the brightest; to find them, use ζ and ε Cephei as guides. ε Cephei is just within the same binocular field as β Lacertae.

The only object of any note is the open cluster NGC7243 (C16), which forms an equilateral triangle with α and β and is just within binocular range. A bright nova (CP) flared up in Lacerta in 1936, and reached magnitude 1.9, but it faded quickly, and is now below the 15th magnitude.

BL Lacertae, which is too faint to be of interest to the user of a small telescope, was once regarded as an ordinary run-of-the-mill variable star, and was given the appropriate designation, but when its spectrum was examined it was found to be something much more dramatic, and is more akin to a quasar. It has given its name to the whole class of such objects (see page 199), which are conventionally known as 'BL Lacs'.

CASSIOPEIA

BRIGHTEST STARS

No.	Star	R.A.			Dec.			Mag.	Spectrum	Proper name
		h	m	s	°	'	"			
27	γ	00	56	42	+60	43	00	2.2v	B0p	
18	α	00	40	30	+56	32	15	2.2v?	K0	Shedir
11	β	00	09	11	+59	08	59	2.27	F2	Chaph
37	δ	0	25	49	+60	14	07	2.68	A5	Ruchbah
45	ε	01	54	24	+63	40	13	3.38	B3	Segin
24	η	00	49	06	+57	48	58	3.44	G0	Achird

Also above magnitude 4.3: ζ (3.67), ι (3.98), ϰ (4.16); next comes θ (Marfak) (4.33). ϱ is an irregular variable which can at times exceed magnitude 4.3, but is usually nearer 4.8.

VARIABLES

Star	R.A.		Dec.		Range	Type	Period	Spectrum
	h	m	°	'	(mags)		(d)	
ϱ	23	54.4	+58	30	4.1–6.2	?	–	F
γ	00	56.7	+60	43	1.6–3.3	Irregular	–	Bp
α	00	40.5	+56	22	2.1–2.5?	Suspected	–	K
R	23	58.4	+51	24	4.7–13.5	Mira	431	M
SU	02	52.0	+68	53	5.7–6.2	Cepheid	1.95	F
RZ	02	48.9	+69	38	6.2–7.7	Algol	1.19	A

DOUBLES

Star	R.A.		Dec.		P.A.	Sep.	Mags	
	h	m	°	'	°	"		
η	00	49.1	+57	49	315	12.6	3.4, 7.5	binary, 480y
ι	02	29.1	+67	24	232	2.4	4.9, 6.9	binary, 840y

CLUSTERS

M	C	NGC	R.A.		Dec.		Mag.	Dimensions	Type
			h	m	°	'		'	
52		7654	23	24.2	+61	35	7.3	13	open cluster
103		581	01	33.2	+60	42	7.4	6	open cluster
	10	663	01	46.0	+61	15	7.1	16	open cluster
	13	457	01	19.1	+58	20	6.4	13	open cluster round φ Cas

CEPHEUS

BRIGHTEST STARS

No.	Star	R.A.			Dec.			Mag.	Spectrum	Proper name
		h	m	s	°	'	"			
5	α	21	18	35	+62	35	08	2.44	A7	Alderamin
35	γ	23	39	21	+77	37	57	3.21	K1	Alrai
8	β	21	28	39	+70	33	39	3.23v	B2	Alphirk
21	ζ	22	10	51	+58	12	05	3.35	K1	
3	η	20	45	17	+61	50	20	3.43	K0	

Also above magnitude 4.3: ι (3.52), ε (4.19), θ (4.22), ν (4.29), ξ (4.29).
The two famous variables can exceed magnitude 4 at maximum, δ and μ.

VARIABLES

Star	R.A.		Dec.		Range	Type	Period	Spectrum
	h	m	°	'	(mags)		(d)	
δ	22	29.2	+58	25	3.5–4.4	Cepheid	5.37	F-G
μ	21	43.5	+58	47	3.4–5.1	Irregular	–	M
T	21	09.5	+68	29	5.2–11.3	Mira	388	M
VV	21	56.7	+63	38	4.8–5.4	Eclipsing	7430	M+B
W	22	36.5	+58	26	7.0–9.2	Semi-regular	Long	K-M
S	21	35.2	+78	37	7.4–12.9	Mira	487	N

DOUBLES

Star	R.A.		Dec.		P.A.	Sep.	Mags	
	h	m	°	'	°	"		
ϰ	20	08.9	+77	43	122	7.4	4.4, 8.4	
β	21	28.7	+70	34	249	13.3	3.2, 7.9	
δ	22	29.2	+58	25	191	41.0	var, 7.5.	
ο	23	18.6	+68	07	220	2.9	4.9, 7.1	binary, 796y
ξ	22	03.8	+64	38	277	7.7	4.4, 6.5	binary, 3800y

CAMELOPARDALIS

The brightest star is β; R.A. 05h 03m, 25s 1, dec. +60° 26' 32", mag. 4.03. The only other stars above mag. 4.3 are 7 (4.21) and α (4.29).

LACERTA

A small, obscure constellation. The brightest star is α: R.A. 22h 31m 17s.3, dec. 150° 16' 17", mag. 3.77. The only other star above mag. 4.3 is 2 (4.13).

CLUSTER

C	NGC	R.A.		Dec.		Mag.	Dimensions	Type
		h	m	°	'		'	
16	7243	22	15.3	+49	53	6.4	21	open cluster, about 40 stars

Boötes, Corona Borealis, Coma Berenices

Magnitudes

–1

0

1

2

3

4

5

Variable star

Galaxy

Planetary nebula

Gaseous nebula

Globular cluster

Open cluster

Boötes. A large northern constellation, said to represent a herdsman who invented the plough drawn by two oxen – for which service to mankind he was rewarded with a place in the heavens.

The whole area is dominated by Arcturus (α), which is the brightest star in the northern hemisphere of the sky and is one of only four with negative magnitudes (the other three are Sirius, Canopus and α Centauri). It is a light orange K-type star, 36 light years away and with a luminosity 115 times that of the Sun; the diameter is about 30 million kilometres (about 19 million miles). It is too bright to be mistaken, but in case of any doubt it can be located by following through the tail of Ursa Major (α, ζ and η).

Arcturus has the exceptionally large proper motion of 2.3 seconds of arc per year, and as long ago as 1718 Edmond Halley found that its position relative to the background stars had shifted appreciably since ancient times. At the moment it is approaching us at the rate of 5 kilometres (3 miles) per second, but this will not continue indefinitely; in several thousand years' time it will pass by us and start to recede, moving from Boötes into Virgo and dropping below naked-eye visibility in 500,000 years. It is a Population II star belonging to the galactic halo, so its orbit is sharply inclined, and it is now cutting through the main plane of the Galaxy.

In 1860 a famous variable star observer, Joseph Baxendell, found a magnitude 9.7 star in the field of Arcturus, at a position angle of 250 degrees and a separation of 25 minutes of arc. Within a week it had disappeared, and has never been seen again, though it is still listed in catalogues as T Boötis. It may have been a nova or recurrent nova, and there is always the chance that it will reappear, so amateur observers make routine checks to see whether it has done so.

ε (Izar), the second brightest star in the constellation, is a fine double; the primary is an orange K-type star, while the companion looks rather bluish by contrast. No doubt the two stars have a common origin, but the revolution period must be immensely long. The primary is 400 times more luminous than the Sun, so it is more powerful than Arcturus, but it is also farther away – around 210 light years. The semi-regular variable W Boötis is in the same binocular field with ε, and is easily recognizable because of its orange-red hue.

ζ is a binary, with almost equal components (magnitudes 4.5 and 4.6) and an orbital period of 123 years, but the separation is never more than 1 second of arc, so a telescope of at least 13-centimetre (5-inch) aperture is needed to split it. There are no Messier objects in Boötes, and in fact no nebular objects with integrated magnitude as bright as 10.

One constellation which is still remembered, even though it is no longer to be found on our maps, is Quadrans Muralis (the Mural Quadrant), added to the sky by Bode in 1775. The nearest brightish star to the site is β Boötis (Nekkar), magnitude 3.5. Like all the rest of Bode's groups, Quadrans was later rejected, but it so happens that the meteors of the early January shower radiate from there, which is why we call them the Quadrantids. For this reason alone, there might have been some justification for retaining Quadrans.

Corona Borealis is a very small constellation, covering less than 180 square degrees of the sky (as against over 900 square degrees for Boötes), but it contains far more than its fair share of interesting objects. The brightest star, α or Alphekka (also known as Gemma), is second magnitude, and is actually an eclipsing binary with an unusually small range; the main component is 50 times as luminous as the Sun, while the fainter member of the pair has just twice the Sun's power. The real separation between the two is less than 30 million kilometres (19 million miles), so they cannot be seen separately. The distance from the Solar System is some 78 light years.

η is a close binary, with an average separation of 0.5 second of arc and yellowish components of magnitudes 5.6 and 5.9; it is a binary with a period of 41.6 years, and is a useful

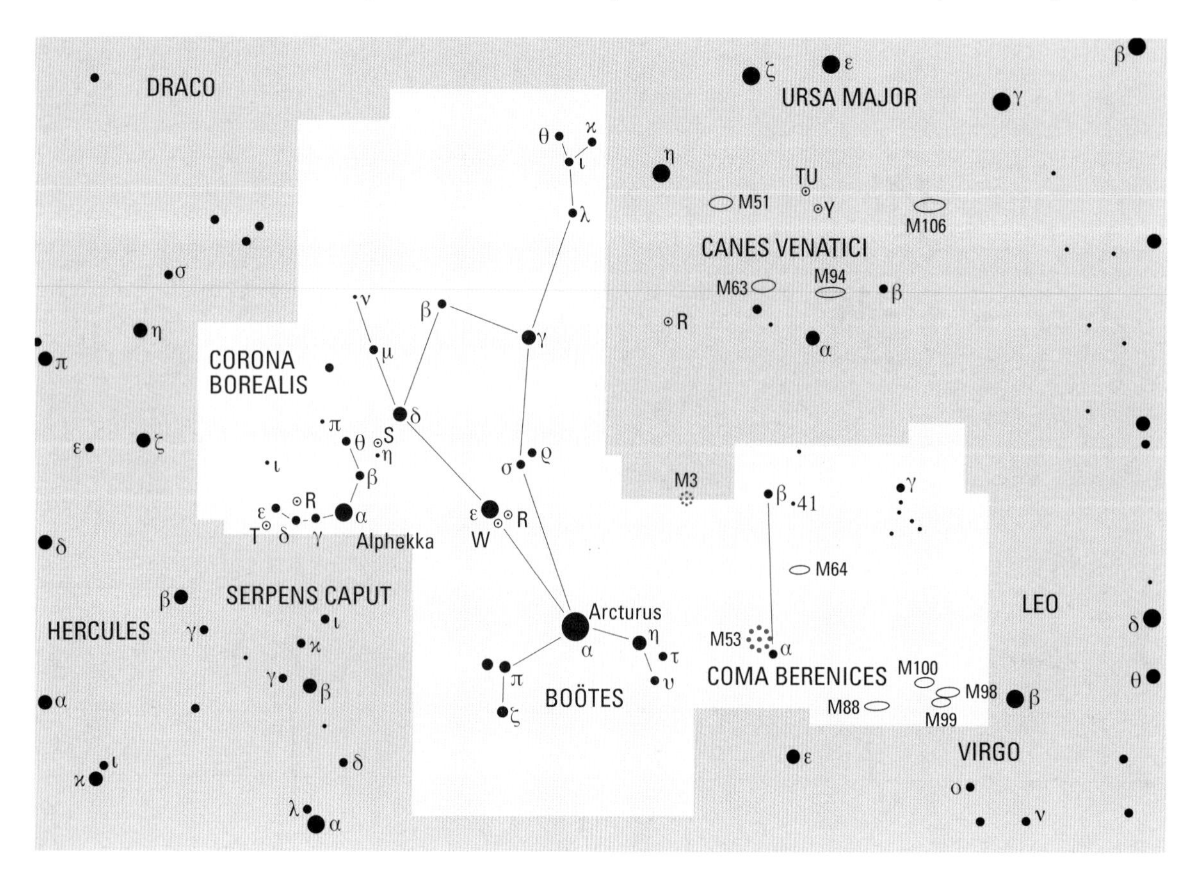

◄ **The map is dominated by Arcturus (α Boo),** the brightest star in the northern hemisphere of the sky; it is sufficiently close to the celestial equator to be visible from every inhabited country, and is at its best during evenings in northern spring (southern autumn). The Y-formation made up of Arcturus, ε and γ Boo, and Alphekka (α CrB) is distinctive. The rejected constellation of Quadrans is now included in Boötes, near β; it is from here that the meteors of early January radiate, which is why they are known as the Quadrantids.

test object for telescopes with apertures of around 13 centimetres (5 inches). There are optical companions at 58 seconds of arc (magnitude 12.5) and 215 seconds of arc (magnitude 10.0). ζ and σ are both easy doubles, while β is a spectroscopic binary, and is also a magnetic variable of the same type as Cor Caroli (α Canum Venaticorum).

Inside the bowl of the crown lies the celebrated variable R Coronae Borealis, which periodically veils itself behind clouds of soot in its atmosphere. Usually it is on the brink of naked-eye visibility, but it shows sudden, unpredictable drops to minimum. At its faintest it fades below magnitude 15, so it passes well out of the range of small telescopes; on the other hand there may be long periods when the light remains almost steady, as happened between 1924 and 1934. It is much the brightest member of its class, and of the rest only RY Sagittarii approaches naked-eye visibility.

R Coronae Borealis is a splendid target for binocular observers. Generally, binoculars will show two stars in this part of the bowl, R and a star (M) of magnitude 6.6. If you examine the area at low power and see only one star instead of two, you may be sure that R Coronae Borealis has 'taken a dive'.

Outside the bowl, near ε, is the Blaze Star, T Coronae Borealis, which is normally around tenth magnitude, but has shown two outbursts during the past century and a half; in 1866, when it reached magnitude 2.2 (equal to α, Alphekka), and again in 1946, when the maximum magnitude was about 3. On neither occasion did it remain a naked-eye object for more than a week, but it is worth keeping a watch on it, though if these sudden outbursts have any periodicity there is not likely to be another until around 2026.

Spectroscopic examination has shown that T Coronae Borealis is, in fact, a binary, made up of a hot B-type star together with a cool red giant. It is the B-star which is the site of the outbursts, while the red giant seems to be irregularly variable over a range of about a magnitude, causing the much smaller fluctuations observed when the star is at minimum. Other recurrent novae are known, but only the Blaze Star seems to be capable of becoming really prominent. Also in the constellation is S Coronae Borealis, a normal Mira variable which rises to the verge of naked-eye visibility when it is at its maximum.

Mythologically, Corona Borealis is said to represent a crown given by the wine-god Bacchus to Ariadne, daughter of King Minos of Crete.

Coma Berenices is not an original constellation – it was added to the sky by Tycho Brahe in 1690 – but there is a legend attached to it. When the King of Egypt set out upon a dangerous military expedition, his wife Berenice vowed that if he returned safely she would cut off her lovely hair and place it in the Temple of Venus. The king returned; Berenice kept her promise, and Jupiter placed the shining tresses in the sky.

Coma Berenices gives the impression of a vast, dim cluster. It abounds in galaxies, of which five are in Messier's list. Of these, the most notable is M64, which is known as the Black-Eye Galaxy because of a dark region in it north of the centre – though this feature cannot be seen using any telescope with an aperture of below around 25 centimetres (10 inches). There is also a globular cluster, M53, close to α Comae Berenices which is an easy telescopic object. β and its neighbour 41 act as good guides to the globular cluster M3, which lies just across the border of Canes Venatici and is described with Star Map 1.

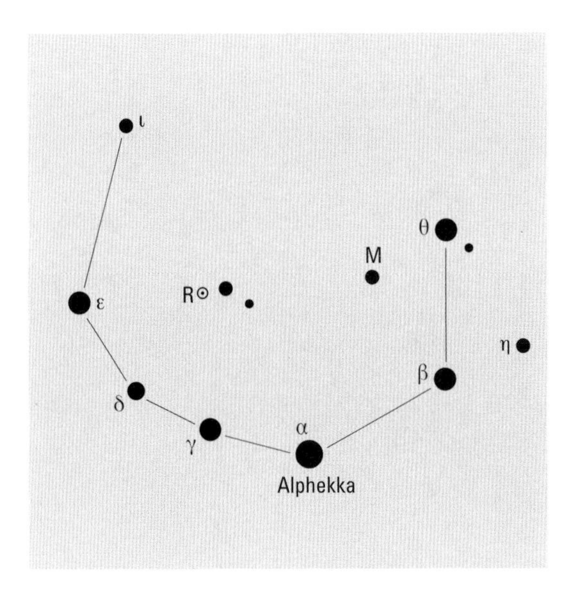

▲ **R Coronae Borealis** is a splendid target for binocular observers. It is found within the bowl of Corona Borealis. If only one star is visible there, it is type M with a magnitude of 6.6, and R CrB has taken one of its periodic 'dives'.

BOÖTES

BRIGHTEST STARS

No.	Star	R.A. h	m	s	Dec. °	'	"	Mag.	Spectrum	Proper name
16	α	14	15	40	+19	10	57	−0.04	K2	Arcturus
36	ε	14	44	59	+27	04	27	2.37	K0	Izar
8	η	13	54	41	+18	23	51	2.68	G0	
27	γ	14	32	05	+38	13	30	3.03	A7	Seginus
49	δ	15	15	30	+33	18	53	3.47	G8	Alkalurops
42	β	15	01	57	+40	23	26	3.50	G8	Nekkar

Also above magnitude 4.3; ϱ (3.58), ζ (3.78), θ (4.05), υ (4.06), λ (4.18).

VARIABLES

Star	R.A. h	m	Dec. °	'	Range (mags)	Type	Period (d)	Spectrum
R	14	37.2	+26	44	6.2–13.1	Mira	223	M
W	14	43.4	+26	32	4.7–5.4	Semi-regular	450	M

DOUBLES

Star	R.A. h	m	Dec. °	'	P.A. °	Sep. "	Mags
K	14	13.5	+51	47	236	13.4	4.6, 6.6
ι	14	16.2	+51	22	033	38.5	4.9, 7.5
π	14	40.7	+16	25	108	5.6	4.9, 5.8
μ	15	24.5	+37	23	171	108.3	4.3, 7.0
ε	14	45.0	+27	04	339	2.8	2.5, 4.9

CORONA BOREALIS

BRIGHTEST STARS

No.	Star	R.A. h	m	s	Dec. °	'	"	Mag.	Spectrum	Proper name
5	α	15	34	41	126	42	53	2.23	A0	Alphekka

The 'crown' is made up of α together with ε (4.15), δ (4.63), γ (3.84), β (3.68) and θ (4.14).

VARIABLES

Star	R.A. h	m	Dec. °	'	Range (mags)	Type	Period (d)	Spectrum
R	15	48.6	+28	09	5.7–15	R Coronae	–	F8p
S	15	21.4	+31	22	5.8–14.1	Mira	360	M
T	15	59.5	+25	55	2.0–10.8	Recurrent nova	–	M1Q

DOUBLE

Star	R.A. h	m	Dec. °	'	P.A. °	Sep. "	Mags	
η	15	23.2	+30	17	030	1.0	5.8, 5.9	
ζ	15	39.4	+36	38	305	6.3	5.1, 6.0	
σ	16	14.7	+33	52	234	7.0	5.6, 6.6	binary, 1000y.

COMA BERENICES

The brightest star in this vast, dim cluster is β; R.A. 13h 11m 52s, dec. 127° 52' 41", mag. 4.26. Then come α (Diadem) (4.32) and γ (4.35).

CLUSTERS AND GALAXIES

M	NGC	R.A. h	m	Dec. °	'	Mag.	Dimensions '	Type
53	5024	13	12.9	+18	10	7.7	12.6	globular cluster
64	4826	12	56.7	+21	41	8.5	9.3 × 5.4	Sb galaxy (Black-Eye)
88	4501	12	32.0	+14	25	9.5	6.9 × 3.9	SBb galaxy
98	4192	12	13.8	+14	54	10.1	9.5 × 3.2	Sb galaxy
99	4254	12	18.8	+14	25	9.8	5.4 × 4.8	Sc galaxy
100	4321	12	22.9	+15	49	9.4	6.9 × 6.2	Sc galaxy

Leo, Cancer, Sextans

Leo was the mythological Nemean lion which became one of Hercules' many victims, but in the sky the Lion is much more imposing than his conqueror, and is indeed one of the brightest of the Zodiacal constellations. The celestial equator cuts its southernmost extension, and Regulus (α), at the end of the sickle, is so close to the ecliptic that it can be occulted by the Moon and planets – as happened on 7 July 1959, when Venus passed in front of it. On that occasion the fading of Regulus before the actual occultation, when the light was coming to us by way of Venus' atmosphere, provided very useful information about the planet's atmosphere itself (of course, this was well before any interplanetary spacecraft had been launched to investigate the atmosphere of Venus more directly).

Regulus is a normal white star, some 78 light years away and around 150 times as luminous as the Sun. It is a wide and easy double; the companion shares Regulus' motion through space, so presumably the two have a common origin. The companion is itself a very close double, difficult to resolve partly because of the faintness of the third star and partly because of the glare from the brilliant Regulus.

About 20 minutes of arc north of Regulus is the dwarf galaxy Leo I, a member of the Local Group, about 750,000 light years from us. It was discovered photographically as long ago as 1950, but even giant telescopes are hard pressed to show it visually, because its surface brightness is so low; it is also one of the smallest and least luminous galaxies known. Even feebler is another member of the Local Group, Leo II, which lies about 2° north of δ (Zosma).

There is a minor mystery associated with Denebola (β). All observers up to and including Bayer, in 1603, ranked it as being first magnitude, equal to Regulus, but it is now almost a whole magnitude fainter. Yet it is a perfectly normal type A Main Sequence star, 39 light years away and 17 times as luminous as the Sun – not at all the kind of star expected to show a slow, permanent change. It is probable that there has been a mistake in recording or interpretation; all the same, a doubt remains, and naked-eye observers may care to check on it to see if there are any detectable fluctuations. The obvious comparison star is γ (Algieba), which is of virtually the same brightness and can often be seen at the same altitude above the horizon.

γ is a magnificent double, easily split with a very small telescope. The primary is orange, and the G-type companion usually looks slightly yellowish. The main star is 180 times as luminous as the Sun, and the companion is the equal of at least 50 Suns; the distance from us is 126 light years. Two other stars, some distance away, are not genuinely connected with the bright pair.

Two fainter stars (not on the map), 18 Leonis (magnitude 5.8) and 19 Leonis (6.5), lie near Regulus and are easily identified with binoculars. Forming a group with them is the Mira variable R Leonis, which can reach naked-eye brightness when at maximum and seldom falls below the tenth magnitude. Like most stars of its type, it is very red, and is a suitable target for novice observers, particularly since it is so easy to find.

There are five Messier galaxies in Leo. M65 and M66, which lie more or less between θ (Chort) and ι, can be seen with binoculars, and are only 21 minutes of arc apart, so they are in the same field of a low-power telescope. Both are spiral galaxies; M66 is actually the brighter of the two, though M65 is often regarded as the easier to see. Unfortunately, both are placed at an unfavourable angle to us, so that the full beauty of the spiral forms is lost. They are around 35 million light years away, and form a true pair. Another pair of spirals, M95 and M96, lies between ϱ and θ; close by is the elliptical galaxy M105, which is an easy object. Leo contains many additional galaxies, and the whole area is worth sweeping.

Before leaving Leo, it is worth mentioning Wolf 359, which lies at R.A. 10h 56m.5, dec. +07° 00′ 52″. Apart from

Magnitudes
-1
0
1
2
3
4
5
Variable star
Galaxy
Planetary nebula
Gaseous nebula
Globular cluster
Open cluster

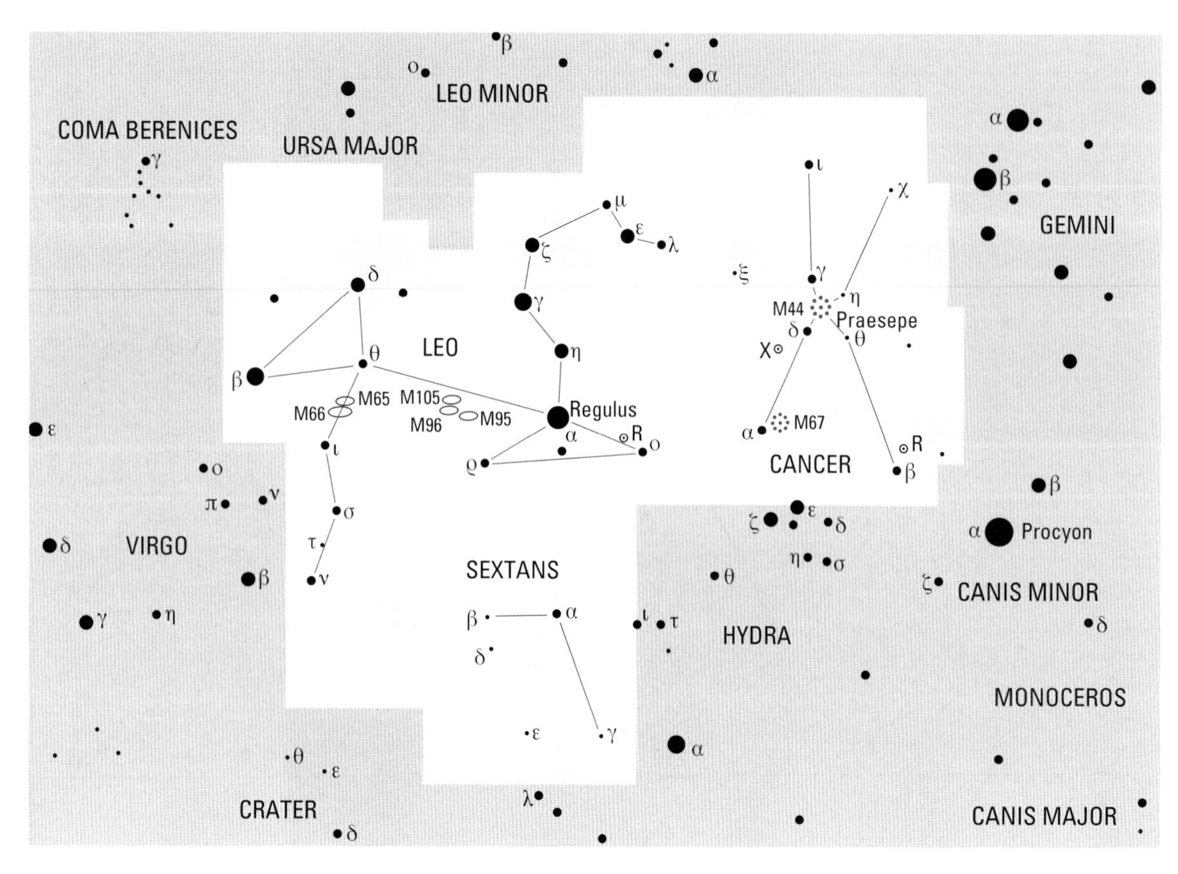

◀ Two Zodiacal constellations are shown here, Leo and Cancer: Leo is large and prominent, Cancer decidedly obscure. Both are at their best during evenings in northern spring (southern autumn). Leo is distinguished by the 'sickle', of which Regulus (α) is the brightest member, while Cancer contains Praesepe (M44), one of the finest open clusters in the sky. Sextans is very barren and obscure. The celestial equator crosses this map, and passes through the southernmost part of the constellation of Leo.

Barnard's Star and the members of the α Centauri group, Wolf 359 is the closest of our stellar neighbours, at a mere 7.6 light years; even so, its apparent magnitude is only 13.5, so it is by no means easy to identify. It is one of the feeblest red dwarfs yet to be discovered, and its luminosity is less than 1/60,000 of that of our own Sun.

Cancer, the celestial Crab, which according to legend met an untimely fate when Hercules trod upon it, looks a little like a dim and ghostly version of Orion. It is easy to locate, since it lies almost directly between the Twins (Castor and Pollux, α and β Geminorum) and Regulus (α Leonis); it is of course in the Zodiac, and wholly north of the equator. ζ (Tegmine) is, in fact, a triple system; the main pair is easy to resolve, and the brighter component is itself a close binary, with a separation which never exceeds 1.2 seconds of arc.

The semi-regular variable X Cancri, near δ, is worth finding because of its striking red colour. As it never fades below magnitude 7.5, it is always within binocular range, and its colour makes it stand out at once. R Cancri, near β, is a normal Mira variable which can rise to almost magnitude 6 at maximum.

The most interesting objects in Cancer are the open clusters, M44 (Praesepe) and M67. Praesepe is easily visible without optical aid, and has been known since very early times; Hipparchus, in the second century BC, referred to it as 'a little cloud'. It was also familiar to the Chinese, though it is not easy to decide why they gave it the unprepossessing nickname of 'the Exhalation of Piled-up Corpses'. Because it is also known as the Manger, the two stars flanking it, δ and γ, are called the Asses. Yet another nickname for Praesepe is the Beehive.

Praesepe is about 577 light years away. It contains no detectable nebulosity, so star formation there has presumably ceased, and since many of the leading stars are of fairly late spectral type, it may be assumed that the cluster is fairly old – about 730 million years. Because Praesepe covers a wide area – the apparent diameter is well over 1° – it is probably best seen with binoculars or else with a very low-power eyepiece. The real diameter is of the order of 10 to 15 light years, though, as with all open clusters, there is no sharp boundary.

M67 is on the fringe of naked-eye visibility, and is easily found; it lies within 2° of α (Acubens). It contains at least 200 stars; the French astronomer Camille Flammarion likened it to 'a sheaf of corn'. Its main characteristic is its great age.

Most open clusters lose their identity before very long, cosmically speaking, because they are disrupted by passing stars, but M67 lies almost 1500 light years away from the main plane of the Galaxy, so it moves in a comparatively sparsely populated region and there is little danger of this happening. Consequently, M67 has retained much of its original structure. Its age is estimated as 4000 million years, slightly less than that of the Sun. Despite its great distance, it is easy to resolve into stars, and in appearance it is not greatly inferior to Praesepe. The distance from the Earth is around 2700 light years, and the real diameter is about 11 light years.

Sextans (originally Sextans Uraniae, Urania's Sextant) is one of the groups formed by Hevelius, but for no obvious reason, since it contains no star brighter than magnitude 4.5 and no objects of immediate interest to the telescope user, though there are several galaxies with integrated magnitudes of between 9 and 12.

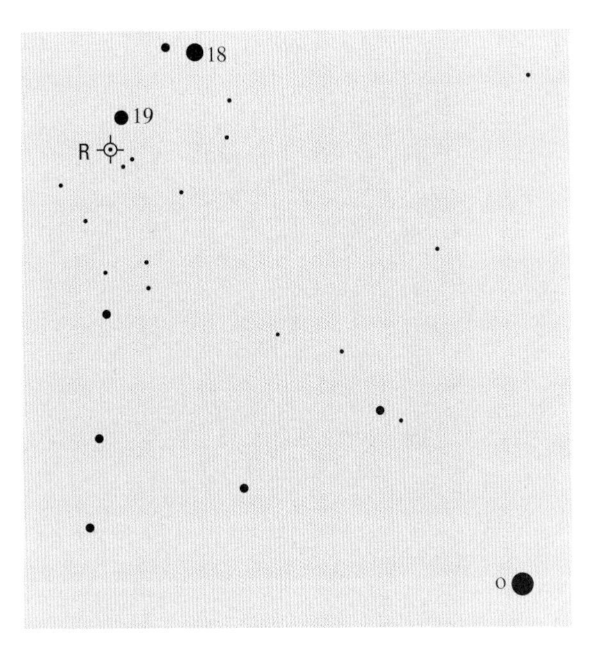

◄ **R Leonis.** The bright star is ο Leo, at mag 3.8. The comparison st for R Leo are 18 Leo, mag. 5.8, and 19 Leo, mag. 6.4.

LEO

BRIGHTEST STARS

No.	Star	R.A. h	m	s	Dec. °	′	″	Mag.	Spectrum	Proper name
32	α	10	08	22	+11	58	02	1.35	B7	Regulus
41	γ	10	19	58	+19	50	30	1.99	K0+G7	Algieba
94	β	11	49	04	+14	34	19	2.14	A3	Denebola
68	δ	11	14	06	+20	31	26	2.56	A4	Zosma
17	ε	09	45	51	+23	46	27	2.98	G0	Asad Australis
70	θ	11	14	14	+15	25	46	3.34	A2	Chort
36	ζ	10	16	41	+23	25	02	3.44	F0	Adhafera

Also above mag. 4.3: η (3.52), ο (Subra) (3.52), ϱ (3.85), μ (3.88), ι (3.94), σ (4.05) and ν (4.30).

VARIABLE

Star	R.A. h	m	Dec. °	′	Range (mags)	Type	Period (d)	Spectrum
R	09	47.6	+11	25	4.4–11.3	Mira	312	M

DOUBLES

Star	R.A. h	m	Dec. °	′	P.A. °	Sep. ″	Mags	
α	10	08.4	+11	58	307	176.9	1.4, 7.7	
τ	11	27.9	+02	51	176	91.1	4.9, 8.0	
γ	10	20.0	+19	51	{AB 124	4.3	2.2, 3.5	binary, 619y
					{AC 291	259.9	9.2	
					AD 302	333.0	9.6	

γ is a fine binary with an orange primary; it is out of binocular range, but a small telescope will split it. The more distant companions are optical.

GALAXIES

M	NGC	R.A. h	m	Dec. °	′	Mag.	Dimensions ′	Type
65	3623	11	18.9	+13	05	9.3	10.0 × 3.3	Sb galaxy
66	3627	11	20.2	+12	59	9.0	8.7 × 4.4	Sb galaxy
95	3351	10	44.0	+11	42	9.7	7.4 × 5.1	SBb galaxy
96	3368	10	46.8	+11	49	9.2	7.1 × 5.1	Sb galaxy
105	3379	10	47.8	+12	35	9.3	4.5 × 4.0	E1 galaxy

CANCER

A dim Zodiacal constellation. The brightest star is β (Altarf); R.A. 08h 16m 30s.9, dec. +09° 11′ 08″, magnitude 3.52. The other stars making up the dim 'pseudo-Orion' pattern are δ (Asellus Australis) (3.84), γ (Asellus Borealis) (4.66), α (Acubens) (4.25), ι (4.02) and χ (5.14).

VARIABLES

Star	R.A. h	m	Dec. °	′	Range (mags)	Type	Period (d)	Spectrum
R	08	16.6	+11	44	6.1–11.8	Mira	362	M
X	08	55.4	+17	14	5.6–7.5	Semi-reg.	195	N

DOUBLE

Star	R.A. h	m	Dec. °	′	P.A. °	Sep. ″	Mags	Binary, 1150 y.
ζ (Tegmine)	08	12.2	+17	39	088	5.7	5.0, 6.2	A is a close binary, and there is a 9.7-mag. third component, P.A. 108°, sep. 288″.

CLUSTERS

M	NGC	R.A. h	m	Dec. °	′	Mag.	Dimensions ′	Type
44	2632	08	40.1	+19	59	3.1	95	open cluster (Praesepe)
67	2682	08	50.4	+11	49	6.9	30	open cluster

SEXTANS

The brightest star is α: R.A. 10h 07m 56s.2, dec. −00° 22′ 18″, mag. 4.49. There is nothing here of interest to the user of a small telescope.

Virgo, Libra

Virgo is one of the largest of all the constellations, covering almost 1300 square degrees of the sky, though it has only one first-magnitude star (α, Spica) and two more above third magnitude. Mythologically it represents the Goddess of Justice, Astraea, daughter of Jupiter and Themis; the name Spica is said to mean 'the ear of wheat' which the Virgin is holding in her left hand. The main stars of Virgo make up a Y-pattern, with Spica at the base and γ at the junction between the 'stem' and the 'bowl'. The bowl, bounded on the far side by β Leonis, is crowded with galaxies.

Spica can be found by continuing the curve from the tail of Ursa Major through Arcturus (α Boötis); if sufficiently prolonged it will reach Spica, which is in any case brilliant enough to be really conspicuous. It is an eclipsing binary, with a very small magnitude range from 0.91 to 1.01; the components are only about 18 million kilometres (11 million miles) apart. Around 80 per cent of the total light comes from the primary, which is more than ten times as massive as the Sun and is itself intrinsically variable, though the fluctuations are very slight indeed. The distance from us is 257 light years, and the brighter component has a luminosity of more than 13,000 times that of the Sun. Like Regulus (α Leonis), Spica is so close to the ecliptic that it can at times be occulted by the Moon or a planet; the only other first-magnitude stars similarly placed are Aldebaran (α Tauri) and Antares (α Scorpii).

The bowl of Virgo is formed by ε, δ, γ, η and β. The last two are much fainter than the rest; early catalogues made them equal to the others, but one must be very wary of placing too much reliance on these old records, and neither star seems to be of the type expected to show long-term changes in brightness. Of the other stars in the main pattern, δ (Minelauva) is a fine red star of type M; Angelo Secchi, the great Italian pioneer of astronomical spectroscopy, nicknamed it Bellissima because of its beautifully banded spectrum. It is 147 light years away, and 130 times as luminous as the Sun. ε (Vindemiatrix or the Grape-Gatherer) is of type G; it is 104 light years away, and its luminosity is 75 times greater than that of the Sun.

γ has three accepted names; Arich, Porrima and Postvarta. It is a binary whose components are identical twins; the orbital period is 168.9 years, and a few decades ago the separation was great enough to make Arich one of the most spectacular doubles in the sky. The two components were are their closest in 2005, with a separation of no more than 0.3 seconds of arc; by late 2007 the separation had opened out to nearly one second of arc. The orbit is eccentric, and the real separation between the components ranges from 10,500 million kilometres (6520 million miles) to only 450 million kilometres (280 million miles). Arich is relatively near, at 36 light years.

There are no bright variables in Virgo, but there is one much fainter star, W Virginis, which is worthy of special mention. Its position is R.A. 13h 23m.5, declination −03° 07′, less than 4° away from ζ, but it is not likely to be of much interest to the user of a small telescope; the magnitude never rises above 9.5, and drops to 10.6 at minimum. The period is 17.3 days. Originally W Virginis was classed as a Cepheid, but it belongs to Population II, and short-period stars of this sort are considerably less luminous than classical Cepheids. It was this which led Edwin Hubble to underestimate the distance of the M31; at the time he had no way of knowing that there are two kinds of short-period variables with very different period–luminosity relationships. For a while the less luminous stars were called Type II Cepheids, but they are now known officially as W Virginis stars. The brightest of them, and the only member of the class easily visible with the naked eye, is ϰ Pavonis, (Star Map 21); unfortunately it is too far south in the sky to be seen from Britain or any part of Europe. W Virginis itself peaks at 1500 times the luminosity of the Sun, but this is not very much compared with δ Cephei, which at its brightest is 6000 times as luminous as the Sun.

Magnitudes
−1
0
1
2
3
4
5
Variable star
Galaxy
Planetary nebula
Gaseous nebula
Globular cluster
Open cluster

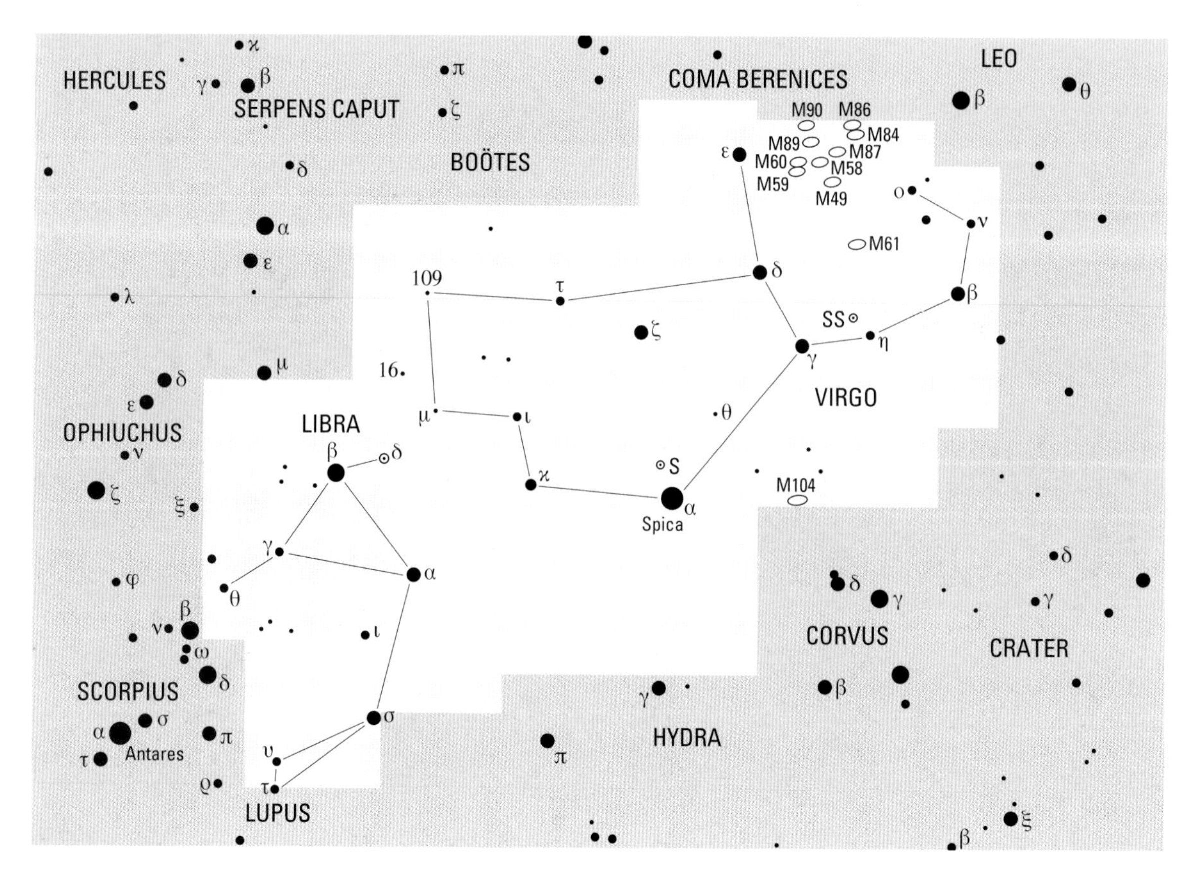

◄ **The constellations** in this map are best seen during evenings around April to June. Both Virgo and Libra are in the Zodiac, and Virgo is crossed by the celestial equator; the 'Y' of Virgo is unmistakable, and the 'bowl' of the Y is crowded with rather faint galaxies. Libra adjoins Virgo to the one side and Scorpius to the other.

Of course, the main feature of Virgo is the cluster of galaxies which spreads into the adjacent constellations of Leo and Coma. The average distance of the cluster members is between 40 and 60 million light years, and since there are thousands of systems it makes our Local Group seem very puny. In Virgo there are no fewer than 11 Messier objects, plus many more galaxies with integrated magnitudes of 12 or brighter.

Pride of place must go to the giant elliptical M87, discovered by Messier himself in 1781. A curious jet, several thousands of light years long, issues from it, and it is attended by many globular clusters – perhaps as many as a thousand. M87 is a very strong radio source, and is known to radio astronomers as Virgo A or as 3C274; it is also a source of X-rays, and it is clear that tremendous activity is going on there. There is considerable evidence that in the heart of the galaxy there is a super-massive black hole.

M104 (near Spica) is different; it is an Sb spiral distinguished by the dark dust lane which crosses it and gives it the nickname of the Sombrero Galaxy. It too is associated with a wealth of globular clusters. With a telescope with an aperture of more than 30 centimetres (12 inches) the dust lane is not hard to see, and there is nothing else quite like it, so it is also a favourite target for astrophotographers.

M49, which makes up an equilateral triangle with δ and ε, is another giant elliptical, strictly comparable with M87 except that it is not a strong radio source. All the other Messier galaxies in the area are fairly easy to locate, as are various others not on his list; this is a favourite hunting ground for observers who concentrate on searching for supernovae in external systems.

Libra, the Scales or Balance, adjoins Virgo and is one of the least conspicuous of the Zodiacal groups. It is the only constellation of the Zodiac named after an inanimate object, but originally it was the Scorpion's Claws; some early Greek legends link it, rather vaguely, with Mochis, the inventor of weights and measures. Its main stars (α, β, γ and σ) make up a distorted quadrilateral; σ has been filched from Scorpius, and was formerly known as γ Scorpii.

α Librae – Zubenelgenubi, the Southern Claw – makes a very wide pair with the magnitude-5.2 star, 8 Librae; the separation is so great that the pair can be well seen with binoculars. The brighter member of the pair is a spectroscopic binary, 72 light years away and 31 times as luminous as the Sun. Of rather more interest is β Librae or Zubenelchemale, the Northern Claw. It is 160 light years away and 130 times as luminous as the Sun; the spectral type is B8, and it has often been said to be the only single star with a decidedly greenish hue. T. W. Webb, a famous nineteenth-century English observer, referred to its 'beautiful pale green' colour. This is certainly an exaggeration, and most people will call it white, but it is worth examining. There have been suggestions that it is yet another star to have faded in historic times, and there is evidence that Ptolemy ranked it as first magnitude, but – as with other similar cases – the evidence is very slender.

There is not much else of interest in Libra, and there are no Messier objects. There is, however, one eclipsing binary of the Algol type. This is δ, which makes up a triangle with 16 Librae (magnitude 4.5) and β; its range is from magnitude 4.8 to 6.1, so it is never conspicuous, and at minimum sinks to the very limit of naked-eye visibility. With low-power binoculars it is in the same field as β, and this is probably the best way to locate it.

VIRGO

BRIGHTEST STARS

No.	Star	R.A.			Dec.			Mag.	Spectrum	Proper name
		h	m	s	°	′	″			
67	α	13	25	11.5	−11	09	41	0.98	B1	Spica
29	γ	12	41	39.5	−01	26	57	2.6	F0 F0	Arich
47	ε	13	02	10.5	+10	57	33	2.83	G9	Vindemiatrix
79	ζ	13	34	41.5	−00	35	46	3.37	A3	Heze
43	δ	12	55	36.1	+03	23	51	3.38	M3	Minelauva

Also above magnitude 4.3: β (Zavijava) (3.61), 109 (3.72), μ (Rijl al Awwa) (3.88), η (Zaniah) (3.89), ν (4.03), ι (Syrma) (4.08), ο (4.12), ϰ (4.19) and τ (4.26).

DOUBLES

Star	R.A.		Dec.		P.A.	Sep.	Mags
	h	m	°	′	°	″	
γ	12	41.7	−01	27	287	3.0	3.5, 3.5
θ (Apami-Atsa)	13	09.9	−05	32	343	7.1	4.4, 9.4

GALAXIES

M	NGC	R.A.		Dec.		Mag.	Dimensions	Type
		h	m	°	′		′	
49	4472	12	29.8	+08	00	8.4	8.9 × 7.4	E3 galaxy
58	4579	12	37.7	+11	49	9.8	5.4 × 4.4	SB galaxy
59	4621	12	42.0	+11	39	9.8	5.1 × 3.4	E3 galaxy
60	4649	12	43.7	+11	33	8.8	7.2 × 6.2	E1 galaxy
61	4303	12	21.9	+04	28	9.7	6.0 × 5.5	Sc galaxy
84	4374	12	25.1	+12	53	9.3	5.0 × 4.4	E1 galaxy
86	4406	12	26.2	+12	57	9.2	7.4 × 5.5	E3 galaxy
87	4486	12	30.8	+12	24	8.6	7.2 × 6.8	E1 galaxy (Virgo A)
89	4552	12	35.7	+12	33	9.8	4	E0 galaxy
90	4569	12	36.8	+13	10	9.5	9.5 × 4.7	Sb galaxy
104	4594	12	40.0	−11	37	8.3	8.9 × 4.1	Sb galaxy (Sombrero)

LIBRA

BRIGHTEST STARS

No.	Star	R.A.			Dec.			Mag.	Spectrum	Proper name
		h	m	s	°	′	″			
27	β	15	17	00.3	−09	22	58	2.61	B8	Zubenelchemale
9	α	14	50	52.6	−16	02	30	2.75	A3	Zubenelgenubi
20	σ	15	04	04.1	−25	16	55	3.29	M4	Zubenalgubi

Also above magnitude 4.3: υ (3.58), τ (3.66), γ (Zubenelhakrabi) (3.91), ι (4.15).
σ Librae was formerly included in Scorpius, as γ Scorpii.

VARIABLE

Star	R.A.		Dec.		Range	Type	Period	Spectrum
	h	m	°	′	(mags)		(d)	
δ	15	01.1	−08	31	4.9–5.9	Algol	2.33	B

DOUBLE

Star	R.A.		Dec.		P.A.	Sep.	Mags
	h	m	°	′	°	″	
α	14	50.9	−16	02	314	231.0	2.8, 5.2

▼ **The Sombrero Galaxy, M101,** in Virgo in an image taken by the Antu unit of the VLT. It is seen almost edge on, but its spiral structure is just visible. It is about 65 million light years away.

Hydra, Corvus, Crater

Magnitudes

–1

0

1

2

3

4

5

Variable star

Galaxy

Planetary nebula

Gaseous nebula

Globular cluster

Open cluster

Hydra, with an area of 1303 square degrees, is the largest constellation in the sky; it gained that distinction when the old, unwieldy Argo Navis was dismembered. As a matter of casual interest, the only other constellations with areas of more than 1000 square degrees are Virgo (1294), Ursa Major (1280), Cetus (1232), Eridanus (1138), Pegasus (1121), Draco (1083) and Centaurus (1060). At the other end of the scale comes Crux, which is a mere 68 square degrees.

Despite its size, Hydra is far from conspicuous, since there is only one star above third magnitude and only ten above fourth. The only reasonably well defined pattern is the 'head', made up of ζ, ε, δ and η; it is easy to find, more or less between Procyon (α Canis Minoris) and Regulus (α Leonis), but there is nothing in the least striking about it. Mythologically, Hydra is said to represent the multi-headed monster which became yet another of Hercules' victims, but many lists relegate it to the status of a harmless watersnake.

The only bright star, α (Alphard), is prominent enough. The Twins, Castor and Pollux (α and β Geminorum), point to it, but in any case it is readily identifiable simply because it lies in so barren an area; there are no other bright stars in the region, and Alphard has been nicknamed the Solitary One. It has a K-type spectrum, and is obviously reddish. It is 180 light years away, and 430 times as luminous as the Sun.

During the 1830s Sir John Herschel, son of the discoverer of Uranus, went to the Cape of Good Hope to survey the far-southern stars. On the voyage home he made some observations of Alphard, and concluded that it was decidedly variable. This has never been confirmed, and today the star is regarded as being constant; however, it may be worth watching – though it is awkward to estimate with the naked eye because of the lack of suitable comparison stars. If there are any fluctuations, they cannot amount to more than a few tenths of a magnitude.

There is, however, one interesting variable in the constellation. This is R Hydrae, close to γ and therefore rather inconveniently low for observers in Britain, Europe and the northern United States. At its maximum it can attain fourth magnitude, and never falls below tenth, so it is always an easy object; but it is not a typical Mira star, because there seems no doubt that the period has changed during the last couple of centuries. It used to be around 500 days; by the 1930s it had fallen to 425 days, and the latest official value is 390 days, so we seem to be dealing with a definite and probably permanent change in the star's evolutionary cycle. Observations are of value, because there is no reason to assume that the shortening in period has stopped. U Hydrae, which forms a triangle with ν and μ, is a semi-regular variable which is worth finding because, like virtually all stars of spectral type N, it is intensely red.

ε is a multiple system. The two main components are easy to resolve; the primary is an extremely close binary with an orbital period of 15 years, and there is a third – and probably a fourth – star sharing a common motion in space. β is also double, but is a difficult test for a 15-centimetre (6-inch) telescope. Though given the second Greek letter, β is below fourth magnitude, and so more than a magnitude fainter than γ, ζ or ν.

There are three Messier objects in Hydra. M48 is an open cluster on the edge of the constellation, close to the boundary with Monoceros (in fact the fourth-magnitude ζ is the best guide to it). It is just visible with the naked eye, but it is not too easy to identify. M68, discovered by Pierre Méchain in 1780, is a globular cluster about 39,000 light years away, lying almost due south of β Corvi and more or less between γ and β Hydrae. M83, south of γ and near the border between Hydra and Centaurus, is a fine face-on spiral galaxy, about 55 million light years away; it is easy to locate with a telescope with an aperture of 10 centimetres (4 inches) or more, and is a favourite photographic target, but it is best seen from the southern hemisphere. Several supernovae have been seen in it in recent years.

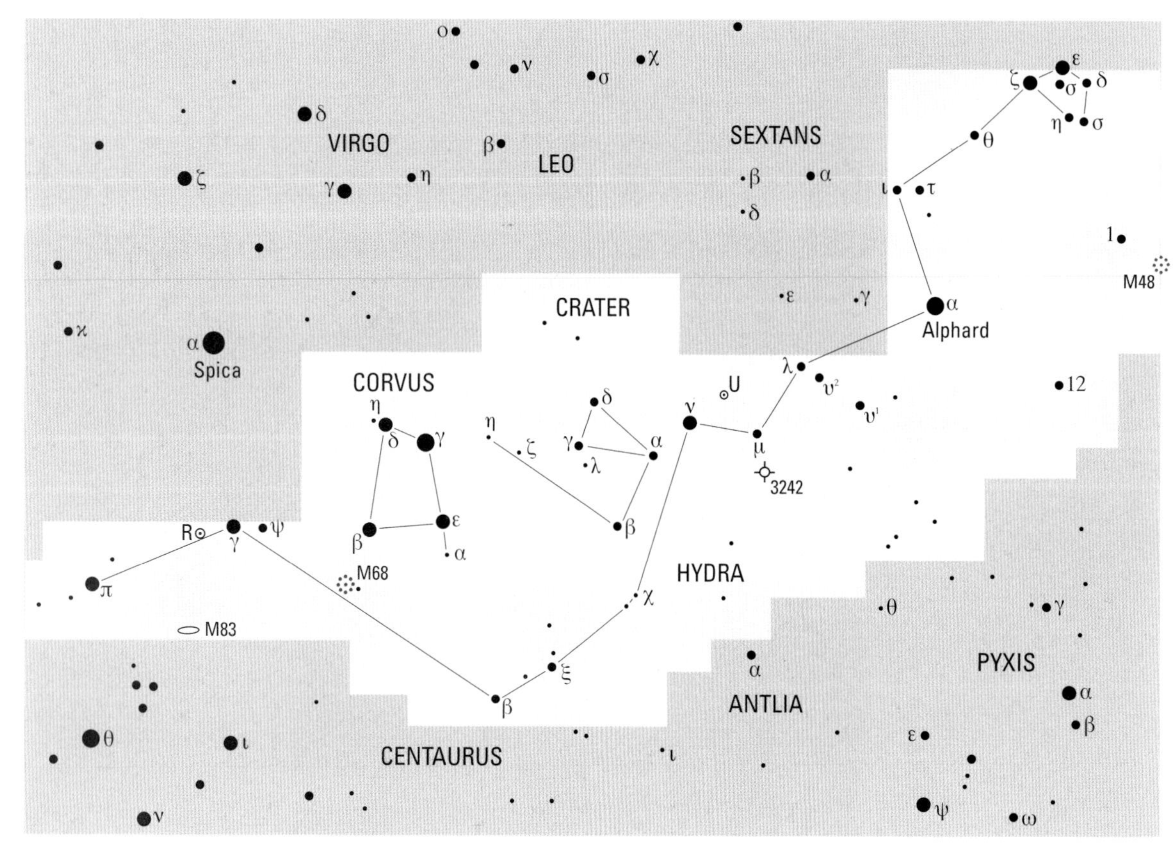

◄ **This map** shows a decidedly barren region. Hydra is the largest of all the constellations, but contains only one fairly bright star, α (Alphard). The 'head' lies near Cancer, the 'tail' extends to the south of Virgo. Corvus is fairly prominent, though none of its stars is as bright as second magnitude; Crater is very obscure.

There is also a planetary nebula, NGC3242 (C59), which has been nicknamed the Ghost of Jupiter. There is a relatively bright oval ring, and a hot 12th-magnitude central star; the whole nebula is said to show a bluish-green colour, and it is certainly well worth locating. It forms a triangle with ν and μ, so it is well placed for northern-hemisphere observers.

Two now rejected constellations border Hydra. Noctua, the Night Owl, was sited close to γ (a rather dim globular cluster, NGC5694, lies here), while Felis, the Cat, nestled against the watersnake's body south of λ. Generally speaking, these small and obscure constellations tend to confuse the sky maps, but in some ways it is sad to think that we have said good-bye to the Owl and the Pussycat!

Corvus is one of the original constellations listed by Ptolemy. According to legend, the god Apollo became enamoured of Coronis, mother of the great doctor Aesculapius, and sent a crow to watch her and report on her behaviour. Despite the fact that the crow's report was decidedly adverse, Apollo rewarded the bird with a place in the sky.

Corvus is easy to identify, because its four main stars, γ (Minkar), β (Kraz), δ (Algorel) and ε, between magnitudes 2.5 and 3, make up a quadrilateral which stands out because there are no other bright stars nearby. The constellation is remarkably devoid of interesting objects. Curiously, the star lettered α (Alkhiba) is more than a magnitude fainter than the four which make up the quadrilateral.

Crater, the Cup, said to represent the wine goblet of Bacchus, is so dim and obscure that it is rather surprising to find that it was one of Ptolemy's original 48 constellations. It is not hard to identify, close to ν Hydrae, but there is nothing of immediate interest in it.

HYDRA

BRIGHTEST STARS

No.	Star	R.A. h	m	s	Dec. °	'	"	Mag.	Spectrum	Proper name
30	α	09	27	35	−08	39	31	1.98	K3	Alphard
46	γ	13	18	55	−23	10	17	3.00	G5	
16	ζ	08	55	24	+05	56	44	3.11	K0	
	ν	10	49	37	−16	11	37	3.11	K2	
49	π	14	06	22	−26	40	56	3.27	K2	
11	ε	08	46	46	+06	25	07	3.38	G0	

Also above magnitude 4.3: ξ (3.54), λ (3.61), υ (4.12), δ (4.16), β (4.28), η (4.30).

VARIABLES

Star	R.A. h	m	Dec. °	'	Range (mags)	Type	Period (d)	Spectrum
R	13	29.7	−23	17	4.0–10.0	Mira	390	M
U	10	37.6	−13	23	4.8–5.8	Semi-reg.	450	N

DOUBLES

Star	R.A. h	m	Dec. °	'	P.A. °	Sep. "	Mags	
ε	08	46.8	+06	25	281	2.8	3.8, 6.8	A is a close binary
β	11	52.9	−33	54	008	0.9	4.7, 5.5	difficult test

CLUSTERS, GALAXIES AND NEBULAE

M	C	NGC	R.A. h	m	Dec. °	'	Mag.	Dimensions '	Type
48		2548	08	13.8	−05	48	5.8	54	open cluster
68		4590	12	39.5	−26	45	8.2	12	globular cluster
83		5236	13	37.0	−29	52	8.2	11.3 × 10.2	Sc galaxy
	59	3242	10	24.8	−18	38	8.6	0.3 × 0.2	planetary nebula (Ghost of Jupiter)

CORVUS

BRIGHTEST STARS

No.	Star	R.A. h	m	s	Dec. °	'	"	Mag.	Spectrum	Proper name
4	γ	12	15	48	−17	32	31	2.59	B8	Minkar
9	β	12	34	23	−23	23	48	2.65	G5	Kraz
7	δ	12	29	52	−16	30	55	2.95	B9	Algorel
2	ε	12	10	07	−22	37	11	3.00	K2	

Also above magnitude 4.3: α (Alkhiba) (4.02).

CRATER

A small, dim group. The brightest stars are α (Alkes) and γ, each of magnitude 4.08. Alkes lies at R.A. 10h 59m 46s, dec. 218° 17' 56".

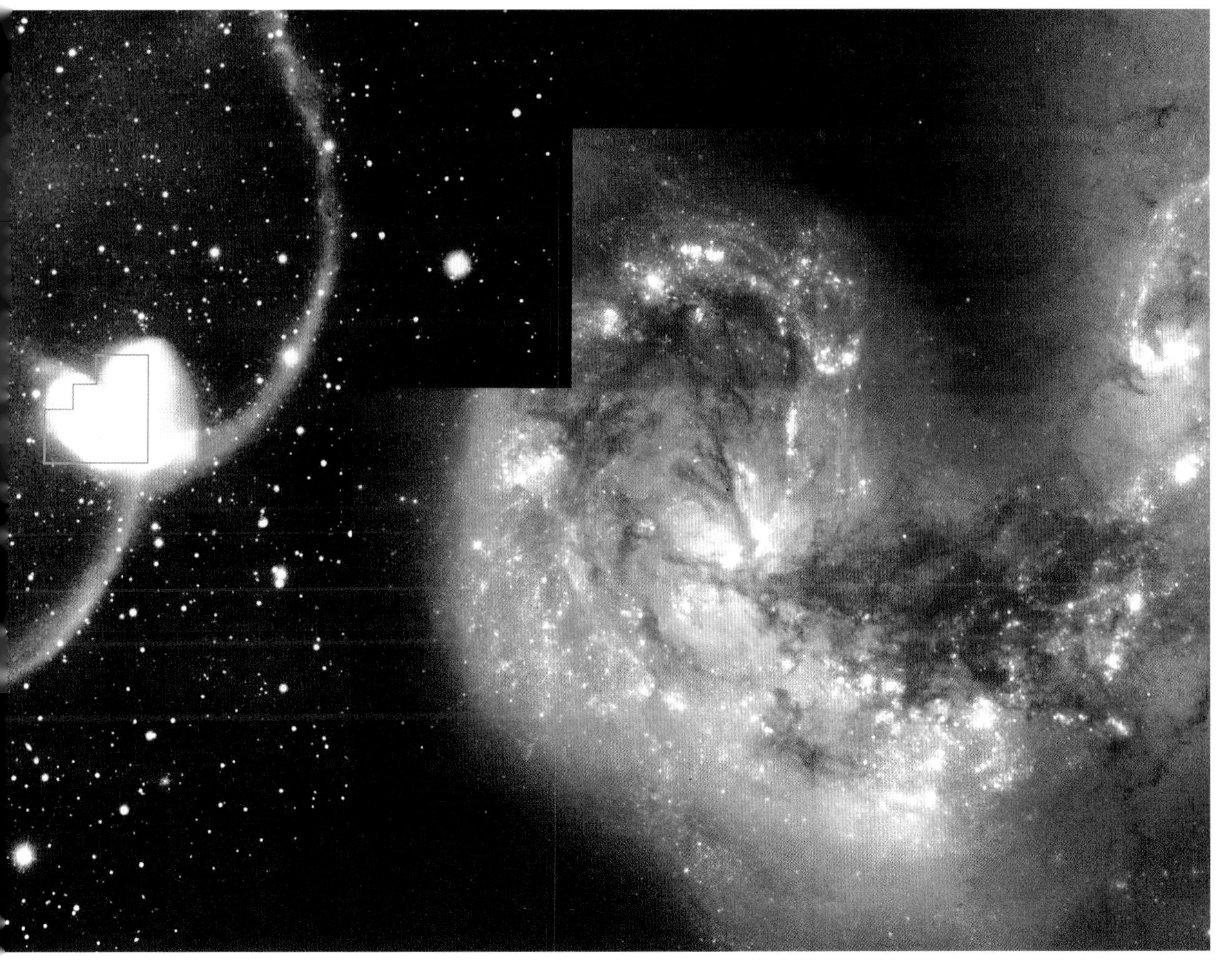

◀ **The Antennae: colliding galaxies in Corvus, NGC4038 and 4039 (C60 and 61).** (Left) Ground-based view, showing long tails of luminous matter formed by the gravitational tidal forces of the encounter 63 million light years away. (Right) Hubble Space Telescope view: the cores of the galaxies are the orange blobs left and right of image centre, criss-crossed by filaments of dark dust. A wide band of chaotic dust, called the overlap region, stretches between the cores of the two galaxies. The sweeping spiral-like patterns, traced by bright blue star clusters, show the result of energetic star birth triggered by the collision.

Lyra, Cygnus, Aquila, Scutum, Sagitta,

Magnitudes
- −1
- 0
- 1
- 2
- 3
- 4
- 5

Variable star

Galaxy

Planetary nebula

Gaseous nebula

Globular cluster

Open cluster

Lyra is a small constellation, but it contains a wealth of interesting objects. α (Vega) is the brightest star in the northern hemisphere of the sky apart from Arcturus, and is distinguished by its steely-blue colour; it is 26 light years away, and 52 times as luminous as the Sun. During 1983, observations made from the Infra-Red Astronomical Satellite (IRAS), showed that Vega is associated with a cloud of cool material which may be planet forming, though it would certainly be premature to claim that any planets actually exist there. Vega's tenth-magnitude companion, at a separation of 60 seconds of arc, merely happens to lie in almost the same line of sight; there is no real connection.

β (Sheliak) is an eclipsing binary with alternate deep and shallow minima; it is the prototype star of its class. Its variations are very easy to follow, because the neighbouring γ (3.24) makes an ideal comparison star; when β is faint there are other comparison stars in ϰ (4.3), δ (also 4.3), and ζ (4.4). R Lyrae is a semi-regular variable, very red in colour, with a rough period of 46 days; useful comparison stars are η and θ, both of which are listed as magnitude 4.4, though I find θ to be appreciably the brighter of the two.

Close to Vega lies ε, a splendid example of a quadruple star. Keen-eyed people can split the two main components, while a 7.6-centimetre (3-inch) telescope is powerful enough to show that each component is again double. It is worth using binoculars to look at the pair consisting of δ^1 and δ^2; here we have a good colour contrast, because the brighter star is an M-type red giant and the fainter member is white. ζ is another wide, easy double.

M57, the Ring Nebula, is the most famous of all the planetary nebulae, though not actually the brightest. It is extremely easy to find, since it lies between β and γ, and a small telescope will show it. The globular cluster M56 is within binocular range, between γ Lyrae and β Cygni; it is very remote, at a distance of over 45,000 light years. Mythologically, Lyra represents the Lyre which Apollo gave to the great musician Orpheus.

Cygnus, the Swan, said to represent the bird into which Jupiter once transformed himself while upon a clandestine visit to the Queen of Sparta, is often called the Northern Cross for obvious reasons; the X-pattern is striking. The brightest star, α (Deneb), is an exceptionally luminous supergiant, at least 160,000 times brighter than the Sun, and 2500 light years away. γ (Sadr), the central star of the X, is of type F8, and equal to 23,000 Suns. One member of the pattern, β (Albireo), is fainter than the rest and also farther away from the centre, so it rather spoils the symmetry; but it compensates for this by being probably the loveliest coloured double in the sky. The primary is golden yellow, the companion vivid blue; the separation is over 34 seconds of arc, so almost any small telescope will show both stars. It is an easy double; so too is the dim 61 Cygni, which was the first star to have its distance measured.

There are several variable stars of note. χ is a Mira variable, with a period of 407 days and an exceptionally large magnitude range; at maximum it may rise to 3.3, brighter than its neighbour η, but at minimum it sinks to below 14, and since it lies in a rich area it is then none too easy to identify. χ is one of the strongest infra-red sources in the sky. U Cygni (close to the little pair consisting of o^1 and o^2) and R Cygni (in the same telescopic field as θ, magnitude 4.48) are also very red Mira variables.

P Cygni, close to γ, has a curious history. In 1600 it flared up from obscurity to third magnitude; since 1715 it has hovered around magnitude 5. It is very luminous and remote, and is also known to be unstable. It is worth monitoring, because there is always the chance of a new increase in brightness; good comparison stars are 28 (4.9) and 29 Cygni (5.0)

The Milky Way flows through Cygnus, and there are conspicuous dark rifts, indicating the presence of obscuring dust. There are also various clusters and nebulae. The open cluster M29 is in the same binocular field as γ and P Cygni and though it is sparse it is not hard to identify. M39, near ϱ, is also loose and contains about 30 stars.

NGC7000 (C20) is known as the North America Nebula. It is dimly visible with the naked eye in the guise of a slightly brighter portion of the Milky Way, and binoculars show it well as a wide region of diffuse nebulosity; photographs show that its shape really does bear a marked resemblance to that of

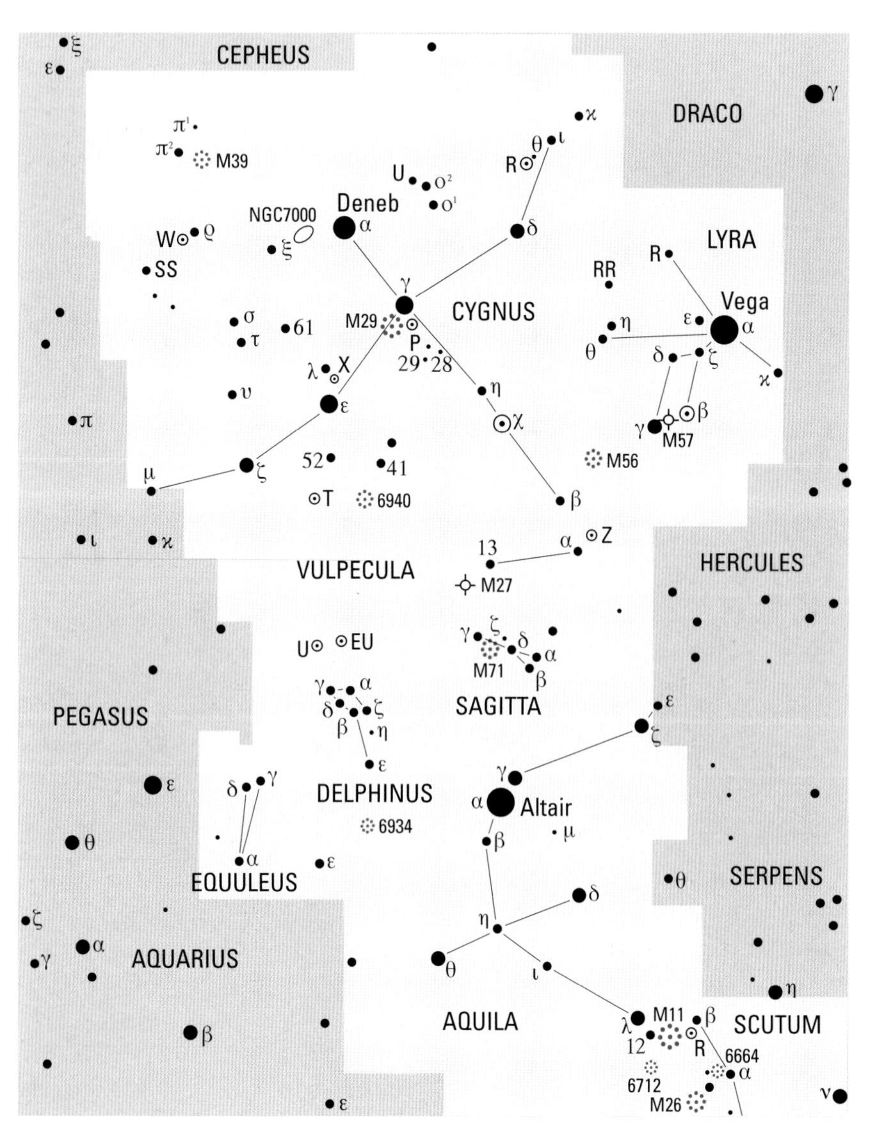

◀ **This map** is dominated by three bright stars: Deneb (α Cyg), Vega (α Lyr and Altair (α Aql). Because they are so prominent during summer evenings in the northern hemisphere, I once referred to them as 'the Summer Triangle', and the name has come into general use, though it is unofficial and is inappropriate in the southern hemisphere. The Milky Way crosses the area, which is very rich. All three stars of the 'Triangle' can be seen from most inhabited countries, though from New Zealand Deneb and Vega are always very low.

Vulpecula, Delphinus, Equuleus

LYRA

BRIGHTEST STARS

No.	Star	R.A.			Dec.			Mag.	Spectrum	Proper name
		h	m	s	°	'	"			
3	α	18	36	56	+38	47	01	0.03	A0	Vega
14	γ	18	58	56	+32	41	22	3.24	B9	Sulaphat
10	β	18	50	05	+33	21	46	3.3 (max)	B7	Shelik

Also above magnitude 4.3: ε (3.9) (combined magnitude); R (3.9, max).

VARIABLES

Star	R.A.		Dec.		Range	Type	Period	Spectrum
	h	m	°	'	(mags)		(d)	
β	18	50.1	+33	22	3.3–4.3	β Lyrae	12.94	B+A
R	18	55.3	+43	57	3.9–5.0	Semi-reg.	46	M
RR	19	25.5	+42	47	7.1–8.1	RR Lyrae	0.57	A–F

DOUBLES

Star	R.A.		Dec.		P.A.	Sep.	Mags	
	h	m	°	'	°	"		
ε	18	44	+39	40	AB+CD 173	207.7	4.7, 5.1	
					ε^1 = AB 357	2.8	5.0, 5.1	quadruple star
					ε^2 = CD 094	2.3	5.2, 5.5	
ζ	18	44.8	+37	36	150	43.7	4.3, 5.9	
β	18	50.1	+33	22	149	45.7	var, 8.6	

CLUSTER AND NEBULA

M	NGC	R.A.		Dec.		Mag.	Dimensions	Type
		h	m	°	'		'	
56	6779	19	16.6	+30	11	8.2	7.1	globular cluster
57	6720	18	53.6	+33	02	9.7	1.1 × 2.5	planetary nebula (Ring)

CYGNUS

BRIGHTEST STARS

No.	Star	R.A.			Dec.			Mag.	Spectrum	Proper name
		h	m	s	°	'	"			
50	α	20	41	26	+45	16	49	1.25	A2	Deneb
37	γ	20	22	13	+40	15	24	2.20	F8	Sadr
53	ε	20	46	12	+33	58	13	2.46	K0	Gienah
18	δ	19	44	58	+45	07	51	2.87	A0	
6	β	19	30	43	+27	57	35	3.08	K5	Albireo
64	ζ	21	12	56	+30	13	37	3.20	G8	

Also above magnitude 4.3: ξ (3.72); τ (3.72), κ (3.77), ι (3.79), o^1 (3.79), η (3.89), ν (3.94), o^2 (3.98), 41 (4.01), ρ (4.02), 52 (4.22), σ (4.23), π^2 (4.23). The curious variable P Cygni has been known to reach magnitude 3, but is usually nearer 5. At some maxima the red Mira variable χ Cygni can reach magnitude 3.3, but most maxima are considerably fainter than this.

VARIABLES

Star	R.A.		Dec.		Range	Type	Period	Spectrum
	h	m	°	'	(mags)		(d)	
χ	19	50.6	+32	55	3.3–14.2	Mira	407	S
P	20	17.8	+38	02	3–6	Recurrent nova	–	Pec.
R	19	36.8	+50	12	5.9–14.2	Mira	426	M
U	20	19.6	+47	54	5.9–12.1	Mira	462	N
X	20	43.4	+35	35	5.9–6.9	Cepheid	16.4	F–G
W	21	36.0	+45	22	5.0–7.6	Semi-reg.	126	M
SS	21	42.7	+43	35	8.4–12.4	SS Cyg (U Gem)	±50	A–G

DOUBLES

Star	R.A.		Dec.		P.A.	Sep.	Mags	
	h	m	°	'	°	"		
β	19	30.7	+27	58	054	34.4	3.1, 5.1	yellow, blue
δ	19	45.0	+45	07	225	2.4	2.9, 6.3	binary, 828 years
61	21	06.9	+38	45	148	29.9	5.2, 6.0	

CLUSTERS AND NEBULAE

M	C	NGC	R.A.		Dec.		Mag.	Dimensions	Type
			h	m	°	'		'	
29		6913	20	23.9	+38	32	6.6	7	open cluster
39		7092	21	32.2	+48	26	4.6	32	open cluster
	20	7000	20	58.8	+44	20	6.0	120 × 100	nebula (North America)

AQUILA

BRIGHTEST STARS

No.	Star	R.A.			Dec.			Mag.	Spectrum	Proper name
		h	m	s	°	'	"			
53	α	19	50	47	+08	52	06	0.77	A7	Altair
50	γ	19	46	15	+10	36	48	2.72	K3	Tarazed
17	ζ	19	05	24	+13	51	48	2.99	B9	Dheneb
65	θ	20	11	18	−00	49	17	3.23	B9	
30	δ	19	25	30	+03	06	53	3.36	F0	
16	λ	19	06	15	−04	52	57	3.44	B9	Althalimain

Also above magnitude 4.3: β (Alshain) (3.71), ε (4.02), 12 (4.02). The Cepheid variable η rises to 3.5 at maximum.

VARIABLES

Star	R.A.		Dec.		Range	Type	Period	Spectrum
	h	m	°	'	(mags)		(d)	
η	19	52.5	+01	00	3.5–4.4	Cepheid	7.2	F–G
R	19	06.4	+08	14	5.5–12.0	Mira	284	M

SCUTUM

The brightest star is α: R.A. 18h 35m 12s.1, dec. −08° 14' 39", mag. 3.85. Also above magnitude 4.3: β (4.22).

VARIABLE

Star	R.A.		Dec.		Range	Type	Period	Spectrum
	h	m	°	'	(mags)		(d)	
R	18	47.5	−05	42	4.4–8.2	RV Tauri	140	G–K

CLUSTERS

M	NGC	R.A.		Dec.		Mag.	Dimensions	Type
		h	m	°	'		'	
11	6705	18	51.1	−06	16	5.8	14	open cluster (Wild Duck)
26	6694	18	45.2	−09	24	8.0	15	open cluster
	6664	18	36.7	−08	13	7.8	16	open cluster (EV Scuti cl.)
	6712	18	53.1	−08	42	8.2	7	globular cluster

SAGITTA

The only two stars above magnitude 4.3 are γ and δ. The brightest, γ, lies at R.A. 19h 58m 45s.3, dec. 119° 29' 32"; magnitude 3.47. The magnitude of δ is 3.82.

CLUSTERS AND NEBULAE

M	NGC	R.A.		Dec.		Mag.	Dimensions	Type
		h	m	°	'		'	
71	6838	19	53.8	+18	47	8.3	7.2	globular cluster

VULPECULA

The brightest star is α: R.A. 19h 28m 42s.2, dec. 124° 39' 24", magnitude 4.44.

VARIABLES

Star	R.A.		Dec.		Range	Type	Period	Spectrum
	h	m	°	'	(mags)		(d)	
T	20	51.5	+28	15	5.4–6.1	Cepheid	4.4	F2G
Z	19	21.7	+25	34	7.4–9.2	Algol	2.45	B1A

DOUBLE

Star	R.A.		Dec.		P.A.	Sep.	Mags
	h	m	°	'	°	"	
α - 8	19	28.7	+24	40	028	413.7	4.4, 5.8

CLUSTER AND NEBULA

M	NGC	R.A.		Dec.		Mag.	Dimensions	Type
		h	m	°	'		'	
27	6853	19	59.6	+22	43	7.6	5.9 × 13.1	planetary nebula (Dumbbell)
	6940	20	34.6	+28	18	6.3	31	open cluster

DELPHINUS

The brightest star is β: R.A. 20h 37m 32s.8, dec. 14° 35' 43", mag. 3.54. Also above magnitude 4.3: α (3.77), γ (3.9 combined magnitude), ε (4.03).

VARIABLES

Star	R.A.		Dec.		Range	Type	Period	Spectrum
	h	m	°	'	(mags)		(d)	
U	20	45.5	+18	05	5.6–8.9	Semi-regular	110	M
EU	20	37.9	+18	16	5.8–6.9	Semi-regular	59	M

DOUBLE

Star	R.A.		Dec.		P.A.	Sep.	Mags
	h	m	°	'	°	"	
γ	20	46.7	+16	07	268	9.6	4.5,5.5

CLUSTER

C	NGC	R.A.		Dec.		Mag.	Dimensions	Type
		h	m	°	'		'	
47	6934	20	34.2	+07	24	8.9	5.9	globular (cluster

EQUULEUS

The only star above mag. 4.3 is α (Kitalpha), R.A. 21h 15m 49s.3, dec. 05° 14' 52", mag. 3.92.

DOUBLE

Star	R.A.		Dec.		P.A.	Sep.	Mags	
	h	m	°	'	°	"		
ε	20	59.1	+04	18	AB 285	1.0	6.0, 6.3	binary, 101 years
					AB+C 070	10.7	7.1	
					AD 280	74.8	12.4	

the North American continent. It is nearly 500 light years in diameter, and may owe much of its illumination to Deneb. It lies in the same field as reddish ξ Cygni.

Aquila, the Eagle, commemorates the bird sent by Jupiter to fetch a shepherd boy, Ganymede, who was destined to become the cup bearer to the gods. α (Altair), at a distance of 16.6 light years, is the closest of the first-magnitude stars apart from α Centauri, Sirius and Procyon; it is ten times as luminous as the Sun, and is known to be rotating so rapidly that it must be egg-shaped. It is flanked by two fainter stars, γ (Tarazed) and β (Alshain). γ is an orange K-type star, much more powerful than Altair but also much more remote.

η is a Cepheid variable. The range is from magnitude 3.4 to 4.4, so δ and θ make ideal comparison stars; when η is near minimum, a useful comparison star is ι (4.0).

Scutum is not an original constellation; it was one of Hevelius' inventions, and was originally Scutum Sobieskii, Sobieski's Shield. The variable R Scuti is the brightest member of the RV Tauri class, and is a favourite binocular target; there are alternate deep and shallow minima, with occasional periods of irregularity. Of the two Messier open clusters, much the more striking is the fan-shaped M11, which has been nicknamed the Wild Duck cluster and is a glorious sight in any telescope; it contains hundreds of stars, and is easily identified, being close to λ and 12 Aquilae.

Sagitta, the Arrow – Cupid's Bow – is distinctive; the main arrow pattern is made up of the two bright stars δ and γ, together with α and β (each magnitude 4.37). There is one Messier object, M71, which was formerly classified as an open cluster, but is now thought to be a globular, though it is much less condensed than other systems of this type. It lies a little less than halfway from γ to δ.

Vulpecula was originally Vulpecula et Anser, the Fox and Goose, but the goose has long since vanished from the maps. The constellation is very dim, but is redeemed by the presence of M27, the Dumbbell Nebula, probably the finest of all planetary nebulae. There is no problem in finding this with binoculars; it is close to γ Sagittae, which is the best guide to it. A moderate power will reveal its characteristic shape. Like all planetaries it is expanding; the present diameter is of the order of two and a half light years.

Delphinus is one of Ptolemy's original constellations. It honours the dolphin which carried the great singer Arion to safety when he had been thrown overboard by the crew of the ship which was carrying him home after winning all the prizes in a competition. Delphinus is a compact little group – unwary observers have been known to confuse it with the Pleiades. Its two leading stars have curious names: α is Svalocin, β is Rotanev. These names were given by one Nicolaus Venator, and the association is obvious enough.

γ is a wide, easy double, and the two red semi-regular variables U and EU are good binocular objects. Near them an interesting nova, HR Delphini, flared up in 1967 and was discovered by the English amateur G. E. D. Alcock; it reached magnitude 3.7, and remained a naked-eye object for months. Its present magnitude is between 12 and 13, and as this was also the pre-outburst value it is unlikely to fade much farther.

Equuleus represents a foal given by Mercury to Castor, one of the Heavenly Twins. It is so small and dim that it is surprising to find it in Ptolemy's original list, but the little triangle made up of α, δ (4.49) and γ (4.69) is not hard to identify, between Delphinus and β Aquarii. ε is a triple star, but otherwise Equuleus contains nothing of interest.

Hercules

Hercules is a very large constellation; it covers 1225 square degrees, but it is not particularly rich. The best guide to it is the second-magnitude α Ophiuchi (Rasalhague), which is bright enough to be prominent and is also rather isolated. Not far away is α Herculis (Rasalgethi), which is some way away from the other main stars of the constellation. The main part of Hercules lies inside the triangle bounded by Alphekka (α Coronae Borealis), Rasalhague and Vega (α Lyrae). With its high northern declination, part of it is circumpolar from the latitudes of Britain and the northern United States. Its main features are the red supergiant Rasalgethi, a wide and easy binary (ζ), and two spectacular globular clusters (M13 and M92).

In 1759 William Herschel discovered that Rasalgethi is variable. At that time only four variables had been found – Mira Ceti, Algol (β Persei), χ Cygni and R Hydrae – so the discovery was regarded as very important. Certainly there is no doubt about the fluctuations; it is said that the extreme range is from magnitude 3.0 to 4.0, though for most of the time the star remains between 3.1 and 3.7. Officially it is classed as a semi-regular with a rough period of 90 to 100 days, but this period is by no means well marked. The variations are slow, but can be followed with the naked eye; suitable comparison stars are ϰ Ophiuchi (3.20), δ Herculis (3.14) and γ Herculis (3.75). β Herculis (2.71), the brightest star in the constellation, is always considerably superior to Rasalgethi.

The distance of Rasalgethi is 218 light years; its spectral type is M. What makes it so notable is its vast size. It may be even larger than Betelgeux, in which case its diameter exceeds 400 million kilometres (250 million miles). It is relatively cool – the surface temperature is well below 3000°C – and its outer layers, at least, are very rarefied. It is a very powerful emitter of infra-red radiation.

Rasalgethi is also a fine double. The companion is of magnitude 5.3, and since the separation is not much short of 5 seconds of arc a small telescope will resolve the pair. The companion is often described as vivid green, though this is mainly, if not entirely, in contrast with the redness of the primary. The companion is itself an extremely close binary, with a period of 51.6 days, and there is every reason to believe that both stars are enveloped in a huge, rarefied cloud. Rasalgethi is indeed a remarkable system.

δ has an eighth-magnitude companion at a separation of 9 seconds of arc (position angle 236 degrees), but this is an optical pair; there is no connection between the two components, and the secondary lies well in the background. δ itself is an ordinary A-type star, 35 times as luminous as the Sun and 91 light years away.

Of more interest is ζ (Rutilicus) which is a fine binary; its double nature was discovered by William Herschel in 1782. The magnitudes are 2.9 and 3.5; the period is only 34.5 years, so both separation and position angle change quickly. In 1994, the separation was 1.6 seconds of arc, so this is a very wide, easy pair. The primary is a G-type subgiant, 31 light years away and rather more than five times as luminous as the Sun.

68 (u) Herculis is an interesting variable of the β Lyrae type. The secondary minimum takes the magnitude down to 5.0 and the deep minimum to only 5.3, so the star is always within binocular or even naked-eye range. Both components are B-type giants, so close that they almost touch; as with β Lyrae, each must be pulled out into the shape of an egg. If the distance is around 600 light years, as

HERCULES

BRIGHTEST STARS

No.	Star	R.A. h	m	s	Dec. °	′	″	Mag.	Spectrum	Proper name
27	β	16	30	13	+21	29	22	2.77	G8	Kornephoros
40	ζ	16	41	17	+31	36	10	2.81	G0	Rutilicus
64	α	17	14	39	+14	23	25	3.0 (max)	M5	Rasalgethi
65	δ	17	15	02	+24	50	21	3.14	A3	Sarin
67	π	17	15	03	+36	48	33	3.16	K3	
86	μ	17	46	27	+27	43	15	3.42	G5	

Also above magnitude 4.3: η (3.53), ξ (3.70), γ (3.75), ι (3.80), ο (3.83), 109 (3.84), θ (3.86), τ (3.89), ε (3.92), 110 (4.19), σ (4.20), 95 (4.27).

VARIABLES

Star	R.A. h	m	Dec. °	′	Range (mags)	Type	Period (d)	Spectrum
α	17	14.6	+14	23	3–4	Semi-reg.	±100?	M
30 (g)	16	28.6	+41	53	5.7–7.2	Semi-reg.	70	M
68 (u)	17	17.3	+35	06	4.6–5.3	β Lyrae	2.05	B1B

DOUBLES

Star	R.A. h	m	Dec. °	′	P.A. °	Sep. ″	Mags	
α	17	14.6	+14	23	107	4.7	var, 5.4	binary, 3600y red, green
ζ	16	41.3	+31	36	089	1.6	2.9, 3.5	binary, 34.5y
ϱ	17	23.7	+37	09	316	4.1	4.6, 5.6	

CLUSTERS

M	NGC	R.A. h	m	Dec. °	′	Mag.	Dimensions ′	Type
13	6205	16	41.7	+36	28	5.9	16.6	globular cluster
92	6341	17	17.1	+43	08	5.5	11.2	globular cluster

seems possible, each star must be well over 100 times as luminous as the Sun.

On 13 December 1934 the English amateur J. P. M. Prentice discovered a bright nova in Hercules, near ι and not far from Draco's head. It rose to magnitude 1.2, so it remains the brightest nova to have appeared in the northern hemisphere of the sky since Nova Aquilae in 1918. During its decline it was strongly green for a while, and was also unusual inasmuch as it remained a naked-eye object for several months. It has now faded back to around its pre-outburst magnitude of 15, and has been found to be an eclipsing binary with the very short period of 4 hours 39 minutes. Both components are dwarfs – one white, one red.

M13, which lies rather more than halfway between ζ and η, is the brightest globular cluster north of the celestial equator; its only superiors are ω Centauri and 47 Tucanae. M13 is just visible with the naked eye on a clear night, but it is far from obvious, and it is not surprising that it was overlooked until Edmond Halley chanced upon it in 1714 – describing it as 'a little patch, but it shows itself to the naked eye when the sky is serene and the Moon absent'. William Herschel was more enthusiastic about it: 'A most beautiful cluster of stars exceedingly condensed in the middle and very rich. Contains about 14,000 stars.' In fact this is a gross underestimate; 500,000 would be closer to the truth.

Like all globulars, M13 is a long way away. Its distance has been given as 22,500 light years and its real diameter perhaps 160 light years, with a condensed central region 100 light years across. It lies well away from the main plane of the Milky Way, so it has not been greatly disturbed by the concentration of mass in the centre of the Galaxy, and is certainly very old indeed. Binoculars give good views of it, and even a small telescope will resolve the outer parts into stars.

The second globular, M92, lies directly between η and ι. It is on the fringe of naked-eye visibility – very keen-eyed observers claim that they can glimpse it – and telescopically it is not much inferior to M3. It is rather farther away, at a distance of 37,000 light years, and in most respects it seems to be similar to M13, though it contains a larger number of variable stars.

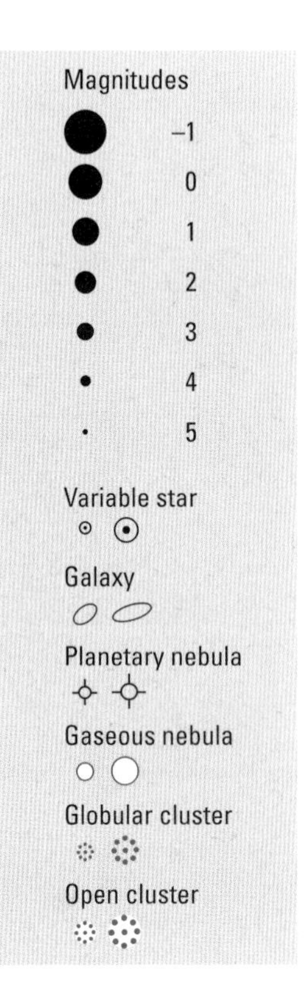

▼ **The area** enclosed in the quadrilateral formed by imaginary lines joining Arcturus (α Boo), Vega (α Lyr), Altair (α Aql) and Antares (α Sco) is occupied by three large, dim constellations: Hercules, Ophiuchus and Serpens. The region is best seen during evenings in the northern summer (southern winter), but there are no really distinctive patterns. Although Hercules is extensive it has no star much brighter than third magnitude.

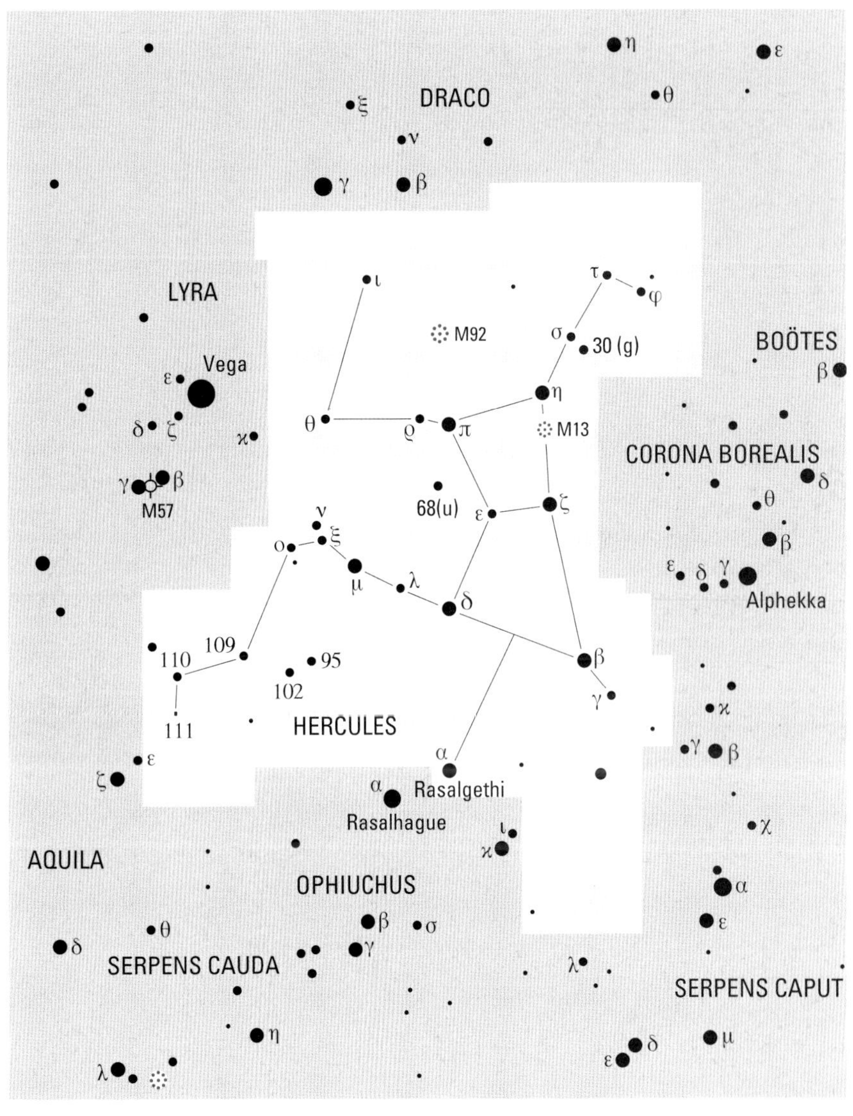

Ophiuchus, Serpens

Magnitudes
−1
0
1
2
3
4
5
Variable star
Galaxy
Planetary nebula
Gaseous nebula
Globular cluster
Open cluster

Ophiuchus is another large constellation, covering 948 square degrees of the sky, but it is confusingly intertwined with the two parts of Serpens. It straddles the equator; of its brighter stars $\varkappa$ is at declination $+9°$ and θ almost $-25°$. There is no distinctive pattern, and the only star bright enough to be really prominent is α (Rasalhague), which is 62 light years away and 67 times as luminous as the Sun. Rasalhague is much closer and less powerful than its neighbour Rasalgethi (α Herculis), even though it looks a full magnitude the brighter of the two. It is also different in colour; Rasalhague is white, while Rasalgethi is a red supergiant.

Of the other leaders of Ophiuchus, δ is of type M, and its redness contrasts well with that of the nearby ε, which is only slightly yellowish. The two are not genuine neighbours; δ is 140 light years away, ε only just over 100. It is worth looking at them with binoculars, because they are in the same field.

On the very edge of the constellation close to σ Scorpii is a wide double, ϱ Ophiuchi (which is not shown on this map) which lies close to a very rich region which is a favourite photographic target. η is a binary with components which are not very unequal (the primary is no more than half a magnitude brighter than the secondary), but the separation is less than one second of arc, so it is a good test for telescopes with apertures of around 25 centimetres (10 inches).

AQUILA
HERCULES
CORONA BOREALIS
OPHIUCHUS
IC4665
SERPENS CAPUT
M5
M14
M12
M10
RS
LIBRA
SCUTUM
M16
SERPENS CAUDA
M107
M9
SAGITTARIUS
M19
M62
Antares
SCORPIUS
LUPUS

The most interesting variable here is RS Ophiuchi, which is a recurrent nova and the only member of the class, apart from the 'Blaze Star', T Coronae Borealis, which can flare up to naked-eye visibility – as it did in 1898, 1933, 1958, 1967, 1985 and 2006; the usual magnitude is rather below 12. It is at least 3000 light years away. Ophiuchus was the site of the last observed galactic supernova, Kepler's Star of 1604, which for a while outshone Mars and Jupiter. It is a pity that it appeared before telescopes came into common use.

Another interesting object in the constellation is Munich 15040, better known as Barnard's Star because it was discovered, in 1916, by the American astronomer Edward Emerson Barnard. It is not easy to locate, because its magnitude is below 9. It is the closest of all stars apart from the members of the α Centauri system, and has the greatest proper motion known, so in fact in only 190 years or so it will shift against its background by a distance equal to the apparent diameter of the Full Moon. It is an extremely feeble red dwarf, and irregularities in its motion have led to the belief that it is attended by at least one companion which is of planetary rather than stellar mass, though definite proof is still lacking. The guide to it is the magnitude-4.6 star 66 Ophiuchi. Incidentally, this was the region where an astronomer named Poczobut, in 1777, tried to introduce a new constellation, Taurus Poniatowski, or Poniatowski's Bull; it also included 70 Ophiuchi (a well-known but rather close binary), together with 67 and 68. Not surprisingly, it has been deleted from current maps.

Ophiuchus contains no fewer than seven globular clusters on Messier's list. All of them are reasonably bright, so they are favourite objects for users of binoculars or wide-field telescopes. M12 is interesting because it is less condensed than most globulars, and is therefore easier to resolve; it may be compared with its neighbour M10, which is much more concentrated.

Mythologically speaking, Ophiuchus is identified with Aesculapius, son of Apollo and Coronis, whose skill in medi-

◀ **These constellations** are relatively hard to identify, particularly as they are so confused. In mythology Ophiuchus was the Serpent-bearer (a former name for it was Serpentarius) and Serpens was the reptile with which he was struggling – and which he has apparently pulled in half! The only bright star in the region is α Oph (Rasalhague). Ophiuchus extends into the Zodiac between Scorpius and Sagittarius, and the major planets can pass through it.

▶ **The globular cluster M12** in Ophiuchus, photographed by John Fletcher using a 25-cm (10-in) reflector. M12 is easier to resolve into individual stars than other globular clusters such as M10.

cine was legendary. He was able even to restore the dead to life, and this angered Pluto, the God of the Underworld, whose realm was starting to become depopulated. Jupiter reluctantly disposed of Aesculapius by striking him with a thunderbolt, but then relented sufficiently to transport him to the sky.

Serpens. Of the two halves of Serpens, the head (Caput) is much more conspicuous, and there is one fairly bright star, the reddish α (Unukalhai), which is magnitude 2.65, and 88 light yearsaway; its luminosity is 90 times that of the Sun. The head is formed by a little triangle of stars; ϰ (magnitude 4.09), β (3.67) and γ (3.85). Directly between β and γ is the Mira variable R Serpentis, which can rise to naked-eye visibility when at its best but which becomes very faint at minimum. Like almost all members of its class, it is very red. Its period is only nine days under a year, so when it peaks at times when it is above the horizon only during the hours of daylight, the maxima are virtually unobservable for several consecutive years. The date of maximum in 1994 was 25 March.

M5, which lies some way from Unukalhai, is one of the finest globular clusters in the entire sky; only ω Centauri, 47 Tucanae, M13 in Hercules and M22 in Sagittarius are brighter. It is very evident in binoculars, and has been known since 1702, when Gottfried Kirch discovered it. M5 is easy to resolve; the distance is 27,000 light years, and, unlike M13, it is particularly rich in variable stars.

The Serpent's body (Cauda) is less prominent, and the brightest star, η (Alava), is only magnitude 3.26; its distance is 52 light years and it is 17 times more luminous than the Sun). However, Cauda does contain two objects of note. One is M16, the Eagle Nebula, in which is embedded the cluster NGC6611. It is on the fringe of the constellation, adjoining Scutum, and the guide star to it is the magnitude-4.7 γ Scuti. The Eagle Nebula is a large, diffuse area of nebulosity, while the cluster is reasonably well marked. The two are not difficult to locate, but of course photographs are needed to bring out their full glory; there is a mass of detail, and there are areas of dark nebulosity together with small, circular 'globules', which will eventually condense into stars.

The other main feature of Cauda is θ (Alya), a particularly wide and easy double. The components are identical twins, each of magnitude 4.5 and of spectral type A5. With the naked eye, θ appears as a single star of magnitude 3.4, but good binoculars will show both components. They make up a genuine binary system, but they are a long way apart – perhaps 900 times the distance between the Earth and the Sun, so the revolution period is immensely long. The distance from us is just over 100 light years. It may be said that θ Serpentis is one of the best 'demonstration doubles' in the entire sky. To find it, follow through the line of θ, η and δ Aquilae; if this is prolonged for an equal distance beyond, it will reach θ Serpentis.

OPHIUCHUS

BRIGHTEST STARS

No.	Star	R.A. h	m	s	Dec. °	'	"	Mag.	Spectrum	Proper name
55	α	17	34	56	+12	33	36	2.08	A5	Rasalhague
35	η	17	10	22	−15	43	30	2.43	A2	Sabik
13	ζ	16	37	09	−10	34	02	2.56	O9.5	Han
1	δ	16	14	21	−03	41	39	2.74	M1	Yed Prior
60	β	17	43	28	+04	34	02	2.77	K2	Cheleb
27	ϰ	16	57	40	+09	22	30	3.20	K2	
2	ε	16	18	19	−04	41	33	3.24	G8	Yed Post
42	θ	17	22	00	−24	59	58	3.27	B2	
64	ν	17	59	01	−09	46	25	3.34	K0	

Also above magnitude 4.3: 72 (3.73), γ (3.75), λ (Marfik) (3.82), 67 (3.97), 70 (4.03), φ (4.28), 45 (4.29).

VARIABLES

Star	R.A. h	m	Dec. °	'	Range (mags)	Type	Period (d)	Spectrum
χ	16	27.0	−18	27	4.2–5.0	Irregular	–	B
U	17	16.5	+01	13	5.9–6.6	Algol	1.68	B+B
X	18	38.3	+08	50	5.9–9.2	Mira	334	M+K
RS	17	50.2	−06	43	5.3–12.3	Recurrent nova	–	O+M

DOUBLES

Star	R.A. h	m	Dec. °	'	P.A. °	Sep. "	Mags	
ϱ	16	25.6	−23	27	344	3.1	5.3, 6.0	
η	17	10.4	−15	43	247	0.5	3.0, 3.5	binary, 64 years difficult test

CLUSTERS

M	NGC	R.A. h	m	Dec. °	'	Mag.	Dimensions '	Type
9	6333	17	19.2	−18	31	7.9	9.3	globular cluster
10	6254	16	57.1	−04	06	6.6	15.1	globular cluster
12	6218	16	47.2	−01	57	6.6	14.5	globular cluster
14	6402	17	37.6	−03	15	7.6	11.7	globular cluster
19	6273	17	02.6	−26	16	7.1	13.5	globular cluster
62	6266	17	01.2	−30	07	6.6	14.1	globular cluster
107	6171	16	32.5	+03	13	8.1	10.0	globular cluster
	6633	18	27.7	+06	34	4.6	27	open cluster
	IC4665	17	46.3	+05	43	4.2	41	open cluster

SERPENS

BRIGHTEST STARS

No.	Star	R.A. h	m	s	Dec. °	'	"	Mag.	Spectrum	Proper name
(Caput)										
24	α	15	44	16	+06	25	32	2.65	K2	Unukalhai
(Cauda)										
58	η	18	21	18	−02	53	56	3.26	K0	Alava
63	θ	18	56	13	+04	12	13	3.4 (combined)	A5+A5	Alya

Also above magnitude 4.3: Caput: μ (3.54), β (3.67), ε (3.71), δ (3.80), γ (3.85), ϰ (4.09); Cauda: ξ (3.54), ο (4.26).

VARIABLE (caput)

Star	R.A. h	m	Dec. °	'	Range (mags)	Type	Period (d)	Spectrum
R	15	50.7	+15	08	5.1–14.4	Mira	356	M

DOUBLES

Star	R.A. h	m	Dec. °	'	P.A. °	Sep. "	Mags	
δ	16	34.8	+10	32	177	4.4	4.1, 5.2	binary, 3168y (in Caput)
θ	18	56.2	+04	12	104	22.3	4.5, 4.5	(inCauda)

CLUSTERS AND NEBULAE

M	NGC	R.A. h	m	Dec. °	'	Mag.	Dimensions '	Type
5	5904	15	18.6	+02	05	5.8	17.4	globular cluster in Caput
16	6611	18	18.8	−13	47	6.4	35 × 28	nebula (Eagle) and cluster (NGC 6611) in Cauda

Scorpius, Sagittarius, Corona Australis

Magnitudes
-1
0
1
2
3
4
5
Variable star
Galaxy
Planetary nebula
Gaseous nebula
Globular cluster
Open cluster

Scorpius is often, incorrectly, referred to as Scorpio. The leader is Antares (α), which is sufficiently close to the celestial equator to reach a reasonable altitude over Britain and the northern United States, though the extreme southern part of the Scorpion does not rise in these latitudes; the southernmost bright star, θ (Sargas), has a declination of almost −43°. Antares is generally regarded as the reddest of the first-magnitude stars, though its colour is much the same as that of Betelgeux. It is interesting to compare these two red supergiants. Antares is 600 light years away, and 12,000 times as luminous as the Sun, so it has only about half the power of Betelgeux; it is slightly variable, but the fluctuations, unlike those of Betelgeux, are too slight to be noticed with the naked eye.

Antares has a companion which looks slightly greenish by comparison. Both are enveloped in a huge cloud of very rarefied material, detected at infra-red wavelengths. The brightest part of the cloud associated with ρ Ophiuchi lies less than four degrees to the north-north-west.

The long chain of stars making Scorpius is striking; it ends in the 'sting', where there are two bright stars close together – λ (Shaula) and υ (Lesath). They give the impression of being a wide double, but there is no true association, because Lesath is 1570 light years away, Shaula only 275. Lesath is extremely luminous, and could match 9000 Suns, so it rivals Antares; Shaula – only just below first magnitude as seen from Earth – is 1300 times as luminous as the Sun. Both Shaula and Lesath are hot and bluish-white. Antares is flanked by τ and σ, both of which are above third magnitude. μ and ζ, farther south, look like naked-eye doubles – again a line of sight effect. The separation between the two stars of ζ is nearly 7 minutes of arc; the fainter of the pair is 2500 light years away, more remote than its brighter, orange neighbour.

The scorpion's head is made up of β (Graffias or Akrab), ν and ω. β is a fine double, so wide virtually any telescope splits it; the primary is a spectroscopic binary.

Scorpius is crossed by the Milky Way, and there are many fine star fields. There are also four clusters in Messier's list. M6 and M7 are among the most spectacular open clusters in the sky, though they are inconveniently far south from Britain and the northern United States. Both are easily seen with the naked eye, and can be resolved with binoculars; M7, the brighter of the two, was described by Ptolemy as 'a nebulous cluster following the sting of Scorpius'. Because of its large size, it is best seen with a very low magnification. M6, the Butterfly Cluster, is also very prominent, and is farther away: 1300 light years, as against 800 light years for M7. Another bright open cluster is NGC6124, which forms a triangle with the μ and ζ pairs. It can be detected with binoculars without difficulty.

The other two Messier objects are globular clusters. M4 is easy to locate: it is in the same binocular field as Antares, less than 2° to the west. It is just visible with the naked eye, and binoculars show it well; it is one of the closest globulars. No more than 7500 light years away, it is very rich in variable stars. M80 is not as prominent, but can be located easily between Antares and β. It is 36,000 light years away. It looks much smaller; it is also more compact, with a diameter of perhaps 50 light years. It is relatively poor in variable stars, but in 1860 a bright nova was seen in it, rising to seventh magnitude; in the lists it is given as T Scorpii. It soon faded away, but in case it is a recurrent nova, M80 is worth monitoring.

Sagittarius is exceptionally rich, and its glorious star clouds hide our view of that mysterious region at the centre of the Galaxy. Since Sagittarius is the southernmost of the Zodiacal constellations, it is never well seen from Britain or the northern United States; part of it never rises at all. The brightest stars are ε (Kaus Australis) and σ (Nunki), with α and β only just above fourth magnitude.

β (Akrab), in the far south of the constellation, is an easy double, and makes up a naked-eye pair with its neighbour. ζ (Ascella) is a very close, difficult binary. The components are almost equal, and the revolution period is 21 years; the sepa-

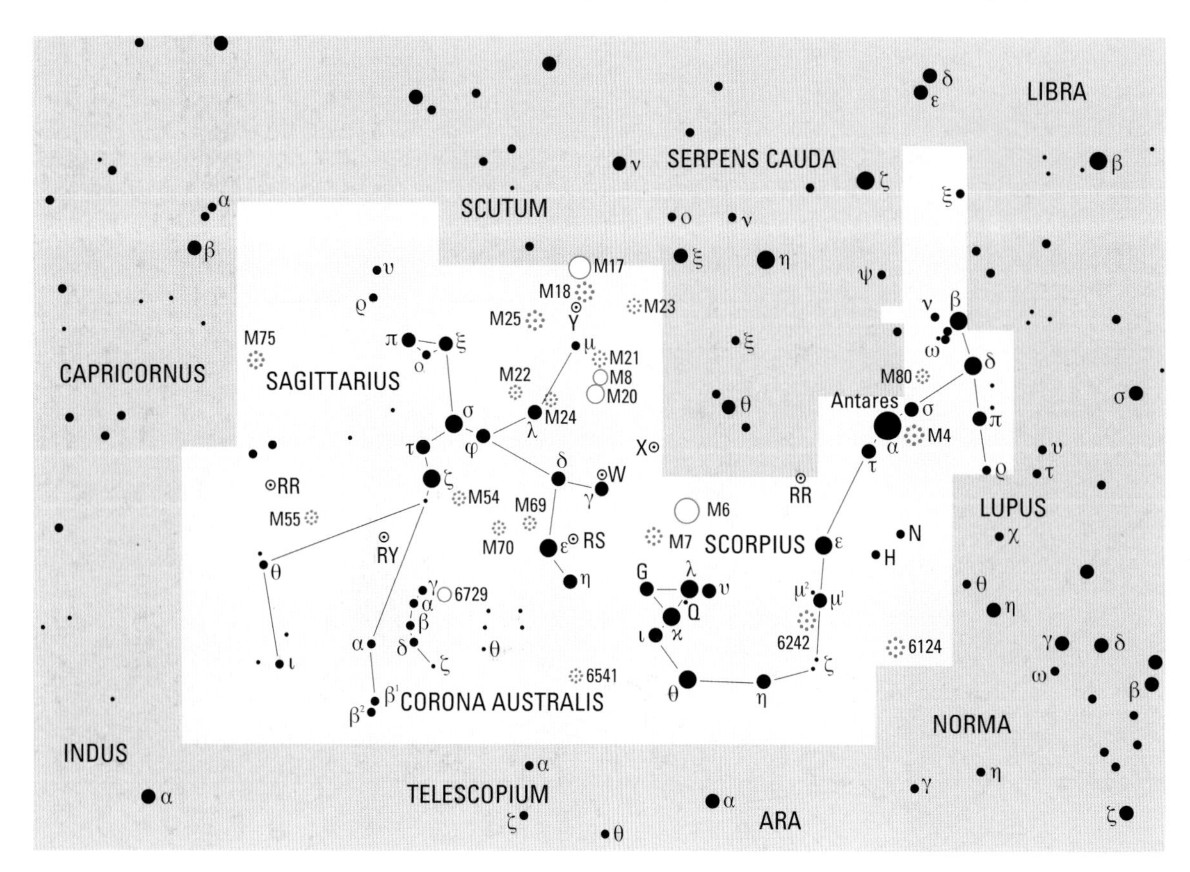

◀ **The two southernmost constellations** of the Zodiac. From the British Isles or the northern United States, parts of Scorpius and Sagittarius never rise. Antares (α Sco) can be seen, and is at its best during summer evenings. From southern countries Scorpius passes overhead, and rivals Orion for the title of the most glorious constellation in the sky, while the star clouds in Sagittarius hide our view of the centre of the Galaxy. Scorpius adjoins Libra, which was once known as the Scorpion's Claws, and the star formerly known as γ Sco has been transferred, so it is now σ Lib.

ration is only about 0.3 of a second of arc, so a telescope with an aperture of at least 38 centimetres (15 inches) is needed to resolve the pair. Users of telescopes of around this size will find it a useful test object.

There are comparatively few bright variables, but RY Sagittarii, again in the southern part of the constellation, is an R Coronae Borealis star; usually it is of magnitude 6, but falls to 15. Unfortunately, its southern declination makes it an awkward target for British and North American observers.

Sagittarius abounds in clusters and nebulae. There are three superb galactic nebulae. M20 (the Trifid) and M8 (the Lagoon) are not far from λ and μ; they can be seen in binoculars as whitish patches. Photography brings out their vivid clouds. Close by is the open cluster M21, which is easy to resolve. M17, in the northern section, and known variously as the Omega, Swan or Horseshoe Nebula, is magnificent. Of the globular clusters, M22 in particular is very fine, and was the first member of the class to be discovered (by Abraham Ihle, as long ago as 1665).

There is a great deal to be seen in Sagittarius, and one has to be systematic about it; for example, once μ has been identified it is not hard to move on to M25, M17, M21, M20 and M18, though care must be taken not to confuse them. Incidentally, M24 is not a true nebular object at all, but merely a star cloud in the Milky Way – though it does contain an open cluster, NGC6603, which lies in its northern part. When the star clouds in the area are high above the horizon, sweeping along them with binoculars or a low-power telescope will give breathtaking views.

Corona Australis, or Corona Austrinus, is one of Ptolemy's original constellations, but no legends appear to be attached to it. It is small, with no stars above fourth magnitude, but the little semi-circle consisting of γ, α, β, δ and θ is distinctive enough, close to the relatively obscure α Sagittarii. γ is a close binary, and makes a useful test object. NGC6541 is a globular cluster, just detectable with binoculars; it lies some way from the main pattern, between θ Coronae Australis and θ Scorpii. The variable nebula NGC6729 surrounds the erratic variable R Coronae Australis (do not confuse it with R Coronae Borealis). The changes in the nebula mimic the fluctuations of the star, but this is faint, and below the range of small telescopes.

SCORPIUS

BRIGHTEST STARS

No.	Star	R.A. h	m	s	Dec. °	′	″	Mag.	Spectrum	Proper name
2	α	16	29	24	−26	25	55	0.96	M1	Antares
35	λ	17	33	36	−37	06	14	1.63	B2	Shaula
	θ	17	37	19	−42	59	52	1.87	F0	Sargas
26	ε	16	50	10	−34	17	36	2.29	K2	Wei
7	δ	16	00	20	−22	37	18	2.32v	B0	Dschubba
	ϰ	17	42	29	−39	01	48	2.41	B2	Girtab
8	β	16	05	26	−19	48	19	2.64	B0+B2	Graffias
34	υ	17	30	46	−37	17	45	2.69	B3	Lesath
23	τ	16	35	53	−28	12	58	2.82	B0	
20	σ	16	21	11	−25	35	34	2.85	B1	Alniyat
6	π	15	58	51	−26	06	50	2.89	B1	
	ι^1	17	47	35	−40	07	37	3.03	F2	
	μ^1	16	51	52	−38	02	51	3.04	B1	
	G	17	49	51	−37	02	36	3.21	K2	
	η	17	12	09	−43	14	21	3.33	F2	

Also above magnitude 4.3: μ2 (3.57), ζ2 (3.62), ϱ (3.88), ω1 (3.96), ν (4.00), ξ (4.16), H (4.16), N (4.23), Q (4.29).

VARIABLE

Star	R.A. h	m	Dec. °	′	Range (mags)	Type	Period (d)	Spectrum
RR	16	55.6	−30	35	5.0–12.4	Mira	279	M

DOUBLES

Star	R.A. h	m	Dec. °	′	P.A. °	Sep. ″	Mags	
ξ	16	04.4	−11	22	051	7.6	4.8, 7.3	A is double
β	16	05.4	−19	48	021	13.6	2.6, 4.9	A is double
σ	16	21.2	−25	36	273	20.0	2.9, 8.5	
α	16	29.4	−26	26	273	2.7	1.2, 5.4	binary, 878y, red, green

CLUSTERS

M	NGC	R.A. h	m	Dec. °	′	Mag.	Dimensions ′	Type
4	6121	16	23.6	−26	32	5.9	26.3	globular cluster.
6	6405	17	40.1	−32	13	4.2	50	open cluster (Butterfly)
7	6475	17	53.9	−34	49	3.3	80	open cluster
80	6093	16	17.0	−22	59	7.2	8.9	globular cluster
	6124	16	25.6	−40	40	5.8	29	open cluster
	6242	16	55.6	−39	30	6.4	9	open cluster

SAGITTARIUS

BRIGHTEST STARS

No.	Star	R.A. h	m	s	Dec. °	′	″	Mag.	Spectrum	Proper name
20	ε	18	24	10	−34	23	05	1.85	B9	Kaus Australia
34	σ	18	55	16	−26	17	48	2.02	B3	Nunki
38	ζ	19	02	37	−29	52	49	2.59	A2	Ascella
19	δ	18	20	59	−29	49	42	2.70	K2	Kaus Meridonalis
22	λ	18	27	58	−25	25	18	2.81	K2	Kaus Borealis
41	π	19	09	46	−21	01	25	2.89	F2	Albaldah
10	γ	18	05	48	−30	25	26	2.99	K0	Alnasr
	η	18	17	37	−36	45	42	3.11	M3	
27	φ	18	45	39	−26	59	27	3.17	B8	
40	τ	19	05	56	−27	40	13	3.32	K1	

Also above magnitude 4.3: ξ^2 (3.51), ο (3.77), μ (Polis) (3.86), ϱ (3.93), β^1 (Arkab) (3.93), α (Rukbat) (3.97), ι (4.13), β^2 (4.29).

VARIABLES

Star	R.A. h	m	Dec. °	′	Range (mags)	Type	Period (d)	Spectrum
X	17	47.6	−27	50	4.2–4.8	Cepheid	7.01	F
W	18	05.0	−29	35	4.3–5.1	Cepheid	7.59	F2G
RS	18	17.6	−34	06	6.0–6.9	Algol	2.41	B2A
Y	18	21.4	−18	52	5.4–6.1	Cepheid	5.77	F
RY	19	16.5	−33	31	6.0–15	R Coronæ	–	Gp
RR	19	55.9	−29	11	5.6–14.0	Mira	335	M

DOUBLES

Star	R.A. h	m	Dec. °	′	P.A. °	Sep. ″	Mags	
η	18	17.6	−36	46	105	3.6	3.2,7.8	
β^1	19	22.6	−44	28	077	28.3	3.9,8.0	wide naked-eye pair with β^2

CLUSTERS, STAR CLOUD AND NEBULAE

M	NGC	R.A. h	m	Dec. °	′	Mag.	Dimensions ′	Type
8	6523	18	03.8	−24	23	6.0	90 × 40	nebula (Lagoon)
17	6618	18	20.8	−16	11	7.0	46 × 37	nebula (Omega)
18	6613	18	19.0	−17	08	6.9	9	open cluster
20	6514	18	02.6	−23	02	7.5	29 × 27	nebula (Trifid)
21	6531	18	04.6	−22	30	5.9	13.0	open cluster
22	6656	18	36.4	−23	54	5.1	24.0	globular cluster
23	6494	17	56.8	−19	01	5.5	27	open cluster
24	6603	18	16.9	−18	29	4.5	90	star cloud
25	IC4725	18	31.6	−19	15	4.6	31.0	open cluster round
28	6626	18	24.5	−24	52	6.9	11.0	globular cluster
54	6715	18	55.1	−30	29	7.7	9.1	globular cluster
55	6809	19	40.0	−30	58	6.9	19.0	globular cluster
69	6637	18	31.4	−32	21	7.7	7.1	globular cluster
70	6681	18	43.2	−32	18	8.1	7.8	globular cluster
75	6864	20	06.1	−21	55	8.6	6.0	globular cluster

CORONA AUSTRALIS

The brightest stars are α (Meridiana) and β, each 4.11; the position of α is R.A. 19h 09m 28s.2, dec. 237° 54′ 16″. The other stars making up the little semi-circle are γ (4.21), δ (4.59) and ζ (4.75).

DOUBLES

Star	R.A. h	m	Dec. °	′	P.A. °	Sep. ″	Mags	
ϰ	18	33.4	−38	44	359	21.6	5.9, 5.9	
γ	19	06.4	−37	04	109	1.3	4.8, 5.1	binary, 12y, good test

CLUSTERS AND NEBULAE

NGC	R.A. h	m	Dec. °	′	Mag.	Dimensions ′	Type
6541	18	08.0	−43	42	6.6	13.1	globular cluster
6729	19	01.9	−36	57	var	1 (var) 40	variable nebula around R Coronae Australis

Andromeda, Triangulum, Aries, Perseus

Magnitudes
- −1
- 0
- 1
- 2
- 3
- 4
- 5

Variable star

Galaxy

Planetary nebula

Gaseous nebula

Globular cluster

Open cluster

Andromeda is a large, prominent northern constellation, commemorating the beautiful princess who was chained to a rock on the seashore to await the arrival of a monster, though fortunately the dauntless hero Perseus was first on the scene. Andromeda adjoins Perseus to one side and Pegasus to the other; why Alpheratz was transferred from the Flying Horse to the Princess remains a bit of a mystery.

The three leading stars of Andromeda are all magnitude 2.1. Their individual names are often used; α is Alpheratz, β is Mirach and γ is Almaak. Their distances are respectively 72, 88 and 121 light years; their luminosities are 96, 115 and 95 times that of the Sun. Alpheratz is an A-type spectroscopic binary; Mirach is orange-red, with colour that is very evident in binoculars. It has been suspected of slight variability. Almaak is a particularly fine double, with a K-type orange primary and a hot companion which is said to look slightly blue-green by contrast. The pair can be resolved with almost any telescope, and the companion is a close binary, making a useful test for a telescope with an aperture of about 25 centimetres (10 inches). δ, between Alpheratz and Mirach, is another orange star of type K.

R Andromedae, close to the little triangle of θ (4.61), σ (4.62) and ϱ (5.18), is a Mira variable which can at times rise above sixth magnitude, and is readily identifiable because it is exceptionally red. The trick is to locate it when it is near maximum, so that the star field can be memorized and the variable followed down to its minimum – though if you are using a small telescope you will lose it for a while, since it drops down to almost 15th magnitude.

Of course the most celebrated object in the constellation is M31, the Andromeda Galaxy. It can just be seen with the naked eye when the sky is dark and clear, and the Persian astronomer Al-Sûfi called it 'a little cloud'. It lies at a narrow angle to us, which is a great pity; if it were face on to us it would indeed be glorious. The modern value for its distance is 2.5 million light years, though if the Cepheid standard candles have been slightly underestimated, as is possible, this value may have to be revised slightly upwards. It is a larger system than ours, and has two dwarf elliptical companions, M32 and NGC205, which are easy telescopic objects.

It has to be admitted that M31 is not impressive when seen through a telescope, and photography is needed to bring out its details. Novae have been seen in it, and there has been one supernova, S Andromedae of 1885, which reached sixth magnitude – though it was not exhaustively studied, simply because nobody was aware of its true nature; at that time it was still believed that M31, like other spirals, was a minor feature of our own Galaxy.

The open cluster NGC752, between γ Andromedae and β Trianguli, is within binocular range, though it is scattered and relatively inconspicuous. It is worth seeking out the planetary nebula NGC7662 (C22), close to the triangle made up of γ, ϰ and τ; a 25-centimetre (10-inch) telescope shows its form, though the hot central star is still very faint.

Triangulum is one of the few constellations which merits its name; the triangle made up of α, β and γ is distinctive even though only β is as bright as third magnitude. There is one reasonably bright Mira variable, R Trianguli, some way from γ, but the main object of interest is M33, the spiral galaxy which lies some way from α in the direction of Andromeda, and is just south of a line joining α Trianguli to β Andromedae. It is looser than M31, but placed at a better angle to us. Some observers claim to be able to see it with the naked eye; binoculars certainly show it, but it can be elusive telescopically, because its surface brightness is low. It is much less massive than our Galaxy.

Aries. According to legend, this constellation honours a flying ram which had a golden fleece and was sent by Mercury to rescue the two daughters of the King of Thebes, who were about to be assassinated by their wicked stepmother.

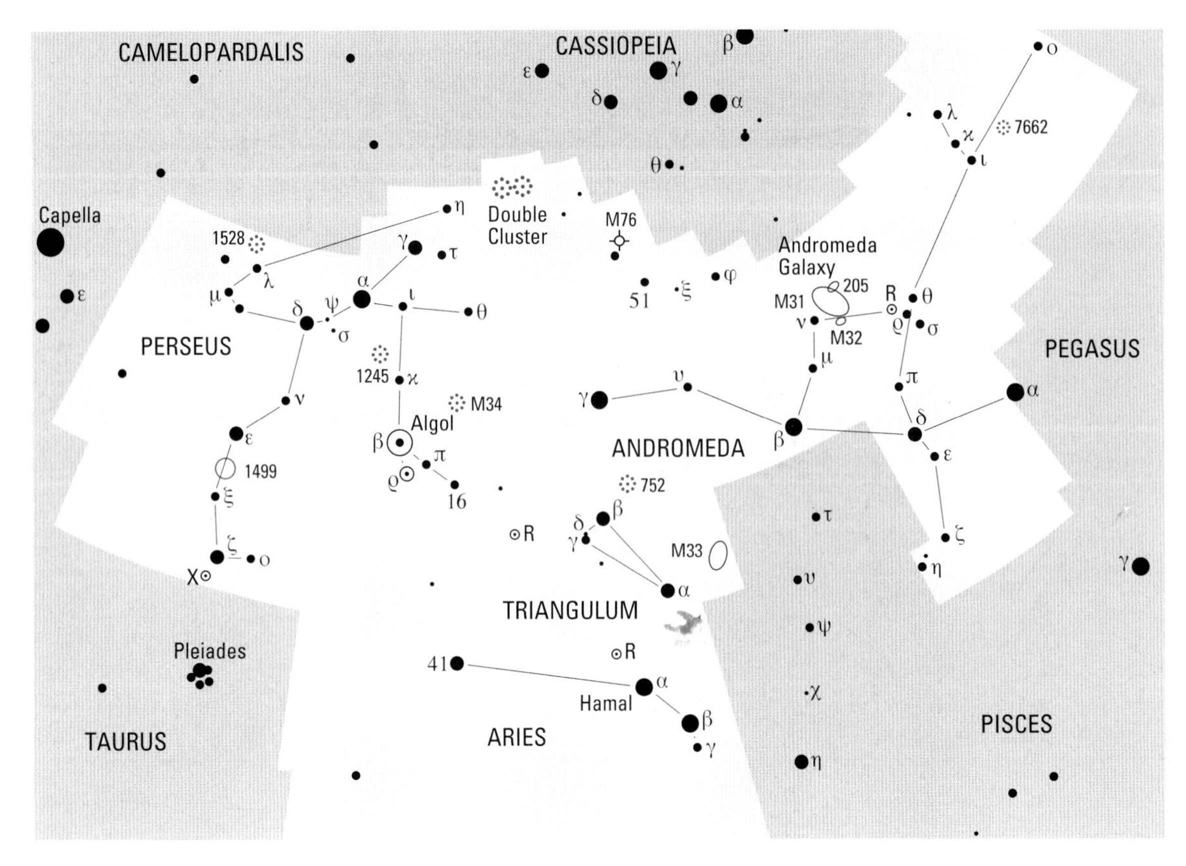

◀ **The constellations** in this map are best seen during evenings in northern autumn (southern spring), though it is true that the northernmost parts of Perseus and Andromeda – as well as Capella (α Aur) – are circumpolar from the British Isles or the northern United States and are always very low from Australia and New Zealand. Andromeda adjoins the Square of Pegasus, and indeed Alpheratz (α And) is one of the four stars of the square. υ And is known to be attended by three planets. Aries is, of course, in the Zodiac, though precession has now carried the vernal equinox across the border of Pisces.

Aries is fairly distinctive, with a small trio of stars (α, β, γ) of which α (Hamal) is reddish and of second magnitude. γ (Mesartim) is a wide, easy double with equal components. Binoculars will not split it, but almost any small telescope will do so.

Perseus. The gallant hero is well represented in the sky, and has an easily identified shape. The constellation is immersed in the Milky Way, and is very rich. There are no first-magnitude stars, but the leader, α (Mirphak), is not far below. It is of type F, 620 light years away and 6000 times as luminous as the Sun.

β (Algol) is the prototype eclipsing binary, and one of the most famous stars in the sky; it lies in the head of Medusa, the Gorgon, who had been decapitated by Perseus but whose glance could still turn any living creature to stone. Algol's period is 2 days 20 hours, 48 minutes and 56 seconds; the primary eclipse is only about 72 per cent total, but is enough to drop the apparent magnitude from 2.1 to 3.4. The secondary minimum, when the fainter component is hidden, amounts to less than a tenth of a magnitude.

The main component (A) is of type B, and is a white star 100 times as luminous as the Sun, with a diameter of 4 million kilometres (2.5 million miles). The secondary (B) is of type G; it is about three times as luminous as the Sun, and is about 5.5 million kilometres (3.4 million miles) across, so that it is larger though less massive than the primary. The true separation is around 10.5 million kilometres (6.5 million miles), so the components are too close together to be seen separately; there is also a third star in the system, well away from the eclipsing pair.

We can work out a good deal about the evolution of the Algol system. Originally the secondary (B) was more massive than its partner, so it swelled out and left the Main Sequence earlier. As it expanded, its gravitational grip on its own outer layers weakened, and material was captured by the other star (A), which eventually became the more massive of the two. The process is still going on. The system is a source of radio waves, from which we can tell that a stream of material is making its way from B to A. This is what is termed mass transfer, and is of the utmost importance in all studies of binary systems.

The fluctuations of Algol are easy to follow with the naked eye; the times of mimima are given in almanacs or in monthly astronomy periodicals such as *Sky & Telescope*. Suitable comparison stars are ζ (2.85), ε (2.89) and ϰ (3.80), as well as γ Andromedae (2.14). Mirphak is rather too bright. Avoid using ϱ Persei, which is a red semi-regular variable with a range of magnitude from 3 to 4, and a very rough period which may range between 33 and 55 days.

ζ is highly luminous (15,000 times as powerful as the Sun) and is the senior member of a 'stellar association', made up of a group of hot, luminous stars which are moving outwards from a common centre and presumably had a common origin. In the same binocular field with ζ are o (3.83) and the irregular variable X Persei, which has a range of between magnitudes 6 and 7 and is of special note because it is an X-ray emitter.

M34, an open cluster near Algol, can be seen with binoculars. However, it pales in comparison with NGC 869 and 885, which make up the Sword Handle. They are easy to locate – γ and δ, in the W of Cassiopeia, point to them – and can be seen with the naked eye; telescopes show a wonderful pair of clusters in the same low-power field. They rank among the most beautiful sights in the stellar sky.

ANDROMEDA

BRIGHTEST STARS

No.	Star	R.A. h	m	s	Dec. °	'	"	Mag.	Spectrum	Proper name
21	α	00	08	23	+29	05	26	2.06	A0p	Alpheratz
43	β	01	09	44	+35	37	14	2.06	M0	Mirach
57	γ	02	03	54	+42	19	47	2.14	K2+A0	Almaak
31	δ	00	39	20	+30	51	40	3.27		

Also above magnitude 4.3: 51 (3.57), σ (3.6v), λ (3.82), μ (3.87), ξ (4.06), υ (4.09) ϰ (4.14), φ (4.25), ι (4.29)

VARIABLES

Star	R.A. h	m	Dec. °	'	Range (mags)	Type	Period (d)	Spectrum
R	00	24.0	+38	35	5.8–14.9		409	S

DOUBLES

Star	R.A. h	m	Dec. °	'	P.A. °	Sep. "	Mags	
γ	02	03.9	+42	20	063	9.8	2.3, 4.8	B is a binary, 61y; 5.5, 6.3; sep 0".5

CLUSTERS, GALAXIES AND NEBULAE

M	C	NGC	R.A. h	m	Dec. °	'	Mag.	Dimensions	Type
31		224	00	42.7	+41	16	3.5	178 × 63	Sb galaxy, great.
32		221	00	42.7	+40	52	8.2	7.6 × 5.8	E2 galaxy, M31 companion
		205	00	40.4	+41	41	8.0	17.4 × 9.8	E6 galaxy, M31 companion
		752	01	57.8	+37	41	5.7	50	open cluster
	22	7662	23	25.9	+42	33	9.2	0.3 × 0.3	planetary nebula

TRIANGULUM

BRIGHTEST STARS

No.	Star	R.A. h	m	s	Dec. °	'	"	Mag.	Spectrum	Proper name
4	β	02	09	32	+34	59	14	3.00	A5	

Also above magnitude 4.3: α (Rasalmothallah) (3.41), γ (4.01).

VARIABLES

Star	R.A. h	m	Dec. °	'	Range (mags)	Type	Period (d)	Spectrum
R	02	37.0	+34	16	5.4–12.6	Mira	266.5	

GALAXY

M	NGC	R.A. h	m	Dec. °	'	Mags	Dimensions '	Type
33	598	01	33.9	+30	39	5.7	62 × 39	Sc galaxy

ARIES

BRIGHTEST STARS

No.	Star	R.A. h	m	s	Dec. °	'	"	Mag.	Spectrum	Proper name
13	α	02	07	10	+23	27	45	2.00	K2	Hamal
6	β	01	54	38	+20	48	29	2.64	A5	Sheratan

Also above magnitude 4.3: 41 (c) (Nair al Butain) (3.63), γ (Mesartim) (3.9) (combined magnitude).

VARIABLES

Star	R.A. h	m	Dec. °	'	Range (mags)	Type	Period (d)	Spectrum
R	00	24.0	+38	35	5.8–14.9		409	S

DOUBLES

Star	R.A. h	m	Dec. °	'	P.A. °	Sep. "	Mags
γ	01	53.5	+19	18	000	7.8	4.8, 4.8

PERSEUS

BRIGHTEST STARS

No.	Star	R.A. h	m	s	Dec. °	'	"	Mag.	Spectrum	Proper name
33	α	03	24	19	+49	51	40	1.80	F5	Mirphak
26	β	03	08	10	+40	57	21	2.12 (max)	B8	Algol
44	ζ	03	54	08	+31	53	01	2.85	B1	Atik
45	ε	03	57	51	+40	00	37	2.89	B0.5	
23	γ	03	04	48	+53	30	23	2.93	G8	
39	δ	03	42	55	+47	47	15	3.01	B5	
25	ϱ	03	05	10	+38	50	25	3.2 (max)	M4	Gorgonea Terti

Also above magnitude 4.3: η (Miram) (3.76), ν (3.77), Î (Misam) (3.80), ε (3.83), τ (Kerb) (3.85), υ (Nembus) (4.04), ξ (Menkib) (4.04), ι (4.05), φ (4.07), θ (4.12), μ (4.14), ψ (4.23), 16 (4.23), 16 (4.23), λ (4.29).

VARIABLES

Star	R.A. h	m	Dec. °	'	Range (mags)	Type	Period (d)	Spectrum
β	03	08.2	+40	57	2.1–3.4	Algol	2.87	B+G
ϱ	03	05.2	+38	50	3.2–4.2	Semi-reg.	33/55	M
X	03	55.4	+31	03	6.0–7.0	Irregular	-	X-ray source

DOUBLES

Star	R.A. h	m	Dec. °	'	P.A. °	Sep. "	Mags
η	02	50.7	+55	54	300	28.3	3.3,8.5

CLUSTERS AND NEBULAE

M	NGC	R.A. h	m	Dec. °	'	Mag.	Dimensions '	Type
34	1039	02	42.0	+42	47	5.2	35	open cluster
76	650-1	01	42.4	+51	34	12.2	2.7 × 1.8	planetary nebula

Pegasus, Pisces

Magnitudes
-1
0
1
2
3
4
5

Variable star

Galaxy

Planetary nebula

Gaseous nebula

Globular cluster

Open cluster

Pegasus forms a square, though one of its main stars, Alpheratz, has been stolen by the neighbouring constellation to become α Andromedae. The stars in the Square of Pegasus are not particularly bright; Alpheratz is second magnitude, the others are between 2.5 and 3. However, the pattern is easy to pick out because it occupies a decidedly barren region of the sky. On a clear night, try to count the number of stars you can see inside the square first with the naked eye, and then with binoculars. The answer can be somewhat surprising.

Three of the stars in the square are hot and white. α Pegasi (Markab) is of type B9, 100 light years away and 75 times as luminous as the Sun. γ (Algenib), which looks the faintest of the four, is also the most remote (520 light years) and the most powerful (equal to 1300 Suns); the spectral type is B. The fourth star, β (Scheat), is completely different. It is an orange-red giant of type M, and the colour is evident even with the naked eye; binoculars bring it out well, and the contrast with its neighbours is striking. Moreover, it is variable. It has a fairly small range, from magnitude 2.3 to 2.8, but the period – around 38 days – is more marked than with most other semi-regular stars. The changes can be followed with the naked eye, α and β make good comparison stars.

When making estimates of this kind, allowance has to be made for what is termed extinction, the dimming of a star through atmospheric absorption which naturally increases at lower altitudes above the horizon (see table, opposite above). The right ascensions of β and α are about the same, and the difference in declination is about 13°. Suppose that β is at an altitude of 32°; it will be dimmed by 0.2 of a magnitude. If α is directly below (as it may be to northern-hemisphere observers; in southern latitudes the reverse will apply) its altitude will be 32 minus 13 = 19°, and it will be dimmed by 0.5 of a magnitude. If the two stars look equally bright, α (magnitude 2.4) will actually be brighter by 0.3 magnitude, so β will be 2.7. Try to find a comparison star at an altitude equal to that of the variable. This is unimportant with telescopic variables; extinction will not change noticeably over a telescopic or binocular field of view.

It is also interesting to compare the real luminosities of the stars in the square. Absolute magnitude is the apparent magnitude a star would have at a standard distance of 10 parsecs (32.6 light years). The values for the four are: Alpheratz −0.1, Markab +0.2, Scheat −1.4 (rather variable), and Algenib −3.0, so the last would dominate the scene.

The other leading star of Pegasus is ε, which is well away from the square and is on the border of Equuleus. It is a K-type orange star, 520 light years away and 4500 times as luminous as the Sun. It has been strongly suspected of variability, and naked-eye estimates are worthwhile; α is a good comparison, though in general ε should be slightly but detectably the brighter of the two.

The globular cluster M15, close to ε, was discovered in 1746 by the Italian astronomer G. F. Maraldi. To find it, use θ and ε as guides. It is just below naked-eye visibility, but binoculars show it as a fuzzy patch; it has an exceptionally condensed centre, and is very rich in variable stars. It is also very remote, at a distance of over 49,000 light years. The real diameter cannot be far short of 100 light years.

Pisces is one of the more obscure Zodiacal constellations, and consists mainly of a line of dim stars running along south of the Square of Pegasus. Mythologically its associations are rather vague; it is sometimes said to represent two fish into which Venus and Cupid once changed themselves in order to escape from the monster Typhon, whose intentions were anything but honourable.

α, magnitude 3.79, has three proper names: Al Rischa, Kaïtain and Okda. It is a binary, not difficult to split with a small telescope; both components have been suspected of slight variability in brightness and colour, but firm evidence is lacking. ζ is another easy double, and here too slight variability has been suspected.

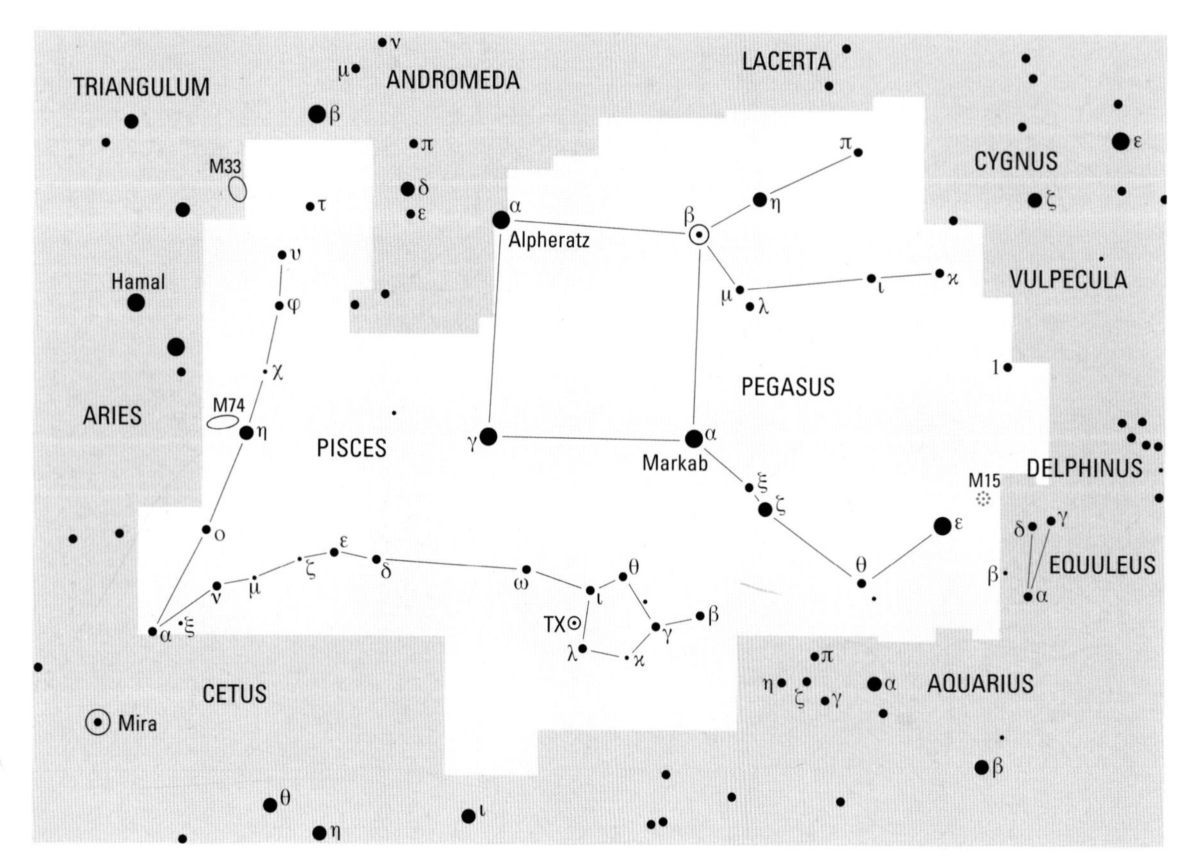

◀ **Pegasus** is the most prominent constellation of the evening sky during northern autumn (southern spring). The four main stars – one of which has been illogically transferred to Andromeda – make up a square, which is easy enough to identify even though maps tend to make it seem smaller and brighter than it really is. In fact the brightest star in Pegasus, ε, is some way from the square. 51 Peg (just outside the square), mag. 5.5, was the first star found to be attended by a planet. Pisces is a very dim Zodiacal constellation occupying the area between Pegasus and Cetus.

EXTINCTION TABLE

Altitude above horizon, °	Dimming in magnitude.
1	3.0
2	2.5
4	2.0
10	1.0
13	0.8
15	0.7
17	0.6
21	0.4
26	0.3
32	0.2
43	0.1

Above this altitude, extinction can be neglected.

◀ **The globular cluster M15** in Pegasus, photographed by Bernard Abrams using a 25-cm (10-in) reflector. It can be found near to ε.

The exceptionally red N-type semi-regular variable TX (19) Piscium is worth locating. It is easily found, near the 'circlet' made up of ι, θ, γ and λ; as it never falls below magnitude 7.7 it is always within binocular range, and its hue is almost as strong as that of the famous Garnet Star, μ Cephei.

The galaxy M74, discovered by Méchain in 1780, is one of the less massive spirals in Messier's catalogue; it can be seen with a 7.6-centimetre (3-inch) telescope, but can be rather elusive. It lies within a couple of degrees of η (Matar). There is a fairly well defined nucleus, but the spiral arms are so loose and faint that even Sir John Herschel mistook it for a globular cluster. The distance is around 26 million light years.

One object in Pisces which is worth finding, though requiring at least a 25-centimetre (10-inch) telescope, is the white dwarf Wolf 28, better known as Van Maanen's Star; it was discovered in 1917 by the Dutch astronomer Adriaan van Maanen. Its position is R.A. 00h 46m.5, dec. +05° 09′, about 2° south of δ. Its visual magnitude is 12.4, and it is one of the dimmest stars known, with a luminosity only about 1/6000 of that of the Sun. The diameter is about the same as that of the Earth, but in mass it is equal to the Sun, so the density must be about a million times that of water; if you could go there and stand on the surface, you would find that your weight has been increased by about 10 million times. The proper motion amounts to nearly 3 seconds of arc per year, and the distance is less than 14 light years, so this is one of the nearest known of all white dwarfs.

PEGASUS

BRIGHTEST STARS

No.	Star	R.A. h	m	s	Dec. °	′	″	Mag.	Spectrum	Proper name
8	ε	21	44	11	+09	52	30	2.38	K2	Enif
53	β	23	03	46	+28	04	58	2.4 (max)	M2	Scheat
54	α	23	04	45	+15	12	19	2.49	B9	Markab
88	γ	00	13	14	+15	11	01	2.83	B2	Algenib
44	η	22	43	00	+30	13	17	2.94	G2	Matar
42	ζ	22	41	27	+10	49	53	3.40	B8	Homan
48	μ	22	50	00	+24	36	06	3.48	K0	Sadalbari

Also above magnitude 4.3: θ (Biham) (3.53), ι (3.76), I (4.08), ϰ (4.13), ξ (Al Suud al Nujam) (4.19), π (4.29). Alpheratz (α Andromedae) was formerly included in Pegasus, as δ Pegasi.

VARIABLE

Star	R.A. h	m	Dec. °	′	Range (mags)	Type	Period (d)	Spectrum
β	23	03.8	+28	05	2.4–2.8	Semi-reg.	38	M

CLUSTER

M	NGC	R.A. h	m	Dec. °	′	Mag.	Dimensions ′	Type
15	7078	21	30.0	+12	10	6.3	12.3	globular cluster

PISCES

The brightest star is η (Alpherg), R.A. 01h 31m 29s, dec.+15° 20′ 45″ mag. 3.62. Also above magnitude 4.3: γ (3.69), α (Al Rischa) (3.79), ω (4.01), ι (4.13), o (Torcular) (4.26), θ (4.28), ε (4.28).

VARIABLE

Star	R.A. h	m	Dec. °	′	Range (mags)	Type	Period (d)	Spectrum
TX	23	46.4	+03	29	6.9–7.7	Irregular	-	N

DOUBLES

Star	R.A. h	m	Dec. °	′	P.A. °	Sep. ″	Mags	
α	02	02.0	+02	46	279	1.9	4.2, 5.1	binary, 933y
ζ	01	13.7	+07	35	063	23.0	5.6, 6.5	

GALAXY

M	NGC	R.A. h	m	Dec. °	′	Mag.	Dimension ′	Type
74	628	01	36.7	+15	47	9.2	10.2 × 9.5	Sc galaxy

Capricornus, Aquarius, Piscis Austrinus

Magnitudes

–1

0

1

2

3

4

5

Variable star

Galaxy

Planetary nebula

Gaseous nebula

Globular cluster

Open cluster

Capricornus has been identified with the demigod Pan, but the mythological association is decidedly nebulous, and the pattern of stars certainly does not recall the shape of a goat, marine or otherwise. Neither can it be said that there is a great deal of interest here, even though the constellation covers over 400 square degrees of the sky. δ is the only star above third magnitude; it is about 49 light years away, and some 13 times as luminous as our own Sun.

β is one of the less powerful naked-eye stars, and is not much more than twice the luminosity of the Sun, though its distance is not known with any certainty and may be less than the official value given in the Cambridge catalogue. It has a sixth-magnitude companion which is within binocular range and is itself a very close double. The bright star appears to be a spectroscopic triple, so β Capricorni is a very complex system indeed.

α^1 and α^2 (Algedi) make up a wide pair, easily separable with the naked eye, but there is no genuine association. The brighter star, α^2, is 109 light years away, while the fainter component, α^1, is very much in the background at a distance of 690 light years; both are of type G, but while the more remote star is well over 930 times as powerful as the Sun, the closer member of the pair could equal no more than 43 Suns. This is a classic example of an optical pair.

There are no notable variables in Capricornus, but there is one Messier object, the globular cluster M30, which lies close to ζ (which is, incidentally, a very luminous G-type giant). M30 was discovered by Messier himself in 1764, and described as 'round; contains no star'. It is in fact a small globular with a brightish nucleus; it is 41,000 light years away, and has no characteristics of special note.

Aquarius, with an area of almost 1000 square degrees, is larger than Capricornus, but it is not a great deal more conspicuous. It is known as the Water-bearer, but its mythological associations are vague, though it has sometimes been identified with Ganymede, the cup bearer to the Olympian gods. Its main claim to fame is that it lies in the Zodiac. Most of it is in the southern hemisphere of the sky.

Both α and β are very luminous and remote G-type giants. The most interesting star is ζ, which is a fine binary with almost equal components; both are F-type subgiants about 100 light years away, with a real separation of at least 15,000 million kilometres (over 9000 million miles). This is an excellent test object for a telescope of around 7.6-centimetre (3-inch) aperture.

There is a distinctive group of stars between Fomalhaut, in Piscis Austrinus, and α Pegasi (Markab). The three stars labelled 'ψ Aquarii' are close together, with χ and φ nearby; several of them are orange, and they have often been mistaken for a very loose cluster, though they are not really associated with each other.

R Aquarii is a symbiotic or Z Andromedae type variable. It is made up of a cool red giant together with a hot subdwarf – both of which seem to be intrinsically variable. The whole system is enveloped in nebulosity, and the smaller star seems to be pulling material away from its larger, less dense companion. R Aquarii is none too easy to locate, but users of larger telescopes will find that it repays study.

M2 is a particularly fine globular cluster, forming a triangle with α and β Aquarii. Some people claim that they can see it with the naked eye; with binoculars it is easy. It was discovered by G. F. Maraldi as long ago as 1746, and is very remote, at around 55,000 light years. Its centre is not as condensed as with most globulars, and the edges are not hard to resolve.

M72 is another globular, discovered by Méchain in 1780; it is 62,000 light years away, and comparatively 'loose'. It is one of the fainter objects in Messier's list, and is none too easy to locate; it lies between θ Capricorni and ε Aquarii (not shown), and is surprisingly difficult to resolve into individual stars.

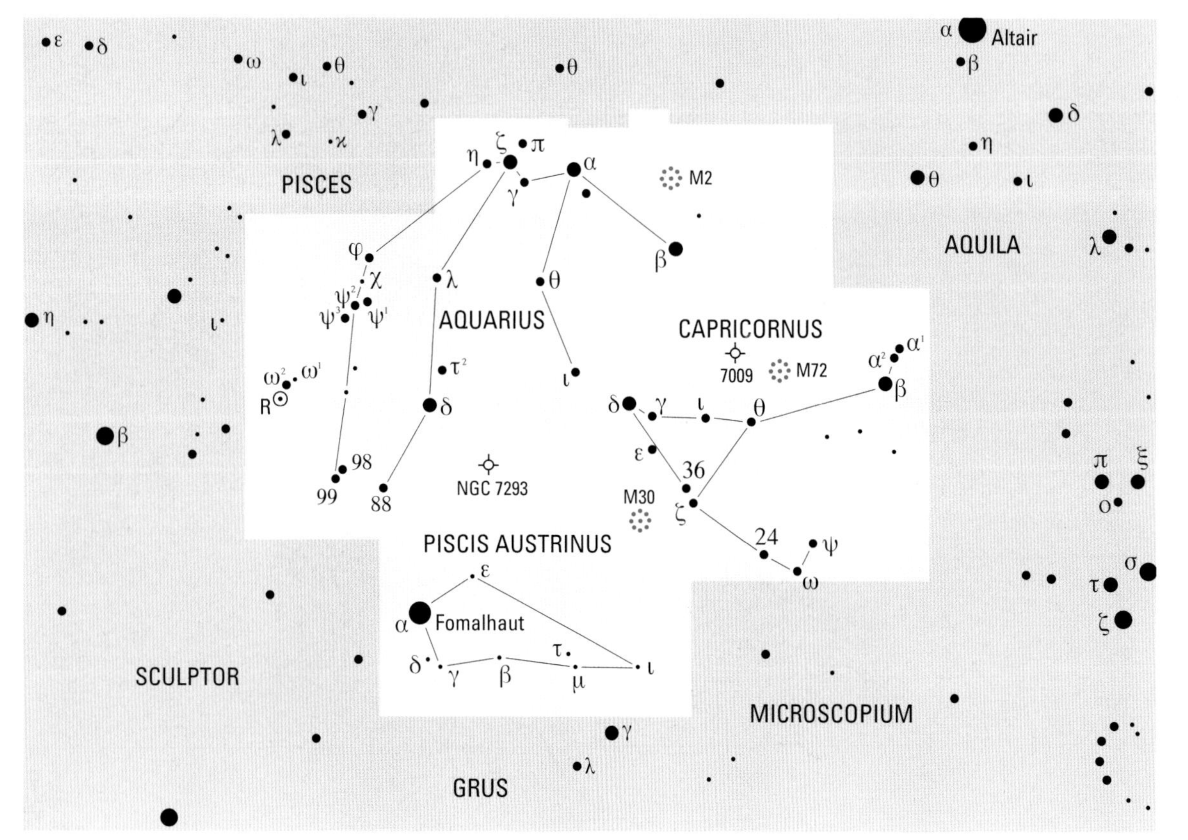

◀ **The two Zodiacal constellations** of Capricornus and Aquarius occupy a wide area, but contain little of immediate interest, and together with Pisces and Cetus they give this whole region a decidedly barren look. Fomalhaut (α PsA), is the southernmost of the first-magnitude stars to be visible from the British Isles or the northern United States; northern observers, who never see it high up, do not always appreciate how bright it really is. The celestial equator just passes through the northernmost part of Aquarius.

◀ **Globular cluster M2** in Aquarius, photographed by Doug Williams and N. A. Sharp with the 0.9-m (36-in) telescope of Kitt Peak National Observatory.

M73, less than 2° from ν Aquarii (magnitude 4.51), is not a real cluster at all, even though it has been given an NGC number; it is made up of a few disconnected stars below tenth magnitude.

There are two interesting planetary nebulae in Aquarius. NGC7009 (C55), the Saturn Nebula, is about one degree west of ν, and is a beautiful object in large telescopes, with a prominent belt of obscuring material. It is about 3900 light years away, and around half a light year in diameter.

NGC7293 (C63), the Helix Nebula, is the largest and the brightest of all the planetaries, and is said to be visible in binoculars as a faint patch, though a telescope is needed to show it clearly because it lies so close to ν. When photographed, the Helix is seen to be not unlike the Ring Nebula in Lyra (M57), but the central star is only 13th magnitude.

Piscis Austrinus, or Piscis Australis, the Southern Fish, is a small though ancient constellation, apparently not associated with any myth or legend. The only star above the fourth magnitude it contains is Fomalhaut (α), which is the southernmost of the first-magnitude stars which are visible from the latitudes of the British Isles (from north Scotland it barely rises). It is easy to find by using β and α Pegasi, in the square, as pointers; but beware of confusing it with Diphda (β Ceti), which is roughly aligned with the other two stars of the square, Alpheratz (α Andromedae) and γ Pegasi. However, Diphda is a magnitude fainter than Fomalhaut.

Fomalhaut is a pure white star, 22 light years away and 13 times more luminous than the Sun; it is therefore one of our closer stellar neighbours. In 1983, the Infra-Red Astronomical Satellite (IRAS) found that it is associated with a cloud of cool matter which may be planet forming; as with Vega (β Lyrae), β Pictoris and other such stars, we cannot certainly claim that a planetary system exists there, but neither can we rule it out, The constellation contains nothing else of particular note, though β is a wide and easy optical double.

CAPRICORNUS

BRIGHTEST STARS

No.	Star	R.A. h	m	s	Dec. °	′	″	Mag.	Spectrum	Proper name
49	δ	21	47	02	−16	07	38	2.87	A5	Deneb al Giedi
9	β	20	21	00	−14	46	53	3.08	F8	Dabih

Also above magnitude 4.3: α² (Al Giedi) (3.57), γ (Nashira) (3.68), ζ (Yen) (3.74), θ (4.07), ω (4.11), φ (4.14), α¹ (4.24), ι (4.28).

DOUBLE

Star	R.A. h	m	Dec. °	′	P.A. °	Sep. ″	Mags	
α	20	18.1	−12	33	291	377.7	3.6, 4.2	naked-eye pair
β	20	21.0	−14	47	267	205.0	3.1, 6.0	

CLUSTER

M	NGC	R.A. h	m	Dec. °	′	Mag.	Dimensions ′	Type
30	7099	21	40.4	23	11	7.5	11.0	globular cluster

AQUARIUS

BRIGHTEST STARS

No.	Star	R.A. h	m	s	Dec. °	′	″	Mag.	Spectrum	Proper name
22	β	21	31	33	−05	34	16	2.91	G0	Sadalsuud
34	α	22	05	47	−00	19	11	2.96	G2	Sadalmelik
76	δ	22	54	39	−15	49	15	3.27	A2	Scheat

Also above magnitude 4.3: ζ (3.6 combined magnitude), 88 (c2) (3.66), λ (3.74), ε (Albali), (3.77), γ (3.84), 98 (b1) (3.97), τ (4.01), η (4.02), θ (Ancha) (4.16), φ1(4.21), τ (4.22), ι (4.27).

VARIABLE

Star	R.A. h	m	Dec. °	′	Range (mags)	Type	Period (d)	Spectrum
R	23	43.8	−15	17	5.8–12.4	Symbiotic	387	M+pec

DOUBLE

Star	R.A. h	m	Dec. °	′	P.A. °	Sep. ″	Mags	
ζ	2	28.8	−00	01	200	2.0	4.3, 4.5	binary, 856y

CLUSTERS AND NEBULAE

M	C	NGC	R.A. h	m	Dec. °	′	Mag.	Dimensions ′	Type
2		7089	21	33.5	−00	49	6.5	12.9	globular cluster
72		6981	20	53.5	−12	32	9.3	5.9	globular cluster
	63	7293	22	29.6	−20	48	6.5	12.8	planetary nebula (Helix)
	55	7009	21	04.2	−11	22	8.3	0.3 × 1.7	planetary nebula (Saturn)

M73 (NGC 6994), R.A. 20h 58.9, dec, −12° 38′, is an asterism of four stars.

PISCIS AUSTRINUS

BRIGHTEST STAR

No.	Star	R.A. h	m	s	Dec. °	′	″	Mag.	Spectrum	Proper name
24	α	22	57	39	−29	37	20	1.16	A3	Fomalhaut

Also above mag. 4.3: ε (4.17), δ (4.21), β (Fum el Samakah) (4.29).

DOUBLE

Star	R.A. h	m	Dec. °	′	P.A. °	Sep. ″	Mags	
β	22	51.5	−32	21	172	30.3	4.4, 7.9	(optical)

Cetus, Eridanus (northern), Fornax

Cetus is a vast constellation, covering 1232 square degrees. Mythologically it is said to represent the sea-monster which was sent to devour the Princess Andromeda, but which was turned to stone when Perseus showed it the Gorgon's head.

The brightest star, β (Diphda), can be found by using Alpheratz (α Andromedae) and Algenib (γ Pegasi), in the Square of Pegasus, as pointers. It is an orange K-type star, 68 light-years away and 75 times as luminous as the Sun. It has been strongly suspected of variability, and is worth monitoring with the naked eye, though the lack of suitable comparison stars makes it awkward to estimate. θ and η lie close together; θ is white, and η, with a K-type spectrum, rather orange. In the same binocular field there is a faint double star, 37 Ceti.

τ Ceti is of special interest. It is only 11.9 light years away and about one-third as luminous as the Sun, with a K-type spectrum. It is one of the two nearest stars which can be said to be at all like the Sun (ε Eridani is the other), and was once regarded as a promising candidate as the centre of a planetary system, so efforts have been made to 'listen out' for signals from it which might be interpreted as artificial – so far with a total lack of success. The flare star UV Ceti lies less than 3° south-west of τ.

The 'head' of Cetus is made up of α, γ, μ, ξ and δ. α (Menkar) is an M-type giant, 130 light years away and 132 times as luminous as the Sun; it too has been suspected of slight variability. It is a binary with a very long revolution period; it is fairly easy to split with a small telescope.

Mira (o Ceti) is the prototype long-period variable, and has been known to exceed second magnitude at some maxima, though at others it barely rises above 4. It is visible with the naked eye for only a few weeks every year, but when at its best it alters the whole aspect of that part of the sky. It was the first variable star to be identified, and is also the closest of the M-type giants; its distance is 420 light years, and it is over 100 times as powerful as the Sun. The average period is 331 days. Mira's diameter is around 680 million kilometres (425 million miles) but it swells and shrinks. It is a binary; the companion, of around magnitude 10, is the flare star VZ Ceti.

M77 is a massive Seyfert galaxy, and is a strong radio source. It lies near δ, and is not hard to locate, but its nucleus is so bright compared with the spiral arms that with low or even moderate magnifications it takes on the guise of a rather fuzzy star. The distance is about 52 million light years. NGC247 (C62), close to β, is fairly large, but has a low surface brightness, and is also placed at an unfavourable angle to us, so the spiral form is not well displayed. All the other galaxies found in Cetus are considerably fainter.

Eridanus. Only the northern part of this immensely long constellation is shown here; the rest sprawls down into the far south of the sky (see Star Map 22). In mythology Eridanus represents the River Po, into which the reckless youth Phaethon fell after he had obtained permission to drive the Sun-chariot for one day – with the result that the Earth was set on fire, and Jupiter had to call a reluctant halt to the proceedings by striking Phaethon with a thunderbolt.

There is not a great deal to see in the northern part of the constellation. There are only two stars above the third magnitude; β (Kursa), which is close to Rigel (β Orionis) of type A, 96 light years away and 83 times as luminous as the Sun, and γ (Zaurak), which is type M, 114 light years away and 120 times as luminous as the Sun. It is worth looking at the δ–ε pair. δ (Rana) is fairly close, at a distance of 29 light years, and is a K-type star only 2.6 times as luminous as the Sun; next to it is ε, at a distance of 10.7 light years, which, with τ Ceti, is one of the two nearest stars to bear any resemblance to the Sun. The Infra-Red Astronomical Satellite (IRAS) found that it is associated with cool material, and it is known to be the centre of a planetary system. It

Magnitudes
–1
0
1
2
3
4
5
Variable star
Galaxy
Planetary nebula
Gaseous nebula
Globular cluster
Open cluster

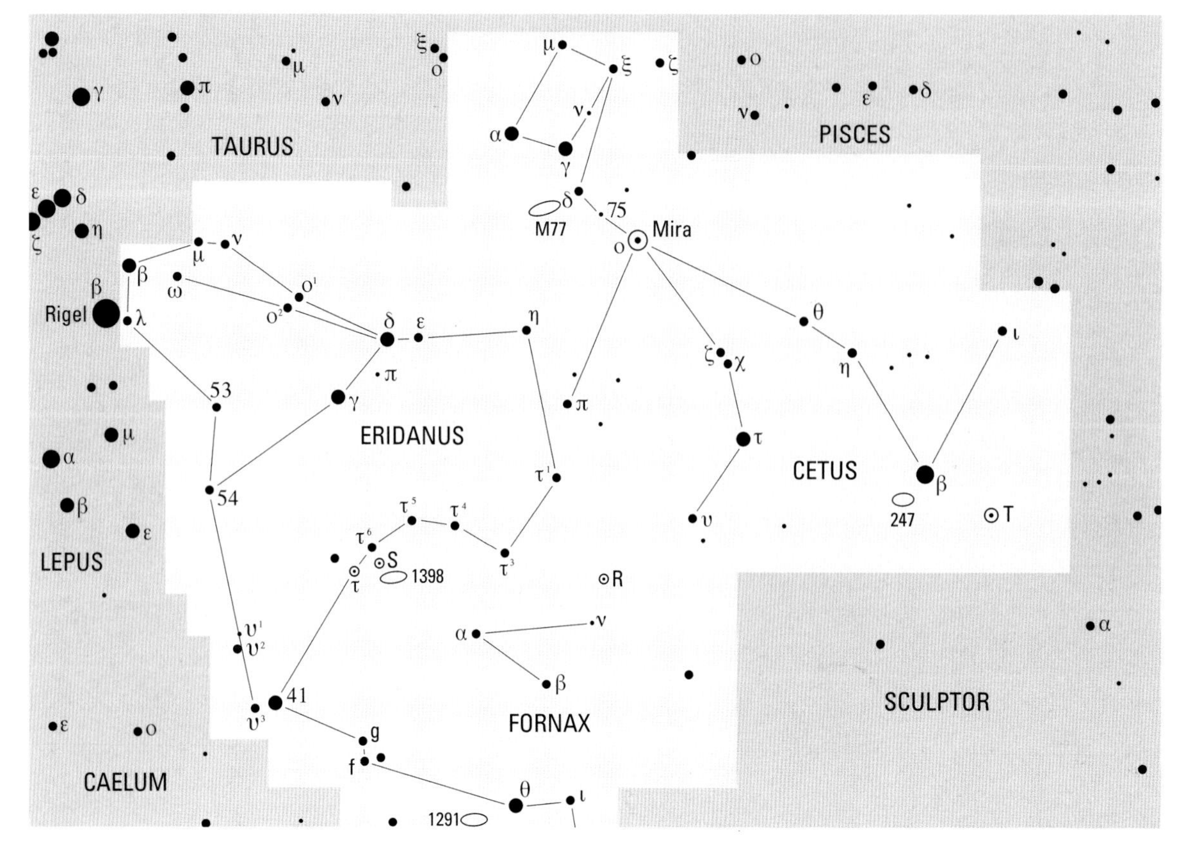

◀ **The constellations** in this map are seen to advantage during evenings in late autumn (northern hemisphere) or late spring (southern). Cetus is large but rather faint, though the 'head', containing α (Menkar) is not hard to identify. Eridanus is so immensely long that not all of it can be conveniently shown on one map; the southern part is contained in Star Map 22, the south polar region – the 'river'ends with the brilliant Achernar (α), which cannot be seen from anywhere farther north than Cairo.

is smaller and less luminous than τ Ceti, and is in fact the feeblest star visible with the naked eye apart from the far-southern ε Indi. Its absolute magnitude is 6.1, so from the standard distance of ten parsecs (32.6 light years) it could not easily be seen without any optical aid.

The two Omicrons, o^1 (Beid) and o^2 (Keid), lie side by side, but are not connected. Beid is 277 light years away and well over 150 times as luminous as the Sun, while Keid is a complex system only 16 light years away. It is a wide, easy binary; the secondary is itself a binary consisting of a feeble red dwarf, of exceptionally low mass (no more than 0.2 that of the Sun) together with a white dwarf whose diameter is about twice that of the Earth and which seems to be associated with a third body of substellar mass.

Fornax is a 'modern' group whose name has been shortened. It was added to the sky by Lacaille in 1752; it was originally Fornax Chemica, the Chemical Furnace.

It is marked by a triangle of inconspicuous stars: α (magnitude 3.87), β (4.46) and ν (4.69). It is crowded with galaxies, but all these are inconveniently faint for users of small telescopes. α is a wide double.

CETUS

BRIGHTEST STARS

No.	Star	R.A. h	m	s	Dec. °	'	"	Mag.	Spectrum	Proper name
16	β	00	43	35	−17	59	12	2.04	K0	Diphda
92	α	03	02	17	+04	05	23	2.54	M2	Menkar
31	η	01	08	35	−10	10	56	3.45	K2	
86	γ	02	43	18	+03	14	09	3.47	A2	Alkaffaljidhina
52	τ	01	44	04	−15	56	15	3.50	G8	

Also above magnitude 4.3: ι (3.56), θ (3.60), ζ (3.73), υ (4.00), δ (4.07), π (4.25), μ (4.27), ξ^2 (4.28).
The variable Mira (ο Ceti) has been known to rise to magnitude 1.6, but most maxima are much fainter than this.

VARIABLES

Star	R.A. h	m	Dec. °	'	Range (mags)	Type	Period (d)	Spectrum
ο	02	19.3	−02	58	1.6–10.1	Mira	331	M
T	00	21.8	−20	03	5.0–6.9	Semi-reg.	159	M

DOUBLES

Star	R.A. h	m	Dec. °	'	P.A. °	Sep. "	Mags
χ	01	49.6	−10	41	250	183.8	4.9, 6.9
γ	02	43.3	+03	14	294	2.8	3.8, 7.3
66	02	12.8	−02	24	234	16.5	5.7, 7.5
ο	02	19.3	−02	58	085	0.3	var, 9.5v

GALAXIES

M	C	NGC	R.A. h	m	Dec. °	'	Mag.	Dimensions '	Type
77		1068	02	42.7	−00	01	8.8	6.9 × 5.9	SBp galaxy (Seyfert galaxy)
	62	247	00	47.1	−20	46	8.9	20.0 × 7.4	SAB galaxy

ERIDANUS

BRIGHTEST STARS

No.	Star	R.A. h	m	s	Dec. °	'	"	Mag.	Spectrum	Proper name
67	β	05	07	51	−05	05	11	2.79	A3	Kursa
34	γ	03	58	02	−13	30	31	2.95	M0	Zaurak

Also above magnitude 4.3: δ (Rana) (3.54), τ^4 (Angetenar) (3.69), ε (3.73), υ^2 (Theemini) (3.82), 53 (Sceptrum) (3.87), η (Azha) (3.89), ν (3.93), μ (4.09), o^1 (4.02), o^3 (Beid) (4.04), λ (4.27); o^2 (Keid), close to o^1, is of magnitude 4.43.
ζ (Zibal), now of magnitude 4.80, is one of the few stars to be strongly suspected of fading during historic times, though the evidence is inconclusive.

DOUBLE

Star	R.A. h	m	Dec. °	'	P.A. °	Sep. "	Mags	
o^2	04	15.2	−07	39	107	82.8	4.9, 9.5	B is double

FORNAX

The brightest star in Fornax is α: R.A. 03h 12m 04s.2, dec. 228° 59' 13", mag. 3.87.
There are no other stars above magnitude 4.3.

DOUBLE

Star	R.A. h	m	Dec. °	'	P.A. °	Sep. "	Mags	
α	03	12.1	−28	59	298	4.0	4.0, 7.0	binary 314y

▼ **M77** is a type Sb spiral galaxy in the constellation Cetus. A Seyfert galaxy, it shows broad and strong emission lines from high velocity gas in the galaxy's inner regions. Cetus A, a strong radio source, is situated in the nucleus of the galaxy.

Orion, Canis Major, Canis Minor,

Magnitudes
- −1
- 0
- 1
- 2
- 3
- 4
- 5

Variable star

Galaxy

Planetary nebula

Gaseous nebula

Globular cluster

Open cluster

Orion, the Hunter, is generally regarded as the most splendid of all the constellations. The two leaders are very different from each other; though lettered β, Rigel is the brighter, and is particularly luminous, since it could match 40,000 Suns and is some 750 light years away. If it were as close to us as Sirius, its magnitude would be −10, and it would be one-fifth as brilliant as the Full Moon. It has a companion star, which is above magnitude 7, and would be easy to see if it were not so overpowered by Rigel: even so it has been glimpsed with a 7.6-centimetre (3-inch) telescope under good conditions. The companion is itself a close binary, with a luminosity 150 times that of the Sun. α (Betelgeux) has an official magnitude range of from 0.4 to 0.9, but it seems definite that at times it can rise to 0.1, almost equal to Rigel. Good comparison stars are Procyon (α Canis Minoris) and Aldebaran (α Tauri), but allowance must always be made for extinction (see table on page 247). The apparent diameter of Betelgeux is greater than for most other stars beyond the Sun, and modern techniques have enabled details to be plotted on its surface.

The other stars of the main pattern are γ (Bellatrix), κ (Saiph) and the three stars of the Belt, δ (Mintaka), ε (Alnilam) and ζ (Alnitak). Bellatrix is 900 times as luminous as the Sun; all the others outshine the Sun by more than 20,000 times, and are over 1000 light years away. Indeed, Saiph is not much less powerful than Rigel, but is even more remote, at 2200 light years. Mintaka is an eclipsing binary with a very small range (magnitude 2.20 to 2.35); both it and Alnitak have companions which are easy telescopic objects.

σ, in the Hunter's Sword, is a famous multiple, and of course θ, the Trapezium, is responsible for illuminating the wonderful Orion Nebula, M42. M43 (an extension of M42) and M78 (north of the belt) are really only the brightest parts of a huge nebular cloud which extends over almost the whole of Orion. Other easy doubles are ι and λ.

The red semi-regular variable W Orionis is in the same binocular field with π^6 (magnitude 4.5), the southernmost member of a line of stars which, for some strange reason, are all lettered π. It has an N-type spectrum, and is always within binocular range; its colour makes it readily identifiable, and it is actually redder than Betelgeux, though the hue is not so striking because the star is much fainter. U Orionis, on the border of Orion and Taurus, is a Mira variable which rises to naked-eye visibility at maximum; it is a member of a well-marked little group lying between τ Tauri and η Geminorum.

Canis Major, Orion's senior dog, is graced by the presence of Sirius, which shines as much the brightest star in the sky even though it is only 26 times as luminous as the Sun; it is a mere 8.6 light years away, and is the closest of all the brilliant stars apart from α Centauri. Though it is pure white, with an A-type spectrum, the effects of the Earth's atmosphere make it flash various colours. All stars twinkle to some extent, but Sirius shows the effect more than any others simply because it is so bright. The white dwarf companion, Sirius B, would be easy to see if it were not so overpowered; the revolution period is 50 years. It is roughly the size of the Earth, but is as massive as the Sun.

ε (Adhara), δ (Wezen), η (Aludra) and o^2 are all very hot and luminous; Wezen, indeed, could match 50,000 Suns, and is over 1800 light years away. It is not easy to appreciate that of all the bright stars in Canis Major, Sirius is much the least powerful. Adhara, only just below the official 'first magnitude', has a companion which is easy to see with a small telescope.

There are two fine open clusters in Canis Major. M41 lies in the same wide field with the reddish ν^2, forming a triangle with ν^2 and Sirius; it is a naked-eye object, and can be partly resolved with binoculars. NGC2362, round the hot, luminous star τ (magnitude 4.39), is 3500 light years away, and seems to be a very young cluster; with a low power it looks almost stellar, but higher magnification soon resolves it. In the same low-power field is a β Lyrae eclipsing binary, UW Canis Majoris, which is an exceptionally massive system. According to one estimate the masses of the two components are 23 and 19 times that of the Sun, so that they rank as cosmic heavyweights. The total luminosity of the system is at least 16,000 times that of the Sun.

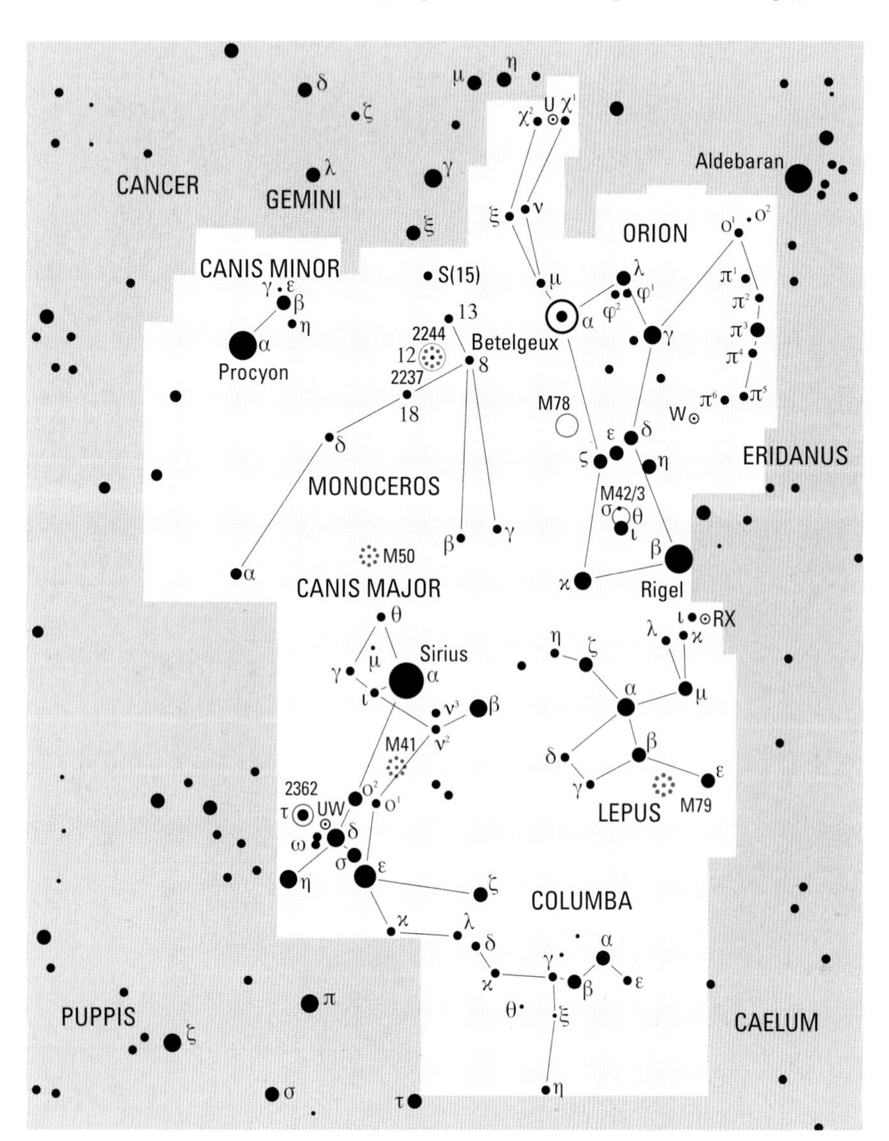

◀ **Orion** is probably the most magnificent of all the constellations, and since it is crossed by the celestial equator it is visible from every inhabited country (though from the station at the South Pole, Rigel is permanently above the horizon and Betelgeux never!). Orion is a superb guide to other groups; the Belt stars point southwards to Sirius and northwards to Aldebaran. Orion and his retinue are prominent in the evening sky throughout the northern winter (southern summer). The stars in the southernmost part of this map do not rise over Britain.

Monoceros, Lepus, Columba

ORION

BRIGHTEST STARS

No.	Star	R.A.	Dec.	Mag.	Spectrum	Proper name
		h m s	° ' "			
19	β	05 14 32	−08 12 06	0.12	B8	Rigel
58	α	05 55 10	+07 24 26	0.1–0.9	M2	Betelgeux
24	γ	05 25 08	+06 20 59	1.64	B2	Bellatrix
46	ε	05 36 13	−01 12 07	1.70	B0	Alnilam
50	ζ	05 50 45	−01 56 34	1.77	O9.5	Alnitak
53	ϰ	05 47 45	−09 40 11	2.06	B0	Saiph

VARIABLES

Star	R.A.	Dec.	Range	Type	Period	Spectrum
	h m	° '	(mags)		(d)	
U	05 55.8	+20 10	4.8–12.6	Mira	372	M
W	05 05.4	+01 11	5.9–7.7	Semi-reg	12	N

DOUBLES

Star	R.A.	Dec.	P.A.	Sep.	Mags
	h m	° '	°	"	
λ	05 36.1	+09 56	043	4.4	3.6,5.5
ι	05 35.4	−05 55	141	11.3	2.8,6.9
β	05 14.5	−08 12	202	9.5	0.1,6.8

CLUSTERS AND NEBULAE

M	NGC	R.A.	Dec.	Mag.	Dimensions	Type
		h m	° '		'	
M42	1976	05 35.4	−05 27	5	66 × 60	Orion Nebula
M43	1982	05 35.6	−05 16	7	20 × 15	Extension of M42
M78	2068	05 46.7	+00 03	8	8 × 8	Nebula

CANIS MAJOR

BRIGHTEST STARS

No.	Star	R.A.	Dec.	Mag.	Spectrum	Proper name
		h m s	° ' "			
9	α	06 45 09	−16 42 58	21.46	A1	Sirius
21	ε	06 58 38	−28 58 20	1.50	B2	Adhara
25	δ	07 08 23	−26 23 36	1.86	F8	Wezen
2	η	07 24 06	−29 18 11	2.44	B5	Aludra

VARIABLES

Star	R.A.	Dec.	Range	Type	Period	Spectrum
	h m	° '	(mags)		(d)	
UW	07 18.4	−24 34	4.0–5.3	β Lyrae	4.39	O7

DOUBLE

Star	R.A.	Dec.	P.A.	Sep.	Mags
	h m	° '	°	"	
α	06 45.1	−16 43	005	4.5	21.5, 8.5

CLUSTERS

M	NGC	R.A.	Dec.	Mag.	Dimensions	Type
		h m	° '		'	
41	2287	06 47.0	−20 44	4.5	38	Open cluster
	2362	07 17.8	−24 57	4	8	Open cluster

CANIS MINOR

BRIGHTEST STARS

No.	Star	R.A.	Dec.	Mag.	Spectrum	Proper name
		h m s	° ' "			
10	α	07 39 18	+05 13 30	0.38	F5	Procyon
3	β	07 27 09	+08 17 21	2.90	B8	Gomeisa

MONOCEROS

The brightest star is β: R.A. 06h 28m 49s, dec. 207° 01' 58", mag. 3.7.

DOUBLE

Star	R.A.	Dec.	P.A.	Sep.	Mags
	h m	° '	°	"	
ε	06 23.8	+04 36	027	13.4	4.5, 6.5
S (15)	06 41.0	+09 54	AB 213	2.8	4.7v, 7.5

CLUSTERS AND NEBULAE

M	NGC	R.A.	Dec.	Mag.	Dimensions	Type
		h m	° '		'	
50	2323	07 03.2	−08 20	5.9	16	open cluster
	2237	06 32.3	+05 03	~6	80 × 60	nebula (Rosette)
	2244	06 32.4	+04 52	5	24	open cluster

LEPUS

BRIGHTEST STARS

No.	Star	R.A.	Dec.	Mag.	Spectrum	Proper name
		h m s	° ' "			
11	α	05 32 44	−17 49 20	2.58	F0	Ameb
9	β	05 28 15	−20 45 35	2.84	G2	Nihal
2	μ	05 12 56	−16 12 20	3.31	B9	

VARIABLES

Star	R.A.	Dec.	Range	Type	Period	Spectrum
	h m	° '	(mags)		(d)	
RX	05 11.4	−11 51	5.0–7.0	Irregular	–	M

DOUBLES

Star	R.A.	Dec.	P.A.	Sep.	Mags
	h m	° '	°	"	
ϰ	05 13.2	−12 56	358	2.6	4.5,7.4
β	05 28.2	−20 46	330	2.5	2.8,7.3
γ	05 44.5	−22 27	350	96.3	3.7,6.3

CLUSTER

M	NGC	R.A.	Dec.	Mag.	Dimensions	Type
		h m	° '		'	
79	1904	05 24.5	−24 33	9.9	8.7	globular cluster

COLUMBA

BRIGHTEST STARS

Star	R.A.	Dec.	Mag.	Spectrum	Proper name
	h m s	° ' "			
α	05 39 39	−34 04 27	2.64	B8	Phakt
β	05 50 57	−35 46 06	3.12	K2	Wazn

Also above magnitude 4.3: δ (3.85), ε (3.87), η (3.96).

Canis Minor, the Little Dog, includes Procyon (α), 11.4 light years away and 10 times as luminous as the Sun. Like Sirius, it has a white dwarf companion, but the dwarf is so faint and so close in that it is a very difficult object. The revolution period is 40 years. The only other brightish star in Canis Major is β (Gomeisa), which makes a pretty little group with the much fainter ε, η and γ.

Monoceros is not an original constellation; it was created by Hevelius in 1690, and although it represents the fabled unicorn there are no legends attached to it. Much of it is contained in the large triangle bounded by Procyon, Betelgeux and Saiph. There are no bright stars, but there are some interesting doubles and nebular objects, and the constellation is crossed by the Milky Way. β is a fine triple; William Herschel, who discovered it in 1781, called it 'one of the most beautiful sights in the heavens'. S Monocerotis is made up of a whole group of stars, together with the Cone Nebula (NGC2264), which is elusive but not too hard to photograph. The open cluster NGC2244, round the star 12 Monocerotis (magnitude 5.8), is easy to find with binoculars; surrounding it is the Rosette Nebula, NGC2237, which is 2600 light years away and over 50 light years across. Photographs show the dark dust lanes and globules which give it such a distinctive appearance. M50 is an unremarkable open cluster near the border between Monoceros and Canis Major.

Lepus, the Hare, is placed here because it represents an animal which Orion is said to have been particularly fond of hunting. Of the two leaders, α (Arneb) is an F-type supergiant, 950 light years away and 6800 times as luminous as the Sun; β (Nihal) is of type G, 316 light years away and 600 Sun-power. γ is a wide, easy double. R Leporis, nicknamed the Crimson Star, is a Mira variable making a triangle with ϰ (4.36) and μ; it can reach naked-eye visibility, and can be followed with binoculars for parts of its cycle. It is cool by stellar standards – hence its strong red colour – but is 1000 light years away, and at least 500 times more powerful than the Sun.

M79, discovered by Méchain in 1780, is a globular cluster at a distance of 43,000 light years; it lies in line with α and β. It is not too easy to find with binoculars, but a small telescope will show it clearly.

Columba (originally Columba Noae, Noah's Dove) contains little of immediate interest, but the line of stars south of Orion, of which α (Phakt) and β (Wazn) are the brightest members, makes it easy to identify. μ, magnitude 5.16, is one of three stars which seem to have been 'shot out' of the Orion nebulosity, and are now racing away from it in different directions; the other two are 53 Arietis and AE Aurigae. μ Columbae is of spectral type O9.5, so that it is certainly very young; it has the high proper motion of 0.025 of a second of arc per year.

Taurus, Gemini

Taurus is a large and conspicuous Zodiacal constellation, representing the bull into which Jupiter once changed himself for thoroughly discreditable reasons. It has no well-defined pattern, but it does contain several objects of special interest.

α (Aldebaran), in line with Orion's Belt, is an orange-red star of type K0, 65 light years away and 140 times as luminous as the Sun. It looks very similar to Betelgeux (α Orionis), though it is not nearly so remote or powerful; it makes a good comparison for Betelgeux, though generally it is considerably the fainter of the two. The stars of the open cluster of the Hyades (C41) extend from it in a sort of V-formation, but there is no true association; Aldebaran is not a cluster member, and merely happens to lie about halfway between the Hyades and ourselves – which is rather a pity, since its brilliant orange light tends to drown the fainter stars. The leading Hyades are γ (3.63), ε (3.54), δ (3.76) and θ (3.42). The cluster was not listed by Messier, presumably because there was not the slightest chance of confusing it with a comet.

Because the Hyades are so scattered, they are best seen with binoculars. σ consists of two dim stars close to Aldebaran; δ makes up a wide pair with the fainter 64 Tauri, of magnitude 4.8; and θ is a naked-eye double, made up of a white star of magnitude 3.4 and a K-type orange companion of magnitude 3.8. The colour contrast is striking in binoculars. Here, too, we are dealing with a simple line of sight effect; the white star is the closer to us by 15 light years, though undoubtedly the two have condensed out of the same nebula which produced all the rest of the Hyades.

Messier did include the Pleiades in his catalogue, and gave them the number 45. Of course, they have been known since very early times; they are referred to by Homer and Hesiod, and are mentioned three times in the Bible. The leader, η (Alcyone), is third magnitude; then follow Electra (17), Atlas (27), Merope (23), Maia (20), Taygeta (19), Celaeno (16), Pleione (28) and Asterope (21/22). This makes nine, though the cluster is always nicknamed the Seven Sisters. However, Pleione is close to Atlas, and is an unstable shell star which varies in light, while Celaeno (magnitude 5.4) and Asterope (5.6) are easy to overlook. On the next clear night, see how many separate stars you can see in the cluster without optical aid; if you can manage a dozen, you are doing very well indeed. Binoculars show many more, and the total membership of the cluster amounts to several hundreds. The average distance of the stars is just over 400 light years.

The Pleiades are at their best when viewed under very low magnification. The leading stars are hot and bluish-white, and the cluster – unlike the Hyades – is certainly very young; there is considerable nebulosity, so star formation is presumably still going on. This nebulosity is very difficult to see through a telescope, but is surprisingly easy to photograph.

The other nebular object is M1, the Crab Nebula, which is the remnant of the supernova of 1054. It can be glimpsed with powerful binoculars, close to the third-magnitude ζ; a telescope shows its form, but photography is needed to bring out its intricate structure. It is expanding, and inside is a pulsar which powerful equipment can record as a faint, flickering object – one of the few pulsars to be identified optically.

λ is an Algol-type variable, easy to follow with the naked eye; good comparison stars are γ, o, ξ and μ. The real separation of the components is of the order of 14 million kilometres (nearly 9 million miles), so they cannot be seen separately; eclipses of the primary are 40 per cent total. The distance is 326 light years; λ is much more luminous than Algol, but is also much farther away. The only other Algol stars to exceed magnitude 5 at maximum are Algol itself, δ Librae and the far-southern ζ Phoenicis.

Magnitudes
−1
0
1
2
3
4
5
Variable star
Galaxy
Planetary nebula
Gaseous nebula
Globular cluster
Open cluster

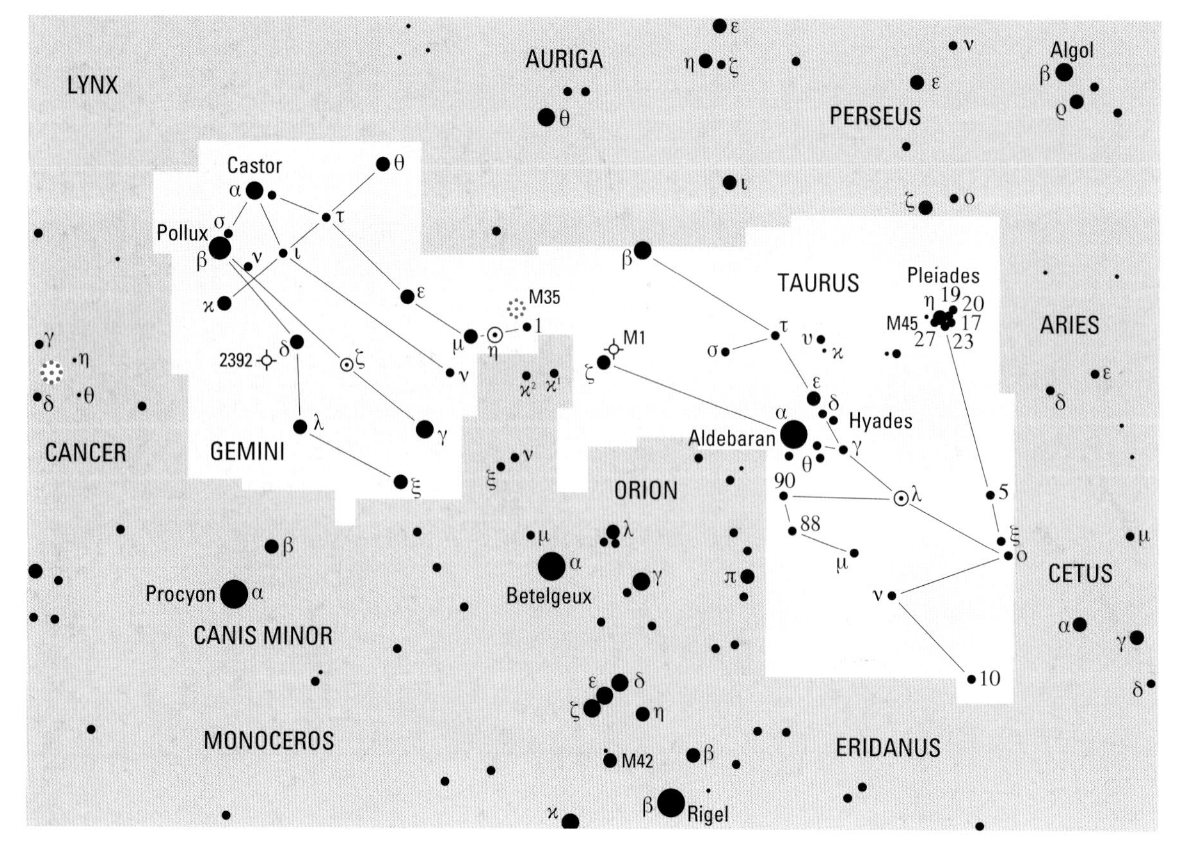

◀ **These two large**, Zodiacal constellations form part of Orion's retinue, and are thus good targets during evenings in northern winter (southern summer). Taurus contains the two most famous open clusters in the sky, the Pleiades and the Hyades, while the 'Twins', Castor and Pollux, make an unmistakable pair. The Milky Way flows through Gemini, and there are many rich star fields.

Of the other leading stars in Taurus, ζ (Alheka) is a highly luminous B-type giant, 490 light years away and 1300 times as powerful as the Sun. β (Alnath) is very prominent, and has been transferred from Auriga to Taurus – which seems illogical, as it belongs much more naturally to the Auriga pattern. It is 130 light years from us, and can equal 470 Suns.

Gemini. The Heavenly Twins, Castor and Pollux, make up a striking pair. Pollux (β) is the brighter; it is 34 light years away as against 53 light years for Castor (α), and it is an orange K-type star, outshining the Sun by over 30 times. Castor is a fine binary with a revolution period of 420 years; though the separation is less than it was a century ago, it is still a suitable target for small telescopes. Each component is a spectroscopic binary, and there is a third member of the system, YY Geminorum, which is an eclipsing binary.

There are two notable variables in Gemini. ζ is a typical Cepheid, with a period of 10.15 days; this is almost twice the period of δ Cephei itself, and ζ Geminorum is correspondingly the more luminous, since at its peak it is well over 5000 times as luminous as the Sun. η (Propus) is a red semi-regular with an extreme range of magnitude 3.1 to 3.9, and a rough period of around 233 days; a good comparison star is μ, which is of the same spectral type (M3) and the same colour. Also in the constellation is U Geminorum, the prototype dwarf nova. Stars of this type are known either as U Geminorum stars or as SS Cygni stars; U Geminorum is much the fainter of the two, since its 'rest' magnitude is only 14.9 and it never reaches magnitude 8. The average interval between outbursts is just over 100 days.

M35 is a very conspicuous cluster close to η and μ. It is 2850 light years away, and was discovered by de Chéseaux in 1746; Messier called it 'a cluster of very small stars'. It is worth seeking out NGC2392, the Eskimo Nebula, which is a planetary nebula lying between ϰ and λ; the central star is tenth magnitude. The Eskimo is decidedly elusive, but photographs taken with larger telescopes show its curious 'face'. Like all planetaries it is expanding, and has now reached a diameter of more than half a light year. It was William Herschel who first called these objects 'planetary nebulae', because he thought that their disks made them look like planets – but the name could hardly be less appropriate.

▼ **The Pleiades cluster** in Taurus, photographed by Bernard Abrams using a 25-cm (10-in) reflector. They have been known since ancient times; Messier included them as 45.

TAURUS

BRIGHTEST STARS

No.	Star	R.A. h	m	s	Dec. °	'	"	Mag.	Spectrum	Proper name
87	α	04	35	55	+16	30	33	0.85	K5	Aldebaran
112	β	05	26	17	+28	36	27	1.65	B7	Al Nath
25	η	03	47	29	+24	06	18	2.87	B7	Alcyone
123	ζ	05	37	39	+21	08	33	3.00	B2	Alheka
35	λ	04	00	41	+12	29	15	3.4 (max)	B3	
78	θ^2	04	28	40	+15	52	15	3.42	A7	

Also above magnitude 4.3: ε (Ain) (3.54), ο (3.60), 27 (Atlas) (3.63), γ (Hyadum Primus) (3.63), 17 (Electra) (3.70), ξ (3.74), δ (3.76), θ' (3.85), 20 (Maia) (3.88), ν (3.91), 5 (4.11), 23 (Merope) (4.18), ϰ (4.22), 88 (4.25), 90 (4.27), 10 (4.28), μ (4.29), υ (4.29), 19 (Taygete) (4.30), τ (4.28), δ3 (4.30). β (Al Nath) was formerly included in Auriga, as γ Aurigæ.

VARIABLES

Star	R.A. h	m	Dec. °	'	Range (mags)	Type	Period (d)	Spectrum
λ	04	00.7	+12	29	3.3–3.8	Algol	3.95	B1A
BU (Pleione)	03	49.2	+24	08	4.8–5.5	Irregular	–	Bp
T	04	22.0	+19	32	8.4–13.5	T Tauri	–	G–K
SU	05	49.1	+19	04	9.0–16.0	R Coronæ	–	G0p

DOUBLES

Star	R.A. h	m	Dec. °	'	P.A. °	Sep. "	Mags	
θ	04	28.7	+15	32	346	337.4	3.4, 3.8	naked-eye
σ	04	39.3	+15	55	193	431.2	4.7, 5.1	naked-eye
K167	04	25.4	+22	18	173	339	4.2, 5.3	naked-eye

CLUSTERS AND NEBULAE

M	C	NGC	R.A. h	m	Dec. °	'	Mag.	Dimensions '	Type
1		1952	05	34.5	+22	01	10	6 × 4	supernova remnant (Crab)
45			03	47.0	+24	07	1.2	110	open cluster (Pleiades)
	41		04	27	+16	00	1	330	open cluster (Hyades)

GEMINI

BRIGHTEST STARS

No.	Star	R.A. h	m	s	Dec. °	'	"	Mag.	Spectrum	Proper name
78	β	07	45	19	+28	01	34	1.14	K0	Pollux
66	α	07	34	36	+31	53	18	1.58	A0	Castor
24	γ	06	37	43	+16	23	57	1.93	A0	Alhena
13	μ	06	22	58	+22	30	49	2.88	M3	Tejat
27	ε	06	43	56	+25	07	52	2.98	G8	Mebsuta
7	η	06	14	53	+22	30	24	3.1 (max)	M3	Propus
31	ξ	06	45	17	+12	53	44	3.36	F5	Alzirr

Also above magnitude 4.3: δ (Wasat) (3.53), ϰ (3.57), λ (3.58), θ (3.60), ζ (Mekbuda) (3.7 max), ι (3.79), υ (4.06), ν (4.15), 1 (4.16), ϱ (4.18), σ (4.28).

VARIABLES

Star	R.A. h	m	Dec. °	'	Range (mags)	Type	Period (d)	Spectrum
η	06	14.9	+22	30	3.1–3.9	Semi-regular	1233	M
ζ	07	04.1	+20	34	3.7–4.1	Cepheid	10.15	F–G

DOUBLES

Star	R.A. h	m	Dec. °	'	P.A. °	Sep. "	Mags	
η	06	14.9	+22	30	266	1.4	3v, 8.8	binary, 470y
α	07	34.6	+31	53	AB 088	2.5	1.9, 2.9	binary 420y
					AC 164	72.5	8.8	

CLUSTERS AND NEBULAE

M	NGC	R.A. h	m	Dec. °	'	Mag.	Dimensions '	Type
35	2168	06	08.9	+24	20	5	28	open cluster
	2392	07	29.2	+20	55	10	0.2 × 0.7	planetary nebula (Eskimo)

Auriga, Lynx

Auriga, the Charioteer, is a brilliant northern constellation, led by Capella (α). In mythology it honours Erechthonius, son of Vulcan, the blacksmith of the gods; Erechthonius became King of Athens, and invented the four-horse chariot.

Capella is the sixth brightest star in the entire sky, and is only 0.05 of a magnitude less bright than Vega (α Lyrae). It and Vega are on opposite sides of the north celestial pole, so when Capella is high up Vega is low down, and vice versa; from Britain, neither actually sets, and Capella is near the zenith or overhead point during evenings in winter. It can be seen from almost all inhabited countries, though it is lost from the extreme southern tip of New Zealand and southern South America.

Capella is yellow, like the Sun, but is a yellow giant rather than a dwarf – or, rather, two giants, because it is a very close binary. One component is 50 times as luminous as the Sun, and the other 80 times; the distance between them is not much more than 100 million kilometres (60 million miles).

Its distance from us is 42 light years. The second star of Auriga, β (Menkarlina), is also a spectroscopic binary, and is actually an eclipsing system with a very small magnitude range. The components are more or less equal, and a mere 12 million kilometres (7.5 million miles) apart; both are of type A.

Of course, the most intriguing objects in Auriga are the two eclipsing binaries ε and ζ. It is sheer chance that they lie side by side, because they are at very different distances – 520 light years for ζ; as much as 4600 light years for ε. The third member of the trio of the Haedi or Kids, η Aurigae, is a useful comparison; the magnitude is 3.17.

It is worth keeping a close watch on ε, because even during long intervals between eclipses it seems to fluctuate slightly. The catalogues give its normal magnitude as 2.99, in which case it appears very slightly but perceptibly brighter than η. All three Kids are in the same low-power binocular field, and this is probably the best way to make estimates of ε; ζ is much fainter, and the only really useful comparison star is ν, which is at magnitude 3.97.

Of the other main stars of the constellation, ι (Hassaleh) and the rather isolated δ, are reddish, with K-type spectra. θ is white, and has two companions; the closer pair makes up a slow binary system, while the more remote member of the group, of magnitude 10.6, merely lies in almost the same line of sight.

Auriga is crossed by the Milky Way, and there are several fine open clusters, of which three are in Messier's list. M36 and M38 were both discovered by Guillaume Legentil in 1749, and M37 by Messier himself in 1764; but there is no doubt all of them had been recorded earlier, because they are all bright.

M36 is easy to resolve, and is 3700 light years away. M37, at about the same distance, is in the same lower-power field as θ, which is a very good way of identifying it; the brightest stars in the cluster form a rough trapezium. M38 is larger and looser, and rather less bright. It lies slightly away from the mid-point of a line joining θ to ι, and within half a degree of it is a much smaller and dimmer cluster, NGC1907.

Note also the Flaming Star Nebula, IC405 (C31), round the irregular variable AE Aurigae – one of the 'runaway stars' which seem to have been ejected from the Orion nebulosity (the others are 53 Arietis and μ Columbae). AE Aurigae illuminates the diffuse nebulosity, which is elusive telescopically though photographs show intricate structure. The distance is of the order of 1600 light years.

Lynx is a very ill defined and obscure northern constellation, created by Hevelius in 1790; it has no mythological associations, and it has been said that only a lynx-eyed observer can see anything there at all. In fact there is one brightish star, α (magnitude 3.13), which is decidedly iso-

Magnitudes
–1
0
1
2
3
4
5
Variable star
Galaxy
Planetary nebula
Gaseous nebula
Globular cluster
Open cluster

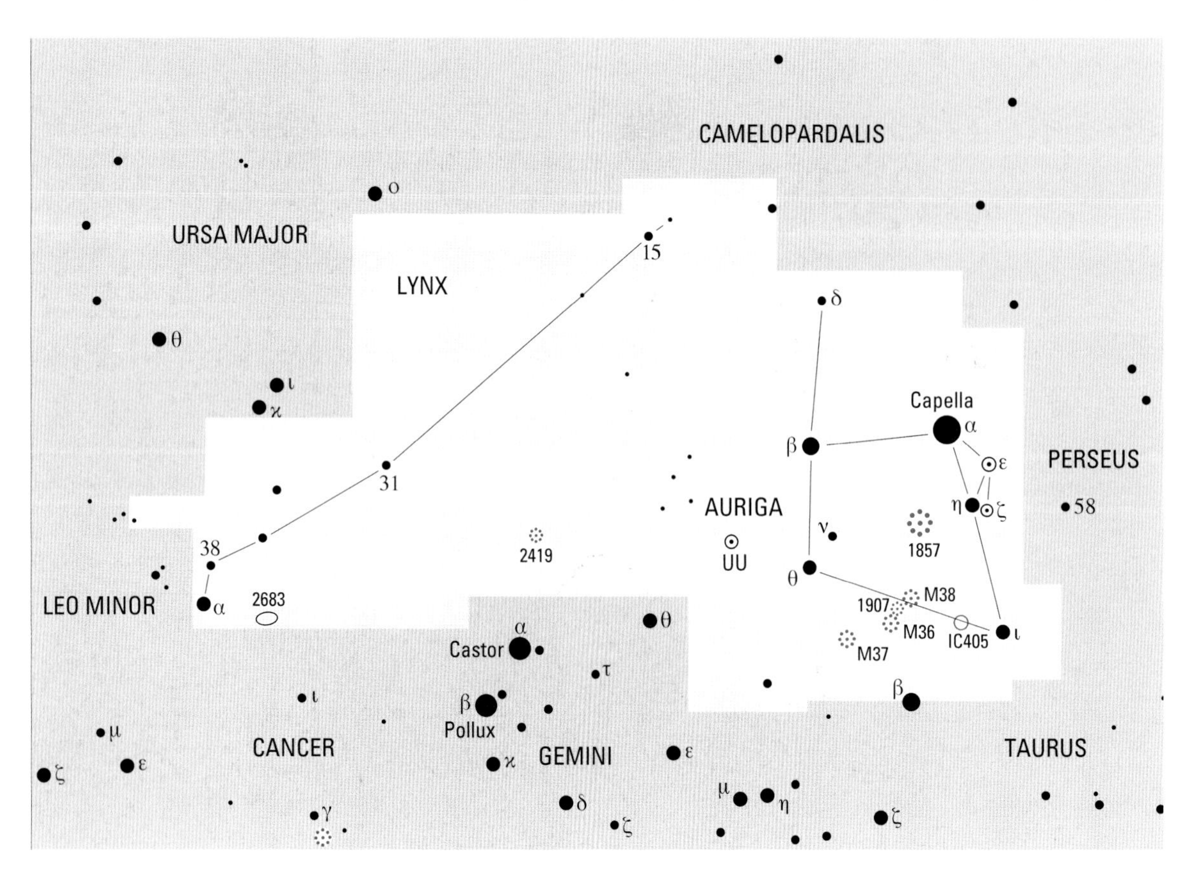

◀ **Capella**, (α Aur), the the sixth brightest star in the sky, is near the zenith during evenings in winter, as seen from the northern hemisphere; this is the position occupied by Vega during summer evenings. From Britain or the northern United States, Capella does not set, though at its lowest it skims the horizon. The Auriga quadrilateral is very easy to identify; a fifth bright star, Alnath, which seems logically to belong to the Auriga pattern, has been transferred to Taurus, and is now β Tau instead of γ Aur.

lated, and forms an equilateral triangle with Regulus (α Leonis) and Pollux (β Geminorum). It is of type M, and obviously red; its distance is 166 light years, and it is 120 times as luminous as the Sun. None of the other stars in Lynx have been given Greek letters, though one of them, 31 Lyncis, has been dignified with a proper name, Alsciaukat.

The globular cluster NGC2419 (C25), about 7° north of Castor (α Geminorum, is faint and none too easy to identify. This is not because it is feeble – on the contrary it is exceptionally large, and must be around 400 light years across – but because it is so far away. The distance has been estimated at around 300,000 light years, and though this may be rather too great it is clear that the cluster is at the very edge of our Galaxy. It may even be escaping altogether, in which case it will become what is termed an intergalactic tramp. It is very rich, and, predictably, its leading stars are red and yellow giants.

We have little direct knowledge of the isolated star systems that lie between the galaxies. There is every reason to believe they exist, but since they will be so much less luminous than full-scale galaxies they will be far less easy to detect.

Indeed, galaxies of very low surface brightness may also be very elusive. Modern electronic techniques used with large telescopes may be able to track these isolated objects, but at the moment we do not know how many of them there are. At least it seems unlikely that NGC2419 will become the only intergalactic tramp.

AURIGA

BRIGHTEST STARS

No.	Star	R.A. h	m	s	Dec. °	′	″	Mag.	Spectrum	Proper name
13	α	05	16	41	+45	59	53	0.08	G8	Capella
34	β	05	59	32	+44	56	51	1.90	A2	Menkarlina
37	θ	05	59	43	+37	12	45	2.62	A0p	
3	ι	04	56	59	+33	09	58	2.69	K3	Hassaleh
7	ε	05	01	58	+43	49	24	2.99v	F0	Almaaz
10	η	05	06	31	+41	14	04	3.17	B3	

Also above magnitude 4.3: δ (3.72), ζ (Sadatoni) (3.75) (max).

VARIABLES

Star	R.A. h	m	Dec. °	′	Range (mags)	Type	Period (d)	Spectrum
ε	05	02.0	+43	49	3.0–3.8	Eclipsing	9892	F
ζ	05	02.5	+41	05	3.7–4.1	Eclipsing	972	K1B
UU	06	36.5	+38	27	5.1–6.8	Semi-reg.	234	N

DOUBLE

Star	R.A. h	m	Dec. °	′	P.A. °	Sep. ″	Mags
θ	05	59.7	+37	13	AB 313	3.6	2.6,7.1
					AC 297	50.0	10.6

CLUSTERS AND NEBULAE

M	C	NGC	R.A. h	m	Dec. °	′	Mag.	Dimensions ′	Type
36		1960	05	36.1	+34	08	6.0	12	open cluster
37		2099	05	52.4	+32	33	5.6	24	open cluster
38		1912	05	28.7	+35	50	6.4	21	open cluster
		1857	05	20.2	+39	21	7.0	6	open cluster
	31	IC405	05	16.2	+34	16	var	30 × 19	nebula (Flaming Star), round AE Aurigæ

LYNX

BRIGHTEST STAR

No.	Star	R.A. h	m	s	Dec. °	′	″	Mag.	Spectrum	Proper name
40	α	09	21	03	+34	23	33	3.13	MO	

Also above magnitude 4.3: 38 (3.92), 31 (Alsciaukat) (4.25).

CLUSTER AND GALAXY

C	NGC	R.A. h	m	Dec. °	′	Mag.	Dimensions ′	Type
25	2419	07	38.2	+38	53	10.4	4.1	globular cluster
	2683	08	52.7	133	25	9.7	9.3 × 2.5	Sc galaxy

▼ **AE Aurigae** is an irregular variable. This spectacular image was obtained by the 0.9-m (36-in) telescope at the Kitt Peak National Observatory, in Arizona. AE Aurigae is the bright blue star in the centre of the image, and it is is surrounded by the diffuse nebula known as the Flaming Star Nebula (IC405, C31). This false-colour image was created by combining images taken at three different wavelengths.

Carina, Vela, Pyxis, Antlia, Pictor, Volans

Magnitudes

- −1
- 0
- 1
- 2
- 3
- 4
- 5

Variable star

Galaxy

Planetary nebula

Gaseous nebula

Globular cluster

Open cluster

Carina, the Keel. We now come to the main constellations of the southern sky, most of which are inaccessible from the latitudes of Britain, Europe or most of the mainland United States. The brightest part of the old constellation of Argo Navis is the Carina (the Keel), which contains Canopus (α), the second brightest star in the sky. It looks half a magnitude fainter than Sirius, but this is only because it is so much more remote. It is 15,000 times as luminous as the Sun, and therefore well over 500 times as luminous as Sirius. The spectral type is F, and this means that in theory it should look slightly yellowish, but to most observers it appears pure white. Its declination is −53°. Over parts of Australia and South Africa it sets briefly, but it is circumpolar from Sydney, Cape Town and the whole of New Zealand.

The second brightest star in Carina is β (Miaplacidus), of type A and 85 times as luminous as the Sun. ε and ι, together with ϰ and δ Velorum, make up the False Cross, which is of much the same shape as the Crux and is often confused with it, even though it is larger and not so brilliant. As with Crux, three of its stars are hot and bluish-white while the fourth – in this case ε Carinae – is red; ε is of type K, 530 light years from us and 6000 times as luminous as the Sun. ι is of type F, very luminous (6800 times more so than the Sun) and over 800 light years away. Its proper name is Tureis, but it has also been called Aspidske.

ZZ Carinae is a bright Cepheid, and R Carinae is one of the brightest of all Mira variables, rising to magnitude 3.9 at some maxima. However, the most interesting variable is η, which has been described earlier. For a while during the nineteenth century it outshone even Canopus; today it is just below naked-eye visibility, but it may brighten again at any time. The associated nebula can be seen with the naked eye; it contains a famous dark mass nicknamed the Keyhole. Telescopically, η looks quite unlike a normal star, and its orange hue is very pronounced. In the future – perhaps tomorrow, perhaps not for a million years – it will explode as a supernova, and it will then provide us with a truly magnificent spectacle.

The cluster IC2602 (C102), round θ Carinae, is very fine; it forms a triangle with β and ι. Also imposing is NGC2516, which lies in line with δ Velorum and ε Carinae in the False Cross; NGC2867, between ι Carinae and ϰ Velorum, is a planetary nebula which is just within binocular range. The whole of Carina is very rich, and there are a great many spectacular star fields.

Vela, the Sails of Argo Navis, are also full of interest, though less striking than Carina. The brightest star is γ (Regor), which is a Wolf–Rayet star of spectral type W, and very hot and unstable. It is a fine, easy double, and there are three fainter companions nearby. δ Velorum has a fifth-magnitude companion which is visible in a very small telescope, and in the same binocular field lies the open cluster NGC2391 (C85), round the 3.6-magnitude star o Velorum. In a low-power telescope, or even in binoculars, the cluster has a vaguely cruciform appearance. Another naked-eye cluster is NGC2547, near γ (Regor).

Pyxis (originally Pyxis Nautica, the Mariner's Compass). A small constellation to the north of Vela. The only object of immediate interest is the recurrent nova T Pyxidis, which is normally of about 14th magnitude, but has flared up to near naked-eye visibility on several occasions. It makes a triangle with α and γ, but in its usual state it is not at all easy to identify.

Antlia, originally Antlia Pneumatica (the Air Pump), was added to the sky by Lacaille in 1752, and seems to be one of the totally unnecessary constellations. It adjoins Vela and Pyxis, and is entirely unremarkable.

Pictor (originally Equuleus Pictoris, the Painter's Easel) is another of Lacaille's constellations. It lies near Canopus; there are no bright stars, but β – which has no individual name – has become famous because of the associated cloud

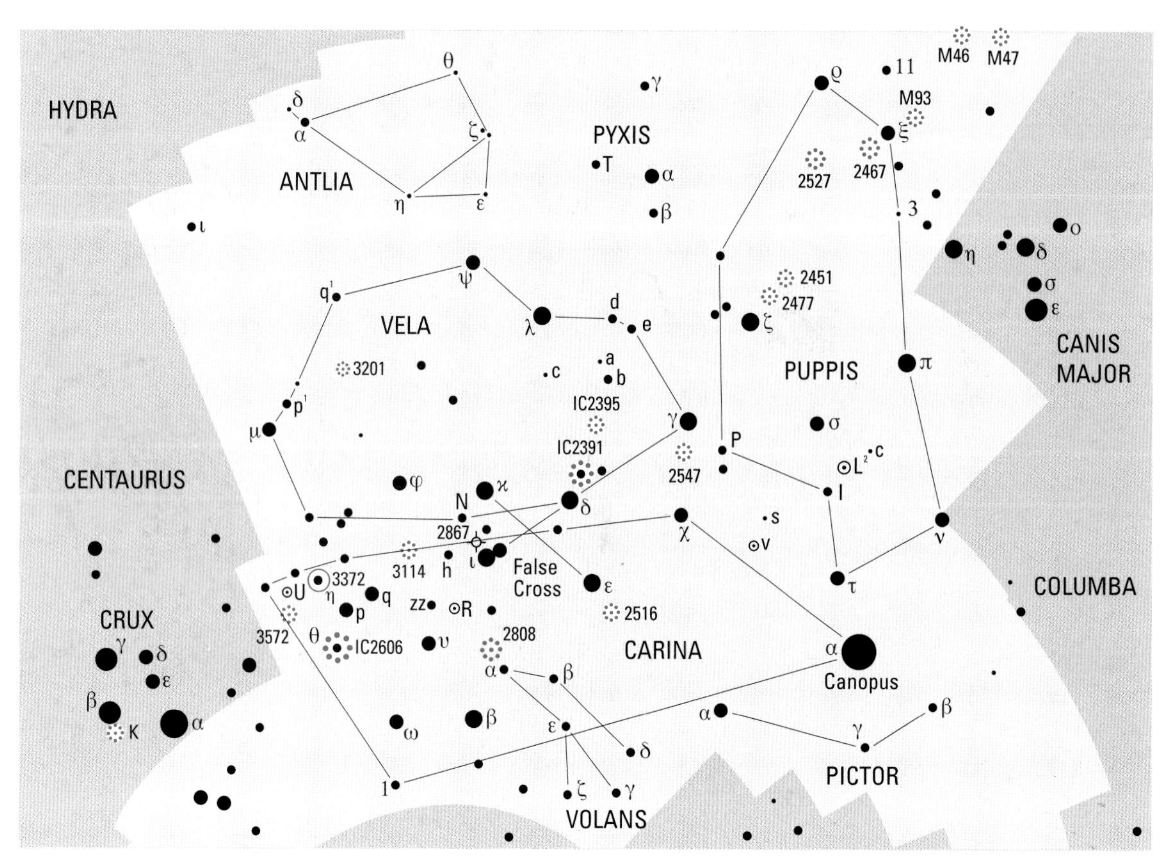

◀ **This region** is well south of the equator, and most of it is invisible from Britain or the northern United States, though part of Puppis can be seen. From southern countries such as Australia, Canopus – the second brightest star in the sky – is near the zenith during evenings around February; it can be seen from Alexandria, but not from Athens – an early proof that the Earth is not flat. Carina, Vela and Puppis were once combined as Argo Navis, the ship Argo; another section formed when Argo was dismembered was Malus (the Mast), part of which survives as Pyxis. The whole region, particularly Carina, is very rich.

Puppis

of cool material which may be planet-forming. It is 78 light-years away, and also 78 times as luminous as the Sun. In 1925 a bright nova, RR Pictoris, flared up, and remained fairly prominent for some time before fading back to obscurity.

Volans (originally Piscis Volans, the Flying Fish). A small constellation which, rather confusingly, intrudes into Carina between Canopus and Miaplacidus. It contains little of interest, though γ is a wide, easy double.

Puppis. Argo Navis' poop, part of which is sufficiently far north to rise in British latitudes though the brightest star, ζ, cannot do so. ζ is a very hot 0-type star, 63,000 times as luminous as the Sun and therefore the equal of Rigel in Orion; it is 2400 light years away. L^2 is a semi-regular variable with a range of magnitude from 3.4 to just below 6; V Puppis is of the β Lyrae type, with a range of about half a magnitude.

There are only three Messier objects in Puppis, because the rest of the constellation never rises over France, where Messier spent all his life. All three are open clusters. M46 and M47 are neighbours, more or less in line with Sirius and Mirzam (α and β Canis Majoris). M93, in the binocular field with ξ Puppis, is fairly bright and condensed. Admiral Smyth, the well-known nineteenth-century amateur astronomer, commented that the arrangement of the brighter stars in M93 reminded him of a starfish. The distance is 3600 light years.

CARINA

BRIGHTEST STARS

Star	R.A. h	m	s	Dec. °	'	"	Mag.	Spectrum	Proper name
α	06	23	57	−52	41	44	20.72	F0	Canopus
β	09	13	12	−69	43	02	1.68	A0	Miaplacidus
ε	08	22	31	−59	30	34	1.86	K0	Avior
ι	09	17	05	−59	16	31	2.25	F0	Tureis
θ	10	42	57	−64	23	39	2.76	B0	
υ	09	47	06	−65	04	18	2.97	A0	
l(ZZ)	09	45	15	−62	30	28	3.3 (max.)	G0	
ϱ	10	32	01	−61	41	07	3.32	B3	
ω	10	13	44	−70	02	16	3.32	B7	
w	10	17	05	−61	19	56	3.40	K5	
q	09	10	58	−58	58	01	3.44	B0	
x	07	56	47	−52	58	56	3.47	B2	

Also above magnitude 4.3: u (3.78), c (3.84), R (3.9 max.), x (3.91), I (4.00), h (4.08).

VARIABLES

Star	R.A. h	m	Dec. °	'	Range (mags)	Type	Period (d)	Spectrum
η	10	45.1	−59	41	−0.8–7.9	Irregular	–	Pec
ZZ	09	45.2	−62	30	3.3–4.2	Cepheid	35.5	F-K
R	09	32.2	−62	47	3.9–10.5	Mira	309	M
U	10	57.8	−59	44	5.7–7.0	Cepheid	38.8	F-G

DOUBLES

Star	R.A. h	m	Dec. °	'	P.A. °	Sep. "	Mags
υ	09	47.1	−65	04	127	5.0	3.1, 6.1

CLUSTERS AND NEBULAE

C	NGC	R.A. h	m	Dec. °	'	Mag.	Dimensions '	Type
102	IC2602	10	43.2	−64	24	2	50	open cluster round θ
96	2516	07	58.3	−60	52	3.8	30	open cluster
	3114	10	02.7	−60	07	4.2	35	open cluster
	3572	11	12.0	−64	52	6.3	14	open cluster
	2808	09	12.0	−64	52	6.3	14	globular cluster
92	3372	10	43.8	−59	52	6		nebula, round η
	2867	09	21.4	−58	19	9.7	0.2	planetary nebula

VELA

BRIGHTEST STARS

Star	R.A. h	m	s	Dec. °	'	"	Mag.	Spectrum	Proper name
γ	08	09	32	−47	20	12	1.78	WC7	Regor
δ	08	44	42	−54	42	30	1.96	A0	Koo She
λ	09	08	00	−43	25	57	2.21	K5	Al Suhail al Wazn
κ	09	22	07	−55	00	38	2.50	B2	Markeb
μ	10	46	46	−49	25	12	2.69	G5	
N	09	31	13	−57	02	04	3.13	K5	

Also above magnitude 4.3: φ (3.54), ψ (3.60), o (3.62), c (3.75), p (3.84),b (3.84), q (3.85), a (3.91), 4 (4.14), x (4.28).

DOUBLES

Star	R.A. h	m	Dec. °	'	P.A. °	Sep. "	Mags	
δ	08	44.7	−54	43	153	2.6	2.1, 5.1	
μ	10	46.8	−49	25	055	2.3	2.7, 6.4.	binary, 116y
γ	08	09.5	−47	20	AB 220	41.2	1.9, 4.2	
					AC 151	62.3	8.2	
					AD 141	93.5	9.1	
					DE 146	1.8	12.5	

CLUSTERS AND NEBULAE

C	NGC	R.A. h	m	Dec. °	'	Mag.	Dimensions '	Type
85	IC2391	08	40.2	−53	04	2.5	50	open cluster (o Velorum)

CLUSTERS AND NEBULAE (CONT.)

C	NGC	R.A. h	m	Dec. °	'	Mag.	Dimensions '	Type
	IC2395	08	41.1	−48	12	4.6	8	open cluster
	2547	08	10.7	−49	16	4.7	20	open cluster
79	3201	10	17.6	−46	25	6.7	18	globular cluster

PYXIS

The brightest star is α: R.A. 08h 43m 35s.5, dec. 233° 11' 11'', mag. 3.68.
Also above magnitude 43: β (3.97), γ (4.01)

VARIABLE

Star	R.A. h	m	Dec. °	'	Range (mags)	Type	Period (d)	Spectrum
T	09	04.7	−32	23	6.3–14.0	Recurrent nova	–	

ANTLIA

The only star brighter than magnitude 4.3 is α: R.A. 10h 27m 09s, dec. 231° 04' 14'', mag. 4.25.

PICTOR

BRIGHTEST STARS

Star	R.A. h	m	s	Dec. °	'	"	Mag.	Spectrum	Proper name
α	06	48	11	−61	56	29	3.27	A5	–

The only other star above magnitude 4.3 is β (3.85). This is the star now known to be associated with a disk of material which may be planet-forming.

VOLANS

The brightest star is γ: R.A. 07h 08m 42s.3, dec. 270° 29' 50'', combined magnitude 3.6.
Also above magnitude 4.3: β (3.77), ~ (3.95), δ (3.90), α (4.00).

DOUBLES

Star	R.A. h	m	Dec. °	'	P.A. °	Sep. "	Mags
γ	07	08.8	−70	30	300	13.6	4.0, 5.9

PUPPIS

BRIGHTEST STARS

Star	R.A. h	m	s	Dec. °	'	"	Mag.	Spectrum	Proper name
ζ	08	03	35	−40	00	12	2.25	O5.8	Suhail Hadar
π	07	17	09	−37	05	51	2.70	K5	
ϱ	08	07	33	−24	18	15	2.81	F6	Turais
τ	06	49	56	−50	36	53	2.93	K0	
υ	06	37	45	−43	11	45	3.17	B8	
σ	07	29	14	−43	18	05	3.25	K5	
ε	07	49	18	−24	51	35	3.34	G3	Asmidiske
ξ	07	13	13	−45	10	59	3.4 (max)	M5	

Also above magnitude 4.3: c (3.59), s (3.73), α (3.82), 3 (3.96), P (4.11), 11 (4.20).

VARIABLES

Star	R.A. h	m	Dec. °	'	Range (mags)	Type	Period (d)	Spectrum
L2	07	13.5	−44	39	3.4–6.2	Semi-reg.	140	M
V	07	58.2	−49	15	4.7–5.2	β Lyræ	1.45	BIB

CLUSTERS

M	C	NGC	R.A. h	m	Dec. °	'	Mag.	Dimensions '	Type
46		2437	07	41.8	−14	49	6.1	27	open cluster
47		2422	07	36.6	−14	30	4.4	30	open cluster
93		2447	07	44.6	−23	52	6.2	22	open cluster
	71	2477	07	52.3	−38	33	5.8	27	open cluster
		2451	07	45.4	−37	58	2.8	45	open cluster
		2527	08	05.3	−28	10	6.5	22	open cluster
		2467	07	52.5	−26	24		14 × 32	open cluster

Centaurus, Crux, Triangulum Australe, Circinus,

Centaurus was one of Ptolemy's original 48 groups. α and β are the Pointers to the Southern Cross; α, the brightest star in the sky apart from Sirius and Canopus, has been known as Toliman, Rigil Kentaurus and Rigil Kent, but astronomers refer to it simply as α Centauri. It is the nearest of the bright stars, and only slightly farther away than its dim red dwarf companion Proxima, which is only of the 11th magnitude and is difficult to identify; it lies 2° from α, and is a feeble flare star.

α itself is a magnificent binary, with components of magnitudes 0.0 and 1.2. The primary is a G-type yellow star rather more luminous than the Sun; the K-type secondary is the larger of the two, but has less than half the Sun's luminosity. The revolution period is 80 years. The apparent separation ranges from 2 to 22 seconds of arc, so the pair is easy to resolve with a small telescope.

β, known as Agena or Hadar, is a B-type star, 530 light years away and 13,000 times the luminosity of the Sun. γ is a binary with almost equal components, but the separation is less than 1.5 seconds of arc, so at least a 10-centimetre (4-inch) telescope is needed to resolve it. The Mira variable R Centauri lies between α and β. At its best it reaches naked-eye visibility.

ω Centauri (NGC5139, C80) is by far the finest globular cluster in the sky. To the naked eye, it is a hazy patch in line with Agena and second-magnitude ε. It is one of the nearer globulars at around 17,000 light years. It probably contains over a million stars, concentrated near the centre of the system within no more than a tenth of a light year.

There are several bright open clusters in Centaurus, notably the two that sit near λ. There is also a remarkable galaxy, NGC5128 (C77), which is crossed by a dark dust lane, and is a fairly easy telescopic object. In 1986 a bright supernova was discovered in it by an Australian amateur astronomer, Robert Evans, using his 32-centimetre (13-inch) reflector.

Crux, the Southern Cross. There can be few people who cannot identify Crux, though it was not accepted as a separate constellation until 1679. One of the four stars, δ, is more than a magnitude fainter than the rest, which rather spoils the symmetry; neither is there a central star to make an X, as with Cygnus in the far north. α, β and δ are hot and bluish-white, while γ is a red giant of type M. α (Acrux) is a wide double, and there is a third star in the same telescopic field. β, of type B, is slightly variable.

The Jewel Box (the Kappa Crucis Cluster, NGC4755, C94), round the striking red supergiant, ϰ Crucis, is one of the loveliest in the sky; its main stars form a triangle. It is 7700 light years away, and the cluster is about 25 light years across; it is believed to be no more than a few million years old, so that by cosmic standards it is a true infant. Close by it is the dark nebula known as the Coalsack, which can be detected with the naked eye as an almost starless region.

Triangulum Australe. The three leaders, α, β and γ, form a triangle; α is identifiable because of its orange-red hue. It is 55 light years away, and 96 times as luminous as the Sun. The globular cluster NGC6025 lies near β and is not far below naked-eye visibility; binoculars show it well.

Circinus was one of Lacaille's additions, lying between the Pointers and Triangulum Australe. α is a wide double; γ is a close binary.

Ara lies between θ Scorpii and α Trianguli Australis. Three of its leading stars, β, ζ and η, are orange K-type giants; R Arae, in the same binocular field with ζ and η, is an Algol-type eclipsing binary which never becomes as faint as the seventh magnitude. Ara contains several brightish clusters, of which the most notable is the globular NGC6397 (C86), close to the β–γ pair. It seems to be no more than 8200 light years away – probably the closest globular cluster. NGC6352 (C81), near α, is considerably brighter even though it is farther away.

Telescopium is a small, dim constellation near Ara. The only object of note is the variable RR Telescopii, less than

Magnitudes

−1
0
1
2
3
4
5

Variable star

Galaxy

Planetary nebula

Gaseous nebula

Globular cluster

Open cluster

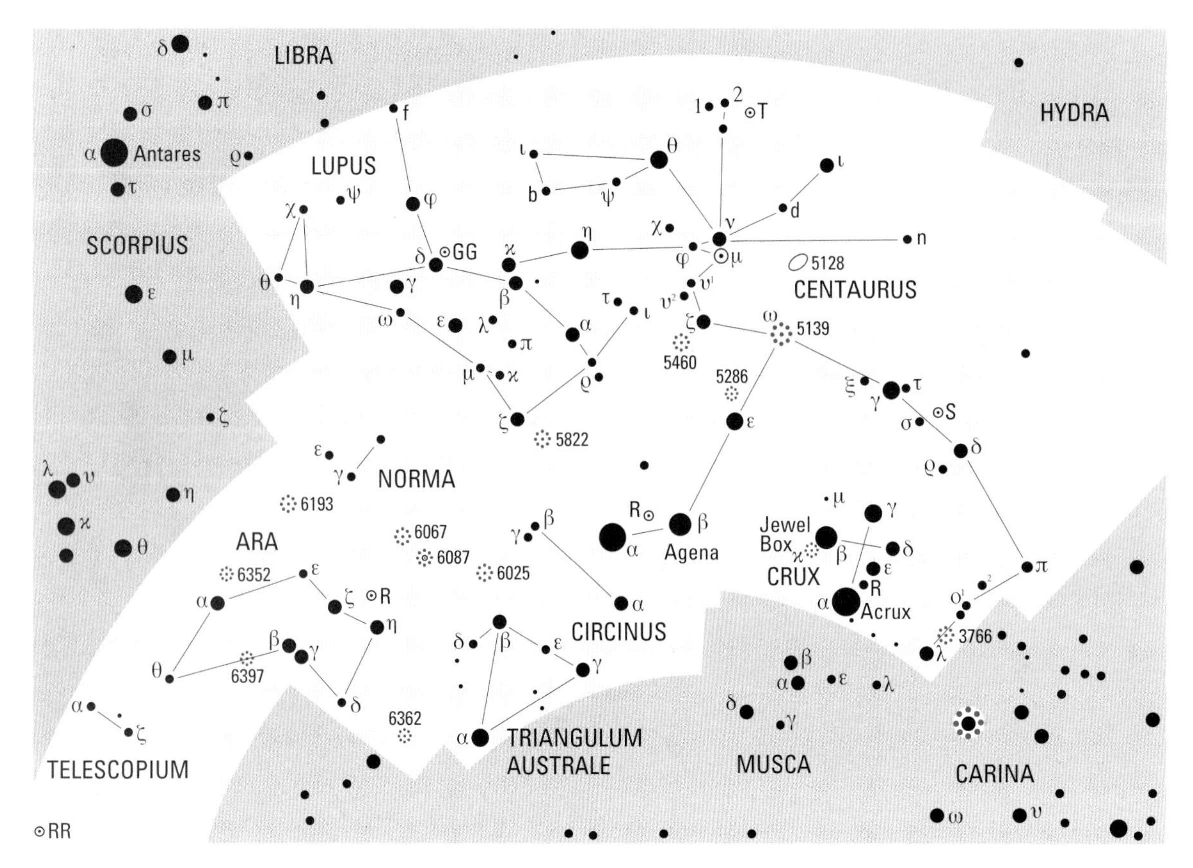

SOUTHERN CROSS

The distances and luminosities of the four main stars in Crux:

	Distance, light years	Luminosity, Sun = 1
α	360	3200–2000
β	460	8200
γ	88	160
δ	260	1300

◀ **The Southern Cross**, Crux, is the smallest constellation in the sky, but one of the most conspicuous, even if it is shaped more like a kite than an X. It is almost surrounded by Centaurus, and the brilliant Pointers, α and β Centauri, show the way to it. Most of Centaurus is too far south to be seen from Europe; from New Zealand, Crux is circumpolar. It is highest during evenings in southern autumn. Centaurus is an imposing constellation; it contains the finest of all globular clusters, ω Centauri.

Ara, Telescopium, Norma, Lupus

4° from α Pavonis. It is very faint indeed, but has flared up to as bright as seventh magnitude.

Norma is an obscure constellation, once known as Quadra Euclidis (Euclid's Quadrant). It adjoins Ara and Lupus, and contains two brightish open clusters; NGC6067, not far from γ, and the NGC6087 (C89), round the Cepheid, S Normae.

Lupus (the Wolf) is an original constellation. It contains a number of brightish stars, though there is no well marked pattern. NGC5722, close to ζ, is an open cluster within binocular range, and ϰ is an easy double. In 1006 a supernova flared up here, and became almost as bright as the Quarter Moon.

CENTAURUS

BRIGHTEST STARS

Star	R.A. h	m	s	Dec. °	'	"	Mag.	Spectrum	Proper name
α	14	39	37	−60	50	02	−0.27	G2IKI	
β	14	03	49	−60	22	22	0.61	B1	Agena
5	14	06	41	−38	22	12	2.06	K0	Haratan
γ	12	21	31	−48	57	34	2.17	A0	Menkent
ε	13	39	53	−53	27	58	2.30	B1	
η	14	35	30	−42	09	28	2.31	B3	
ζ	13	55	32	−47	17	17	2.55	B2	Al Nair al Kentaurus
δ	12	08	21	−50	43	20	2.60	B2	
ι	13	20	36	−36	42	44	2.75	A2	
μ	13	49	37	−42	28	25	3.04 max	B3	
ϰ	14	59	10	−42	06	15	3.13	B2	Ke Kwan
λ	11	35	47	−63	01	11	3.13	B9	
ν	13	49	30	−41	41	16	3.41	B2	

Also above magnitude 4.3: φ (3.83), τ (3.86), υ (3.87), d (3.88), π (3.89), σ (3.91), 65G (4.11), 1 (4.23), n (4.27), 2 (4.19), ξ^2 (4.27).

VARIABLES

Star	R.A. h	m	Dec. °	'	Range (mags)	Type	Period (d)	Spectrum
R	14	16.6	−59	55	5.3–11.8	Mira	546	M
μ	13	49.6	−42	28	3.0–3.5	Irregular	–	B
T	13	41.8	−33	36	5.5–9.0	Semi-reg.	60	K–M
S	12	24.6	−49	26	6.0–7.0	Semi-reg.	65	N

DOUBLES

Star	R.A. h	m	Dec. °	'	P.A. °	Sep. "	Mags	
α	14	39.6	−60	50	215	19.7	0.0, 1.2	binary, 80y
γ	12	41.5	−48	58	353	1.4	2.9, 2.9	binary, 84y

CLUSTERS, GALAXIES AND NEBULAE

C	NGC	R.A. h	m	Dec. °	'	Mag.	Dimensions '	Type
	IC2944	11 36.6.2		−63	02	4.5	15	open cluster (λ Centauri)
	3766	11	36.1	−61	37	5.3	12	open cluster
	5460	14	07.6	−48	19	5.6	25	open cluster
80	5139	13	25.8	−47	29	3.6	36	globular cluster (ω Centauri)
84	5286	13	46.4	−51	22	7.6	9	globular cluster
77	5128	13	25.5	−43	01	7.0	18.2 × 14.3	SOp galaxy (Centaurus A)

CRUX

BRIGHTEST STARS

Star	R.A. h	m	s	Dec. °	'	"	Mag.	Spectrum	Proper name
α	12	26	26	−63	05	56	0.83	B1IB3	Acrux
β	12	47	43	−59	41	19	1.25	B0	
γ	12	31	10	−57	06	47	1.63	M3	
δ	12	15	09	−58	44	55	2.80	B2	

Also above magnitude 4.3: ε (3.59), μ^1 (4.03), ξ (4.04), η (4.15).

DOUBLES

Star	R.A. h	m	Dec. °	'	P.A. °	Sep. "	Mags
α	12	26.6	−63	06	115	4.4	1.4, 1.9
					202	90.1	1.0, 4.9
γ	12	31.2	−57	07	031	110.6	1.6, 6.7
					082	155.2	9.5
μ^1	12	54.6	−57	11	017	34.5	4.0, 5.2

CLUSTER AND NEBULA

C	NGC	R.A. h	m	Dec. °	'	Mag.	Dimensions '	Type
94	I4755	12	53.6	−60	20	4	10	open cluster (ϰ Crucis, Jewel Box)
99		12	53	−63	–	–	400 × 300	dark nebula (Coalsack)

TRIANGULUM AUSTRALE

BRIGHTEST STARS

Star	R.A. h	m	s	Dec. °	'	"	Mag.	Spectrum	Proper name
α	16	48	40	−69	01	39	1.92	K2	Atria
β	16	55	08	−63	25	50	2.85	F5	
γ	15	18	54	−68	40	46	2.89	A0	

Also above magnitude 4.3: δ (3.85), ε (4.03), ε (4.11).

CLUSTER

C	NGC	R.A. h	m	Dec. °	'	Mag.	Dimensions '	Type
95	6025	16	03.7	−60	30	5.1	12	globular cluster

CIRCINUS

BRIGHTEST STARS

Star	R.A. h	m	s	Dec. °	'	"	Mag.	Spectrum	Proper name
α	14	42	28	−64	58	43	3.19	F0	

Also above magnitude 4.3: β (4.07).

DOUBLES

Star	R.A. h	m	Dec. °	'	P.A. °	Sep. "	Mags
α	14	42.5	−64	59	232	15.7	3.2, 8.6

ARA

BRIGHTEST STARS

Star	R.A. h	m	s	Dec. °	'	"	Mag.	Spectrum	Proper name
β	17	25	18	−55	31	47	2.85	K3	
α	17	31	50	−49	52	34	2.95	B3	Choo
ζ	16	58	37	−55	59	24	3.13	K5	
γ	17	25	23	−56	22	39	3.34	B1	

Also above magnitude 4.3: δ (3.62), θ (3.66), η (3.76), ε^1 (4.06).

VARIABLE

Star	R.A. h	m	Dec. °	'	Range (mags)	Type	Period (d)	Spectrum
R	16	39.7	−57	00	6.0–6.9	Algol	4.42	B

CLUSTERS

C	NGC	R.A. h	m	Dec. °	'	Mag.	Dimensions '	Type
82	6193	16	41.3	−48	46	5.2	15	open cluster
81	6352	17	25.5	−48	25	8.1	7.1	globular cluster
	6362	17	31.9	−67	03	8.3	10.7	globular cluster
86	6397	17	40.7	−53	40	5.6	25.7	globular cluster

TELESCOPIUM

The brightest star is α: R.A. 18h 26m 58s.2, dec. 245° 58' 06", mag. 3.51.
Also above magnitude 4.3: ζ (4.13).

VARIABLE

Star	R.A. h	m	Dec. °	'	Range (mags)	Type	Period (d)	Spectrum
RR	20	04.2	−55	43	6.5–16.5	Z Andromedae	–	F5p

NORMA

The only star in Norma above magnitude 4.3 is γ2: R.A. 16h 19m 50s, dec. 250° 09' 20", mag. 4.02.

VARIABLE

Star	R.A. h	m	Dec. °	'	Range (mags)	Type	Period (d)	Spectrum
S	16	18.9	−57	54	6.1–6.8	Cepheid	9.75	F–G

CLUSTERS

C	NGC	R.A. h	m	Dec. °	'	Mag.	Dimensions '	Type
	6067	16	18.9	−54	13	5.6	13	open cluster
89	6087	16	18.9	−57	54	5.4	1	open cluster (S Normae)

LUPUS

BRIGHTEST STARS

Star	R.A. h	m	s	Dec. °	'	"	Mag.	Spectrum	Proper name
α	14	41	56	−47	23	17	2.30	B1	Men
β	14	58	32	−43	08	02	2.68	B2	Ke Kouan
γ	15	35	08	−41	10	00	2.78	B3	
δ	15	21	22	−40	38	51	3.22	B2	
ε	15	22	41	−44	41	21	3.37	B3	
ζ	15	12	17	−52	05	57	3.41	G8	
η	16	00	07	−38	23	48	3.41	B2	

Also above magnitude 4.3: φ (3.56), ϰ (3.72), π (3.89), χ (3.95), ϱ (4.05), λ (4.05), θ (4.23), μ (4.27).

VARIABLE

Star	R.A. h	m	Dec. °	'	Range (mags)	Type	Period (d)	Spectrum
GG	15	18.9	−40	47	5.4–6.0	β Lyrae	2.16	B1A

DOUBLE

Star	R.A. h	m	Dec. °	'	P.A. °	Sep. "	Mags
K	15	11.9	−48	44	144	26.8	3.9, 5.8

CLUSTER

NGC	R.A. h	m	Dec. °	'	Mag.	Dimensions '	Type
5822	15	16.8	−45	39	7	40	open cluster

Grus, Phoenix, Tucana, Pavo, Indus,

Magnitudes
–1
0
1
2
3
4
5
Variable star
Galaxy
Planetary nebula
Gaseous nebula
Globular cluster
Open cluster

Grus, the Crane, is much the most prominent of the four constellations making up the Southern Birds; one way to identify it is to continue the line from α and β Pegasi, in the square, through Fomalhaut. The line of stars running from γ through β and on to ε and ζ really does give some impression of a bird in flight. The little pairs making up δ and μ give the impression of being wide doubles, though both are nothing more than line of sight effects.

Of the two leaders of Grus, α (Alnair) is a bluish-white B-star, 150 light years away and 100 times as luminous as the Sun. β (Al Dhanab) is an M-type giant, 228 light years away and its luminosity is 750 times that of the Sun. The two are almost equally bright, and the contrast between the steely hue of Alnair and the warm orange of Al Dhanab is striking in binoculars – or even with the naked eye. Grus contains a number of faint galaxies, but there is not much of interest here for the user of a small telescope.

Phoenix was the mythological bird which periodically burned itself to ashes, though this did not perturb it in the least and it soon recovered. α (Ankaa) is the only bright star; it is of type K, decidedly orange, lying at a distance of 78 light years. It is 75 times as luminous as the Sun. It makes up a triangle with Achernar (α Eridani and Al Dhanab), which is probably the best way to identify it.

The main object of interest is ζ Phoenicis, which is a typical Algol-type eclipsing binary with a range from magnitude 3.6 to 4.4; the variations are easy to follow with the naked eye, and there are suitable comparison stars in β (3.31), δ (3.95) and η (4.36). Both components are of type B. This is actually the brightest of all stars of its kind apart from Algol itself and λ Tauri.

The interesting variable SX Phoenicis lies less than 7° west of Ankaa. It is a pulsating star of the δ Scuti type, with the remarkably short period of only 79 minutes, during which time the magnitude ranges between 7.1 and 7.5 – though the amplitude is not constant from one cycle to another. The spectrum, too, is variable, sometimes being of type A and at others more like type F. Its distance is no more than 150 light years, and the luminosity is roughly twice that of the Sun. Stars of this type are sometimes known as dwarf Cepheids. SX itself forms a triangle with Ankaa and ι (4.71), but the field is not very easy to identify without a telescope equipped with good setting circles.

Tucana, the Toucan. Though the dimmest of the Southern Birds, Tucana is graced by the presence of the Small Magellanic Cloud (SMC) and two superb globular clusters. The brightest star is α, which is of type K and is decidedly orange; β is a wide double in a fine binocular field. The fainter component is a close binary.

The SMC is very prominent with the naked eye; it is further away than the Large Magellanic Cloud (LMC), but the two are connected by a 'bridge' of material, and are no more than 80,000 light years apart. The SMC contains objects of all kinds, including many short-period variables – in fact it was by studying these, in 1912, that Henrietta Leavitt was able to establish the period–luminosity relationship which has been so invaluable to astronomers. It has been suggested that the SMC may be of complex form, and that we are seeing it almost end on.

Almost silhouetted against the SMC is 47 Tucanae (NGC104), the brightest globular cluster apart from ω Centauri. It has even been claimed that 47 Tucanae is the more spectacular of the two, because it is small enough to be fitted into a moderate-power telescopic field. It is surprisingly poor in variable stars, but there are several of the 'blue stragglers' referred to earlier; photographs taken with the Hubble Space Telescope resolve the cluster right through to its centre. It is about 15,000 light years away. Telescopically, or even with binoculars, it is evident that its surface brightness is much greater than that of the SMC.

NGC362 (C104) is another globular cluster in the same region of the sky. It is close to naked-eye visibility, and tele-

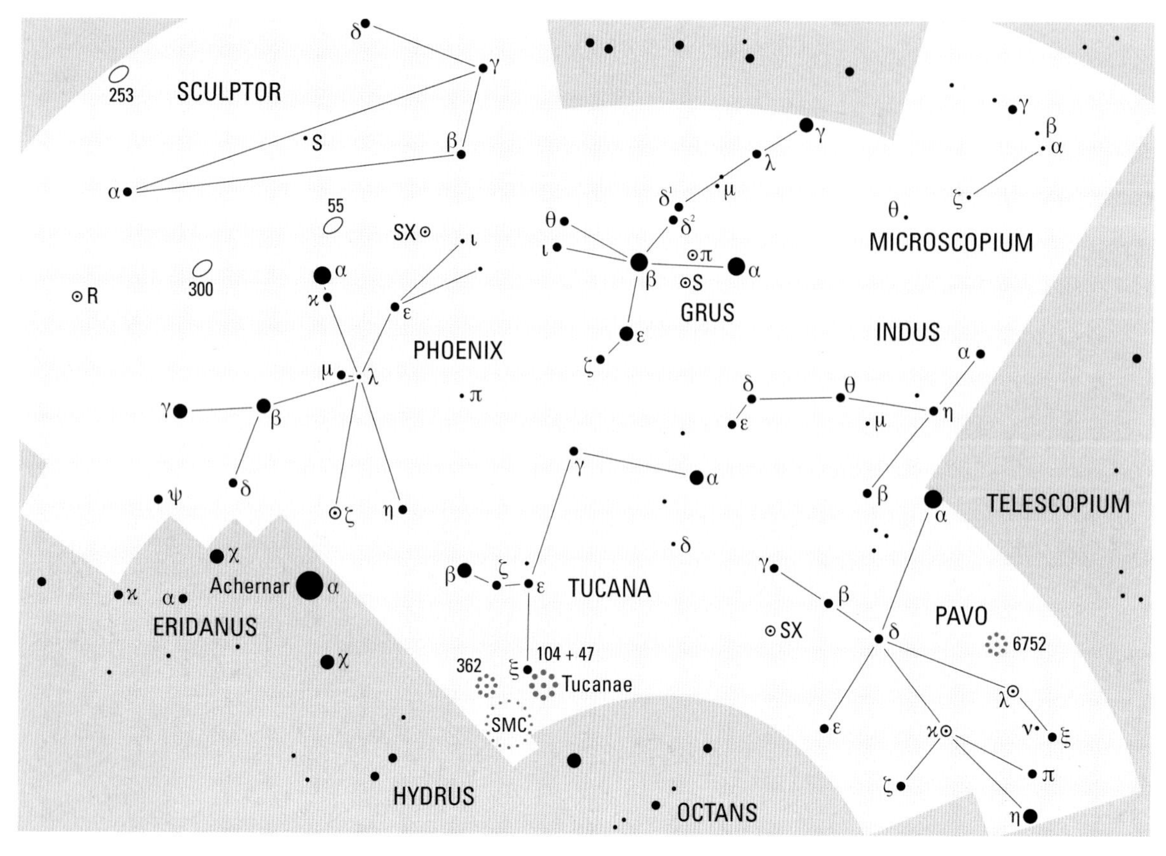

◄ **The region of the 'Southern Birds'** is apt to be somewhat confusing, because only Grus is distinctive, and the other birds are comparatively ill formed – though in Tucana we find the Small Magellanic Cloud together with the splendid globular cluster 47 Tucanae. However, the fact that Achernar (α Eri) lies nearby is a help in identification. The other constellations in this map – Indus, Microscopium and Sculptor – are very obscure.

Microscopium, Sculptor

scopically it is not greatly inferior to 47 Tucanae, though it is less than half the size of the latter.

Pavo, the Peacock, has one bright star, α, which can be found by using α Centauri and α Trianguli Australis as pointers; indeed, this is perhaps the best way of leading into the region of the Southern Birds. α has no proper name, and is rather isolated from the rest of the constellation; it is of type B, 230 light years away and 700 times as luminous as the Sun.

κ Pavonis is a short-period variable with a range of from magnitude 3.9 to 4.7 and a period of just over 9 days; suitable comparison stars are ε (3.96), γ (4.22), ζ (4.01), ξ (4.36) and ν (4.64). (Avoid λ, which is itself a variable of uncertain type; the same comparison stars can be used.) κ is of the W Virginis type and much the brightest member of the class. It was once known as a Type II Cepheid, but W Virginis stars are much less luminous than classical Cepheids with the same period, and κ is no more than about four times as luminous as the Sun; its distance is 75 light-years.

The fine globular cluster NGC6752 (C93) lies not far from λ. It is easy to see with binoculars, and is moderately condensed; the distance is about 20,000 light years. It seems to have been discovered by J. Dunlop in 1828.

Indus is a small constellation created by Bayer in 1603; its brightest star, α, forms a triangle with α Pavonis and Alnair in Grus. There is nothing here of immediate interest for the telescopic observer, but it is worth noting that ε Indi, at magnitude 5.69, is one of the nearest stars – just over 11 light years away – and has only one-tenth the luminosity of the Sun, so it is actually the feeblest star which can be seen with the naked eye. If it could be observed from the standard distance of 10 parsecs (32.6 light years), its apparent magnitude would be 7, and it would be invisible without optical aid. It is of type K, and orange in colour. Despite its low luminosity, it may be regarded as a fairly promising candidate for the centre of a planetary system.

Microscopium is a very dim constellation adjoining Grus and Piscis Austrinus. It was formerly included in the Southern Fish, so γ was known as 1 Piscis Austrini and ε as 4 Piscis Austrini. It contains nothing of special note. α is an easy double (magnitudes 5.0 and 10.0, separation 20.5 seconds, position angle 166 degrees). There are also two Mira variables which can reach binocular visibility at maximum. U ranges from magnitude 7 to 14.4 in a period of 334 days, while S ranges between 7.8 and 14.3 in 209 days. Like most Mira stars they are of spectral type M, and are obviously orange-red.

Microscopium was one of the numerous constellations introduced by Nicolas-Louis de Lacaille in his famous maps of the southern sky in 1752, but frankly it seems unworthy of a separate identity.

Sculptor is another one of Lacaille's groups, originally Apparatus Sculptoris, the Sculptor's Apparatus. It occupies the large triangle bounded by Fomalhaut (α Piscis Austrini), Ankaa (α Phoenicus), and Diphda (β Ceti), but the only objects of interest are the various galaxies.

NGC253 (C65) lies almost edgewise on to us, and lies not far from α, close to the border between Sculptor and Cetus; it is a favourite photographic target. NGC55 (C72) lies near Ankaa on the border between Sculptor and Phoenix, and seems to be one of the nearest galaxies beyond the Local Group, lying at no more than 8 million light years from us. Like NGC253 it is a spiral, seen almost edge on; it is easy to identify and attractive to photograph. The south galactic pole lies in Sculptor, and the whole region is noticeably lacking in bright stars.

GRUS

BRIGHTEST STARS

Star	R.A. h	m	s	Dec. °	′	″	Mag.	Spectrum	Proper name
α	22	08	14	−46	57	40	1.74	B5	Alnair
β	22	42	40	−46	53	05	2.11	M3	Al Dhanab
γ	21	53	56	−37	21	54	3.01	B8	
ε	22	48	33	−51	19	01	3.49	A2	

Also above magnitude 4.3: ι (3.90), δ¹ (3.97), δ² (4.11), ζ (4.12), θ (4.28)

VARIABLES

Star	R.A. h	m	Dec. °	′	Range (mags)	Type	Period (d)	Spectrum
π¹	22	22.7	−45	57	5.4–6.7	Semi-reg.	150	S
S	22	26.1	−48	26	6.0–15.0	Mira	401	M

PHOENIX

BRIGHTEST STARS

Star	R.A. h	m	s	Dec. °	′	″	Mag.	Spectrum	Proper name
α	00	26	17	−42	18	22	2.39	K0	Ankaa
β	01	06	05	−46	43	07	3.31	G8	
γ	01	28	22	−43	19	06	3.41	K5	

Also above magnitude 4.3: ζ (3.6, max), ε (3.88), ι (3.94), δ (3.95).

VARIABLES

Star	R.A. h	m	Dec. °	′	Range (mags)	Type	Period (d)	Spectrum
ζ	01	08.4	−55	15	3.6–4.4	Algol	1.67	B+B
SX	23	46.5	−41	35	6.8–7.5	δ Scuti	0.055	A–F

TUCANA

BRIGHTEST STARS

Star	R.A. h	m	s	Dec. °	′	″	Mag.	Spectrum	Proper name
α	22	18	30	−60	15	35	2.86	K3	

Also above magnitude 4.3: β (3.7, combined), γ (3.99), ξ (4.23).

DOUBLES

Star	R.A. h	m	Dec. °	′	P.A. °	Sep. ″	Mags
β	00	31.5	−62	58	169	27.1	4.4, 4.8; B is a close binary (444y)
κ	01	15.8	−68	53	336	5.4	5.1, 7.3

CLUSTERS AND GALAXY

C	NGC	R.A. h	m	Dec. °	′	Mag.	Dimensions ′	Type
		00	53	−72	50	2.3	280 × 160	galaxy (Small Magellanic Cloud
106	104	00	24.1	−72	05	4.0	30.9	globular cluster (47 Tucanae)
104	362	01	03.2	−70	51	6.6	12.9	globular cluster

PAVO

BRIGHTEST STARS

Star	R.A. h	m	s	Dec. °	′	″	Mag.	Spectrum	Proper name
α	20	25	39	−56	44	06	1.94	B3	
β	20	44	57	−66	12	12	3.42	A5	
λ	18	52	13	−62	11	16	3.4 (max)	B1	

Also above magnitude 4.3: δ (3.56), η (3.62), κ (3.9 max), ε (3.96), ζ (4.01), γ (4.22).

VARIABLES

Star	R.A. h	m	Dec. °	′	Range (mags)	Type	Period (d)	Spectrum
κ	18	56.9	−67	14	3.9–4.7	W Virginis	9.09	F
λ	18	52.2	−62	11	3.4–4.3	Irregular	–	B
SX	21	28.7	−69	30	5.4–6.0	Semi-reg.	50	M

CLUSTER

M	C	NGC	R.A. h	m	Dec. °	′	Mag.	Dimensions ′	Type
	93	6752	19	10.9	−59	59	5.4	20.4	globular cluster

INDUS

BRIGHTEST STARS

Star	R.A. h	m	s	Dec. °	′	″	Mag.	Spectrum	Proper name
α	20	37	34	−47	17	29	3.11	K0	Persian

Also above magnitude 4.3: β (3.65).

MICROSCOPIUM

The brightest star is γ: R.A. 21h 01m 17s.3, dec. −32° 5′ 28″, mag. 4.67. It was formerly known as 1 Piscis Austrinus.

SCULPTOR

The brightest star is α: R.A. 00h 58m 36s.3, dec. 229° 21′ 27″, mag. 4.31.

VARIABLES

Star	R.A. h	m	Dec. °	′	Range (mags)	Type	Period (d)	Spectrum
S	00	15.4	−32	03	5.5–13.6	Mira	365.3	M
R	01	27.0	−32	33	5.8–7.7	Semi-reg.	370	N

GALAXIES

C	NGC	R.A. h	m	Dec. °	′	Mag.	Dimensions ′	Type
72	55	00	14.9	−39	11	8.2	32.4 × 6.5	SB galaxy
65	253	00	47.6	−25	17	7.1	25.1 × 7.4	Sc galaxy
70	300	01	54.9	−37	41	8.7	20.0 × 14.8	Sd galaxy

Eridanus (southern), Horologium, Caelum, Dorado, Reticulum,

Eridanus, the River, stretches down into the far south, ending at Achernar (α), which is the ninth brightest star in the sky; it is 144 light years away, and 1000 times as luminous as the Sun. It can be seen from anywhere south of Cairo; from New Zealand it is circumpolar.

There is a minor mystery attached to Acamar (θ Eridani). Ptolemy ranked it as first magnitude, and seems to have referred to it as 'the last in the River', but it is now little brighter than magnitude 3. It is not likely to have faded, and it is just possible that Ptolemy had heard reports of Achernar, which is not visible from Alexandria – though Acamar can be seen, low over the horizon. Acamar, 55 light years away, is a splendid double, with one component rather brighter than the other; both are white, of type A, and are respectively 50 and 17 times more luminous than the Sun.

Horologium is one of Lacaille's obscure constellations, bordering Eridanus. The only object of any note is the red Mira variable R Horologii, which can rise to magnitude 4.7 at maximum. It is rather isolated, but χ and φ Eridani, near Achernar, point more or less to it.

Caelum is another Lacaille addition; he seems to have had a fondness for sculpture, since Caelum was originally Caela Sculptoris, the Sculptor's Tools. There is nothing of interest here; the constellation borders Columba and Dorado.

Dorado, the Swordfish, once commonly known as Xiphias, lies between Achernar and Canopus (α Carinae). The most notable star is β, which is a bright Cepheid variable; it has a period of over 9 days, and is therefore considerably more luminous than δ Cephei itself. If its distance is correctly given in the Cambridge catalogue, it is 7500 light years away, with a peak luminosity of 200,000 times that of the Sun.

Most of the Large Magellanic Cloud (LMC) lies in Dorado, and here we have the superb Tarantula Nebula (NGC2070, 30 Doradûs), probably the finest in the sky. The LMC, 169,000 light years away, was once classed as an irregular galaxy, but shows clear indications of barred spirality. It remains visible with the naked eye even in moonlight, and is of unique importance to astronomers – which is partly why so many of the latest large telescopes have been sited in latitudes from which it is accessible. There have been various novae in it, and the spectacular supernova of 1987.

Reticulum (the Net) was originally Reticulus Rhomboidalis, the Rhomboidal Net. It is a small but compact group bordering Horoligium, Dorado and Hydrus, not far from Achernar and the LMC. Of its leading stars, β (3.85), γ (4.51), δ (4.56) and ε (4.44) are all orange, with K- or M-type spectra, so that they are quite distinctive. The Mira variable R Reticuli, with a range of from magnitude 6.5 to 14, lies in the same wide field with α, and when near maximum is a binocular object. Close beside it is R Doradûs, just across the boundary of the Swordfish, which is a red semi-regular star always within binocular range.

Hydrus, the Little Snake, is easy enough to find, though it is far from striking. α and β are relatively nearby stars; α is at a distance of 36 light years, and β, at less than 21 light years, is even closer. Since β is a G-type star, only two-and-a-half times as luminous as the Sun, it may well have a system of planets, though we have no proof. It is a just under 13° away from the south celestial pole.

Mensa (the Table) is yet another Lacaille creation, originally Mons Mensae, Table Mountain. It has the unenviable distinction of being the only constellation with no star as bright as fifth magnitude, but at least a small part of the LMC extends into it.

Chamaeleon. Another dim group. The best way to find it is to follow a line from ι Carinae through Miaplacidus (β Carinae) and extend it for some distance. The four leading stars of Chamaeleon, α (4.07), β (4.26), γ (4.11) and δ (4.45), are arranged in a diamond pattern; β lies roughly between Miaplacidus and α Trianguli Australis.

Musca Australis, the Southern Fly, generally known simply as Musca (there used to be a Musca Borealis, in the northern hemisphere, but this has now disappeared from our maps; no doubt somebody has swatted it). There are two bright globular clusters, NGC4833 (C105) near δ and NGC4372 (C108) near γ. They are not easy to locate with binoculars, but are well seen in a small telescope.

Apus was added to the sky by Bayer in 1603, originally under the name of Avis Indica, the Bird of Paradise. To find

Magnitudes
−1
0
1
2
3
4
5
Variable star
Galaxy
Planetary nebula
Gaseous nebula
Globular cluster
Open cluster

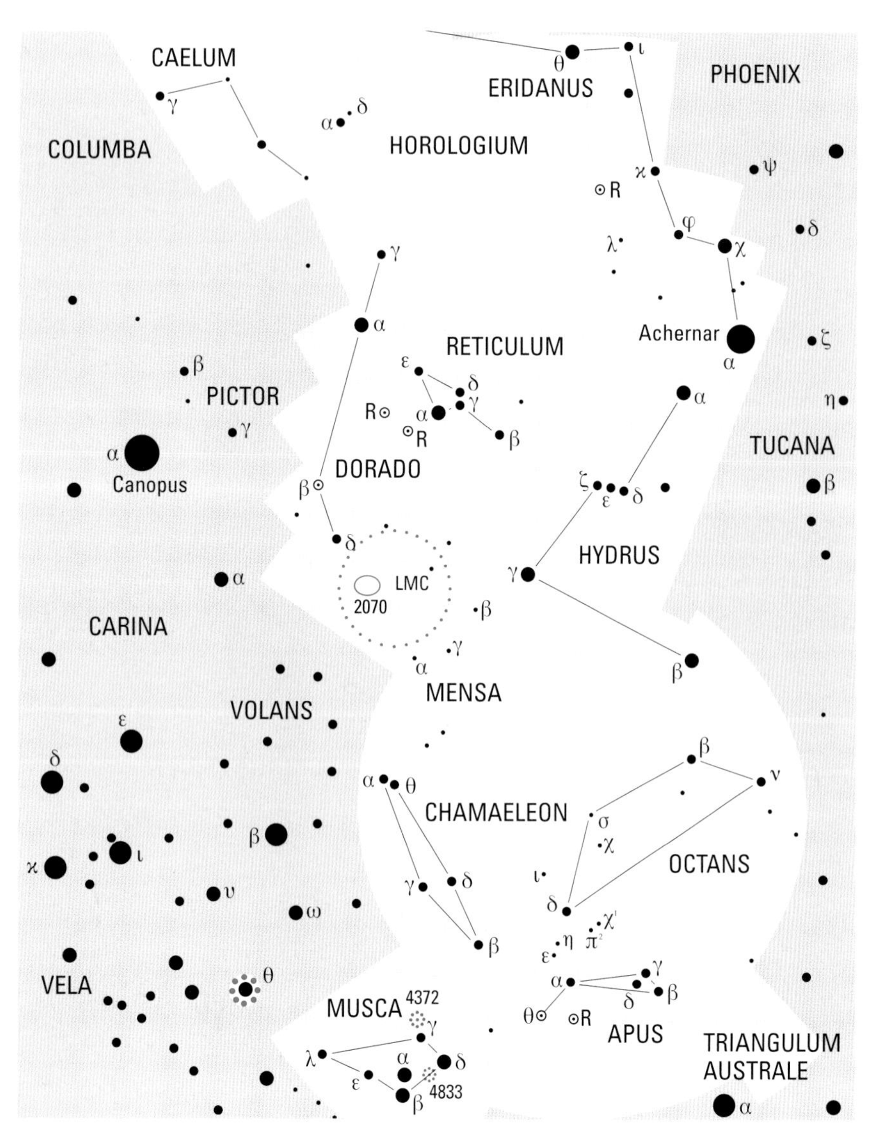

◄ **The region of the south celestial pole** is decidedly barren; if there is any mist, for example, the whole area of the sky will appear completely blank, and even against a dark sky the south polar star, σ Oct, is none too easy to identify. The nearest reasonably bright star to the south celestial pole is β Hyi, but its distance from the pole is almost 13°. The polar area is divided up into small constellations, few of which are easy to locate, but the Large Magellanic Cloud is present, mainly in Dorado but extending across into Mensa. It is also worth noting that Achernar (α Eri) lies fairly close to α Hyi.

Hydrus, Mensa, Chamaeleon, Musca, Apus, Octans

it, take a line from α Centauri through α Circini and continue until you come to α Apodis; the other main stars of the constellation – γ, δ and ε – make up a small triangle. δ is a red M-type star, and has a K-type companion at a separation of 103 seconds of arc. θ, in the same wide field with α, is a semi-regular variable which is generally within binocular range; also in the field is R Apodis, which is below magnitude five and suspected of variability.

Octans lies nearest to the pole. The brightest star in the southernmost constellation is ν, which is a K-type orange giant 75 times as luminous as the Sun. The south polar star, σ Octantis, is only of magnitude 5.5, and is not too easy to locate at first glance. A good method, using 7 × binoculars, is as follows:

Identify α Apodis, as given above. In the same field as α Apodis are two faint stars, ε (5.2) and η (5.0). These point straight to the orange δ Octantis (4.3), which has two dim stars, π^1 and π^2 Octantis, close beside it. Now put δ Octantis at the edge of the field, and continue the line from Apus. χ Octantis (5.2) will be on the far side of the field; centre it, and you will see two more stars of about the same brightness, σ and ι. These three are in the same field, and make up a triangle. The south polar star, σ, is the second in order from δ. Using 12 × binoculars, the three are in the same field with υ (5.7).

σ Octantis is of type F, and is less than seven times as luminous as the Sun, so that it pales in comparison with the northern Polaris. The pole is moving slowly away from it, and the separation will have grown to a full degree by the end of the century.

▼ **Large Magellanic Cloud** (left) and Small Magellanic Cloud (right) are the nearest notable galaxies: the LMC is 169,000 light years away, the SMC slightly farther.

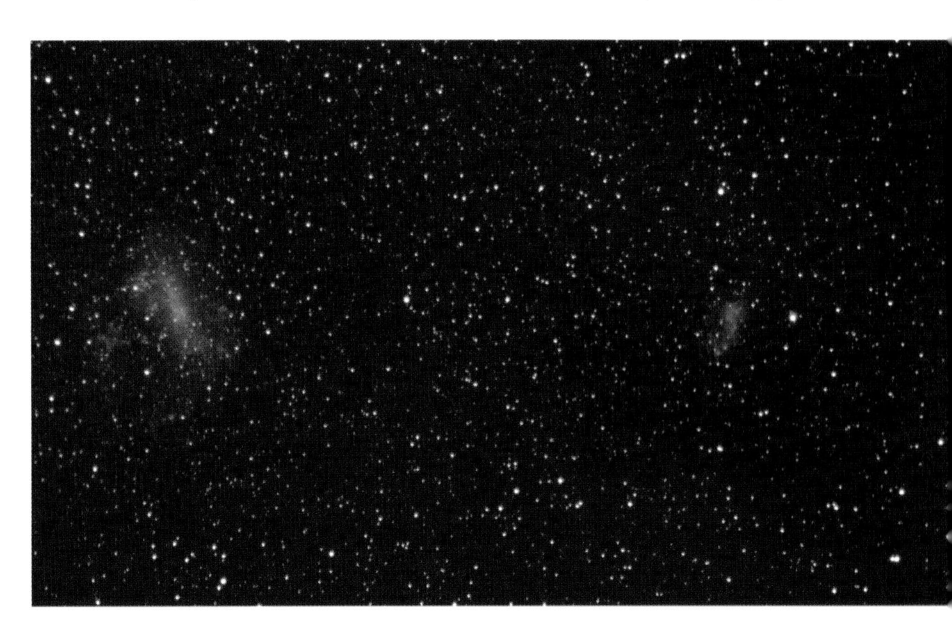

ERIDANUS

BRIGHTEST STARS

Star	R.A. (h m s)	Dec. (° ′ ″)	Mag.	Spectrum	Proper name
α	01 37 43	−57 14 12	0.46	B5	Achernar
θ	02 58 16	−40 18 17	2.92	A3+A2	Acamar

Also above magnitude 4.3: υ^4 (3.56), φ (3.56), χ (3.70), υ^2 (3.82), υ^3 (3.96), ϰ (4.25), ι (4.11), e (4.27), g (4.27).

DOUBLE

Star	R.A. (h m)	Dec. (° ′)	P.A. (°)	Sep. (″)	Mags
θ	02 58.3	−40 18	088	8.2	3.4, 4.5

HOROLOGIUM

The brightest star is α; R.A. 04h 14m 00.0s, dec. −42° 17′ 40″, mag. 3.186.

VARIABLE

Star	R.A. (h m)	Dec. (° ′)	Range (mags)	Type	Period (d)	Spectrum
R	02 53.9	−49 53	4.7–14.3	Mira	404	M

CAELUM

The brightest star is α: R.A. 04h 40m 33s.6, dec. −41° 51′ 50″, mag. 4.45.

DORADO

BRIGHTEST STAR

Star	R.A. (h m s)	Dec. (° ′ ″)	Mag.	Spectrum	Proper name
α	04 34 00	−55 02 42	3.27	A0	

Also above magnitude 4.3: β (3.7 max), γ (4.25).

VARIABLES

Star	R.A. (h m)	Dec. (° ′)	Range (mags)	Type	Period (d)	Spectrum
β	05 33.6	−62 29	3.7–4.1	Cepheid	9.84	F
R	04 36.8	−62 05	4.8–6.6	Semi-reg.	338	M

GALAXY AND NEBULA

C	NGC	R.A. (h m)	Dec. (° ′)	Mag.	Dimensions (′)	Type
		05 24	−69 45	0	650 × 550	galaxy (Large Magellanic Cloud)
103	2070	05 38.7	−69 06	3	40 × 25	nebula (30 Doradûs in Large Magellanic Cloud)

RETICULUM

BRIGHTEST STARS

Star	R.A. (h m s)	Dec. (° ′ ″)	Mag.	Spectrum	Proper name
α	04 14 25	−62 28 26	3.35	G6	

Also above magnitude 4.3: β (3.85).

HYDRUS

BRIGHTEST STARS

Star	R.A. (h m s)	Dec. (° ′ ″)	Mag.	Spectrum	Proper name
β	00 25 46	−77 15 15	2.80	G1	
α	01 58 46	−61 34 12	2.86	F0	
γ	03 47 14	−74 14 20	3.24	M0	

Also above magnitude 4.3: δ (4.09), ε (4.11).

MENSA

The brightest star is α: R.A. 6h 10m 14s.6, dec. −74° 45′ 11″, mag. 5.09.
A small part of the Large Cloud of Magellan extends into Mensa.

CHAMAELEON

The brightest star is α: R.A. 08h 18m 31s.7, dec. −76° 55′ 10″, mag. 4.07.
Also above magnitude 4.3: γ (4.11).

MUSCA

BRIGHTEST STARS

Star	R.A. (h m s)	Dec. (° ′ ″)	Mag.	Spectrum	Proper name
α	12 37 11	−69 08 07	2.69	B3	
β	12 46 17	−68 06 29	3.05	B3	

Also above magnitude 4.3: δ (3.62), λ (3.64), γ (3.87), ε (4.11).

DOUBLE

Star	R.A. (h m)	Dec. (° ′)	P.A. (°)	Sep. (″)	Mags	
β	12 46.3	−68 06	014	1.4	4.7, 5.1	binary; period many centuries

CLUSTERS

C	NGC	R.A. (h m)	Dec. (° ′)	Mag.	Dimensions (′)	Type
105	4833	13 00	−70 53	7.3	13.5	globular
108	4372	12 25.8	−72 40	7.8	18.6	globular cluster

APUS

The brightest star is α: R.A. 14h 47m 51s.6, dec. −79° 02′ 41″, mag. 3.83.
Also above magnitude 4.3: γ (3.89), β (4.24).

VARIABLES

Star	R.A. (h m)	Dec. (° ′)	Range (mags)	Type	Period (d)	Spectrum
θ	14 05.3	−76 48	6.4-8.6	Semi-reg.	119	M

DOUBLE

Star	R.A. (h m)	Dec. (° ′)	P.A. (°)	Sep. (″)	Mags
δ	16 20.3	−78 41	012	102.9	4.7, 5.1

OCTANS

The brightest star is γ: R.A. 21h 41m 29s dec. −77° 23′ 24″, mag. 3.76.
The south polar star is σ: R.A. 20h 15m 1s dec. −89° 08′, mag. 5.46.

VARIABLES

Star	R.A. (h m)	Dec. (° ′)	Range (mags)	Type	Period (d)	Spectrum
δ	22 20.0	−80 26	4.9–5.4	Semi-reg.	55	M

The author's 22-cm (8½-inch) telescope in its weather-proof housing at his home in Selsey – a practical proposition for most amateur astronomers. Having your own observatory helps to ensure that you make the most of your telescope.

THE PRACTICAL ASTRONOMER

Beginner's Guide 1

Few can fail to be stirred by the sight of a star-filled sky seen from a clear, dark location. Even the most astronomically uninterested will, at some time or other, gaze up in awe at the beauty of a thin crescent Moon hanging in the vignette of an evening twilight or perhaps instinctively cry out when a bright meteor streaks across the sky. Astronomical curiosity is present, in varying amounts, in each and every one of us. For those with more than a vague interest, amateur astronomy is the hobby to nurture that curiosity.

With spacecraft visiting the planets and giant mountaintop telescopes fitted with sophisticated imaging equipment, is there really a place left for the amateur astronomer in our light-polluted towns and cities? Fortunately, the answer to this question is a resounding yes. Astronomy is one of the few sciences where amateurs can still contribute useful observations to assist the professionals. At the most basic level, the hobby is also great fun and will deliver many rewards if you're prepared to put the time and effort in.

Astronomy is a large subject, encompassing many diverse disciplines. If you're just starting out, this diversity can appear daunting. Fortunately, the internet and astronomy work well together and a few simple searches will normally provide you with plenty of help and guidance. Rushing out to buy a telescope before you have any idea at what you'll actually be looking at isn't really a good plan. The best starting strategy is to forget about spending any money at all until you have learned some of the basics.

Among the multitude of stars you can see on a moonless clear night, there is a surprising amount of order. The ancients saw animals, figures and objects in the night sky and it is from these beginnings that we now have 88 officially recognized constellations. To start out you need only to identify a small handful of main constellations which will act as signposts to the rest of the sky. A simple star chart or planisphere is ideal for the task.

While it can be said that some patterns are easy to identify, it's also true to say that some are rather obscure. The trick to finding your way around the night sky is a bit like fumbling your way around a darkened room, grabbing onto familiar objects like chairs and tables for support. If you identify a few

▼ **The stirring sight** of a young crescent Moon close to the planet Venus. Pete Lawrence.

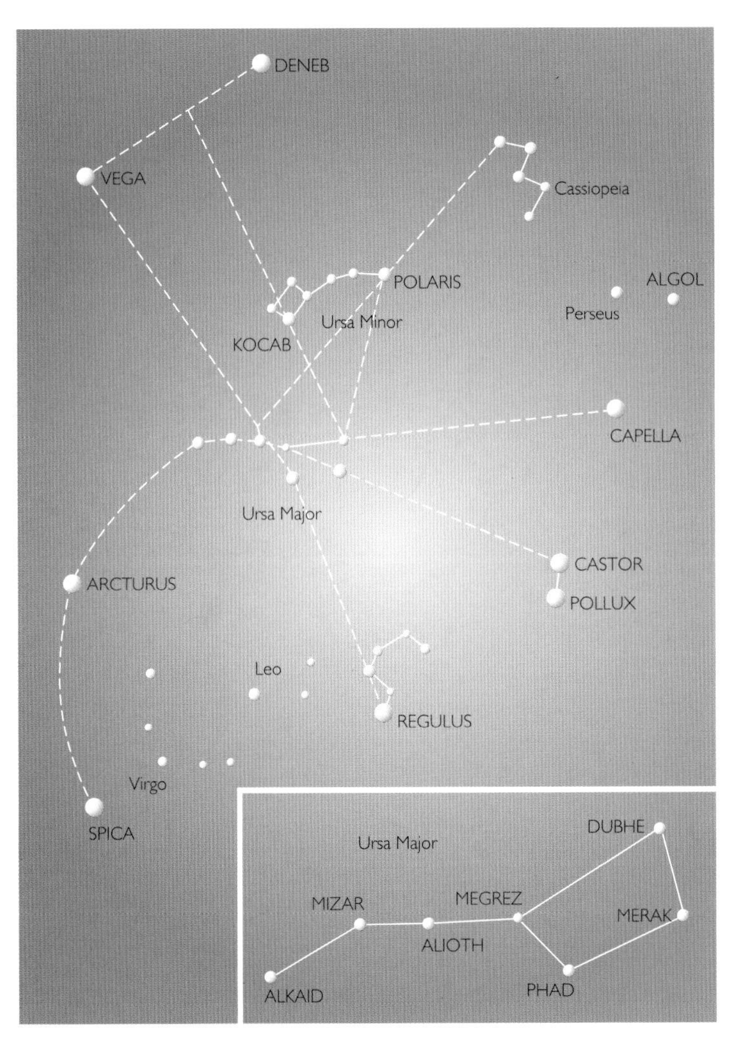

◀ **The easy-to-find stars of Ursa Major** make a useful signpost to the bright stars of other constellations including Perseus, Casseopeia, Leo, Cygnus, Lyra, Gemini, Ursa Minor, Boötis and Virgo.

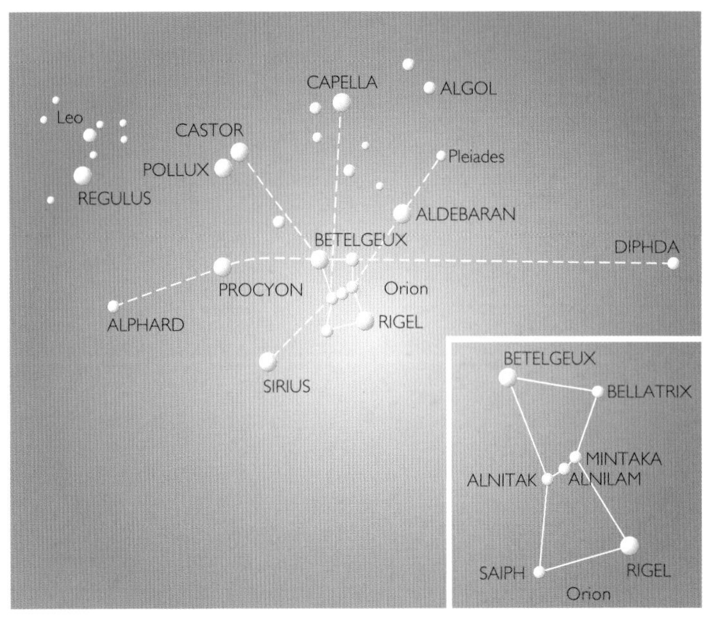

▲ **Orion.** The arrangement of stars in Orion leads to other constellations around including Leo, Canis Major, Taurus, Gemini, Hydra, Cetus, Perseus and Auriga.

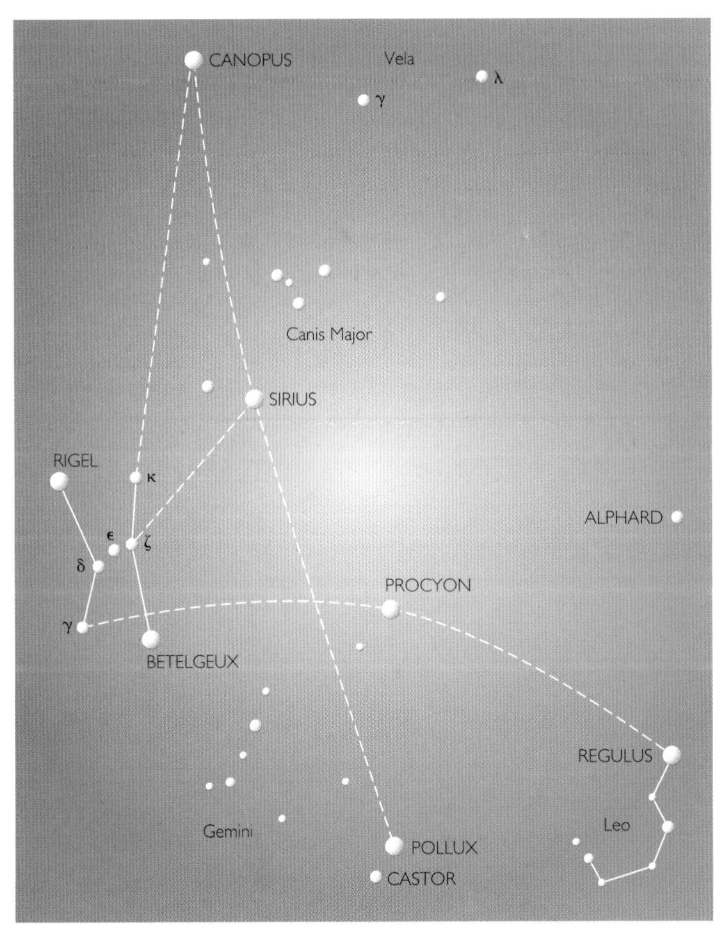

◀ **In the Southern Hemisphere** the stars of Orion can also be used to locate the constellations of Carina and Vela.

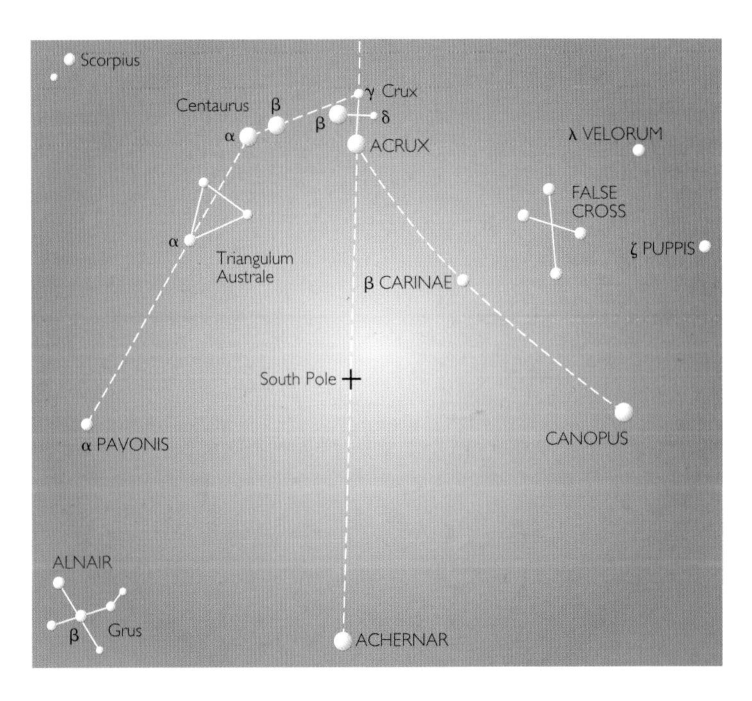

▲ **The South Celestial Pole.** Crux and Achernar can be used to locate the South Celestial Pole and prominent stars in the southern sky, so making finding your way around this region easier.

Beginner's Guide 2

familiar 'signpost' constellations, you can use them as guides to the more obscure areas of the sky. Examples of popular signposts are the Plough, or Big Dipper, and Orion, in the Northern Hemisphere with Crux, Scorpius and Sagittarius in the Southern Hemisphere.

Another useful skill involves learning how to judge scale in the sky. Linear measurements like centimetres and metres have no meaning when describing how far one star appears to be from another. Instead, astronomers use angular measures of degrees, minutes and seconds (° ′ ″). Your hand at arm's length makes a great sky ruler. Hold up your little finger at arm's length and its width is approximately 1°. A fist is about 10° wide while an outstretched hand is about 18°. This is especially useful if you invest in a set of star charts or a planisphere as it can help you to judge exactly how big something will appear in the sky.

Once you have learned the main star patterns, you can try something a little more taxing such as tracking down the brighter planets. The internet has many free websites that list what is visible on a given night and the location of objects such as the planets and the Moon. The constellations act as a framework against which these other objects can be located.

With just your eyes, it is possible to enjoy the night sky in many different ways. For example, many stars in the sky vary in brightness (variable stars) and it is possible, with a little practice, to make very accurate estimates of these variations. Watching meteor showers is another popular naked-eye pastime; observations of meteors can provide valuable insight into the activity profile of a particular stream. Then there are rare events such as bright comets, auroral displays (for those of high northern or southern latitudes) and noctilucent cloud displays, during the summer months from high temperate latitudes, the visibility of which may have links with climate change.

The curse of light pollution is something that is removing the grandeur of the night sky for many of us, especially those who live close to a large city or town centre. Here naked-eye astronomy can be difficult. Even for those without light pollution there will come a time when you will want to delve deeper into the stars. The best advice here is to purchase a good pair of binoculars, an ideal instrument for portable observing.

Binocular sizes are defined by two numbers, for example 10 × 50. The first number indicates magnification while the last gives the diameter of the front lens in millimetres. A magnification of 7 × to 10 × is ideal for a pair of hand held binoculars, anything more and you may struggle to keep the image steady. A lens diameter of 50 millimetres (2 inches) is also a good choice, providing enough light grasp to see many faint objects while keeping the instrument's weight down. Larger magnifications may seem inviting but they can be restrictive

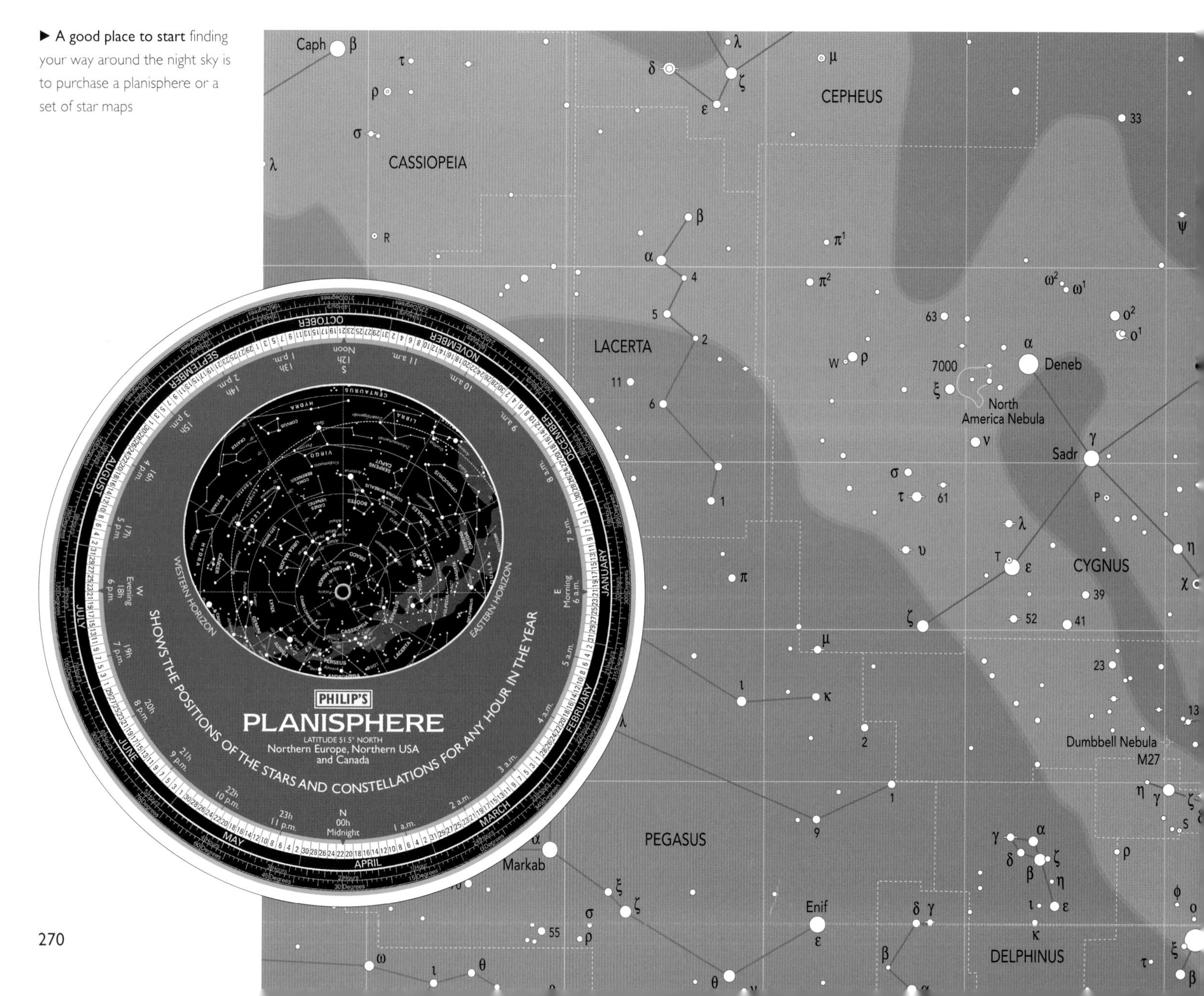

► **A good place to start** finding your way around the night sky is to purchase a planisphere or a set of star maps

and in extreme cases, useless. Larger apertures, too, are also tempting but beyond a certain size, weight becomes an issue requiring some form of mounting such as a tripod to make the binoculars useable for extended periods.

Tripods and binoculars don't always work well together, especially when being used to look at objects which are nearly overhead. Here, specialist binocular mounts can greatly ease the task. Alternatively special mirror stands are available which allow you to view the night sky with the binoculars pointing down rather than up. These may suit those who are prone to backaches or a stiff neck.

Binoculars open up a whole new universe of views. Apart from being able to see familiar naked-eye objects more clearly, the added light grasp and magnification they provide allow you to see fainter and smaller objects such as the larger craters on the Moon, bright galaxies, star clusters and nebulae. If you have an interest in following variable stars or observing double stars, the number that will come within your reach with a simple pair of binoculars is quite staggering. While a fixed low power is ideal for sweeping across wide expanses of the sky at once, it is not ideal for looking at very small objects such as the planets. If, after using binoculars to enhance your general views, you still find yourself yearning to see the disks of the main planets, then there is nothing else for it, the time has come to get a telescope...

▲ **Bright comets** appear infrequently but once seen they are not easily forgotten. This is Comet C/2006 P1 (McNaught) which graced our skies in January 2007.

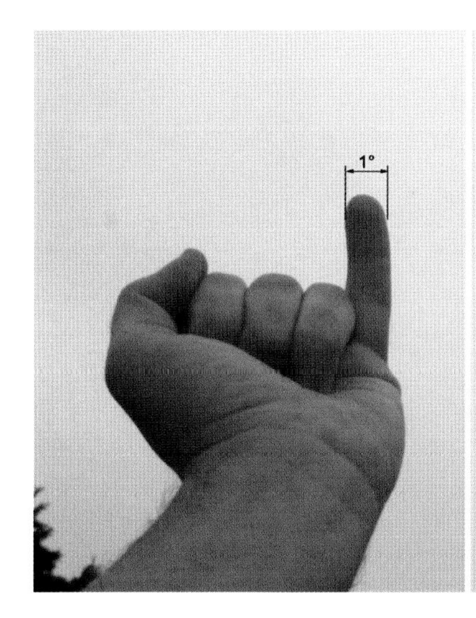

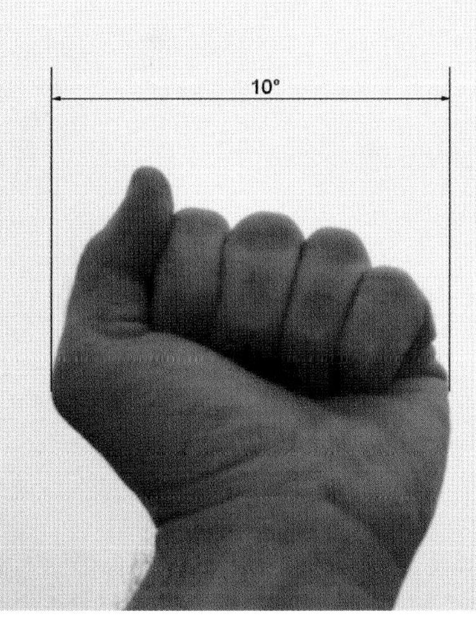

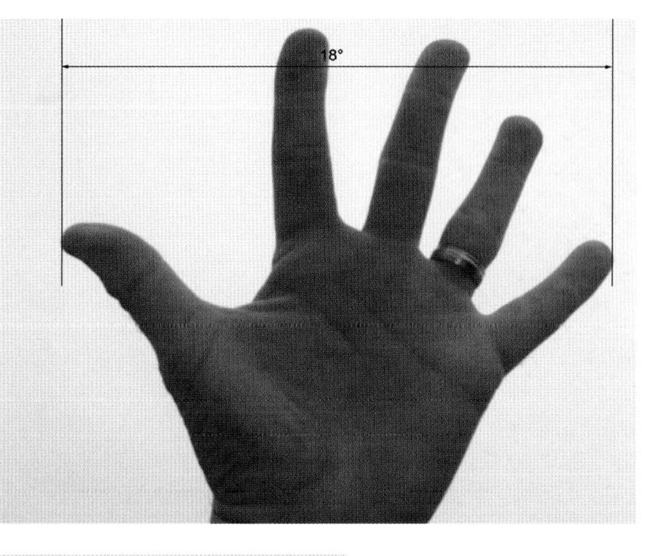

▲ **Angular measurements.** 1° (little finger at arm's length), 10° (fist at arm's length) and 18° (stretched hand at arm's length).

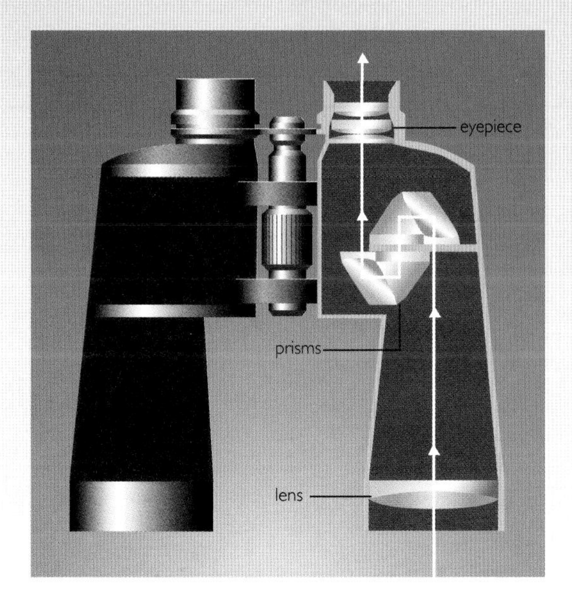

BINOCULARS
Binoculars are graded according to their magnification and their aperture, which is always given in millimetres. Thus a 7 × 50 pair yields a magnification of seven, with each object-glass 50 mm (2 in) in diameter. If only one pair is to be obtained, this type is probably a wise choice, because these binoculars have a wide field and are lightweight enough to be 'handy'.

Telescope Basics

There are two main types of astronomical telescope, the refractor which uses lenses to collect the light and the reflector which uses mirrors. In the case of the refractor, light passes through the front lens (the object-glass or objective) and is brought to a focus at the rear of the telescope. The reflector functions in much the same way except that here the light is reflected off a mirror (the primary) and brought to a focus up the tube. The most common type of reflector, the Newtonian, has a smaller, flat mirror (the secondary) located in the centre of the tube close to the front aperture. This is orientated at 45° to the primary mirror and diverts the light cone from the primary to the side of the tube for viewing. The secondary is held in place by a structure known as a spider. It is this structure that causes the diffraction spikes often seen around bright stars in photographs.

Both have pros and cons. Low cost refractors (achromats) tend to suffer from colour aberration which results in a blue fringe around bright objects. More complex lens designs (apochromats) can reduce this effect but can be expensive.

Although reflectors don't suffer from chromatic aberrations like refractors, they are prone to coma, which affects how stars appear at the edge of the field of view. In addition, the mechanical design of the reflector requires it to be regularly checked to ensure that the optical components are lined up correctly (collimated). Many poor reflector views are caused by bad collimation. There are plenty of tutorials (the internet once again is a good source) to help reflector owners to collimate their instruments properly. Although it can appear daunting to begin with, after a few attempts, the process becomes second nature and allows the instrument to reach its full potential in terms of the views it can deliver.

Reflectors are generally good value for money although small ones are not recommended as their light grasp is rather low. The secondary mirror in a reflector blocks some of the incoming light and reduces contrast. As a consequence, size for size, a refractor will be more effective than a reflector.

Compound or catadioptric telescopes such as Schmidt-Cassegrains and Maksutovs attempt to reduce aberrations such as coma by introducing a corrector plate at the front of the optical tube. The basic design of these instruments is similar to that of the reflector except that generally the secondary sends the focus of the main mirror straight back towards the primary which has a hole in it to allow the light cone to escape through the rear of the telescope. This folded optical design means that a long focal length instrument can be contained in a fairly short tube. The more complex optical nature of catadioptric telescopes can make them expensive to buy.

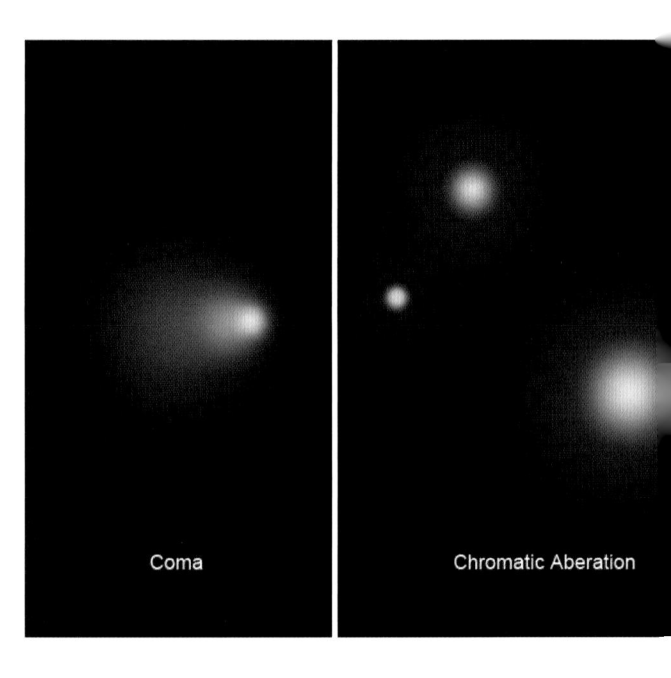

▶ **Aberrations affect the quality of the image** that is presented at the eyepiece. Achromatic refractors suffer from colour fringing around bright objects while short focal length Newtonian reflectors suffer from off-axis coma.

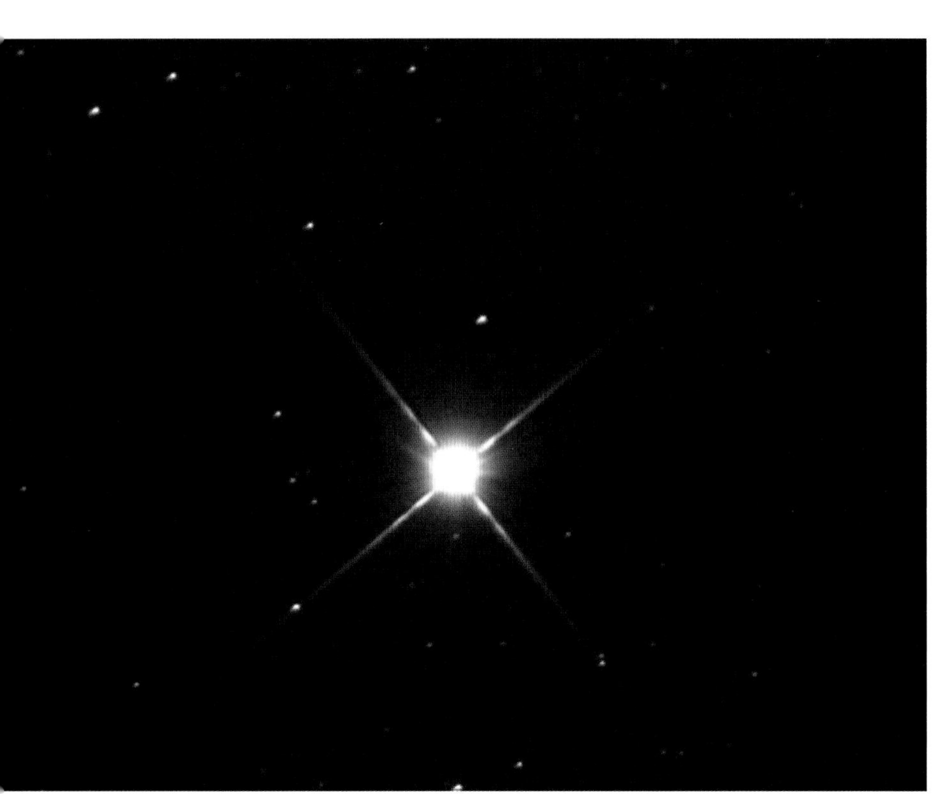

▼ **Betelgeux,** photographed through the author's 38-cm (15-in) reflector. Vanes holding the secondary mirror give rise to the diffraction spikes shown.

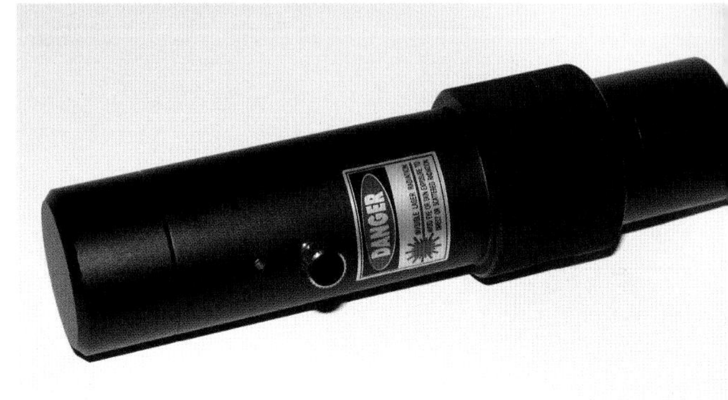

◀ ▼ **Tools such as Cheshire eyepieces** and laser collimators make the task of collimation easier.

Telescope magnifications are adjusted by using eyepieces of different focal lengths according to the equation M=Ft/Fe where Ft is the focal length of the telescope and Fe the focal length of the eyepiece in the same units. A general rule of thumb states that the maximum useable magnification a telescope of given aperture can deliver is 2 × the diameter in millimetres; for example, a 15-centimetre (6-inch) telescope can be used up to 300 ×.

At least three eyepieces are required to make the best use of a telescope. A low power one (30–40 millimetres) will provide the widest views while a mid range one (12–20 millimetres) will probably give the most useful service. A high power eyepiece should be chosen to give close to the maximum useable magnification as discussed above.

A telescope is normally described in terms of its focal ratio; the value you get if you divide its focal length by its aperture using the same units. Those with low focal ratios (f/5 or smaller) are described as fast, delivering lots of light for wide, bright low-magnification views. Those in the range f/8 or higher are described as slow – giving narrow, low contrast, high-magnification views. Mid-way between slow and fast telescopes are those which cover both bases but typically don't excel in either. The speed of a telescope (fast, medium or slow) is the biggest factor in deciding which to purchase in order to view certain objects. Fast telescopes are good for deep-sky objects such as clusters, nebulae and galaxies. Slow telescopes can provide high magnification which is ideal for studying the Moon or planets. Most astronomical objects are faint and the rule of thumb about telescope size is the larger the aperture the better.

Up to a point, optical amplifiers such as Barlow lenses or focal reducers can be used to adjust the effective focal length of a telescope. The new effective focal length of a telescope is obtained by multiplying the natural focal length by the amplifier's power.

Mounts

A telescope mount should be sturdy and strong enough to carry the main telescope tube without vibration or flexure. If the telescope is intended to be used for astrophotography, a precise mount is one of the most critical requirements.

Many telescopes now come with computerized controls which allow you to select an object in the telescope's database and request the computer slew directly to it. Care should be taken here as the most important thing about a mount is its rigidity and suitability for the task. It's all too easy to be won over by the apparent sophistication of the electronics at the expense of the mount's engineering.

Motorized mounts allow you to view your subject without having to adjust the telescope's position. In order to do this the mount either has to be accurately polar aligned (equatorial mount) or computer controlled (alt-az or equatorial) to follow the motion of the target in the sky.

Computer controlled mounts will typically allow for a connection to an external computer. Certain star charting programs can talk to the mount allowing you to pick on object on the computer screen while the program instructs the telescope to point to it in the sky.

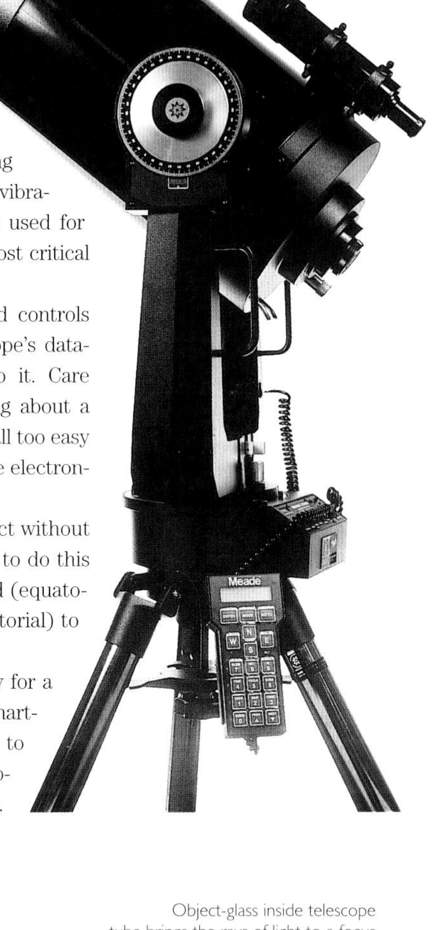

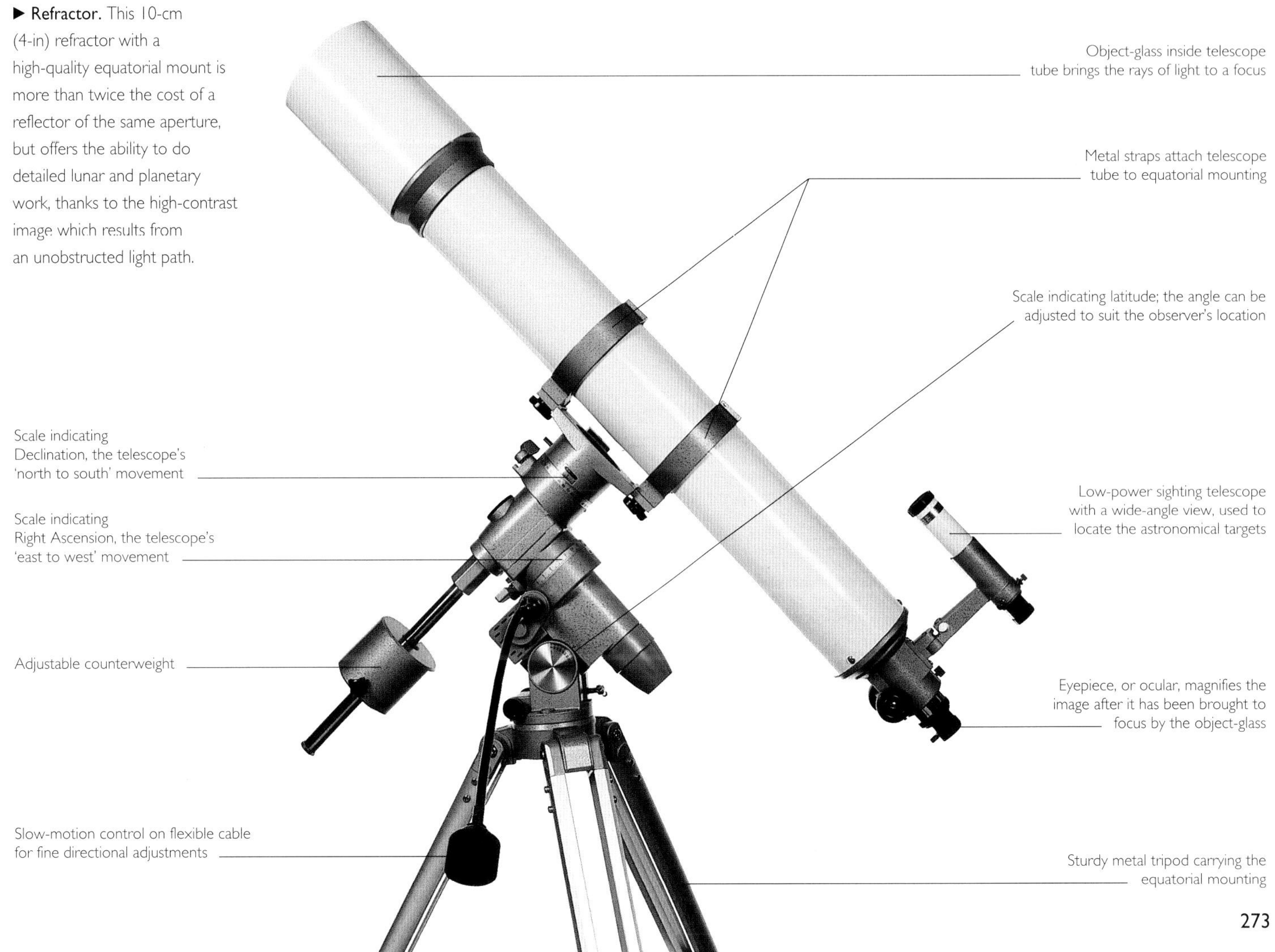

► **Refractor.** This 10-cm (4-in) refractor with a high-quality equatorial mount is more than twice the cost of a reflector of the same aperture, but offers the ability to do detailed lunar and planetary work, thanks to the high-contrast image which results from an unobstructed light path.

Digital Imaging 1

▶ **Three different types** of digital photographic camera ranging from a simple point-and-shoot camera (minimum control) on the left, a 'prosumer' fixed lens camera in the middle and a removable lens DSLR (maximum control) on the right.

With the advent of digital technology, astronomical imaging has undergone something of a revolution in recent years. However, choosing a camera for a specific imaging purpose can be a confusing task, because of the sheer number of models and types available. Astronomically suitable cameras fall into the following general categories; general photographic digital cameras, dedicated astronomical CCD cameras, webcams and high frame rate cameras.

There are three main categories of digital camera; 'point-and-shoot', prosumer and DSLR. Point-and-shoot cameras are inexpensive, have fixed lenses and limited controls. They are suitable for photographing bright scenes such as sunsets, possibly the brighter planets (as dots) and bright objects, such as the Moon, through a telescope.

Prosumer (a contraction of 'professional' and 'consumer') cameras are more expensive than their point-and-shoot cousins, have fixed lenses and offer increased manual control over factors such as exposure and sensor sensitivity. In general prosumer fixed lenses are of a high quality and often offer quite respectable optical zoom values. Digital zoom refers to a method of increasing image scale by software interpolation. As such it has limited application and for astronomical purposes can be ignored. Here the range of potential targets increases to include the brighter stars, brighter deep sky objects, bright comets, artificial satellites and aurorae. An afocally coupled prosumer camera can also produce very respectable images of the Sun (always with the appropriate filters) and Moon.

▼ **M31,** the Andromeda Galaxy captured with a Canon DSLR camera attached to an 80-mm (3.5-in) refractor

The holy grail of photographic digital cameras is the digital single lens reflex camera or DSLR. This design offers through-the-lens focusing and composition, removable lenses and full manual control over the camera's capabilities.

Not all DSLR cameras are created equal and some are quite unsuitable for astronomical imaging purposes. Fortunately, a simple internet search on the camera's name and the term 'astrophotography' is normally enough to return examples showing how good a particular model is for the purpose. If no results are returned the camera will either be new and untested or unsuitable for astrophotography.

There can be many differences between camera models but one of the main considerations for the astrophotographer is how well the camera handles noise. Photographic camera sensors are inherently noisy when used for long exposures and some models handle noise better than others. Static noise, such as that produced from pixels that are stuck in a permanently 'on' state (hot pixels) are relatively easy to deal with by a process called dark frame subtraction. Following a normal exposure, another exposure of equal length is taken with the lens cap on (a dark frame). This reveals the static noise which will also be present in the first image. Subtracting the dark frame from the normal exposure removes the static noise.

▼ **Static noise** (below left) is revealed when an image is taken under the same conditions (exposure, ISO, temperature) as a normal astrophotograph but with the lens cap on. This dark frame can be subtracted from the normal photograph, effectively removing the static noise. Random noise (below right) can be seen in the background between the stars. Stacking techniques are normally employed to remove this type of noise.

▼▼ **As well as chip noise,** images taken through a telescope can be contaminated with effects from the optical system. These effects can be revealed by taking a photograph with the telescope's field of view evenly illuminated (bottom left and right). The resulting flat field is subtracted from the main image which removes the unwanted blemishes.

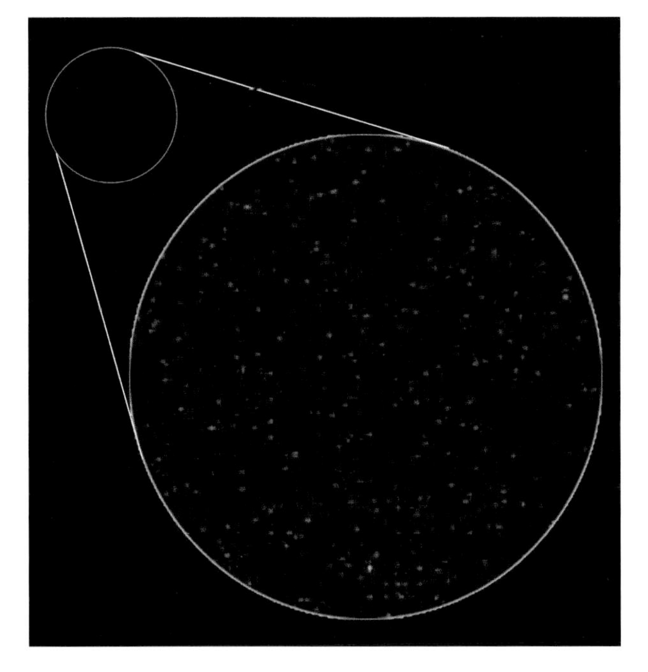

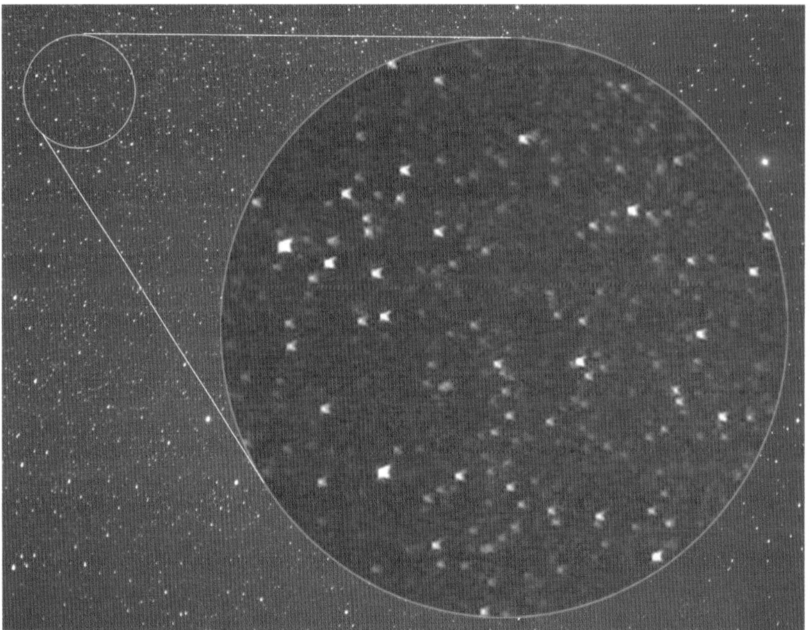

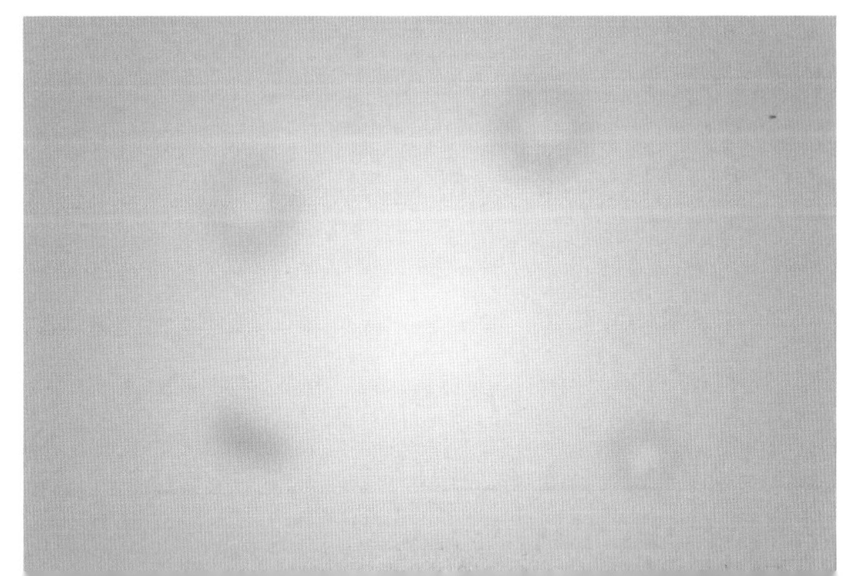

Digital Imaging 2

► **Astronomical CCD cameras** are dedicated imaging devices with cooled imaging chips to reduce thermal noise.

► **A filter wheel** provides a convenient means of changing a filter without having to remove the camera between changes.

Random noise, such as thermally induced background noise, cannot be dealt with in the same way because its location is not predictable, so image stacking techniques are employed to average it out and increase signal strength.

Most random noise comes from the chip being warm (thermally induced noise). Dedicated astronomical CCD cameras have in-built cooling which can cool the chip to as much as 30°C (54°F) below the ambient temperature to reduce this unwanted effect. Although this doesn't remove thermal noise completely, it does reduce it to manageable levels.

Astronomical CCD cameras are available in either one-shot colour or monochrome versions. Colour CCD cameras use mono sensors overlaid with a special filter matrix (a Bayer matrix). After capture, software interprets the values of each pixel as seen through the filter and recreates the colour information in the scene. The Bayer matrix has an effect on the sensitivity of these one-shot colour devices and many purists prefer to work with more sensitive mono cameras.

▼ **High frame rate cameras** allow amateur imagers to capture stunning detail on the Moon and planets

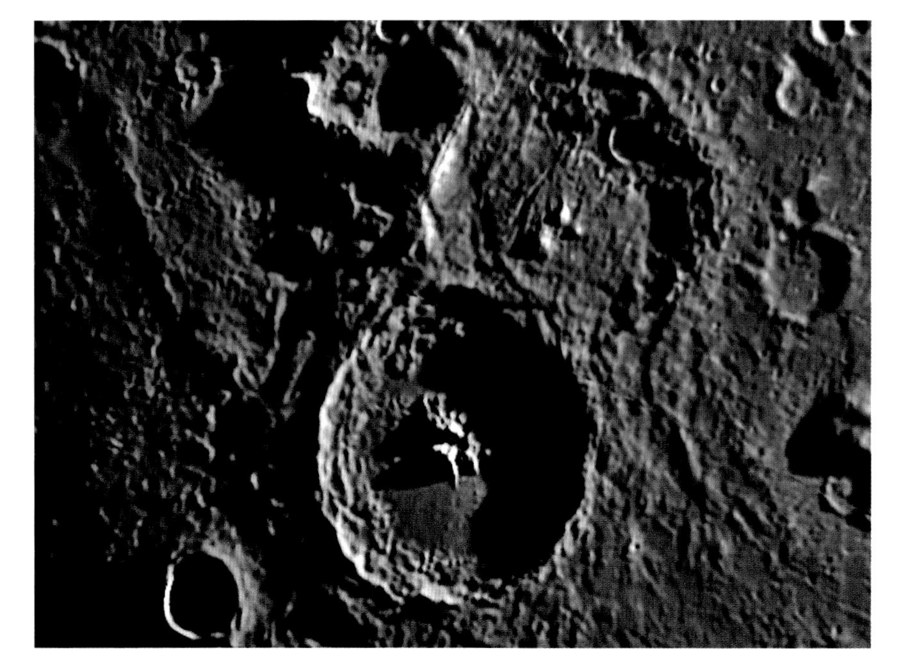

Colour renditions of astronomical targets are still possible with mono cameras through the use of imaging filters. An image is taken in three individual colours (normally red, green and blue – RGB) which are subsequently combined, with graphics software, into a full colour composite. It is common to combine the RGB image with a Luminance image – a high quality mono image that contains sharp spatial information about the target – to produce what is known as an LRGB image.

Specialist narrowband filters can also be used with mono astronomical CCD cameras to produce results that show the light of certain elements glowing within the scene. These can be combined in a similar way to the LRGB images described above to produce beautiful and scientifically useful results. Typical narrowband filters include hydrogen-alpha (Hα), oxygen-III (O-III), hydrogen-beta (Hβ) and sulphur-II (SII).

Nothing will test the accuracy of a telescope set-up more than long exposure deep-sky imaging. Variations in drive and mount quality combined with a need for precise polar alignment mean that the deep-sky imager can really struggle to keep stars looking round and sharp during long exposures. One solution is to hand the control of your telescope mount and drive system to a computer and use a technique known as autoguiding. Here, a star (the guide star) is picked from a star field image on screen. The computer is then instructed to keep that star in the same position on relative to the edges of the chip which it does by issuing commands to the telescope mount to compensate for any drift in the star's position.

Digital cameras and astronomical CCD cameras are typically (but not exclusively) used for imaging deep sky objects such as galaxies, clusters and nebulae. For Solar System targets such as the Sun, Moon and planets, an alternative approach is needed as they are badly affected by the effects of the Earth's atmosphere. As light from these extended objects passes though pockets of air of different temperatures and densities, their light is bent (refracted) by different amounts. To make matters worse still, these pockets can be moving quite rapidly. The net effect (called seeing) makes the target appear to wobble and distort. Amazing advances in Solar System imaging have come about through the use of high frame rate cameras capable of capturing hundreds or even thousands of frames over time scales of seconds. Within the captured frame list there will be a few images that catch the target in moments where the seeing may be generally fair to good. By pulling these good frames out of the frame list and stacking them together, a reasonable quality end result can be produced. Specialist software is used to sort and analyse the frame stack.

It was found, largely through experimentation, that certain types of inexpensive PC webcams were suitable for lunar and planetary imaging. However, beyond certain frame rates (typically 10 frames per second) these cameras use compression techniques to keep the data stream flowing, and this is not suitable for astronomical imaging. New dedicated high frame rate cameras are now becoming established in this area of astrophotography. These devices can be thought of as industrial strength webcams capable of delivering uncompressed capture rates in excess of 60 frames per second – perfect for capturing fleeting moments of good seeing.

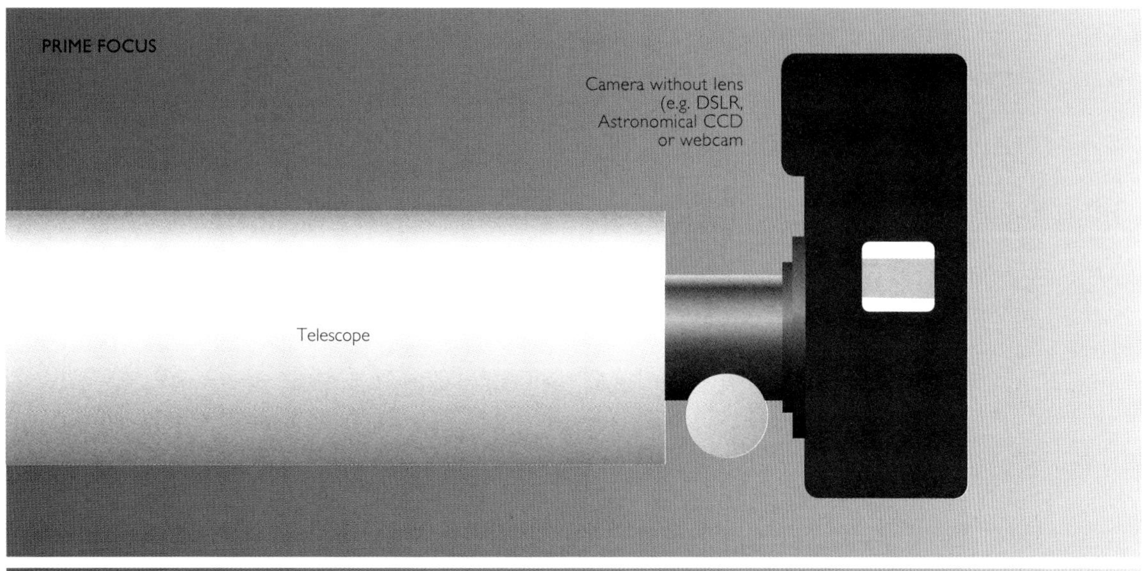

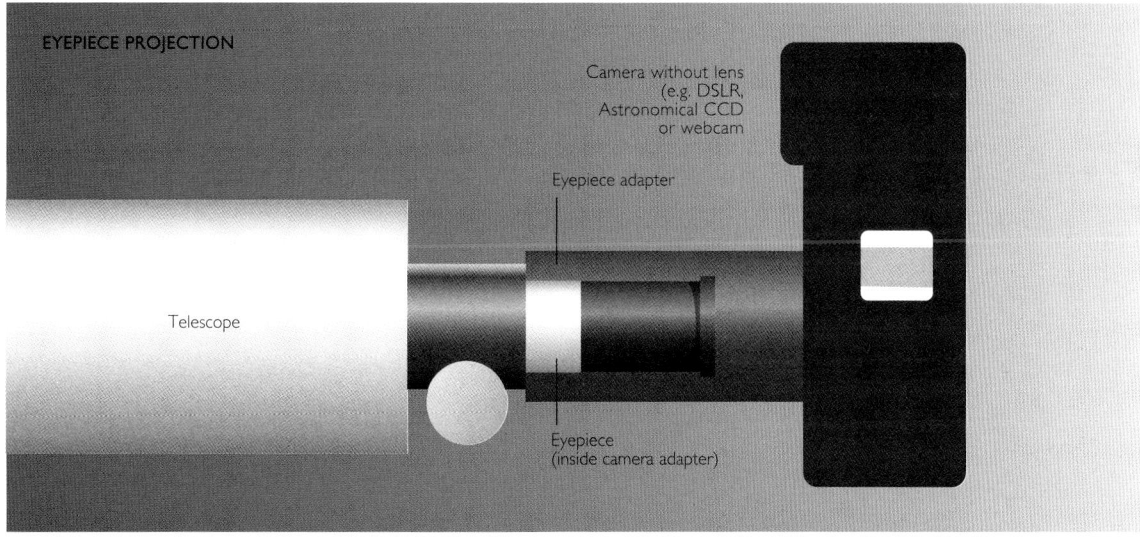

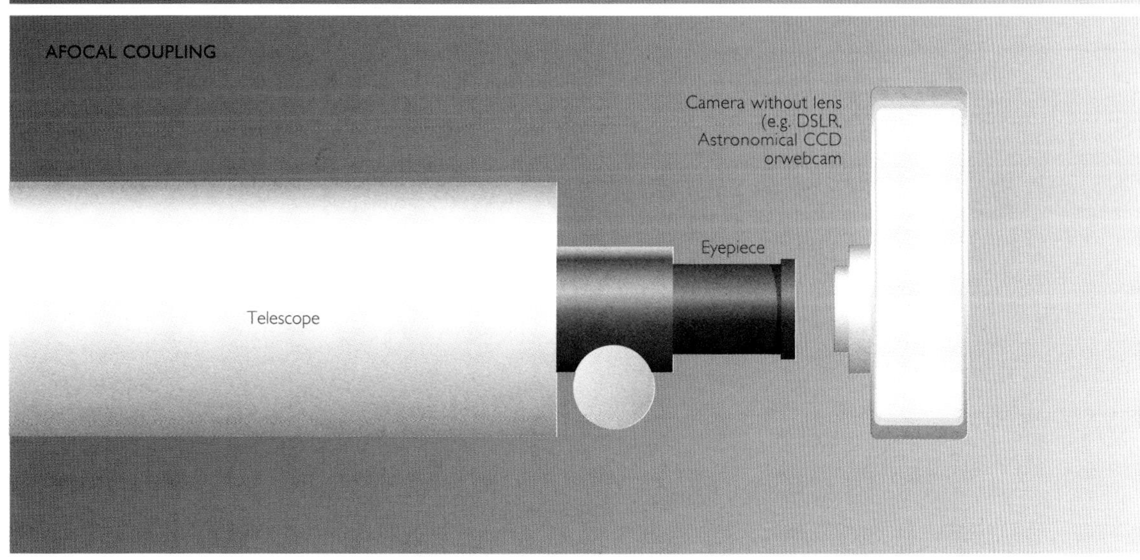

◄ **There are three main methods of coupling a camera** to a telescope. Prime focus is the purest method with the focus of the telescope falling directly onto the sensor in the camera. This method only works with cameras which can have their front lenses removed. Eyepiece projection requires a special adapter into which an eyepiece is inserted. The image from the eyepiece is projected directly onto the camera sensor. Increasing the distance between the eyepiece and the camera increases the magnification (image scale). Finally, afocal coupling applies to fixed lens cameras. Essentially it is the same as holding a regular camera against the eyepiece of a telescope. Specialist afocal coupling platforms are available for a wide variety of point-and-shoot cameras.

Observatories

The best views of the night sky are given on nights when the sky is really dark. These often occur in the depths of winter and can, depending on climate, offer rather uncomfortable viewing conditions. In addition, modern telescopes, especially if they are being used for imaging purposes, can require a significant amount of set-up time. If you find yourself carrying your telescope into and out of your garden each time the sky clears you also run a risk of dropping the instrument and incurring some rather expensive damage. There is a saying in astronomy that the best telescope is the one that gets used. If the sheer effort of setting up is putting you off observing then another solution has to be found and the usual solution is to build an observatory.

Amateur observatories come in all shapes and sizes. Some are complex structures requiring careful planning and a great deal of engineering knowledge. However, this doesn't need to be the case for all sites and a simple run-off shed will normally afford excellent protection and quick set-up on nights when the clouds are absent. Perhaps the most basic form of protection for a semi-permanent set-up is a simple tarpaulin cover. However, care is needed here because it is important that the cover can breath, allowing moisture out from around the telescope. Such covers are not suitable for long-term outdoor storage of telescopes.

A run-off shed is a more robust alternative and need not be a complex affair. The best designs comprise a shed of two parts that split in the middle and move away revealing the telescope within. Simple rails mounted on concrete can ease the process. Other popular designs add a run-off roof to the top of a rectangular box shed. When the time comes, the roof is simply pushed back to leave the shed completely open to the sky. Those with the necessary DIY skills can adapt off-the-shelf DIY shed designs for the purpose. For those who are less adept, run-off roof sheds can now be bought as ready-made kits, albeit at a premium. Whatever design is adopted, care must be taken to ensure that any seals are watertight to prevent wind-driven rain from gaining access to the telescope.

A more complex option, if you have the room and permission to do so, is to construct a dome. If it is done correctly, as well as being perfectly suited to astronomy, a dome can also make a very attractive garden feature. There are many designs of dome currently available from do-it-yourself kits to complete fibre glass dome buildings bought and delivered as a whole.

The larger the dome the more comfortable it will be inside allowing you to fit all manner of ancillary equipment to assist you. Networking technology now makes it possible to control amateur telescopes and imaging equipment from remote locations. In the simplest form this could mean operating a telescope in a cold dome from the comfort of a bedroom or study. Alternatively, if you live under light-polluted skies and are lucky enough to have a telescope permanently mounted out in the country it is perfectly feasible to control the telescope from the comfort of your own home over the telephone network.

If an observatory of your own is out of the question, there is a growing number of publicly available remote observatories. These offer you the ability to book time on a remotely operated telescope in a dark-sky location, which may, for example, be situated on the top of a mountain in a completely different country.

▼ **The author's run-off shed** for his 31-cm (12.5-in) reflector. The shed is in two parts which run back on rails in opposite directions.

▲ **Greg Parker's observatory** in the New Forest, Hampshire, in southern England. The wooden decking gives the base stability, while the upper part rotates manually.

◀ **The Mountain Skies Observatory** built by Curtis MacDonald, near Laramie, Wyoming, USA. It houses a 31-cm (12.5-in) Newtonian reflector with a 12-cm (5-in) refractor on the same mounting. The dome has a wooden framework covered with Masonite and painted with exterior latex. This photograph was taken by moonlight with Comet Hale-Bopp in the background.

Glossary

A

aberration of starlight The apparent displacement of a star from its true position in the sky due to the fact that light has a definite velocity (299,792.5 km/186,000 mi per second). The Earth is moving around the Sun, and thus the starlight seems to reach it 'at an angle'. The apparent positions of stars may be affected by up to 20.5 seconds of arc.

absolute magnitude The apparent magnitude that a star would have if it were observed from a standard distance of 10 parsecs, or 32.6 light years. The absolute magnitude of the Sun is +4.8.

absolute zero The lowest limit of temperature: −273.16°C (−459.688°F). This value is used as the starting point for the Kelvin scale of temperature, so absolute zero = 0 Kelvin.

absorption of light in space Space is not completely empty, as used to be thought. There is appreciable material spread between the planets, and there is also material between the stars; the light from remote objects is therefore absorbed and reddened. This effect has to be taken into account in all investigations of very distant objects.

absorption spectrum A spectrum made up of dark lines against a bright continuous background. The Sun has an absorption spectrum; the bright background or continuous spectrum is due to the Sun's brilliant surface (**photosphere**), while the dark absorption lines are produced by the solar atmosphere. These dark lines occur because the atoms in the solar atmosphere absorb certain characteristic wavelengths from the continuous spectrum of the photosphere.
see also **emission spectrum**

acceleration Rate of change of velocity. Conventionally, increase of velocity is termed acceleration; decrease of velocity is termed deceleration, or negative acceleration.

accretion disk Disk of matter that surrounds an astronomical object and through which material is transferred to that object. In many circumstances, material does not transfer directly from one astronomical object to another. Instead, the material is pulled into equatorial orbit about the object before accreting. Such material transfer systems are known as accretion disks. Accretion disks occur in protostellar clouds, close **binary star** systems and at the centre of galaxies.

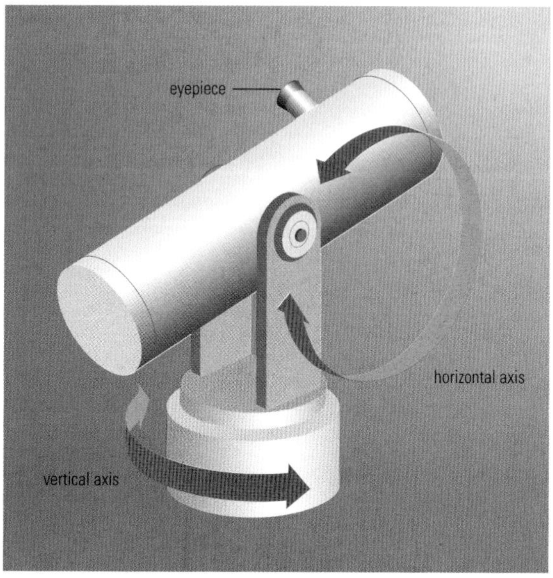

▶ altazimuth mounting This simple form of telescope mount allows free movement in both horizontal and vertical axes, but is not suitable for use with motor drives, unless they are computer controlled.

active region Region of enhanced magnetic activity on the Sun often, but not always, associated with **sunspots** and extending from the solar **photosphere** to the **corona**. Where sunspots occur, they are connected by strong magnetic fields that loop through the **chromosphere** into the low corona (coronal loops). Radio, ultraviolet and X-ray radiation from active regions is enhanced relative to neighbouring regions of the chromosphere and corona. Active regions may last from several hours to a few months. They are the sites of intense explosions, **flares**, which last from a few minutes to hours. The occurrence and location of active regions varies in step with the approximately 11-year **solar cycle**. Loops of gas seen as **filaments** or **prominences** are often suspended in magnetic fields above active regions.

aeropause A term used to denote that region of the **atmosphere** where the air density has become so slight as to be disregarded for all practical purposes. It has no sharp boundary, and is merely the transition zone between 'atmosphere' and 'space'.

airglow The faint natural luminosity of the night sky caused by reactions in the Earth's upper **atmosphere**.

air resistance Resistance to a moving body caused by the presence of **atmosphere**. An artificial **satellite** will continue in orbit indefinitely only if its entire orbit is such that the satellite never enters regions where air resistance is appreciable.

Airy disk The apparent size of a star's disk produced even by a perfect optical system. Since the star can never be focused perfectly, 84 per cent of the light will concentrate into a single disk, and 16 per cent into a system of surrounding rings.

albedo The reflecting power of a planet or other non-luminous body. A perfect reflector would have an albedo of 100 per cent.

altazimuth mounting Telescope mounting that has one axis (altitude) perpendicular to the horizon, and the other (azimuth) parallel to the horizon. An altazimuth (short for 'altitude–azimuth' and also expressed 'alt-az') mounting is much lighter, cheaper and easier to construct than an **equatorial mounting** for the same sized telescope, but is generally not capable of tracking the apparent motion of celestial objects caused by the Earth's rotation. Many amateur instruments with altazimuth mountings can therefore be used for general viewing but are not suitable for long photographic exposures.

Historically, large professional telescopes were built with massive equatorial mountings, which often dwarfed the instrument they held. The lightweight and simple nature of altazimuth mountings, combined with high-speed computers, has led to almost all modern instruments being built with altazimuth mountings. On these telescopes, computers are used to control the complex three-axis motions needed for an altazimuth mount to track the stars. The altitude and azimuth axes are driven at continuously varying rates but, in addition, the field of view will rotate during long photographic exposures, requiring an additional drive on the optical axis to counter field rotation. Some amateur instruments, especially **Dobsonian telescopes**, are now equipped with such three-axis drive systems, controlled by personal computers.

altitude Angular distance above an observer's horizon of a celestial body. The altitude of a particular object depends both on the location of the observer and the time the observation is made. It is measured vertically from 0° at the horizon, along the great circle passing through the object, to a maximum of 90° at the **zenith**. Any object below the observer's horizon is deemed to have a negative altitude. *See also* **azimuth**; **celestial coordinates**

Ångström unit The unit formerly used for measuring the wavelength of light and other electromagnetic vibrations. It is equal to 100 millionth of a cm. Visible light ranges from about 7500 Å (red) down to about 3900 Å (violet).

antenna A conductor, or system of conductors, for radiating or receiving radio waves. Systems of antennae coupled together to increase sensitivity, or to obtain directional effects, are known as antenna arrays, or as radio telescopes when used in **radio astronomy**.

apastron The point in the orbit of a **binary** system where the stars are at their furthest from each other. The closest point is known as the **periastron**.

aperture The clear diameter of the primary lens or mirror of a telescope. The aperture is the single most important factor in determining both the **resolving power** and the light-gathering ability of the instrument.

aphelion Point at which a Solar System body, such as a planet, asteroid or comet, moving in an elliptical orbit, is at its greatest distance from the Sun. For the Earth, this occurs around 4 July, when it lies 152 million km (94 million mi) from the Sun. *See also* **apsides; perihelion**

apogee The point in the orbit of the Moon or an artificial **satellite** at which the body is farthest from the Earth. The closest point is known as the **perigee**.

apsides (sing. apse) The two points in an elliptical **orbit** that are neartest to and farthest away from the primary body. The line joining these two points is the line of apsides. For an object orbiting the Sun the nearest apse is termed the **perihelion** and the farthest apse is the **aphelion**. For the Moon or an artificial satellite orbiting the Earth the apsides are the **perigee** and **apogee**. The components of a binary star system are at **periastron** when they are closest together and at **apastron** when farthest apart. For objects orbiting, say, the Moon or Jupiter, one sometimes sees terms such as periselinium or perijove. The terms pericentre and apocentre (or less commonly periapse and apoapse) are generally used for all cases where the primary is not the Sun, the Earth or a star, or when discussing elliptic motion in general.

arc, degree of One 360th of a full circle (360°).

arc, minute of One 60th of a degree of arc.

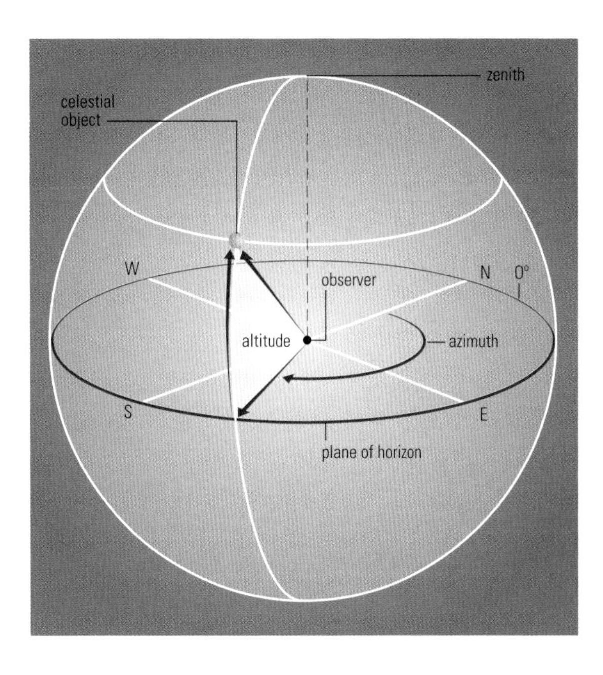

◀ **altitude** The altitude of a celestial object relative to an observer is measured on a scale of 0–90° from the observer's horizon to the zenith – the point directly overhead.

arc, second of One 60th of a minute of arc.

ashen light The faint luminosity of the night side of the planet Venus, seen when Venus is in the crescent phase. It is probably a genuine phenomenon rather than a contrast effect, but its cause is not fully understood.

asteroids The minor planets, most of which move around the Sun between the orbits of Mars and Jupiter. Several thousands of asteroids are known; the largest is Ceres (now reaclassified as a dwarf planet), whose diameter is 1003 km (623 mi). Only one asteroid (Vesta) is ever visible with the naked eye.

astrolabe A disk-shaped early astronomical instrument, for measuring positions on the **celestial sphere**, equipped with sights for observing celestial objects. The classical form, known as the planispheric astrolabe, originated in ancient Greece, attained its greatest refinement in the hands of medieval Arab astronomers and reached the Christian West in about the tenthth century AD. The basic form consists of two flat disks, one of which (the *mater*) is fixed and represents the observer on the Earth. The other disk (the *rete*) is movable and represents the celestial

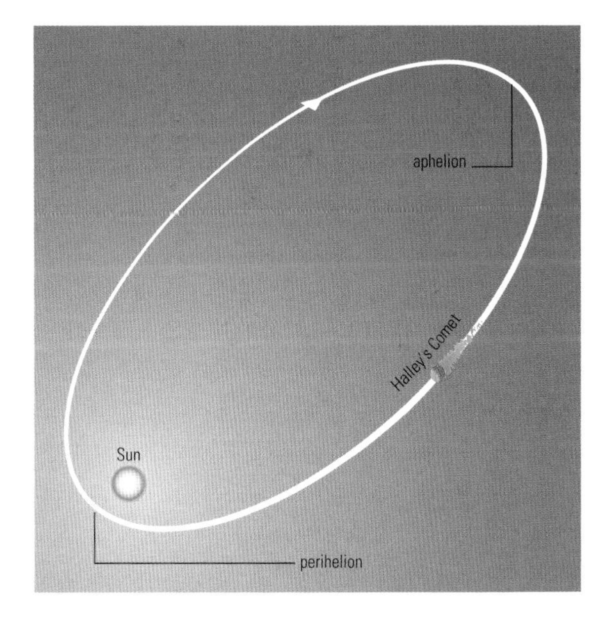

◀ **aphelion** Halley's Comet has a highly eccentric orbit: its aphelion – its greatest distance from the Sun – is 35.295 AU (which is roughly half-way between the mean orbits of Uranus and Pluto), while its perihelion (its closest approach to the Sun) is just 0.587 AU, well within Venus' orbit.

sphere. The **altitudes** and **azimuths** of celestial bodies can be read. A pointer, called an alidade, can be used for measuring altitudes when the instrument is suspended on a length of string.

Given the latitude, the date and the time, the observer can read off the altitude and azimuth of the Sun, the bright stars and the planets, and measure the altitude of a body and find the time. Its function of representing the night sky for a certain latitude at various times is reproduced in the modern **planisphere**. The astrolabe can also be used as an analogue computer for many problems in spherical trigonometry, and it was even used for terrestrial surveying work such as determining the altitude of a tower.

astrology A pseudoscience which claims to link the positions of the planets with human destinies. It has no scientific foundation.

astrometric binary A **binary star** where only one component is seen. This star is observed to 'wobble' in a periodic manner as it moves across the sky, a movement caused by the gravitational pull of the unseen companion on the brighter component.

astronomical unit The distance between the Earth and the Sun. It is equal to 149,597,900 km, usually rounded off to 150 million km (93 million mi).

astrophysics The application of the laws and principles of physics to all branches of astronomy. It has often been defined as 'the physics and chemistry of the stars'.

atmosphere The gaseous mantle surrounding a planet or other body. It can have no definite boundary, but merely thins out until the density of matter in it is no greater than that of surrounding space.

atom The smallest unit of a chemical element which retains its own particular character. (Of the 92 elements known to occur naturally, hydrogen is the lightest and uranium is the heaviest.)

aurorae (northern and southern lights) The Aurora Borealis in the northern hemisphere and the Aurora Australis in the southern. They are glows are caused by charged particles emitted by the Sun exciting molecules in the Earth's upper atmosphere. Because the particles are electrically charged, they tend to be attracted towards the magnetic poles, so aurorae are seen best at high latitudes.

autoguider Electronic device that ensures a telescope accurately tracks the apparent movement of a celestial object across the sky. Although a telescope is driven about its axes to compensate for the rotation of the Earth, it is still necessary to make minor corrections, particularly during the course of a long observation. An autoguider achieves this by using a photoelectric device, such as a quadrant photodiode or a CCD, to detect any drifting of the image. The light-sensitive surface of the detector is divided into four quarters, with light from a bright guide star in the field of view focused at the exact centre. If the image wanders from the centre into one of the four quadrants, an electrical current is produced. By knowing from which quadrant this has emanated, it is possible to correct the telescope drive to bring the image back to the exact centre again.

azimuth The horizontal direction or bearing of a celestial body, reckoned from the north point of the observer's horizon. Because of the Earth's rotation, the azimuth of a body is changing all the time.

B

background radiation Very weak microwave radiation coming from space, continuously from all directions and indicating a general temperature of 3°C above **absolute zero**. It is believed to be the last remnant of the Big Bang, in which the universe was created about 13,700 million years ago. The Cosmic Background Explorer satellite (COBE) has detected slight variations in it.

Baily's beads Brilliant points seen along the edge of the Moon's disk at a total solar eclipse, just before totality and again just after totality has ended. They are caused by the Sun's light shining through valleys between mountainous regions on the limb of the Moon.

barycentre The centre of gravity of the Earth–Moon system. Because the Earth is 81 times more massive than the Moon, the barycentre lies within the terrestrial globe.

binary star Double star in which the two components are gravitationally bound to each other and orbit their common centre of mass. More than half of stars observed are binary or belong to multiple systems, with three or more components. Other double stars are optical doubles, which result when two stars appear to be close because they lie almost on the same line of sight as viewed from Earth, but in reality lie at a vast distance from each other. Binary systems are classified in a variety of ways, including the manner in which their binary nature is known. *See also* **close binary**; **common proper motion binary**; **contact binary**; **eclipsing binary**; **Roche lobe**; **spectroscopic binary**; **spectrum binary**; **visual binary**.

black hole Object that is so dense and has a gravitational field so strong that not even light or any other kind of radiation can escape: its escape velocity exceeds the speed of light. Black holes are predicted by Einstein's theory of General Relativity, which shows that if a quantity of matter

▼ binary star The two components of a binary system move in elliptical orbits around their common centre of gravity (G), which is not half-way between them but nearer to the more massive component. In accordance with Kepler's second law of motion, they do not travel at uniform rates. In (2) they are greatly separated and moving slowly. In (6) they are closer and moving more rapidly.

1 G

2 G

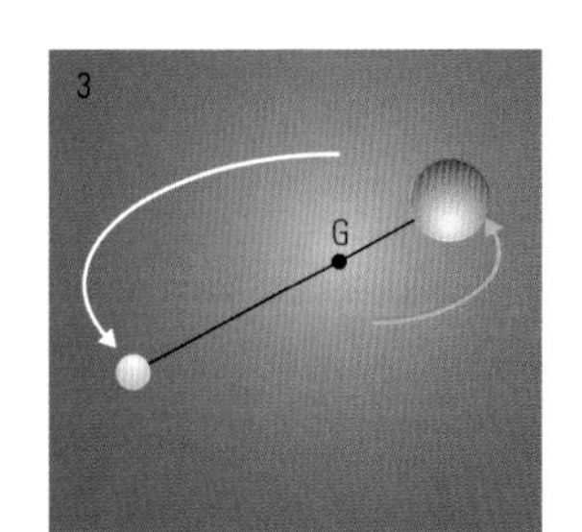

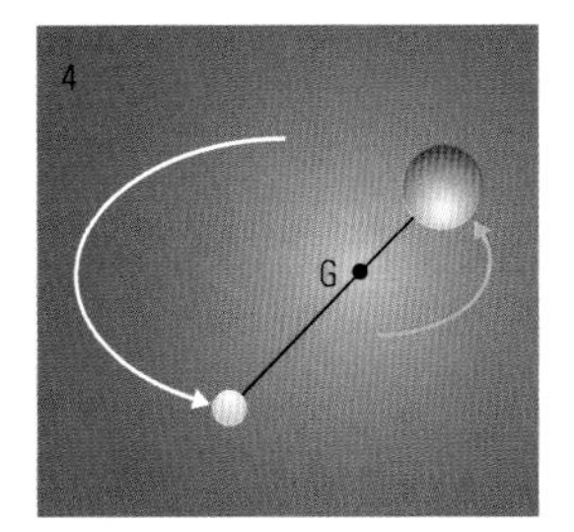

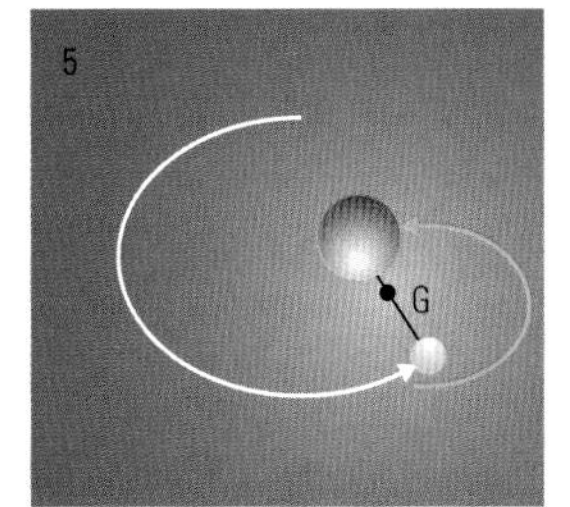

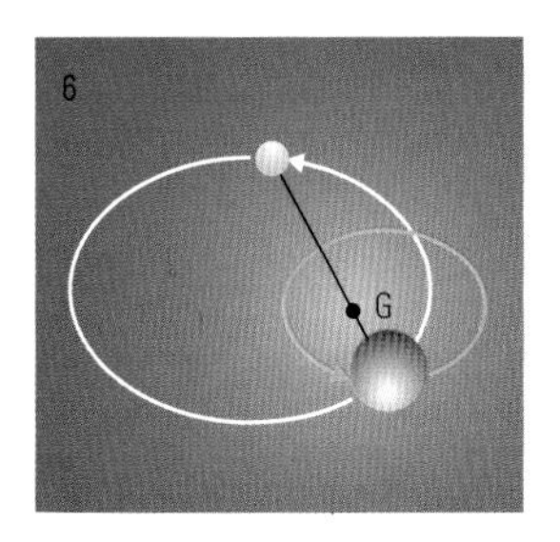

is compressed within a critical radius, no signal can ever escape from it. Thus, although there are many black hole candidates, they cannot be observed directly. Candidates are inferred from the effects they have upon nearby matter. There are three classes of black hole: stellar, primordial (or mini) and supermassive.

Bode's Law An empirical relationship between the distances of the planets from the Sun, discovered by J. D. Titius in 1772 and made famous by J. E. Bode. The law seems to be fortuitous, and without any real scientific basis.

bolide A brilliant **meteor**, which may explode during its descent through the Earth's atmosphere.

bolometer A very sensitive radiation detector, used to measure slight quantities of radiation over a very wide range of wavelengths.

brown dwarf Star with mass between about 0.01 solar mass and 0.08 solar mass; its core temperature does not rise high enough to start thermonuclear reactions. It is luminous, however, because it slowly shrinks in size and radiates away its gravitational energy. As its surface temperature is below the 2500 K lower limit for **red dwarfs**, it is known as a brown dwarf.

C

caldera Large, roughly circular volcanic depression formed by collapse over an evacuated magma chamber. Small calderas (less than 5 km/3 mi in diameter) are common at the crests of terrestrial basaltic and andesitic volcanoes. Calderas as large as 75 km (47 mi) across have formed on Earth during volcanic ash flow eruptions. The summit caldera of the Martian volcano Olympus Mons, the largest volcano in the Solar System, is about 80 km (50 mi) across.

Caldwell catalogue A list of 109 bright nebular objects, none of which is included in the Messier catalogue.

carbon-nitrogen cycle The stars are not 'burning' in the usual sense of the word; they are producing their energy by converting hydrogen into helium, with release of radiation and loss of mass. One way in which this conversion takes place is by a whole series of reactions, involving carbon and nitrogen as catalysts. It used to be thought that the Sun shone because of this process, but modern work has shown that another cycle, the so-called proton-proton reaction, is more important in stars of solar type. The only stars which do not shine because of the hydrogen-into-helium process are those at a very early or relatively late stage in their evolution.

Cassegrain reflector A type of reflecting telescope (see **reflector**) in which the light from the object under study is reflected from the main mirror to a convex secondary, and thence back to the eyepiece through a hole in the main mirror.

celestial coordinates Reference system used to define the positions of points or celestial objects on the **celestial sphere**. A number of systems are in use, depending on the application.

Equatorial coordinates are the most commonly used and are the equivalent of latitude and longitude on the Earth's surface. **Declination** is a measure of an object's angular distance north or south of the **celestial equator**, values north being positive and values south being negative. **Right ascension**, or R.A., equates to longitude and is measured in hours, minutes and seconds eastwards from the **First Point of Aries**. Hour angle and polar distance can also be used as alternative measures. The horizontal coordinate system uses the observer's horizon as its plane of reference, measuring the **altitude** (the angular measure of the object above the horizon) and **azimuth** (bearing measure westwards around the horizon from north). **Ecliptic** coordinates are based are based upon the plane of the ecliptic and use the measures of celestial latitude and celestial longitude. Celestial latitude is measured in degrees north and south of the ecliptic, while celestial longitude is measure in degrees eastwards along the ecliptic from the First Point of Aries. The galactic coordinate system takes the plane of the Galaxy and the galactic centre as its reference points.

celestial sphere An imaginary sphere surrounding the Earth, concentric with the Earth's centre. The Earth's axis indicates the positions of the celestial poles; the projection of the Earth's equator on to the celestial sphere marks the celestial equator.

centrifuge A motor-driven apparatus with long arm, at the end of which is a cage. When people (or animals) are put into the cage, and revolved and rotated at high speeds, it is possible to study effects comparable with the accelerations experienced in spacecraft. Astronauts are given tests in a centrifuge during training.

Cepheid variable An important type of variable star, Cepheids have short periods of from a few days to a few weeks, and are regular in their behaviour. It has been found that the period of a Cepheid is linked with its real luminosity: the longer the period, the more luminous the star. From this it follows that once a Cepheid's period has been measured, its distance can be worked out. Cepheids are luminous stars, and may be seen over great distances; they are found not only in our Galaxy, but also in external galaxies. The name comes from Delta Cephei, the brightest and most famous member of the class.

Charge-Coupled Device (CCD) An electronic imaging device which is far more sensitive than a photographic plate, and is now replacing photography for most branches of astronomical research.

chromatic aberration A defect found in all lenses, resulting in the production of 'false colour'. It is due to the fact that light of all wavelengths is not bent or refracted equally; for example, blue light is refracted more strongly than red, and so is brought to focus nearer the lens. With an astronomical telescope, the object-glass is made up of several lenses composed of different kinds of glass. In this way chromatic aberration may be reduced, although it can never be entirely cured.

chromosphere The part of the Sun's atmosphere lying above the bright surface or **photosphere**, and below the outer **corona.** It is visible with the naked eye only during total solar **eclipses**, when the Moon hides the photosphere; but by means of special instruments it may be studied at any time.

circular velocity The velocity with which an object must move, in the absence of air resistance, in order to describe a circular orbit around its primary.

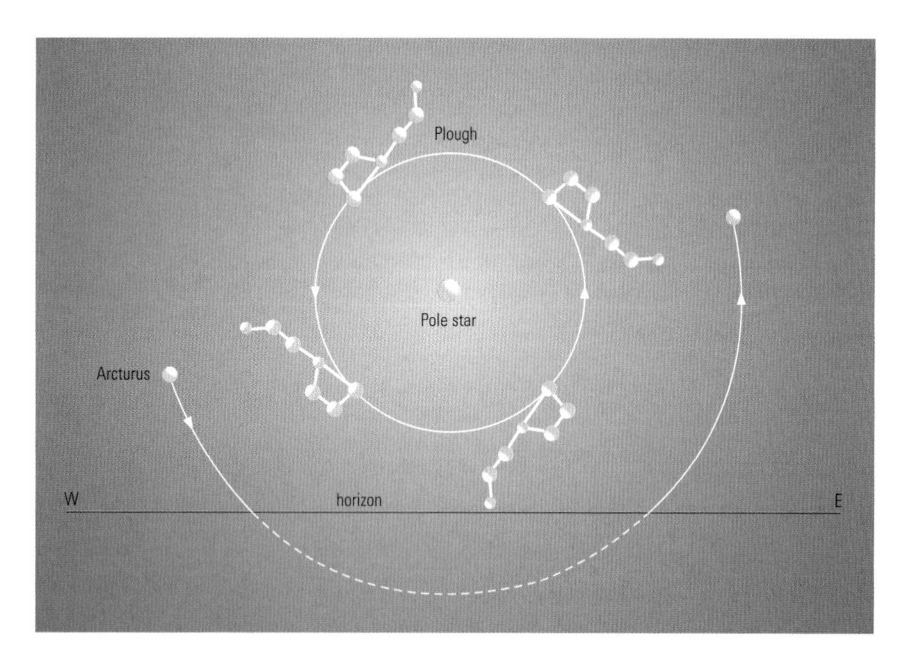

▲ circumpolar star Provided it lies sufficiently close to the celestial pole, a star may describe a complete circle once per sidereal day without disappearing below the observer's horizon. In the example shown, Alkaid, the end star on the Plough's 'handle', is such a circumpolar star, while Arcturus is not.

circumpolar star Star that never sets below the observer's horizon. For a star to be circumpolar at a given latitude its declination must be greater than 90° minus that latitude. For example, if the observer's latitude is 52°, by subtracting 52° from 90° we get 38°. Any star with a declination greater then 38° will therefore be circumpolar for that observer. At the equator, no stars are circumpolar whereas from the poles, all visible stars are circumpolar.

close binary A **binary star** that has an orbital period of less than 30 years, implying that the two components are less than about 10 AU apart. Because of this proximity, most close binaries are **spectroscopic binaries** and/or **eclipsing binaries**. **Mass transfer** occurs at some stage in most close binaries, profoundly affecting the evolution of the component stars. If the two components in a close binary do not fill their **Roche lobes**, the system is a detached binary. In a semidetached binary one star fills its Roche lobe and mass transfer occurs. In a **contact binary** both stars fill their Roche lobes.

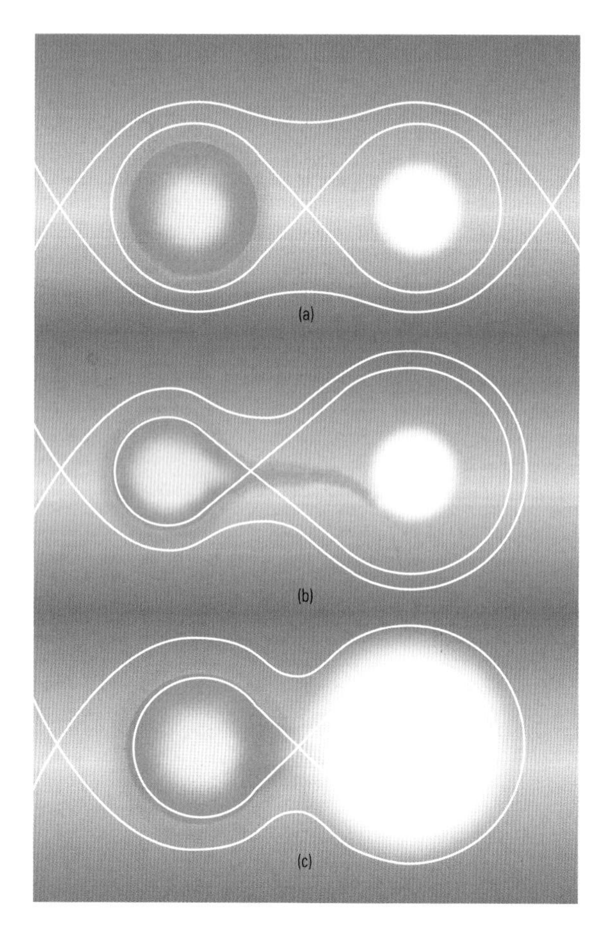

▶ close binary In close binary systems, the two stars may be completely detached (a). Semi-detached systems have one star whose atmosphere fills its Roche lobe, leading to mass transfer to the other component (b). Where both stars fill their Roche lobes, the pair share a common atmosphere and the system becomes a contact binary (c).

clusters, stellar A collection of stars which are genuinely associated. An open cluster may contain several hundred stars, usually together with gas and dust; there is no particular shape to the cluster. Globular clusters contain thousands of stars, and are regular in shape; they are very remote, and lie near the edge of the Galaxy. Both open and globular clusters are also known in external galaxies. Moving clusters are made up of widely separated stars moving through space in the same direction and at the same velocity. (For example, five of the seven bright stars Ursa Minor are members of the same moving cluster.)

collimation Process of aligning the optical elements of a telescope. Certain sealed instruments, such as refractors, are generally collimated at the factory and need never be adjusted, but most telescopes with mirrors do require occasional re-alignment, especially amateur **Newtonian telescopes** built with Serrurier trusses where the secondary and focuser assembly are routinely separated from the primary for transport. The procedure for collimating a Newtonian telescope is usually carried out in two steps, first aligning the primary to direct light to the centre of the secondary, then adjusting the secondary to direct the light cone down the centre of the focuser. The term is also used to describe an optical arrangement of lenses or mirrors used to bring incoming light rays into a parallel beam before they enter an instrument such as a **spectroscope** or X-ray telescope.

colour index A measure of a star's colour and hence of its surface temperature. The ordinary or visual **magnitude** of a star is a measure of the apparent brightness as seen with the naked eye; the photographic magnitude is obtained by measuring the apparent size of a star's image on a photographic plate. The two magnitudes will not generally be the same, because in the old standard plates red stars will seem less prominent than they appear to the eye. The difference between visual and photographic magnitude is known as the colour index. The scale is adjusted so that for a white star, such as Sirius, colour index = 0. A blue star will have negative colour index; a yellow or red star will have positive colour index.

colures Great circles on the **celestial sphere**. The equinoctial colure, for example, is the **great circle** which passes through both celestial poles and also the **First Point of Aries** (vernal equinox), i.e. the point where the **ecliptic** intersects the **celestial equator**.

coma (1) The hazy-looking patch surrounding the nucleus of a **comet**. (2) The blurred haze surrounding the images of stars on a photographic plate, because of optical defects in the equipment used to make the image.

comet A member of the Solar System, moving around the Sun in an orbit which is generally highly eccentric. It is made up of relatively small particles (mainly ices) together with tenuous gas: the most substantial part of the comet is the nucleus, which may be several kilometres in diameter. A comet's tail always points more or less away from the Sun, due to the effects of **solar wind**. There are many comets with short periods, all of which are relatively faint; the only bright comet with a period of less than a century is Halley's. A long-period comet is one for which the interval between successive perihelion returns is greater than 200 years. About 750 long-period comets are known within this category. The most brilliant comets have periods so long that their return cannot be predicted. See also **sun-grazers**.

common proper motion binary A **binary star** whose elements are seen as distinct objects that move across the sky together. No orbital motion is observed because they lie so far apart that their orbital period is very long.

conic section Curve that is obtained by taking a cross-section across a circular cone. The curve will be a circle, ellipse, parabola or hyperbola depending on the angle at which the cross-section is taken. The significance for astronomy is that these curves are also the possible paths of a body moving under the gravitational attraction of a primary body.

conjunction Alignment of two Solar System bodies with the Earth so that they appear in almost the same position in the sky as viewed from Earth. The **inferior planets**, Mercury and Venus, can align in this way either between the Sun and the Earth, when they said to be at *inferior conjunction*, or when they lie on the opposite side of the Sun to the Earth, and are at *superior conjunction*. The superior planets can only come to superior conjunction. The strict definition of conjunction is when two bodies have the same celestial longitude as seen from Earth and because of the inclination of the various planetary orbits to the ecliptic, exact coincidence of position is rare. The term is also used to describe the apparent close approach in the sky of two or more planets, or of the Moon and one or more planets.

constellation A group of stars named after a living or a mythological character, or an inanimate object. The names are highly imaginative, and have no real significance. Neither is a constellation made up of stars that are genuinely associated with one another; the individual stars lie at very different distances from the Earth, and merely happen to be in roughly the same direction in space. The International Astronomical Union currently recognizes 88 separate constellations.

contact binary A **binary star** in which both components fill their **Roche lobes**. *See* **close binary**.

corona The outermost part of the Sun's atmosphere; it is made up of very tenuous gas at a very high temperature, and is of great extent. It is visible to the naked eye only during total solar **eclipses**.

coronagraph A type of **telescope** designed to view the solar **corona** in ordinary daylight; ordinary telescopes are unable to do this, partly because of the sunlight scattered across the sky by the Earth's atmosphere, and partly because of light which is scattered inside the telescope – mainly by particles of dust. The coronagraph was invented by the French astronomer Bernard Lyot.

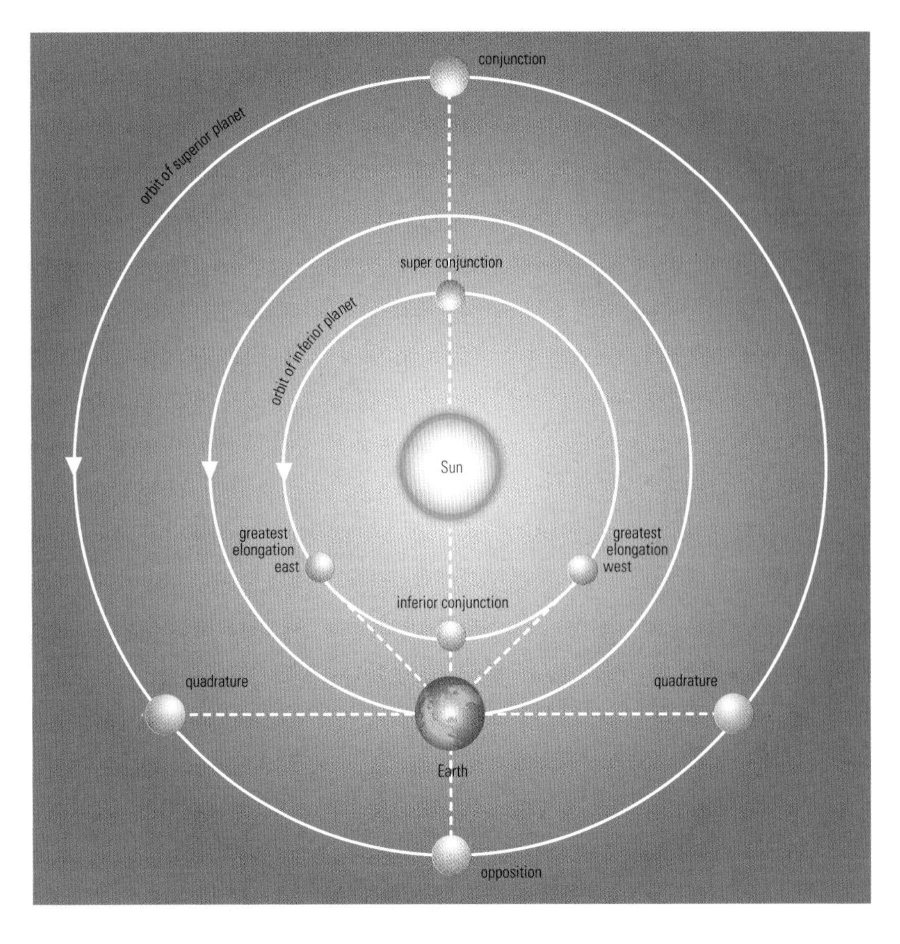

▲ **conjunction** Superior planets come to conjunction with the Sun when on the far side of it, as seen from Earth (and are therefore lost from view). The inferior planets, Mercury and Venus, can undergo conjunction at two stages in their orbit. At superior conjunction, they lie on the far side of the Sun from Earth, while at inferior conjunction they are between the Sun and the Earth. Under certain circumstance, Mercury and Venus can transit across the Sun's disk at inferior conjunction.

coronal mass ejection (CME) Transient ejection into interplanetary space of plasma and magnetic fields from the solar **corona**, seen in sequential images taken with a **coronagraph**. A coronal mass ejection expands away from the Sun at supersonic speeds up to 1200 km/s (750 mi/s), becoming larger than the Sun in a few hours and removing up to 50 million tonnes of material. CMEs are often associated with eruptive **prominences** in the chromosphere, and sometimes with solar **flares** in the lower corona. CMEs are more frequent at the maximum of the **solar cycle**, when they occur at a rate of about 3.5 events per day. They can, however, occur at any time in the solar cycle, and unlike coronal streamers, they are not confined to equatorial latitudes at solar minimum. CMEs are most readily seen when directed perpendicular to the line of sight, expanding outwards from the solar limb; Earth-directed events are seen as diffuse, expanding rings described as halo coronal mass ejections. Earth-directed CMEs can cause intense **geomagnetic storms**, and trigger enhanced auroral activity. Coronal mass ejections produce intense shock waves in the **solar wind**, and accelerate vast quantities of energetic particles. A large CME may release as much as 10^{25} Joule of energy, comparable to that in a solar flare. Like solar flares, CMEs are believed to result from the release of stored magnetic stress.

cosmic rays High-velocity particles reaching the Earth from outer space. The heavy cosmic-ray primaries are broken up when they enter the top part of the Earth's atmosphere, and only the secondary particles reach ground level. There is still

▶ coudé focus In this optical configuration, mirrors direct light from the telescope along the polar axis to a fixed observing position. This has several advantages for observations that require use of heavy or bulky detectors.

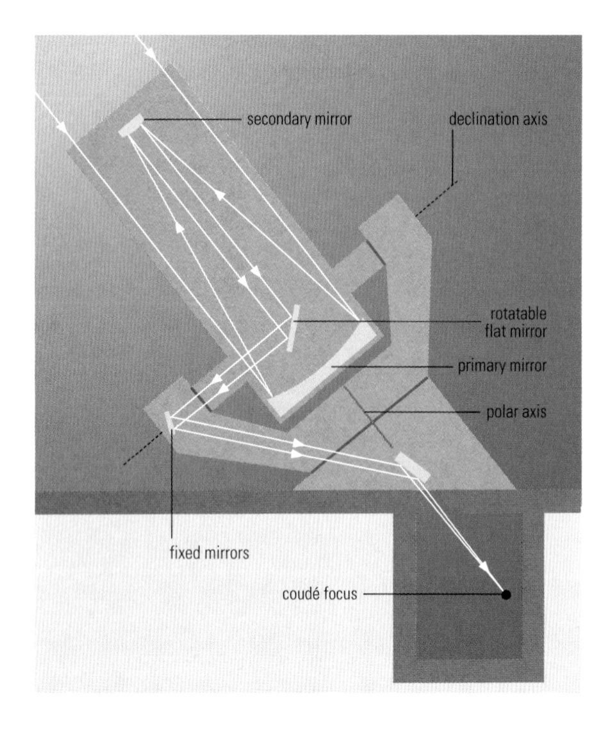

much doubt whether cosmic radiation will prove to be a major hazard for astronauts on long space flights.

cosmology The study of the universe as a whole; its nature, origin, evolution, and the relations between its various parts.

coudé focus Focal point of an equatorially mounted telescope in which the light path is directed along the polar axis to a fixed position that remains stationary, regardless of the orientation of the telescope. Instruments such as high-dispersion **spectroscopes**, which are too large or heavy to be mounted on a moving telescope, may be placed at the coudé focus, which is often located in an adjacent room or even separate floor. A series of auxiliary mirrors is used to direct the converging beam of light from the secondary mirror, down the hollow polar axis of the telescope mount, to the slit of the spectrograph. The word coudé is French for 'elbow', and describes the bending of the light path.

counterglow See **gegenschein**.

Crab Nebula The remnant of a **supernova** observed in 1054; an expanding cloud of gas, approximately 6000 **light years** away, according to recent measurements. It is important because it emits not only visible light but also radio waves and **X-rays**. Much of the radio emission is due to **synchrotron radiation** (that is, the acceleration of charged particles in a strong magnetic field). The Crab Nebula contains a **pulsar**, the first to be identified with an optical object.

culmination The time when a star or other celestial body reaches the observer's **meridian**, so that it is at its highest point (upper culmination). If the body is circumpolar, it may be observed to cross the meridian again 12 hours later (lower culmination). With a non-circumpolar object, lower culmination cannot be observed as, at that point, the object is then below the horizon.

cybernetics The study of methods of communication and control that are common to machines and to living organisms.

▼ differential rotation In a fluid body, such as a star or gas giant planet, the equatorial regions rotate more rapidly than the poles. As shown, a consequence of this is that a set of points lined up on the central meridian will become spread out in longitude over the course of a rotation. Points close to the equator will return to the central meridian earlier than those near the poles.

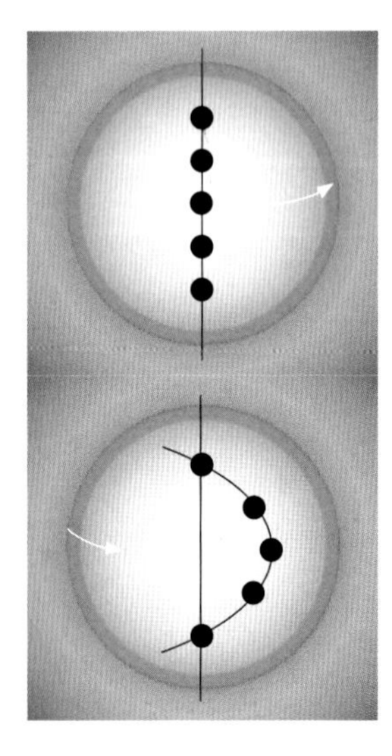

D

day In everyday language, a day is the amount of time it takes for the Earth to spin once on its axis. A sidereal day (see **sidereal time**) is the rotation period measured with reference to the stars (23 hours 56 minutes 4.091 seconds). A solar day is the time interval between two successive noons; the length of the mean solar day is 24 hours 3 minutes 56.555 seconds – rather longer than the sidereal day, since the Sun is moving eastwards along the **ecliptic**. The civil day is, of course, taken to be 24 hours.

declination The angular distance of a celestial body north or south of the celestial equator. It may be said to correspond to latitude on the surface of the Earth.

density The mass of a substance per unit volume. Taking water as one, the density of the Earth is 5.5.

density wave Theoretical explanation for the common occurence of spiral arms in **galaxies**, when any arms that are simply lines of material should wind up in a relatively short cosmic time and lose their identity. In density wave theory spiral arms are patterns through which stars and gase move, with the pattern moving at a different speed. The arms would appear enhanced because the extra mass in them keeps stars and gas there longer than they would be in the absence of the extra density, and star formation is triggered in molecular clouds by the denser environment. This theory probably applies to 'grand design' spiral galaxies, many of which have close companions whose gravitational **perturbations** would excite the density wave.

dichotomy The half-phase of Mercury, Venus or the Moon.

differential rotation Rotation of a gaseous or fluid body at a rate that differs with latitude: the equatorial regions rotate more quickly, and revolve faster, than higher-latitude polar ones. The differential rotation of the Sun persists from the **photosphere** down to the base of its convective zone (*see* **solar interior**). Differential rotation is also shown by the **giant planets**. A solid body like the Earth cannot undergo differential rotation: it must rotate so that the angular velocity and period of rotation is the same everywhere.

diffraction rings Concentric rings surrounding the image of a star as seen in a **telescope**. They cannot be eliminated, since they are due to the wave-motion of light. They are most evident in small instruments.

direct motion Bodies which move around the Sun in the same sense as the Earth are said to have direct motion. Those which move in the opposite sense have retrograde motion. The term may also be applied to satellites of the planets. No planet or asteroid with retrograde motion is known, but there are various retrograde satellites and comets. The terms are also used with regard to the apparent movements of the planets in the sky. When moving eastwards against the stars, the planet has direct motion; when moving westwards, it is retrograde.

diurnal motion The apparent daily rotation of the sky from east to west. It is due to the real rotation of the Earth from west to east.

Dobsonian telescope **newtonian telescope** equipped with a low, stable **altazimuth mounting**; its designer, amateur astronomer John Lowry Dobson (1915–), refers to it as a sidewalk telescope. For amateur instruments, the Dobsonian (or 'Dob') has advantages of economy and is stable enough for useful observation. The design can be applied to large telescopes, and has led to modern amateurs using instruments with far greater apertures – up to a metre (3 ft) or even more – than were common when it first emerged in the 1950s. A Dobsonian requires no great skill to build, and can be made from very simple parts. The basic features of the mounting are a ground board (or baseplate), topped by Teflon pads on which rests a cradle (or rocker-box). A single loose bolt in the centre of the ground board keeps the cradle in place but allows it to turn in azimuth. The sides of the cradle have Teflon-lined semi-circular cut-outs, to accommodate rings bolted to the side of a box or tube containing the primary mirror. The secondary may be in the same tube, or can be attached, with other components, before observing. Typically, the components are all made from lightweight materials, and the instrument can be disassembled into separate modules for added portability.

Doppler effect The apparent change in the wavelength of light caused by the motion of the source relative to the observer. When a light-emitting body is approaching the Earth, more light waves per second enter the observer's eye than would be the case if the object were stationary; therefore, the apparent wavelength is shortened, and the light seems 'too blue'. If the object is receding, the wavelength is apparently lengthened, and the light is 'too red'. For ordinary velocities the actual colour changes are very slight, but the effect shows up in the spectrum of the object concerned. If the dark lines are shifted towards the red or long wave, the object must be receding; and the amount of the shift to the red is a key to the velocity of recession. Apart from the galaxies in our **Local Group**, all external systems show red shifts, and this is the observational proof that the universe is expanding. The Doppler principle also applies to radiations at radio wavelengths.

double star A star which is made up of two components. Some doubles are optical; that is to say, the components are not truly associated, and simply happen to lie in much the same direction as seen from Earth. Most double stars, however, are physically associated or **binary** systems.

drive Mechanism by which a telescope is moved quickly (slewed) from one part of the sky to another, or moved slowly (tracked) to follow celestial objects in their sidereal motion. A tracking motor can be use to turn the telescope about the polar axis of an **equatorial mounting** at the sidereal rate of one revolution per $23^h\ 56^m$. The most advanced telescope drives use computer control. In such systems, the computer rapidly and repeatedly calculates the required position of the telescope, determining for the time given by a quartz clock the hour angle and altitude of the object that it has been instructed to follow. The computer reads the telescope position from encoders attached to the two mounting axes (*see also* **GO TO telescope**). It is possible in principle to attain a pointing accuracy of 10 by these methods. The final adjustment of the telescope's position is made by visual inspection of the field of view, or by measuring the position of a guide star in the field, using a CCD camera. Both professional and large amateur telescopes are now commonly equipped with CCD-based **autoguiders**.

dwarf nova A member of a class of irregular variable stars whose light curves resemble those of **novae**. Their **luminosity** stays the same for long periods, then rapidly increases, and finally returns slowly to normal. Dwarf novae are **close binary** stars in which one component is a **white dwarf** and the other is a subgiant or dwarf.

E

Earthshine The faint luminosity of the night hemisphere of the Moon, caused by the Sun's light reflected on to the Moon from the Earth.

eccentricity (symbol *e*) One of the **orbital elements**, it describes the degree of **elongation** of an elliptical orbit. The eccentricity is obtained by dividing the distance between the foci of the **ellipse** by the length of the major axis. A circular orbit has $e = 0$; a parabolic orbit is the extreme case of an ellipse, with $e = 1$.

eclipses These are of two kinds: solar and lunar. (1) A solar eclipse is caused by the Moon passing in front of the Sun. By coincidence, the two bodies appear almost equal in size. When the alignment is exact, the Moon covers up the Sun's bright disk for a brief period, either totally or partially (never more than about eight minutes; usually much less). When the eclipse is total the Sun's surroundings – the **chromosphere**, **corona** and **prominences** – may be seen with the naked eye (though you should never look directly at the Sun). If the Sun is not fully covered, the eclipse is partial, and the spectacular phenomena of totality are not seen. If the Moon is near its greatest distance from the Earth (see **apogee**) it appears slightly smaller than the Sun, and at central alignment a ring of the Sun's disk is left showing around the body of the Moon; this is an annular eclipse, and again the phenomena of totality are not seen. (2) A lunar eclipse is caused when the Moon passes into the shadow cast by the Earth; it may be either total or partial. Generally, the Moon does not vanish, as some sunlight is refracted on to it by way of the ring of atmosphere surrounding the Earth.

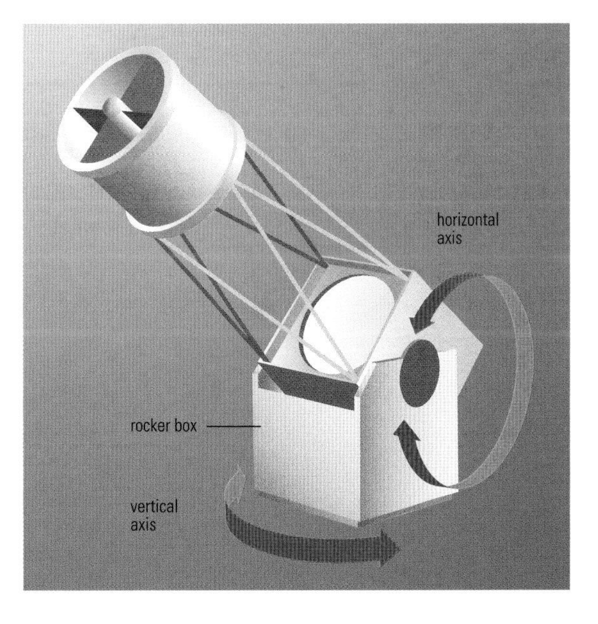

◀ **Dobsonian telescope** This simple altazimuth mounting has enabled many amateur astronomers to construct large-aperture instruments comparatively cheaply. The Dobsonian offers considerable advantages of portability.

eclipsing binary (or eclipsing variable) A **binary star** made up of two components moving around their common centre of gravity at an angle such that, as seen from the Earth, the components mutually eclipse each other. In the case of the eclipsing binary Algol, one component is much brighter than the other; every two and a half days the fainter star covers up the brighter, and the star seems to fade by more than a magnitude.

ecliptic The projection of the Earth's orbit on to the celestial sphere. It may also be defined as 'the apparent yearly path of the Sun against the stars', passing through the constellations of the **Zodiac.** Since the plane of the Earth's orbit is inclined to the equator by 23.5°, the angle between the ecliptic and the celestial equator must also be 23.5°.

ecosphere The region around the Sun in which the temperatures are neither too hot nor too cold for life to exist under suitable conditions. Venus lies near the inner edge of the ecosphere; while Mars is near the outer edge. The ecospheres of other stars will depend upon the luminosities of the stars concerned.

electromagnetic spectrum The full range of what is termed electromagnetic radiation: **gamma-rays**, **X-rays**, **ultra-violet radiation**, visible light, **infra-red radiation** and **radio** waves. Visible light makes up only a very small part of the whole electromagnetic spectrum. Of all forms of radiation, only visible light and some radio waves can pass through the Earth's atmosphere and reach ground level.

electron A fundamental particle carrying unit negative charge of electricity; the orbital components of the **atom**.

electron density The number of free electrons in unit volume of space. A free electron is not attached to any particular atom, but is moving independently.

element A substance which cannot be chemically split up into simpler substances; 92 elements are known to exist naturally on the Earth and all other substances are made up from these fundamental 92. Various extra elements have been made artificially, all of which are heavier than uranium (number 92 in the natural sequence) and most of which are very unstable.

ellipse A **conic section**, so called because it is an intersection of a plane with a cone, with the significance for astronomy that the unperturbed orbits of the planets and satellites are ellipses around a primary body located at one of the foci of the ellipse. The other focus is empty. The longest line that can be drawn through the centre of the ellipse is the major axis; the shortest line is the minor axis. These two axes are at right angles to each other. The two foci lie inside the ellipse on the major axis. The sum of the distances from any point on the ellipse to the two foci is constant, and equals the length of the major axis. The two parameters that are used to describe the size and shape of an ellipse are the semimajor axis *a*, which is equal to half of the major axis, and the **eccentricity** *e*. The distance between the two foci is equal to 2*ae*. The eccentricity can range from 0 for the special case of a circle, where the two foci coincide at the centre of the circle, to almost 1, where the ellipse becomes very elongated, and the part close to either of the foci is virtually a **parabola**. *See also* **Kepler's Laws of Planetary Motion; orbital elements**

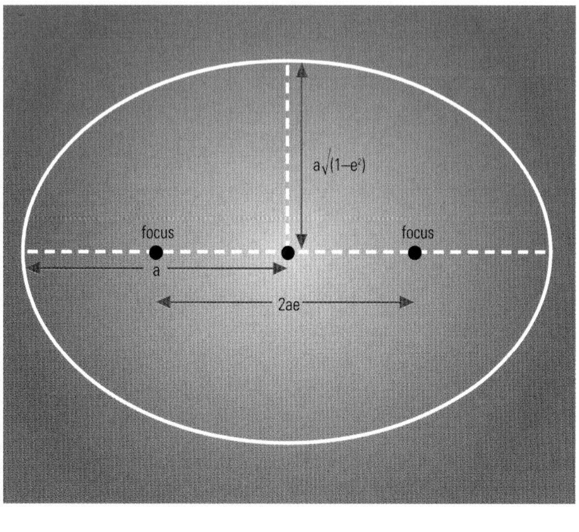

► **ellipse** The principal features of an ellipse are its two foci, its semimajor axis (a), its short (minor) axis and its eccentricity (e). In the case of an ellipse used to define a planetary orbit, the Sun occupies one of the foci while the other is empty. The planets' orbits are reasonably close to circular, but for most comets the orbital ellipse is extremely eccentric, with a very long major axis.

elongation Angular distance between the Sun and a planet, or between the Sun and the Moon, as viewed from Earth. More accurately, the difference in the celestial **longitude** of the two bodies, measured in degrees. An elongation of 0° is called **conjunction**, one of 90° **quadrature** and one of 180° **opposition**. For an inferior planet, the maximum angular distance, east or west, reached during each orbit is known as the greatest elongation. When Mercury and enus are at eastern elongation they may be visible in the evening sky; when at western elongation they may be seen in the morning.

emission spectrum A spectrum consisting of bright lines or bands. Incandescent gases at low density yield emission spectra. *See also* **absorption spectrum**

ephemeris A table giving the predicted positions of a moving celestial body, such as a planet or a comet.

epoch A date chosen for reference purposes in quoting astronomical data. For instance, some star catalogues are given for 'epoch 1950'; by the year 2000 the given positions will have changed slightly because of the effects of **precession**.

equation of time The Sun does not move among the stars at a constant rate, because the Earth's orbit is not circular. Astronomers therefore make use of a *mean* sun, which travels among the stars at a speed equal to the average speed of the real Sun. The interval by which the real Sun is ahead of or behind the mean sun is termed the *equation of time.* It can never exceed 17 minutes; four times every year it becomes zero.

equator, celestial The projection of the Earth's equator on to the **celestial sphere** divides the sky into two equal hemispheres.

equatorial mounting Telescope mounting in which one axis, the polar axis, is parallel to Earth's axis of rotation, and the telescope can be moved about the other axis, the declination axis, which is perpendicular to the polar axis. With an equatorial mount, it is possible to track objects across the sky by using a **drive** mechanism to turn the polar axis at the sidereal rate of one rotation per $23^h\ 56^m$.

Smooth, efficient driving depends on having the telescope carefully balanced by appropriate counterweights to minimize stress on the motor and its gear assembly. Although equatorial mounts were for a long time the standard for large telescopes, developments in computers and automated drive systems have led to modern professional instruments being mounted on **altazimuth mountings**. Similarly, many amateur **GO TO telescopes** are provided with altazimuth mounts.

equinox Twice a year the Sun crosses the celestial equator, once when moving from south to north (about 21 March) and once when moving from north to south (about 22 September). These points are known respectively as the vernal equinox, or **First Point of Aries**, and the autumnal equinox, or **First Point of Libra**. (The equinoxes are the two points at which the ecliptic cuts the celestial equator.)

escape velocity The minimum velocity at which an object must move in order to escape from the surface of a planet, or other body, without being given extra propulsion and neglecting any air resistance. The escape velocity of the Earth is 11 km/s (7 mi/s), or about 40,200 km/h (25,000 mi/h); for the Moon it is only 2.4 km/s (1.5 mi/s); for Jupiter, as much as 60 km/s (37 mi/s).

exosphere The outermost part of the Earth's atmosphere. It is very rarefied, and has no definite upper boundary, since it simply 'thins out' into surrounding space.

eyepiece (ocular) System of lenses in an optical instrument through which the observer views the image. The eyepiece is usually a composite **lens** that magnifies the final real image produced by the optics. There are many different designs of eyepiece, all attempting to achieve the best result for a particular application. Most incorporate an eye lens that is closest to the eye and a field lens that is closest to the **object-glass**. The magnification provided by a telescope is calculated by dividing its focal length by the focal length of the eyepiece used with it. Thus the magnification can be changed simply by changing the eyepiece; astronomical eyepieces come in standard sizes with a simple push-fit to facilitate this. Standard sizes include 20 mm, 26 mm 31 and and 36 mm. When an eyepiece with a shorter focal length is used and the overall magnification increases, the field of view – the area of sky the observer can see through the eyepiece – will decrease. However, some eyepiece designs provide a larger field of view than others, and manufacturers often quote an apparent field of view determined by the acceptance angle of the eyepiece, which is typically around 40°. The telescopic field of view is the acceptance angle of the eyepiece divided by the magnification. Wide-angle eyepieces can be designed to have acceptance angles up to 80°, but this additional width comes at the cost of increased distortion, particularly towards the edges of the field.

F

faculae Bright temporary patches in the Sun's upper photosphere, usually (although not always) associated with **sunspots**. Faculae frequently appear in a position near which a spot or group of spots is about to appear, and may persist for some time – even months – in the region of a group which has disappeared.

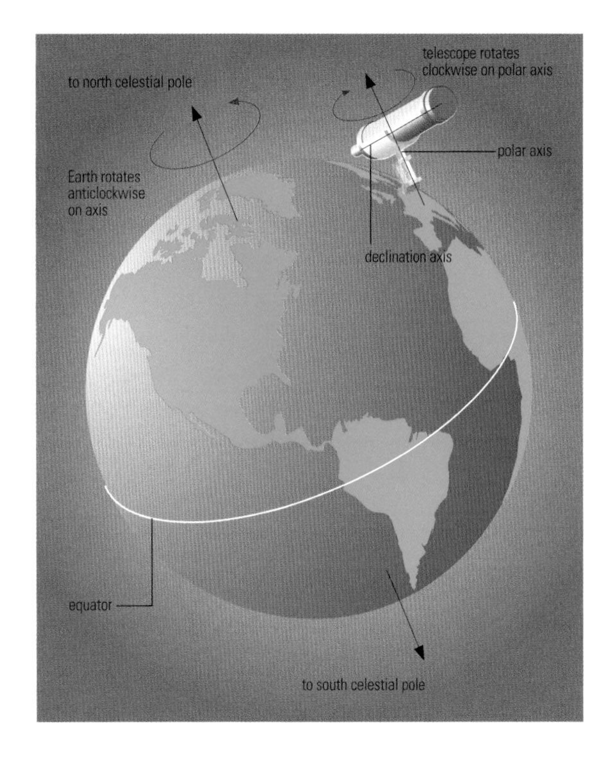

◄ **equatorial mounting** A useful means of mounting a telescope so that it can be driven to follow the apparent motion of the stars due to Earth's rotation, the equatorial is based around two main axes. The polar axis, around which the driving motor turns the telescope, is aligned parallel to Earth's axis of rotation. North and south movement is made along the declination axis.

filament Dark, elongated, ribbon-like structure against the surface of the Sun at certain wavelengths. A filament is a **prominence** seen in projection against the bright solar disk.

First Point of Aries The vernal equinox. The **right ascension** of the vernal equinox is taken as zero, and the right ascensions of all celestial bodies refer to it. See **equinox**.

First Point of Libra The autumnal equinox. See **equinox**.

flares, solar Brilliant outbreaks in the outer part of the Sun's atmosphere, usually associated with active **sunspot** groups. They send out electrified particles which may later reach the Earth, causing magnetic storms and aurorae (see **aurora**); they are also associated with strong outbursts of solar radio emission. It has been suggested that the particles emitted by flares may present a hazard to astronauts who are in space or on the unprotected surface of the Moon.

flare stars Faint red dwarf stars which may brighten up by several magnitudes over a period of a few minutes, fading back to their usual brightness within an hour or so. It is thought that this must be due to intense flare activity in the star's atmosphere. Although the energies involved are much higher than for solar **flares**, it is not yet known whether the entire stellar atmosphere is involved, or only a small area, as in the case of flares on the Sun. Typical flare stars are UV Ceti and AD Leonis.

flash spectrum Just before the Moon completely covers the Sun at a total solar eclipse, the Sun's atmosphere is seen shining by itself, without the usual brilliant background of the **photosphere**. The dark lines in the spectrum then become bright, producing what is termed the flash spectrum. The same effect is seen just after the end of totality.

flocculi Patches on the Sun's surface, seen by instruments based on the principle of the **spectroscope**. Bright flocculi are composed of calcium; dark flocculi are made up of hydrogen.

Focal length The distance between a lens (or mirror) and the point at which the image of an object at infinity is brought to focus. The focal length divided by the aperture of the mirror or lens is termed the *focal ratio*.

Fraunhofer lines The dark lines in the solar spectrum caused by absorption at specific wavelengths, named in honour of the German optician Joseph von Fraunhofer, who studied and mapped them from 1814 onwards.

free fall The normal state of motion of an object in space under the influence of the gravitational pull of a central body; thus the Earth is in free fall around the Sun, while an artificial satellite moving beyond the atmosphere is in free fall around the Earth. While no thrust is being applied, a lunar probe travelling between the Earth and the Moon is in free fall; the same applies for a probe in a transfer orbit between the Earth and another planet. While a vehicle is in free fall, an astronaut will have no apparent 'weight', and will be experiencing zero gravity or weightlessness.

fringe region The upper part of the **exosphere**. Atomic particles in the fringe region have little chance of collision with one another, and to all intents and purposes they travel in free orbits, subject to the Earth's gravitation.

G

g Symbol for the force of gravity at the Earth's surface. The acceleration due to gravity is 9.75 metres per second per second at sea level.

galaxies Systems of stars; our Galaxy contains about 100,000 million stars, but is not exceptional in size. Galaxies are of various shapes; some are spiral, some elliptical, some irregular. The most remote galaxies known are at least 18,000 million light-years away; all, apart from those of our Local Group, are receding from us, so that the entire universe is expanding.

Galaxy, the The Galaxy of which our Sun is a member. Also known as the Milky Way Galaxy.

gamma-rays Extremely short-wavelength electromagnetic radiations. Cosmic gamma-ray sources have to be studied by space research methods.

gegenschein (or counterglow) A very faint glow in the sky, exactly opposite to the Sun; it is very difficult to observe, and has never been satisfactorily photographed. It is due to tenuous matter spread along the main plane of the **Solar System**, so that it is associated with the **zodiacal light**.

▼ gravitational lens (top) Light from a distant quasar may be bent in the gravitational field of a foreground galaxy to produce a distorted or multiple image. In this example, the quasar's light is split into four separate images, each arriving at the telescope along a slightly different path. (bottom) The massive galaxy cluster Abell 2218 in Draco acts as a gravitational lens, producing distorted, arc-shaped images of more distant objects in the same line of sight.

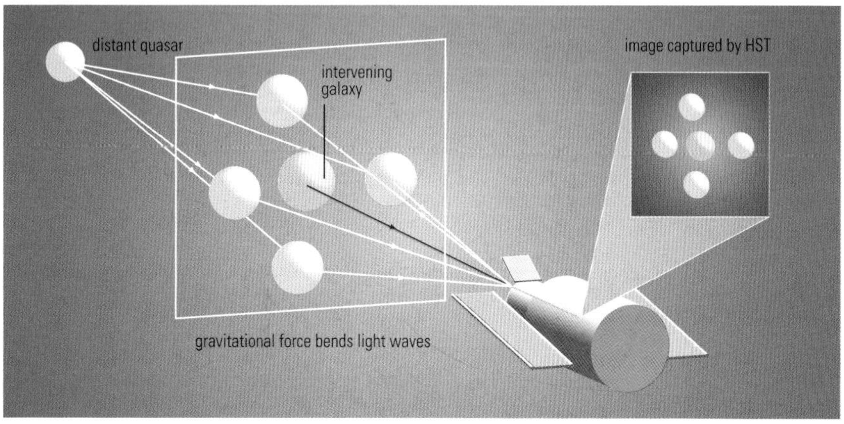

geocentric Relative to the Earth as a centre – or as measured with respect to the centre of the Earth.

geocorona A layer of very tenuous hydrogen surrounding the Earth near the uppermost limit of the atmosphere.

geodesy The science which deals with the Earth's form, dimensions, elasticity, mass, gravitation and allied topics.

geomagnetic storm A sudden disturbance of the Earth's magnetic field, shown by interference with radio communication as well as by variations in the compass needle. It is due to charged particles sent out from the Sun, often associated with solar **flares**.

geophysics The science dealing with the physics of the Earth and its environment. Its range extends from the interior of the Earth out to the limits of the magnetosphere. In 1957–8 an ambitious international programme, the International Geophysical Year (IGY), was organized to undertake intensive studies of geophysical phenomena at the time of a sunspot maximum. It was extended to 18 months, and was so successful that at the next sunspot minimum a more limited but still extensive programme was organized, the International Year of the Quiet Sun (IQSY).

giant planet Term used to describe the planets Jupiter, Saturn, Uranus and Neptune to distinguish them from the smaller, less massive rocky terrestrial planets of the Solar System. They are also known as the Jovian planets or gas giants.

gibbous A phase of the Moon or a planet which is more than half, but less than full.

GO TO telescope Telescope, usually on an **altazimuth mounting**, whose axes are fitted with position encoders that can be directed to find objects by entering simple commands from a handset. On-board computers on GO TO telescopes can store databases of many thousands of objects. Following initial set-up, which usually requires the observer to centre a couple of reference stars in the field of view, the mount will automatically slew the telescope to objects of choice. This greatly eases object location for novice amateur observers.

gravitation The force of attraction which exists between all particles of matter in the universe. Particles attract one another with a force which is directly proportional to the product of their masses and inversely proportional to the square of the distance between them.

gravitational lens Configuration of an observer, an intervening mass and a source of light in which the mass acts as a lens or imaging device for light from the distant source. In 1917, Sir Arthur Eddington's observation of the deflection of starlight around the Sun during a solar eclipse was the very first confirmation of the prediction made in Albert Einstein's Theory of General Relativity that a mass situated in front of a distant light source might bend the light and possibly for images of the distant source. This

confirmation paved the way for more complex situations, in which light from distant galaxies and quasars might be bent enough by foreground galaxies or stars to form images. This consequence was championed by Fritz Zwicky in 1937 and was verified in 1979 with the discovery by Dennis Walsh, Robert Carswell and Ray Weymann of the first of the known 'double quasars' (0957+561).
Many more gravitational lens systems have now been observed, including distant galaxies and quasars lensed by nearer galaxies; these distant galaxies would not be visible were it not for the amplification provided by the gravitational lens. Under certain conditions, a compact lens that lies directly along the line of sight forms multiple images that overlap. The resulting magnified image appears much brighter than the original source. This situation is known as microlensing. Microlensing might be important in quasars when intervening objects (stars or black holes) move across the line of sight to distant quasars and cause an increase of brightness in the quasar. These 'microlensed' outbursts are symmetric in time and frequency independent. Dark matter in our galactic halo might also 'microlens' stars in nearby galaxies, thus revealing their presence.

great circle A circle on the surface of a sphere (such as the Earth, or the celestial sphere) whose plane passes through the centre of the sphere. Thus a great circle will divide the sphere into two equal parts.

green flash (or green ray) When the Sun is setting, the last visible portion of the disk may flash brilliant green for a very brief period. This is due to effects of the Earth's atmosphere, and is best observed over a sea horizon. Venus has also been known to show a green flash when setting.

Greenwich Mean Time (GMT) The time reckoned from the Greenwich Observatory in London, England. It is used as the standard throughout the world. Also known as Universal Time (UT).

Greenwich Meridian The line of longitude which passes through the Airy Transit Circle at Greenwich Observatory. It is taken as longitude zero degrees, and is used as the standard throughout the world.

Gregorian reflector A type of reflecting telescope (see **reflector**) in which the incoming light is reflected from the main mirror on to a small concave mirror placed outside the focus of the main mirror; the light then comes back through a hole in the main mirror and is brought to focus. Gregorian reflectors are not now common.

H

H I and H II regions Clouds of hydrogen in the **Galaxy**. In H I regions the hydrogen is neutral, and the clouds cannot be seen, but they may be studied by radio telescopes by virtue of their characteristic emission at a wavelength of 21 centimetres. In H II regions the hydrogen is ionized (see **ion**), generally in the presence of hot stars. The recombination of the ions and free electrons to form neutral atoms gives rise to the emission of light, by which the H II regions can be seen.

halation ring A ring sometimes seen around a star image on a photograph. It is purely a photographic effect.

halo (1) A luminous ring around the Sun or Moon, caused by ice crystals in the Earth's upper atmosphere. (2) The galactic halo: The spherical star cloud that surrounds the main part of the **Galaxy**.

Hertzsprung-Russell Diagram (H/R Diagram) A diagram in which stars are plotted according to spectral type and luminosity. It is found that there is a well-defined band known as the **Main Sequence** which runs from the upper left of the Diagram (very luminous bluish stars) down to the lower right (faint red stars); there is also a giant branch to the upper right, while the dim, hot **white dwarfs** lie to the lower left. H/R Diagrams have been of the utmost importance in studies of stellar evolution. If colour index is used instead of spectrum, the diagram is known as a colour-magnitude diagram.

Hohmann orbit See **transfer orbit**.

hour angle The time which has elapsed since a celestial body crossed the meridian of the observer.

hour circle A great circle on the celestial sphere which passes through both poles of the sky. The zero hour circle corresponds to the observer's meridian.

Hubble Constant The relationship between the distance of a galaxy and its recessional velocity. Its value is of the order of 70 kilometres per second per megaparsec.

I

inferior planets Mercury and Venus, whose orbits lie closer to the Sun than does that of the Earth. When their **right ascensions** are the same as that of the Sun, so they are lying approximately between the Sun and the Earth, they reach inferior **conjunction**. If the planet's **declination** is also the same as that of the Sun, the result will be a **transit** of the planet.

infra-red radiation Radiation with wavelengths longer than that of red light, but shorter than microwaves. Infra-red sources in the sky are studied either from high-altitude observatories (as at Mauna Kea) or with space techniques. In 1983 the Infra-Red Astronomical Satellite (IRAS) carried out a full survey of the sky in infra-red.

ion An atom which has lost or gained one or more electrons; it has a corresponding positive or negative electrical charge, since in a complete atom the positive charge of the nucleus is balanced out by the combined negative charge of the electrons. The process of producing an ion is termed *ionization.*

ionosphere The region above the **stratosphere**, from about 65–c. 800 km (40–c. 500 mi). Ionization of the atoms in this region (see **ion**) produces layers which reflect radio waves, making long-range communication over the Earth possible. Solar events have effects upon the ionosphere, and produce ionospheric storms; on occasion, radio communication is interrupted.

irradiation The effect which makes brightly lit or self-luminous bodies appear larger than they really are. For example, the Moon's bright crescent appears larger in diameter than the Earth-lit part of the disk.

J

Julian day A count of the days, starting from 12 noon on 1 January 4713 BC. The system was introduced by Scaliger in 1582. The 'Julian' is in honour of Scaliger's father, and has nothing to do with Julius Caesar or the Julian Calendar. Julian days are used by **variable star** observers, and for reckonings of phenomena which extend over very long periods of time.

K

Kepler's Laws of Planetary Motion The three important laws announced by J. Kepler between 1609 and 1618 . They are:
(1) The planets move in elliptical orbits, the Sun being located at one focus of the ellipse, while the other focus is empty.
(2) The radius vector, or imaginary line joining the centre of the planet to the centre of the Sun, sweeps out equal areas in equal times.
(3) The squares of the sidereal periods of the planets are proportional to the cubes of their mean distances from the Sun (Harmonic Law).

kiloparsec 1000 **parsecs**, or 3260 **light years**.

Kirkwood gaps Regions in the belt of **asteroids** between Mars and Jupiter in which almost no asteroids move. The gravitational influence of Jupiter keeps these zones 'swept clear'; an asteroid which enters a Kirkwood region will be regularly perturbed by Jupiter until its orbit has been changed. They were first noted by the American mathematician Daniel Kirkwood.

L

Lagrangian points Points in the orbital plane of two massive bodies at which a third body can remain in equilibrium, so that all three bodies remain in a fixed geometrical configuration. There are five such points, as shown in the diagram. The Lagrangian points L_4 and L_5 form equilateral triangles with M_1 and M_2, and L_1, L_2 and L_3 are collinear with M_1 and M_2. Orbits close to L_4 and L_5 are stable and perform slow oscillations around the equilibrium points. The points L_1, L_2 and L_3 are unstable, and any objects in these orbits would very slowly drift away (although some particular orbits in these vicinities are stable). The **trojan asteroids** are close to the L_4 and L_5 points of Jupiter's orbit around the Sun, with oscillation amplitudes around the equilibrium points of up to 30°. Similar orbits occur in Saturn's satellite system. The tiny satellites Telesto and Calypso are close to the L_4 and L_5 points of Tethys' orbit around Saturn, and Helene is close to the L_4 point of Dione's orbit. The spacecraft SoHO has been placed into a stable orbit close to the L_1 point of the Earth. This point is located between the Earth and the Sun at about 1,500,000 km (930,000 mi) from the Earth, which is about four times the distance of the Moon and one hundredth of the distance to the Sun.

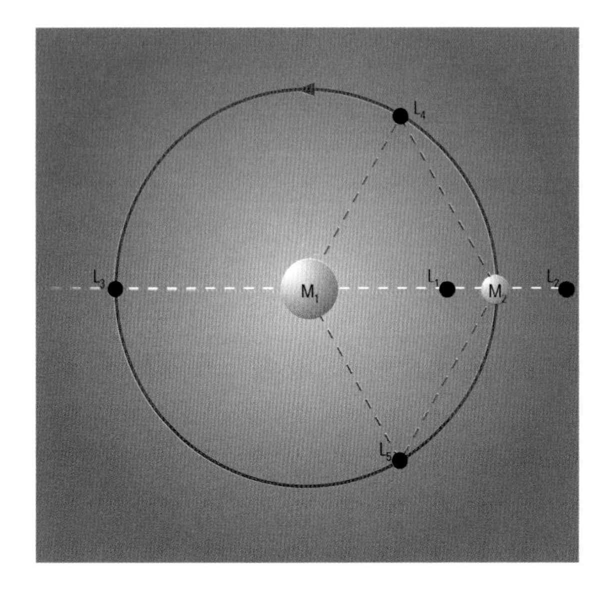

▶ **Lagrangian point** In a system where a body orbits a more massive primary, there are five key positions where the gravity of the two balances out such that a third body can remain there stably. The L_1 position, close to the less massive body on the line towards the primary, is one such Lagrangian point; many spacecraft are stationed around the L_1 point between Earth and the Sun. The L_4 and L_5 (60° ahead of and behind the planet) Lagrangian points of Jupiter's orbit are occupied by Trojan asteroids.

Laser (Light Amplification by the Simulated Emission of Radiation) A device which emits a beam of light made up of rays of the same wavelength (coherent light) and in phase with one another. It can be extremely intense. Laser beams have already been reflected off the Moon.

latitude, celestial The angular distance of a celestial body from the nearest point on the **ecliptic**.

lens Basic optical component made of transparent material through which light passes in order to produce or modify an image. Astronomical lenses range in size and complexity from tiny spherical sapphire lenses which are attached to the ends of fibre optics, through to large **object-glass** lenses which are used to collect and focus light in refracting telescopes. Lenses are also used to shape and direct light as it passes through astronomical instruments for analysis and measurement, but the most common example as far as the amateur astronomer is concerned is the telescope **eyepiece**.

There are two basic types of lens, converging and diverging. Converging lenses are thicker in the centre than at the edges and cause parallel light passing through them to bend towards a common point, that is, to converge. The simplest example of a converging lens is the magnifying lens. Diverging lenses cause parallel light to spread out and to appear to come from a common point behind the lens. Diverging lenses are thicker at the edges than at the centre and the most common example is the spectacle lens used to correct short-sightedness. Converging lenses often have two convex surfaces and so are sometimes called convex lenses. Diverging lenses often have two concave surfaces and so are called concave lenses. However, both types can have one convex surface and one concave so this terminology can be slightly misleading. A lens can produce a focused image because light is bent (refracted) as it passes from air to glass and then from glass back to air. It is the shape of the air–glass surface that determines how the light is bent and what effect this has on the image produced. Surfaces are usually spherical, although more complex shapes are used in order to reduce optical aberrations. Several manufacturers now advertise binoculars with aspheric optics which offer superior performance.

librations, lunar Although the Moon's rotation is captured with respect to the Earth, there are various effects, known as librations, which enable us to examine 59 per cent of the total surface instead of only 50 per cent, although no more than 50 per cent can be seen at any one time. There are three librations: in longitude (because the Moon's orbital velocity is not constant), in latitude (because the Moon's equator is inclined by 6 degrees to its orbital plane), and diurnal (due to the rotation of the Earth).

light year The distance travelled by light in one year. It is equal to 9.4607×10^{12} km, 0.3066 **parsecs** or 63,240 **astronomical units**.

limb The edge of the visible disk of the Sun. Moon, a planet, or the Earth (as seen from space).

Local Group of galaxies The group of which our Galaxy is a member. There are more than two dozen systems, of which the most important are the Andromeda Galaxy, our Galaxy, the Triangulum Galaxy and the two Magellanic Clouds.

longitude, celestial The angular distance from the vernal equinox to the foot of a perpendicular drawn from a celestial body to meet the **ecliptic**. It is measured eastwards along the ecliptic from zero° to 360°.

luminosity Total amount of energy emitted by a star per second in all wavelengths. It is dependent on the radius of the star and its temperature. Luminosity is usually expressed in units of watts.

lunation (synodical month) The interval between successive new moons: 29 days 12 hours 44 minutes. See also **synodic period**.

Lyot filter (monochromatic filter) A device used for observing the Sun's prominences and other features of the solar atmosphere, without the necessity of waiting for a total **eclipse**. It was invented by the French astronomer Bernard Lyot in the 1930s.

M

Mach number The velocity of a vehicle moving in an atmosphere divided by the velocity of sound in the same region. Near the surface of the Earth, sound travels at about 1200 km/h (745 mi/h); so Mach 2 would be $2 \times 1200 = 2400$ km/h (1490 mi/h).

magnetic storm *See* **geomagnetic storm**.

magnetohydrodynamics The study of the interactions between a magnetic field and an electrically conducting fluid. The Swedish scientist H. Alfven is regarded as the founder of magnetohydrodynamics.

magnetosphere Region surrounding a magnetized planet that is occupied by the planetary magnetic field; it acts to control the structure and dynamics of plasma populations and ionized particles within it. A magnetosphere can be considered as the magnetic sphere of influence of the planet. In our Solar System, Mercury, the Earth, Jupiter, Saturn, Uranus and Neptune possess magnetospheres. The size of the magnetosphere is controlled by the interaction of the planetary field with the **solar wind**, which compresses the upstream side while dragging the downwind side out into an extended magnetotail, resembling a windsock. The magnetosphere represents an obstacle to the supersonic solar wind flow from the Sun, with the solar wind having to flow around the magnetosphere. To do this, the wind is slowed, deflected and heated at a bow shock, standing in the flow upstream of the magnetosphere. The shocked solar wind plasma and associated magnetic field downstream of the bow shock is known as a magnetosheath. This region is separated from the magnetosphere proper by a layer of electrical current known as the magnetopause. In the solar direction, the magnetopause occurs at the point where the pressure associated with the planetary magnetic field balances the pressure of the solar wind.

A planetary magnetosphere contains a number of regions with distinct plasma or energetic charged particles. Closest to the planet, these regions include the radiation belts (the **Van Allen belts** at the Earth) and the **plasmasphere**. The radiation belts contain trapped energetic particles captured from the solar wind or originating from collisions between upper atmosphere atoms and high-energy **cosmic rays**. The plasmasphere is populated by the polar wind, an upflow of plasma from the upper **ionosphere**. Farther out, the **plasma sheet** occupies the central portion of the magnetotail and contains a hot plasma of solar wind origin. Magnetic storms and magnetospheric substorms cause global restructuring of these magnetospheric regions and significant energization of the particles within them. In addition, there is significant coupling between the magnetosphere and the ionosphere. At the Earth this coupling is achieved by the flow of large-scale electrical currents from the magnetosphere through the polar ionosphere, which results in heating of the ionosphere, particularly at disturbed times. It is also associated with spectacular displays of the **aurorae**.

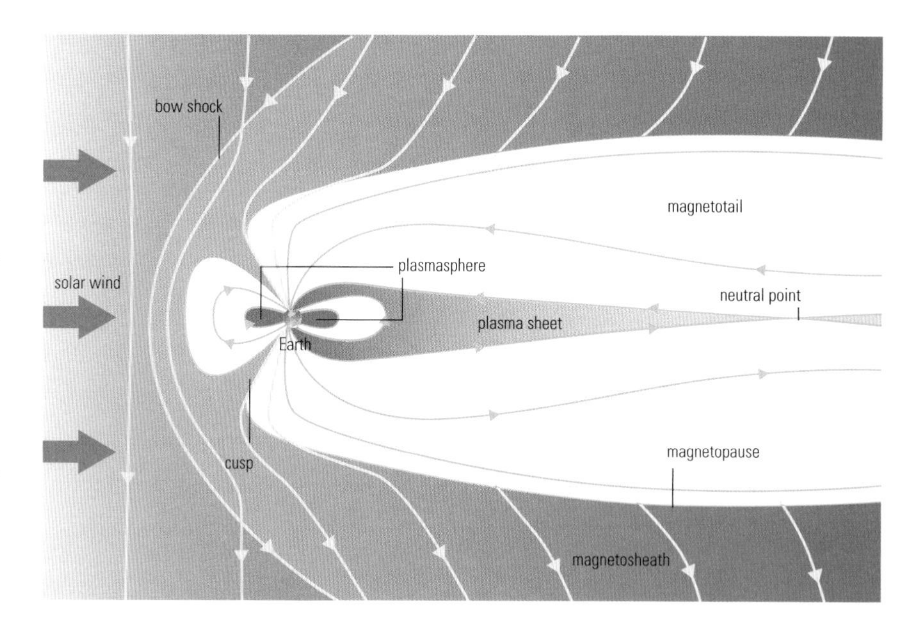

▲ magnetosphere The Earth's moving iron core gives it an active magnetic field that extends well beyond the planet. The arrows indicate the direction of the magnetic field. The magnetosphere protects us from solar radiation.

magnitude This is really a term for 'brightness', but there are several different types. (1) Apparent or visual magnitude: the apparent brightness of a celestial body as seen with the eye. The brighter the object, the lower the magnitude. The planet Venus is of about magnitude −4.5; Sirius, the brightest star, −1.4; the Pole Star, +2; stars just visible with the naked eye, +6; the faintest stars that can be recorded with the world's largest telescopes, below +30. A star's apparent magnitude is no reliable key to its luminosity. (2) Absolute magnitude: the apparent magnitude that a star would have if seen from a standard distance of 10 **parsecs** (32.6 **light years**). (3) Photographic magnitude: the magnitude derived from the size of a star's image on a photographic plate. (4) Bolometric magnitude: this refers to the total radiation sent out by a star, not merely to visible light.

Main Sequence The well-defined band from the upper left to lower right of a **Hertzsprung-Russell Diagram**. The Sun is typical Main Sequence star.

Maser (Microwave Amplification by Simulated Emission of Radiation) The same basic principle as that of the laser, but applied to radio wavelengths rather than to visible light.

mass The quantity of matter that a body contains. It is not the same as weight, which depends upon local gravity; thus on the Moon an astronaut has only one-sixth of his normal weight, but his mass remains unaltered.

mass transfer Process that occurs in **close binary** stars when one star fills its **Roche lobe** and material transfers to the other star through the inner **Lagrangian point**. Material can be transferred directly to the other star but is more usually transferred via an **accretion disk**.

meridian, celestial The **great circle** on the **celestial sphere** which passes through the **zenith** and both celestial poles. The meridian cuts the observer's horizon at the exact north and south points.

Messier numbers Numbers given by the French astronomer Charles Messier to various nebulous objects including open and globular clusters, gaseous nebulae and galaxies. Messier's catalogue contained slightly over a hundred objects. His numbers are still used; thus the Andromeda Galaxy is M31, the Orion Nebula M42, and so on.

meteor Cometary debris; a small particle which enters the Earth's upper atmosphere and burns away, producing the effect known as a shooting star.

meteorite A larger body, which is able to reach ground-level without being destroyed. There is a fundamental difference between meteorites and **meteors**; a meteorite seems to be more nearly related to an **asteroid** or minor planet. Meteorites may be stony (*aerolites*), iron (*siderites*) or of intermediate type. In a few cases meteorites have produced craters; the most famous example is the large crater in Arizona, which is almost 1.5 km (nearly 1 mi) in diameter and was formed in prehistoric times.

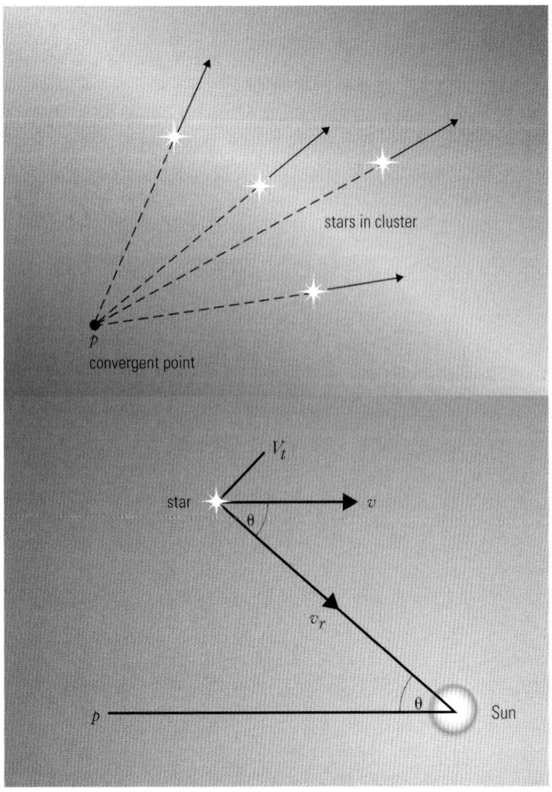

▶ moving cluster The velocity of stars in an open cluster can be determined if they are close enough for their parallax to be measurable over a short period. Their parallax is combined with their line-of-sight rate of motion towards or away from Earth to obtain their speed of movement at right angles to the line of sight. Combined with the measurement of the angle of movement, this allows astronomers to judge their distances.

meteoroids The collective term for meteoritic bodies. It was once thought that they would present a serious hazard to spacecraft travelling outside the Earth's atmosphere, but it now seems that the danger is very much less than was feared, even though it cannot be regarded as entirely negligible.

meteor shower Enhancement of **meteor** activity produced when Earth runs through a **meteor stream**. About 25 readily recognisable meteor showers occur each year, the most prominent being the *Quadrantids*, *Perseids* and *Geminids*. Meteor showers occur at the same time each year, reflecting Earth's orbital position, and its intersection with the orbit of the particular meteor stream whose meteoroids are being swept up. Shower meteors appear to emerge from a single part of the sky, known as the **radiant**. Some showers are active only periodically as a result of uneven distribution of stream meteoroids, the *Giacobinids* perhaps being the best example.

meteor stream Trail of debris, usually from a **comet**, comprised of **meteoroids** that share a common orbit around the Sun. Passage of Earth through such a stream gives rise to a **meteor shower**. Meteor streams may also be associated with some asteroids – most notably 3200 Phaethon, the debris from which produces the *Geminids*.

micrometeorite An extremely small particle, less than 0.1016 mm (0.004 in) in diameter, moving around the Sun. When a micrometeorite enters the Earth's atmosphere, it cannot produce a shooting-star effect, as its mass is too slight. Since 1957, micrometeorites have been closely studied from space probes and artificial satellites.

micron A unit of length equal to 0.001 mm (0.00004 in). There are 10,000 **Ångströms** to one micron. The usual symbol is μ.

midnight Sun The Sun seen above the horizon at midnight. This can occur for some part of the year anywhere inside the Arctic and Antarctic Circles.

Milky Way The luminous band stretching across the night sky. It is due to a line-of-sight effect; when we look along the main plane of our **Galaxy** (that is, directly towards or away from the galactic centre) we see many stars in roughly the same direction. Despite appearances, the stars in the Milky Way are not closely crowded together.

millibar The unit which is used as a measure of atmospheric pressure. The standard atmospheric pressure is 1013.25 millibars (75.97 cm of mercury).

minor planets See **asteroids**.

molecule A stable association of atoms; a group of atoms linked together. For example, a water molecule (H_2O) is made up of two hydrogen atoms and one atom of oxygen.

month (1) Calendar month: the month in everyday use. (2) Anomalistic month: the time taken for the Moon to travel from one **perigee** to the next. (3) Sidereal month: the time taken for the Moon to complete one journey around the **barycentre**, with reference to the stars.

moving cluster A group of stars, often widely scattered, that share the same movement through the Galaxy. The stars in a moving cluster all follow nearly parallel paths. However, because of a perspective effect, they seem to be moving towards, or away from a point known as the convergent point. This allows the distance of the cluster to be determined. Some clusters are very scattered, for example the Ursa Major moving cluster, which includes five stars of the Plough, Sirius, Delta Leonis, Beta Aurigae, Beta Eridani and Alpha Coronae Borealis.

multiple star A star made up of more than two components which are physically associated and orbit their mutual centre of gravity.

N

nadir The point on the celestial sphere that lies immediately below the observer. It is directly opposite to the overhead point or **zenith**.

nanometre (symbol nm) Unit of length equivalent to one thousand millionth (10^{-9}) of a metre. The nanometre is commonly used as a measurement of the wavelength of electromagnetic radiation, having replaced the previously used **Ångström unit**.

Nasmyth focus Focal point of an **altazimuth**-mounted reflecting telescope in which the converging light beam is reflected by means of a third mirror to a point outside the lower end of the telescope tube, one side of the altitude axis. This allows large or heavy instruments, such as spectrographs, to be mounted on a permanent platform which rotates in **azimuth** with the telescope. An arrangement of prisms and mirrors compensates for the rotation of the image as the telescope tracks the object, holding it stationary relative to the instrument. This arrangement was first used by the Scottish engineer James Nasmyth (1808–90) and has been revived with the latest generation of altazimuth-mounted, large professional telescopes. *See also* **coudé focus**

nebula A mass of tenuous gas in space together with what is loosely termed 'dust'. If there are stars in or very near the nebula, the gas and dust will become visible, either because of straightforward reflection or because the stellar radiation excites the material to self-luminosity. If there are no suitable stars, the nebula will remain dark, and will betray its presence only because it will blot out the light of stars lying beyond it. Nebulae are regarded as regions in which fresh stars are being formed out of the interstellar material.

neutrino A fundamental particle which has no mass and no electric charge – which makes them extremely difficult to detect.

neutron A fundamental particle whose mass is equal to that of a **proton**, but which has no electric charge. Neutrons exist in the nuclei of all atoms apart from that of hydrogen.

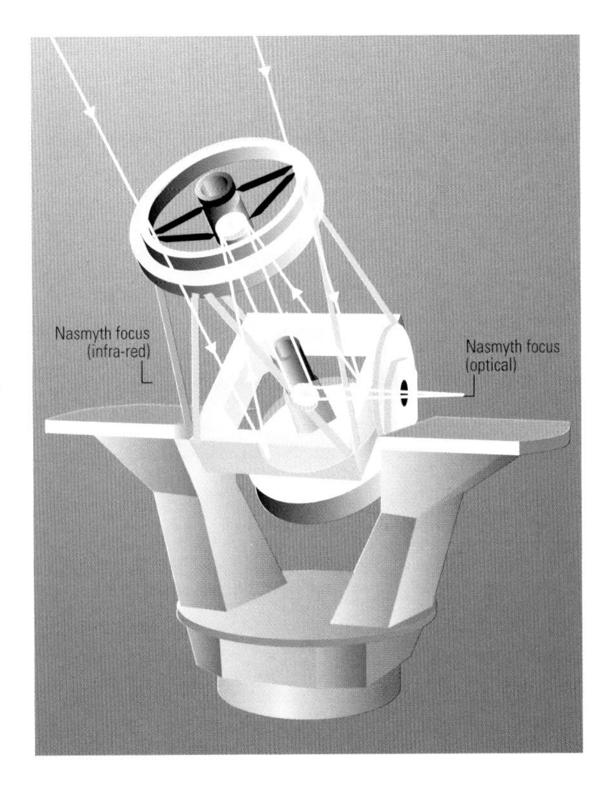

◀ **Nasmyth focus** A number of large, professional telescopes have instruments at their Nasmyth foci. In the William Herschel Telescope there are two foci, one used for infrared and the other for optical observations. They are sited at opposite ends of the declination axis, and the third mirror (the Nasmyth flat) can be moved to direct light to either of them.

neutron star A star made up principally or completely of **neutrons**, so that it will be of low luminosity but almost incredibly high density. Theoretically, a neutron star should represent the final stage in a star's career. The radio sources known as **pulsars** are in fact neutron stars.

Newtonian telescope A **reflecting telescope** having a **paraboloidal** primary mirror, and a diagonal plane mirror positioned to divert the light path through 90° to a focus near the side and towards the upper end of the telescope tube. For apertures less than 150 mm (6 in), simpler spheroidal primaries can be used without significant loss of optical performance. Before the rise of the Schmidt–Cassegrain telescope, the Newtonian was the reflector of choice for many amateur observers.

noctilucent clouds Rare, strange clouds in the **ionosphere**, best seen at night when they continue to catch the rays of the

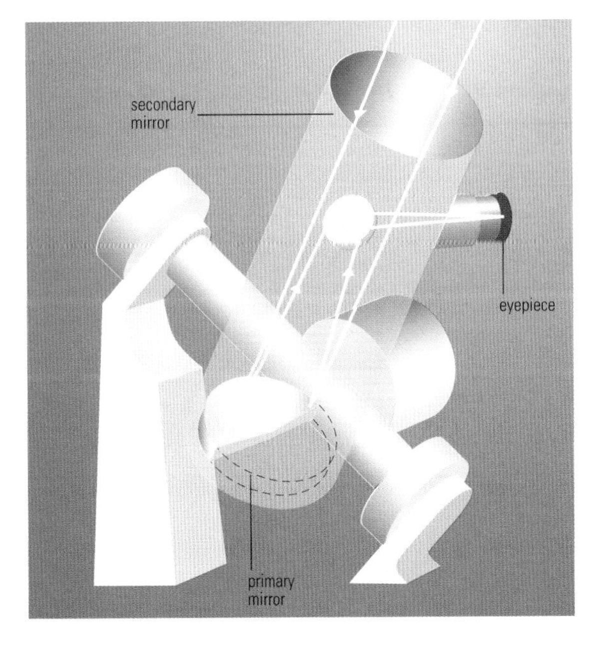

◀ **Newtonian telescope** In this form of reflecting telescope, light is reflected from a parabaloidal primary mirror to a flat secondary mirror (or sometimes a prism), placed diagonally within the tube, and directed to a side-mounted eyepiece. The resulting image is inverted.

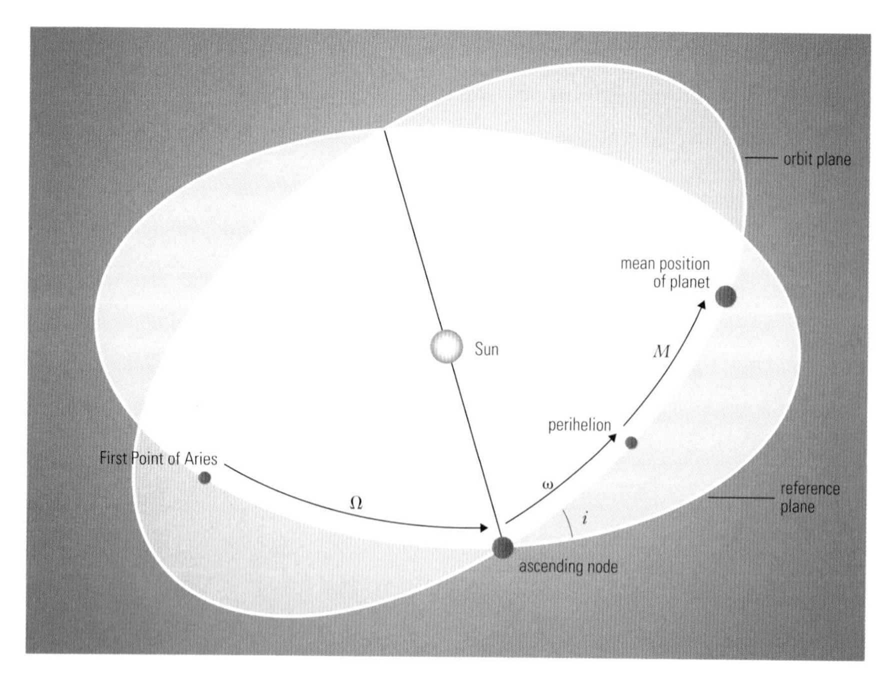

▲ orbital elements The orbit of a planet around the Sun can be defined by a number of characteristics, including the inclination (i) relative to the reference plane of the ecliptic, the longitude (V) of the ascending node measured in degrees from the First Point of Aries, and the argument of perihelion (v) measured in degrees from the ascending node.

Sun, after it has set. They lie at altitudes of greater than 80 km (50 mi), and are noticeably different from normal clouds. It is possible that they are produced by meteoritic dust in the upper atmosphere.

nodes The points at which the orbit of a planet, a comet or the Moon cuts the plane of the **ecliptic**, either as the body is moving from south to north (ascending node) or from north to south (descending node). The line joining these two points is known as the line of nodes.

nova A star which undergoes a sudden outburst, flaring up to many times its normal brilliancy for a while before fading back to obscurity. A nova is a binary system in which one component is a white dwarf; it is the white dwarf which is responsible for the outbursts.

nutation A slight, slow 'nodding' of the Earth's axis, due to the fact that the Moon is sometimes above and sometimes below the ecliptic, and therefore does not always pull on the Earth's equatorial bulge in the same direction as the Sun. The result is that the position of the celestial pole seems to 'nod' by about 9 seconds of arc to either side of its mean position with a period of 18 years 220 days. Nutation is superimposed on the more regular shift of the celestial pole caused by precession.

O

object-glass (objective) The main lens of a refracting telescope (see **refractor**).

obliquity of the ecliptic The angle between the ecliptic and the celestial equator. Its value is 23° 26′ 54″. It may also be defined as the angle by which the Earth's axis is tilted from the perpendicular to the orbital plane.

occultation The covering up of one celestial body by another. Thus the Moon may pass in front of a star or (occasionally) a planet; a planet may occult a star; and there have been cases when one planet has occulted another – for instance, Venus occulted Mars in 1590. Strictly speaking, solar **eclipses** are occultations of the Sun by the Moon.

opposition The position of a planet when it is exactly opposite the Sun in the sky, and so lies due south at midnight. At opposition, the Sun, the Earth and the planet are approximately aligned, with the Earth in the mid position. The **inferior planets** (Mercury and Venus) can never come to opposition.

orbit The path of an artificial or natural celestial body. See also **transfer orbit**.

orbital elements Describing the size, shape and orientation of an **orbit**, and the position of the orbiting body in the orbit at some **epoch**. There are six orbital elements, and if the mass of the primary body is not know (such as in a binary star system) then the orbital period (or the mean motion) is also needed. In addition, the epoch of the elements is usually specified. The elements vary slowly with time as a result of perturbations by other bodies, and the epoch is the time at which they had the particular values specified. The various angles defining an elliptical orbit are shown in the diagram. The point *P* is the position of a fictitious object that orbits with constant angular speed with the same period as the real object, and coincides with the real object at pericentre and apocentre. The orbital elements are:

a the semimajor axis (half the length of the major axis of the ellipse)
e the eccentricity
i the inclination to the reference plane
M the mean anomaly
ω the argument of pericentre
Ω the longitude of the ascending node

Note that M and ω are measured around the orbit plane, and Ω is measured around the reference plane. Two further composite angles are defined, which are measured partly around the reference plane and partly around the orbit plane:

$\varpi = \Omega + \omega$ the longitude of pericentre
$\lambda = \Omega + \omega + M$ the mean longitude

Further quantities that are sometimes used are:

T the time of pericentre passage, that is, the time at which $M = 0$
$q = a(1 - e)$ the distance from the primary at pericentre.

Only six of these quantities are needed to define an orbit. For the planets and natural satellites, it is usual to take as the orbital elements the quantities a, e, i, λ, ϖ, Ω. The reason for choosing the composite angles λ and ϖ rather than M and ω is that for many planets and satellites the inclinations are small, and so the **node** is not a well-defined point (for the Earth the inclination to the ecliptic of date is zero, and so the node is not defined at all). It is preferable, therefore, to measure all angles from the well-defined **equinox**. For asteroids and artificial satellites of the Earth the inclinations are generally fairly large, and so the node is well defined. In this case the orbital elements are usually taken as the quantities a, e, i, M, ω, Ω. In either case, the quantities λ or M vary rapidly as the object moves around its orbit, and so the epoch of the elements is an important quantity.

For comets and newly discovered asteroids the perihelion distance and the time of perihelion passage are usually of particular interest, and so these are taken as two of the orbital elements. The full set used is T, q, e, I, ω, Ω.

As the mean longitude or mean anomaly at epoch is not used, the epoch of the elements is of lesser importance than it is for planetary elements.

ozone Triatomic oxygen (O_3). The ozone layer in the Earth's upper atmosphere absorbs many of the lethal short-wavelength radiations coming from space. Were there no ozone layer, it is unlikely that life on Earth could ever have developed.

P

parabola Open curve, one of the **conic sections**, obtained by cutting a cone in a plane parallel to the side of the cone. It can be regarded as an **ellipse** with only one focus, an infinite major axis and **eccentricity** of 1. A parabolic orbit is used for some of the long-period **comets**. These comets are observable over only a short arc of their orbits near perihelion, and it is often not possible to distinguish between exactly parabolic orbits and extremely elongated elliptical orbits.

paraboloid Three dimensional surface generated by rotating a parabola about its own axis. This shape is commonly used for the primary mirror in **Newtonian telescopes** in preference to a simple spherical surface in order to overcome **spherical aberration** and produce a diffraction-limited image for a point object, such as a star, that is in the centre of the field of view. Incoming light that is not parallel to the axis of the paraboloid (that is, from objects away from the centre of the field) produces images that suffer from **coma**.

parallax, trigonometrical The apparent shift of a body when observed from two different directions. The separation of the two observing sites is called the baseline. The Earth's **orbit** provides a baseline 300 million km (220 million mi) long (since the radius of the orbit is 150 million km (110 million mi); therefore, a nearby star observed at a six-monthly interval will show a definite parallax shift relative to the background of more distant stars. It was in this way that Friedrich Wilhelm Bessel, in 1838, made the first measurement of the distance of a star (61 Cygni). The method is useful for stars out to a distance of about 300 **light years**, beyond which the parallax shifts become too small to detect.

parsec The distance at which a star would show a parallax of one second of arc. It is equal to 3.26 **light years** or 206,265 **astronomical units**. (Apart from the Sun, no star lies within one parsec of us.)

penumbra (1) The comparatively light surrounding parts of a **sunspot**. (2) The area of partial shadow lying to either side of the main cone of shadow cast by the Earth. During lunar **eclipses**, the Moon must move through the penumbra before reaching the main shadow (or **umbra**). Some lunar eclipses are penumbral only.

periastron The point of the orbit of a member of a **binary** system in which the stars are at their closest to each other. The most distant point is termed **apastron**.

perigee The point in the orbit of the Moon or an artificial satellite at which the body is closest to the Earth. The most distant point is the **apogee**.

perihelion The point in the orbit of a member of the **Solar System** in which the body is at its closest to the Sun. The most distant point is the **aphelion** (*see* diagram under aphelion). The Earth reaches perihelion in early January.

periodic times See **sidereal period**.

perturbations, gravitational The disturbances in the orbit of a celestial body produced by the gravitational pulls of other bodies.

phase Extent to which the illuminated hemisphere of a Solar System body, particularly the Moon or an **inferior planet**, is visible as viewed from the Earth. When close to **quadrature**, Mars shows a distinct **gibbous** phase. These bodies exhibit phases because they do not emit any light of their own and only shine by reflected sunlight. The observed phase, which is sometimes expressed as a percentage or decimal fraction, depends on the relative positions of the Sun, Earth and the body in question. When the whole of the illuminated side is visible, the phase is said to be full (100%). When the illuminated side is turned away from the Earth, it is new. Between the two extremes, a gradually increasing or decreasing proportion of the disk is seen. From new, the phases progress through crescent, first quarter, gibbous to full and then through gibbous, last quarter, crescent to new again. *See also* **phases of the moon**

photometry The measurement of the intensity of light. The device now used for accurate determinations of star magnitudes is the photoelectric photometer, which consists of a photoelectric cell used together with a **telescope**. (A photoelectric cell is an electronic device. Light falls upon the cell and produces an electric current; the strength of the current depends on the intensity of the light.)

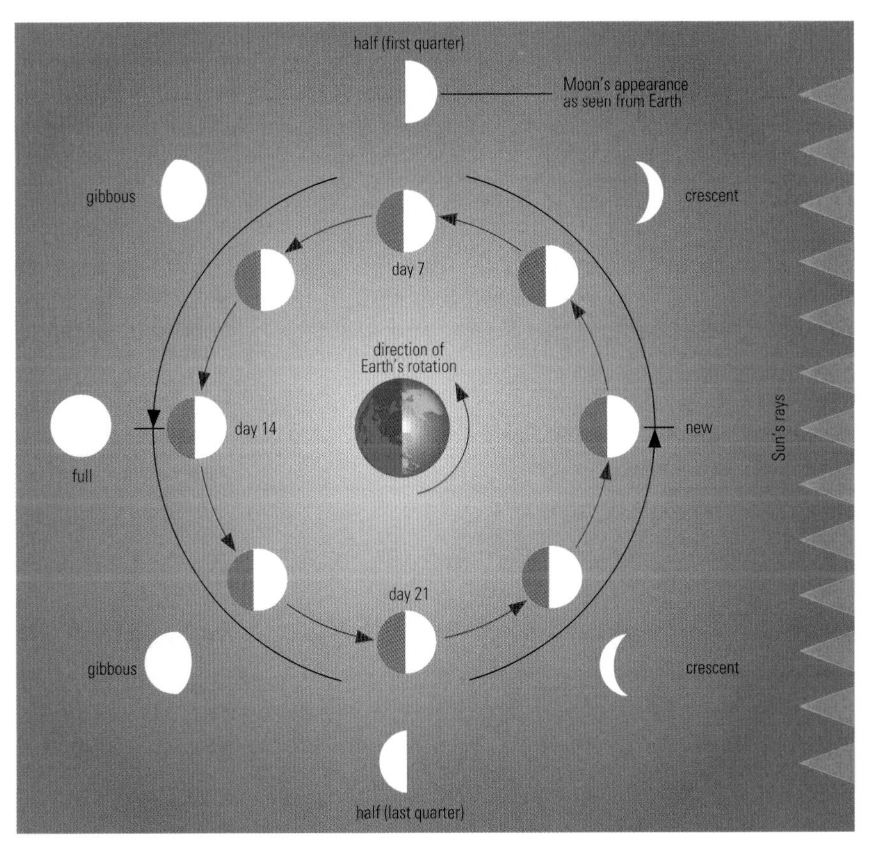

▼ **phase** The changing appearance of the Moon's illuminated disk results from its orbital motion around the Earth. At new (right), the Moon lies in line with the Sun, and the hemisphere presented to Earth is completely dark. At full, the Moon lies opposite the Sun in Earth's sky and is completely illuminated. Between these two extremes, varying degrees of illumination are seen as shown.

photosphere The part of the Sun from which visible light is emitted; more generally, the region of any star that gives rise to visible radiation.

planet A non-luminous body moving round a star.

planetarium An instrument used to show an artificial sky on the inner surface of a large dome, and to reproduce celestial phenomena of all kinds. A planetarium projector is extremely complicated, and is very accurate. The planetarium is an educational device, and has become very popular in recent years. Planetaria have been set up in many large cities all over the world, and are also used in schools and colleges.

planetary nebula A faint star approaching the end of its life and surrounded by an immense 'shell' of tenuous gas. More than 300 are known in our Galaxy. They are so called because their telescopic appearance under low magnification is similar to that of a planet.

planisphere Easy-to-use, portable two-dimensional map of the **celestial sphere**, as seen from a particular latitude, showing which stars are visible at a given date and time. The planisphere comprises two disks, one a star map and the other an overlay containing an oval window. The two are connected at the centre and may be rotated relative to one another in order to select the desired date and time, which are displayed around their outer edges. The oval window in the overlay disk will then show those stars that are currently above the horizon.

plasma A gas consisting of ionized atoms (*see* **ion**) and free electrons, together with some neutral particles. Taken as a whole, it is electrically neutral, and is a good conductor of electricity.

poles, celestial The north and south points of the **celestial sphere**.

populations, stellar There are two main types of star regions. *Population I* areas contain a great deal of interstellar material, and the brightest stars are hot and white; it is assumed that star formation is still in progress. The brightest stars in *Population II* areas are red giants, well advanced in their evolutionary cycle; there are almost no hot, white giant stars, and there is little interstellar material, so that star formation has apparently ceased. Although no rigid boundaries can be laid down, it may be said that the arms of spiral galaxies are mainly of Population I; the central parts of spirals, as well as elliptical galaxies and globular clusters, are mainly of Population II.

position angle The apparent direction of one object with reference to another, such a members of binary star systems, measured from the north point of the main object through east (90°), south (180°) and west (270°).

precession The apparent slow movement of the celestial poles. It is caused by the pull of the Moon and the Sun upon the Earth's equatorial bulge. The Earth behaves rather in the manner of a spinning top which is running down and starting to topple, but the movement is very gradual; the pole describes a circle on the **celestial sphere** – centred on the pole of the ecliptic – which is 47° in diameter and takes 25,800 years to complete. Because of precession, the celestial equator also moves, and this in turn affects the position of the **First Point of Aries** (vernal **equinox**), which shifts westwards along the **ecliptic** by 50 seconds of arc each year. Since ancient times, this motion has taken the vernal equinox out of Aries into the adjacent constellation of Pisces (the Fish). Our present Pole Star will not retain its title indefinitely. In AD 12000, the north polar star will be the brilliant Vega, in Lyra.

prism A glass block having flat surfaces inclined to one another. Light passing through a prism will be split up into the familiar rainbow, since different colours are refracted by different amounts.

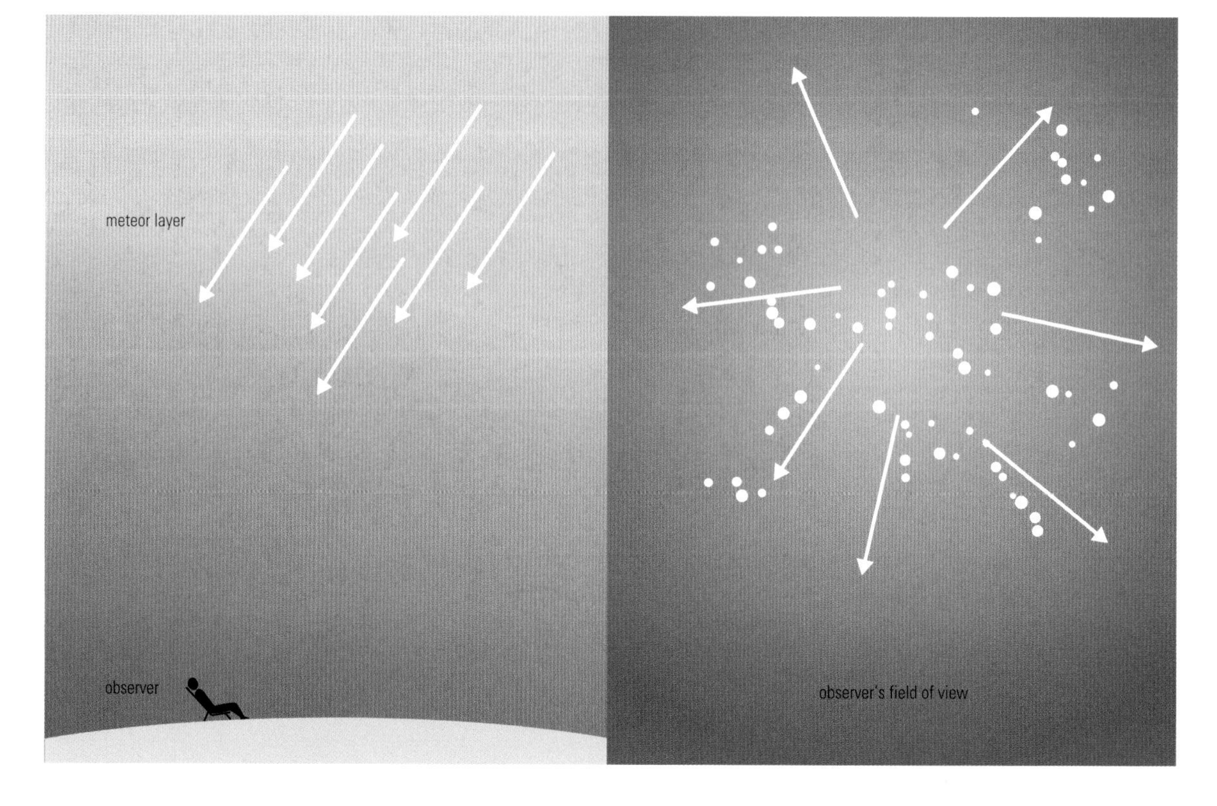

► **radiant** Meteors from a common source, occurring during a shower such as the Perseids, enter the atmosphere along parallel trajectories, becoming luminous at altitudes around 80–100 km (50–60 mi). If an observer on the ground plots the apparent positions of meteors on a chart of the background stars, they will appear to diverge from a single area of sky, the radiant. The radiant effect is a result of perspective.

prominences Masses of glowing gas, chiefly hydrogen, above the Sun's bright surface. They are visible with the naked eye only during total solar **eclipses**, but modern equipment allows them to be studied at any time. They are of two main types, eruptive and quiescent.

proper motion The individual motion of a star on the **celestial sphere**. Because the stars are so remote, their proper motions are slight. The greatest known is that of Barnard's Star (a red dwarf at a distance of 6 **light years**); this amounts to one minute of arc every six years, so that it will take 180 years to move by an amount equal to the apparent diameter of the Moon. The proper motions of remote stars are too slight to be measured at all.

proton A fundamental particle with unit positive electrical charge. The nucleus of a hydrogen atom consists of one proton. See also **neutron**.

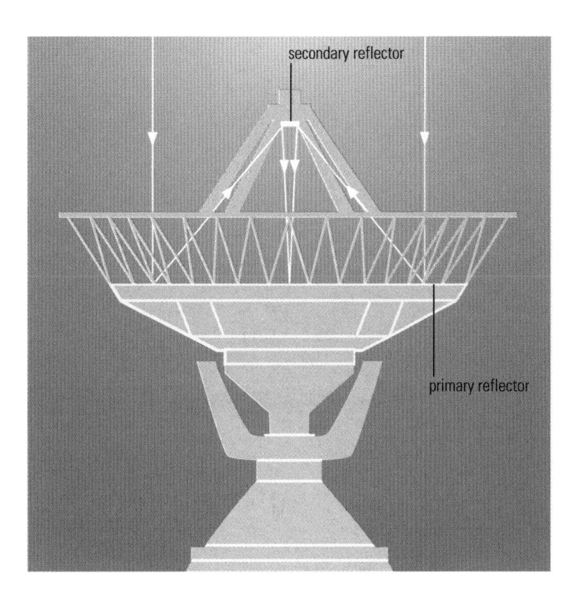

◀ **radio telescope** Just like an optical reflector, a radio telescope depends on collection of electromagnetic radiation by a large aperture, parabolic reflector, often in the form of a dish. Radio waves are brought to a focus on the detector. Most radio telescopes can be steered in altitude and azimuth.

pulsar A **neutron star** radio source which does not emit continuously, but in rapid, very regular pulses. Their periods are short (often much less than one second).

Purkinje effect An effect inherent in the human eye, which makes it less sensitive to light of longer wavelength when the general level of intensity is low. Consider two lights, one red and one blue, which are of equal intensity. If the intensity of both are reduced by equal amounts, the blue light will appear to be the brighter of the two.

Q

quadrature The position of the Moon or a planet when at right angles to the Sun as seen from Earth. Thus the Moon is in quadrature when it is seen at half-phase.

quantum The smallest amount of light-energy which can be transmitted at any given wavelength.

quasar A very remote immensely luminous object, now known to be the core of a very active galaxy – possibly powered by a massive black hole inside it. Quasars are also known as QSOs (Quasi-Stellar Objects). *BL Lacertae objects* are of the same type, though less important.

R

radar astronomy The technique of using radar pulses to study astronomical objects. Most planets and some asteroids have been contacted by radar, and the radar equipment carried in space probes such as Magellan has provided us with detailed maps of the surface of Venus.

radial velocity The towards-or-away movement of a celestial body, measured by the **Doppler effect** in its spectrum. If the spectral lines are red-shifted, the object is receding; if the shift is to the blue, the object is approaching. Conventionally, radial velocity is said to be positive with a receding body, negative with an approaching body.

radiant Circular area of sky, usually taken as a matter of observational convenience to have a diameter of 8°, from which a **meteor shower** appears to emanate. Radiant positions are normally expressed in terms of right ascension and declination. Meteor showers usually take their name from the constellation in which the radiant lies – the Perseids from Perseus, Geminids from Gemini, and so on. The radiant effect, with meteors appearing anywhere in the sky with divergent paths whose backwards projection meets in a single area, is a result of perspective: in reality, shower members have essentially parallel trajectories in the upper atmosphere. Meteors close to the radiant appear foreshortened, whilst those 90° away have the greatest apparent path lengths. Meteor shower radiants move eastwards relative to the sky background by about one degree each day thanks to Earth's orbital motion.
A meteor shower radiant position may be determined from the backwards projection of plotted or photographed trails. The radiant can also be found from parallactic observations of a single meteor recorded at geographical locations separated by a few tens of kilometres.

radio astronomy Astronomical studies carried out in the long-wavelength region of the **electromagnetic spectrum**. The main instruments used are known as **radio telescopes**; they are of many kinds, ranging from 'dishes', such as the 76-m (250-ft) **paraboloid** at Jodrell Bank (Cheshire), to long lines of aerials.

radio telescope Instrument used to collect and measure electromagnetic radiation emitted by astronomical bodies in the radio region of the spectrum. The radio region extends from around 10 mm (0.4 in) to around 10–20 m (33–66 ft). Almost all the types of object studied with optical telescopes have also been observed with radio telescopes. These include the Sun, planets, stars, gaseous nebulae and galaxies. Furthermore, **radio astronomy** has been responsible for the discovery of several new and unsuspected types of astronomical phenomena, such as quasars, pulsars and the cosmic microwave background (*see* **electromagnetic spectrum**). A single radio telescope does not produce a picture directly (unlike an optical telescope); the dish must be scanned forwards and backwards to build it up, and the data then fed through a computer. The computer is also vital for controlling the telescope (which may weigh 1000 tonnes), pointing it accurately at the required position in the sky, moving continually in two coordinates (altitude and azimuth), and correcting for the inevitable deformation of a large dish. The accuracy required is often better than 10″, which becomes millimetres when translated into dish movements.

▶ **refractor** The optical layout of a refractor is shown in this schematic cutaway. Light is collected by the main, objective lens at the front end of the telescope and brought to a focus. The observer views a magnified image of the focused light through a smaller lens, or set of lenses, making up the eyepiece. The telescope shown here is on a German equatorial

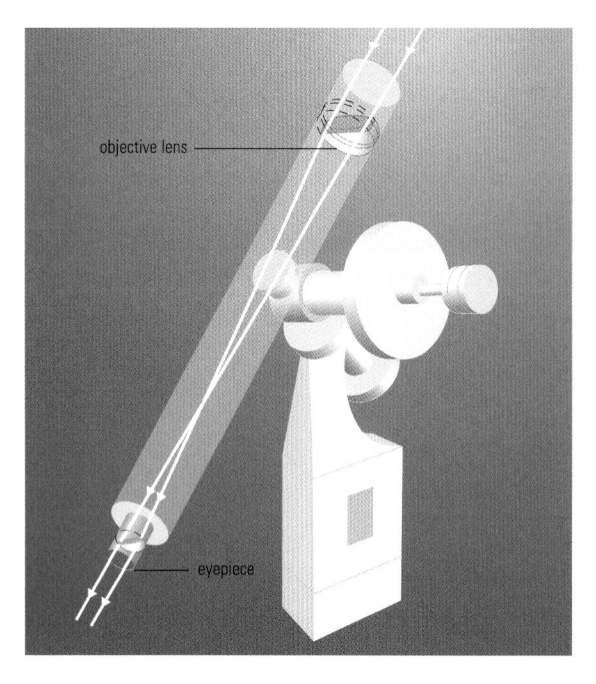

Most radio telescopes are not located inside buildings, so the computers have to correct for weather too, such as high winds, cold and heat. Radio telescopes can also be adversely influenced by ice, rain and lightning.

radio galaxies Galaxies which are extremely powerful emitters of radio radiation.

red dwarf Star at the lower (cool) end of the **Main Sequence** with spectral type K or M. Red dwarfs' surface temperatures are between 2500 and 5000 K and their masses are in the range 0.08 to 0.8 times that of the Sun. Red dwarfs are the most common type of star in our Galaxy, comprising at least 80 per cent of the stellar population.

red shift The **Doppler** displacement of spectral lines towards the red or long-wave end of the spectrum, indicating a velocity of recession. Apart from the members of the **Local Group,** all galaxies show red shifts in their spectra.

reflector (or reflecting telescope) A telescope in which the light is collected by means of a mirror.

refraction The change in direction of a ray of light when passing from one transparent substance into another.

refractor (or refracting telescope) A telescope which collects its light by means of a lens. The light passes through this lens (**object-glass**) and is brought to focus; the image is then magnified by an eyepiece.

▶ **Roche lobe** In a binary star system, the stars (here marked M_1 and M_2) have their own gravitational sphere of influence, defined by the Roche lobe. Dependent on the evolutionary state and relative masses of the partners in a binary system, material from one star, or both, may fill the respective Roche lobes. Overspill of material from a distended, highly evolved star from its Roche lobe into that of a smaller more massive partner is an important driving mechanism in several forms of cataclysmic variable stars.

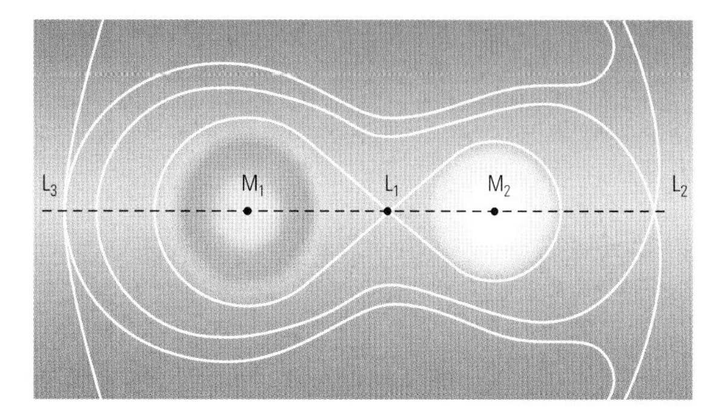

resolving power The ability of a **telescope** to separate objects which are close together; the larger the telescope the greater its resolving power. Radio telescopes (see **radio astronomy**) have poor resolving power compared with optical telescopes.

retardation The difference in the time of moonrise between one night and the next. It may exceed one hour, or it may be as little as a quarter of an hour.

retrograde motion In the **Solar System**, movement in a sense opposite to that of the Earth in its orbit; some **comets**, notably Halley's, have retrograde motion. The term is also used with regard to the apparent movements of planets in the sky; when the apparent motion is from east to west, relative to the fixed stars, the direction is retrograde. The term may be applied to the rotations of planets and satellites. Since Uranus has an axial inclination of more than a right angle, its rotation is technically retrograde; Venus also has retrograde axial rotation.

reversing layer The gaseous layer above the bright surface or **photosphere** of the Sun. Shining on its own, the gases would yield bright spectral lines; but as the photosphere makes up the background, the lines are reversed, and appear as dark absorption or **Fraunhofer lines**. Strictly speaking, the whole of the Sun's **chromosphere** is a reversing layer.

right ascension The right ascension of a celestial body is the time which elapses between the culmination of the **First Point of Aries** and the culmination of the body concerned. For example, Aldebaran in Taurus culminates 4h 33m after the First Point of Aries has done so; therefore the right ascension of Aldebaran is 4h 33m. The right ascensions of bodies in the **Solar System** change quickly. However, the right ascensions of stars do not change, apart from the slow cumulative effect of **precession**.

Roche limit The distance from the centre of a planet, or other body, within which a second body would be broken up by gravitational distortion. This applies only to an orbiting body which has no appreciable structural cohesion, so that strong, solid objects, such as artificial satellites, may move safely well within the Roche limit for the Earth. The Roche limit lies at 2.44 times the radius of the planet from the centre of the globe, so that for the Earth it is about 9170 km (above ground-level. Saturn's ring system lies within the Roche limit for Saturn.

Roche lobe Surface that defines the maximum sizes of stars in a **binary star** system relative to their separation. If both components are well within their Roche lobes, the system is termed a detached binary. If one component fills its Roche lobe, the companion's gravity will pull matter off it. *See* **close binary**

RR Lyrae variables Regular **variable stars** whose periods are very short (between about 70–80 minutes and about 30 hours). They seem to be fairly uniform in luminosity; each is around 100 times as luminous as the Sun. They can therefore be used for distance measures, in the same way as **Cepheid** variables. Many of them are found in star clusters, and they were formerly known as cluster Cepheid variables. No RR Lyrae variable appears bright enough to be seen with the naked eye.

S

Saros A period of 18 years 11.3 days, after which the Earth, Moon and Sun return to almost the same relative positions. Therefore, an **eclipse** of the Sun or Moon is liable to be followed by a similar eclipse 18 years 11.3 days later. The period is not exact, but is good enough for predictions to be made – as was done in ancient times by Greek philosophers.

satellite A secondary body orbiting a primary. The Earth has one satellite (the Moon); Jupiter has 53, Saturn 30, Uranus 23, and Neptune 11, while Mercury and Venus are unattended.

Schmidt camera Type of **reflecting telescope** incorporating a spherical primary mirror and a specially shaped glass correcting plate to achieve a wide field of view combined with fast focal ratio. The conventional paraboloidal primary mirrors of reflecting telescopes only form a perfect point image as long as the incoming parallel light is exactly aligned with the axis. Departure from this causes steadily increasing degradation of the images due to coma. The Schmidt camera, which was invented by Estonian instrument-maker and astronomer Bernhard Schmidt in 1930, overcomes this problem by the use of a spherical rather than a paraboloidal, primary mirror. This eliminates the problem of coma but introduces **spherical abberation**. To correct this, a thin glass plate, known as a 'Schmidt corrector' and with a very shallow profile worked into one or both of its surfaces, is placed at the centre of curvature. The profile is complex in shape and has to be precisely computed to match and counteract the spherical aberration of the mirror, thus producing perfect images over a wide field of view. Because the image is formed on a curved surface, the photographic film or plate must be deformed by clamping it in a suitably shaped holder. Schmidt cameras are used photographically to record detailed images of very large areas of sky. This makes them ideal for observing extensive objects such as comets or certain nebulæ, or for recording large numbers of more compact sources, such as stars or distant galaxies, for statistical studies. Perhaps their most important role is as survey telescopes used for identifying unusual or specific kinds of object, which can then be investigated in more detail with larger, conventional (narrow-field) instruments. The largest Schmidt in use is the 122-cm (48-in) instrument at the Palomar Observatory.

scintillation Twinkling of stars. It is due entirely to the effects of the Earth's atmosphere; a star will scintillate most violently when it is low over the horizon, so that its light is passing through a thick layer of atmosphere. A planet, which shows up as a small disk rather than a point, will generally twinkle much less than a star.

seasons Effects on the climate due to the inclination of the Earth's axis. The fact that the Earth's distance from the Sun is not constant has only a minor effect upon our seasons.

secular acceleration Because of friction produced by the tides, the Earth's rotation is gradually slowing down; the 'day' is becoming longer. The average daily lengthening is only 0.00000002 seconds, but over a sufficiently long period the effect becomes detectable. The lengthening of terrestrial time periods gives rise to an apparent speeding-up of the periods of the Sun, Moon and planets. Another result of these tidal phenomena is that the Moon is receding from the Earth slowly.

seeing The quality of the steadiness and clarity of a star's image. It depends upon conditions in the Earth's atmosphere. From the Moon, or from space, the 'seeing' is always perfect.

seismometer An earthquake recorder. Very sensitive seismometers were taken to the Moon by the Apollo astronauts, and provided interesting information about seismic conditions there.

selenography The study of the Moon's surface.

sextant An instrument used for measuring the altitude of a celestial body above the horizon.

Seyfert galaxies Galaxies with small, bright nuclei. Many of them are radio sources, and show evidence of violent disturbances in their nuclei.

shooting star The luminous appearance caused by a meteor falling through the Earth's atmosphere.

sidereal period The time taken for a planet or other body to make one journey around the Sun (365.2 days in the case of the Earth). The term is also used for a satellite in orbit around a planet. Also known as periodic time.

sidereal time The local time reckoned according to the apparent rotation of the celestial sphere. It is zero hours when the **First Point of Aries** crosses the observer's meridian. The sidereal time for any observer is equal to the right ascension of an object which lies on the meridian at that time. Greenwich sidereal time is used as the world standard (this is, of course, merely the local sidereal time at Greenwich Observatory).

singularity Point in space or spacetime at which the current laws of physics make non-real predictions for the values of some quantities. Thus at the centre of a **black hole**, the density, the force of gravity and the curvature of spacetime are all predicted to be infinite.

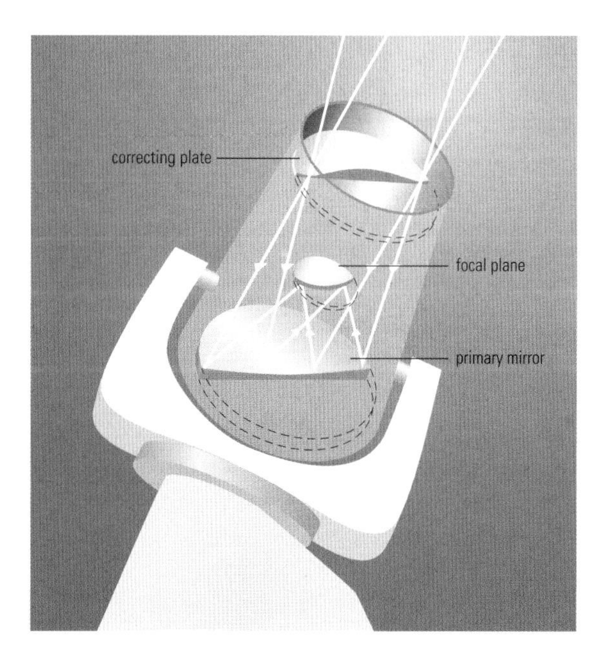

◀ Schmidt camera This type of reflecting telescope gives a wide field of view. They are used only for astrophotography and are useful in survey work.

solar apex The point on the celestial sphere towards which the Sun is apparently travelling. It lies in the constellation Hercules; the Sun's velocity towards the apex is 19 km/s (1 mi/s). The point directly opposite in the sky to the solar apex is termed the solar *antapex*. This motion is distinct from the Sun's rotation around the centre of the Galaxy, which amounts to about 320 km/s (200 mi/s).

solar constant The unit for measuring the amount of energy received on the Earth's surface by solar radiation. It is equal to 1.94 calories per minute per square cm. (A calorie is defined as the amount of heat needed to raise the temperature of 1 g of water by 1°C.)

solar cycle Cyclical variation in solar activity with a period of about 11 years between maxima (or minima). The solar cycle is characterized by waxing and waning of various forms of solar activity, such as **active regions**, **coronal mass ejections**, **flares**, the **solar constant** and **sunspots**. The rise to maximum activity is usually much more rapid than the subsequent decline.

solar flares See **flares, solar**.

solar interior The Sun's internal structure. It comprises three principal regions: a central core, above which is found the radiative zone, and finally the convective zone, from which material rises to the visible surface of the **photosphere**.

solar parallax The trigonometrical parallax of the Sun. It is equal to 8.79 seconds of arc.

Solar System The system made up of the Sun, the planets, satellites, comets, asteroids, meteoroids and interplanetary dust and gas.

solar time, apparent The local time reckoned according to the Sun. Noon occurs when the Sun crosses the observer's meridian, and is therefore at its highest in the sky.

solar wind A steady flow of atomic particles streaming out from the Sun in all directions. It was detected by means of space probes, many of which carry instruments to study it. Its velocity in the neighbourhood of the Earth exceeds 965 km/s (600 mi/s). The intensity of solar wind is enhanced during solar storms. It is the solar wind that causes the tails of comets always to point away from the Sun.

▶ solar interior Not only does the Sun rotate faster at the equator than at the poles but also belts of plasma have been observed moving at different speeds. These belts drift down towards the equator during the solar cycle and, as sunspots appear to form on the boundaries, may be a major factor in the changes in the Sun's magnetic field. There are also deeper currents within the convective zone, which travel from the equator towards the pole.

solstices Times when the Sun is at its northernmost point in the sky (dec. +23.5°, around 22 June), or at its southernmost point (−23.5°, around 22 December). The dates of the solstices vary somewhat, because of the calendar irregularities due to leap years.

spacesuit Equipment designed to allow an astronaut to operate outside the atmosphere.

specific gravity The density of any substance compared with that of an equal volume of water.

spectroheliograph An instrument used for photographing the Sun in the light of one particular wavelength only. If adapted for visual use, it is known as a spectrohelioscope.

spectroscope An instrument used to analyse the light on a star or other luminous object. Astronomical spectroscopes are used in conjunction with telescopes. Without them our knowledge of the nature of the universe would still be very rudimentary.

spectroscopic binary A **binary star** that is too close for its components to be resolved optically; its orbital motion is deduced from periodic shifts in its spectral lines, indicating variable radial velocity. A spectroscopic binary is likely to be a **close binary** with a short orbital period and relatively high radial velocity. Binaries cannot be detected spectroscopically if their orbit is perpendicular to the Earth. If the components have similar brightness, both sets of **spectral lines** are observed, and the system is called a double-lined spectroscopic binary. A composite spectrum binary has a spectrum that consists of two sets of lines from stars of dissimilar spectral type. Systems with only one set of spectral lines observable are known as single-lined spectroscopic binaries. The **radial velocity** of the star (or stars) is determined by measuring the **Doppler shift** of the spectral lines. The individual masses can be determined if the inclination of the orbit is known, for example when the system is also an **eclipsing binary**.

spectrum binary A **binary system** where a **spectrogram** of an apparently single star reveals two sets of spectral lines characteristic of two different types of stars.

speculum The main mirror of reflecting telescope (see **reflector**). Older mirrors were made of speculum metal; modern ones are generally of glass.

spherical aberration The blurred appearance of an image as seen in a telescope, due to the fact that the lens or mirror does not bring the rays falling on its edge and on its centre to exactly the same focus. If the spherical aberration is noticeable, the lens or mirror is of poor quality, and should be corrected.

spicules Jets up to 16,000 km (10,000 mi) in diameter, in the solar chromosphere. Each lasts for 4–5 minutes.

star A self-luminous gaseous body. The Sun is a typical star.

steady-state theory A theory according to which the universe has always existed, and will exist for ever. The theory has now been abandoned by almost all astronomers.

stratosphere The layer in the Earth's atmosphere lying above the **troposphere**. It extends from c. 11–64 km (7–40 mi) above sea level.

sublimation The change of a solid body to the gaseous state without passing through a liquid condition. (This may well apply to the polar caps on Mars.)

sundial An instrument used to show the time, by using an inclined style, or gnomon, to cast a shadow on to a graduated dial. The gnomon points to the celestial pole. A sundial gives apparent time; to obtain mean time, the value shown on the dial must be corrected by applying the **equation of time**.

Sun-grazers These are **comets** which at **perihelion** make very close approaches to the Sun. All the sun-grazers are brilliant comets with extremely long periods.

sunspots Darker patches on the solar photosphere; their temperature is about 4,000°C (as against about 6,000°C for the general photosphere), so that they are dark only by contrast; if they could be seen shining on their own, their surface brilliance would be greater than that of an arc-light. A large sunspot consists of a central darkish area or umbra, surrounded by a lighter area or penumbra, which may be very extensive and irregular. Sunspots tend to appear in groups, and are associated with strong magnetic fields; they are also associated with **faculae** and with solar **flares**. They are most common at the time of solar maximum (approximately every 11 years). No sunspot lasts for more than a few months at most.

supergiant stars Stars of exceptionally low density and great luminosity. Betelgeux in Orion is a typical supergiant.

superior conjunction The position of a planet when it is on the far side of the Sun as seen from Earth.

superior planets The planets beyond the orbit of the Earth in the Solar System: that is to say, all the principal planets apart from Mercury and Venus.

supernova A colossal stellar outburst. A Type I supernova involves the total destruction of the white dwarf component of a **binary** system; a Type II supernova is produced by the collapse of a very massive star. At its peak, a supernova may exceed the combined luminosity of all the other stars of an average galaxy.

synchronous satellite An artificial satellite moving in a west-to-east equatorial orbit in a period equal to that of the Earth's axial rotation (approximately 24 hours): as seen from Earth the satellite appears to remain stationary, and is of great value as a communications relay. Many synchronous satellites are now in orbit.

synchrotron radiation Radiation which is emitted by charged particles moving at relativistic velocities in a strong magnetic field. Much of the radio radiation coming from the **Crab Nebula** is of this type.

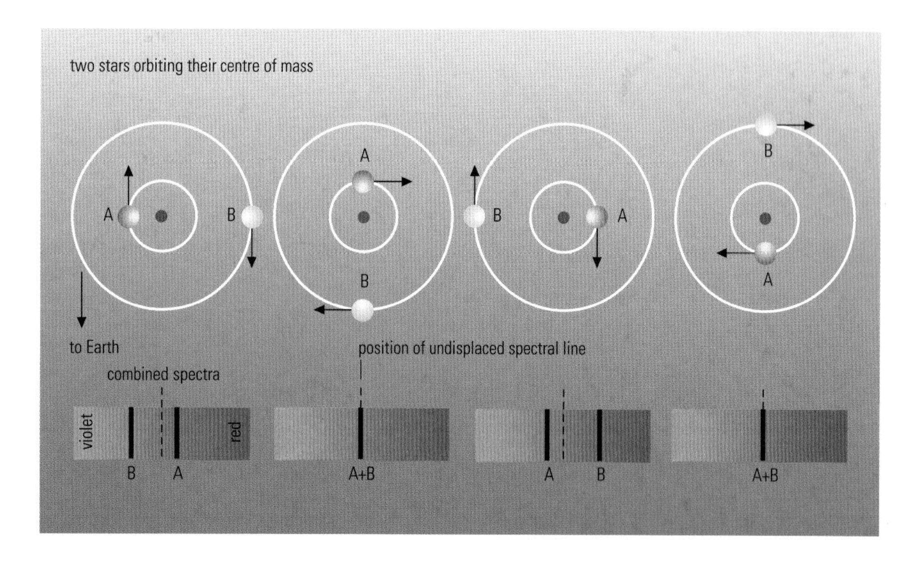

▲ spectroscopic binary Redshift is used to ascertain the relative motions of stars in close binary pairs. When a star is moving away, its spectrum is shifted to the red, whereas the spectrum of a star moving towards us will be blueshifted. When both stars are moving at right angles to our line of sight their spectra are not displaced.

synodic period The interval between successive oppositions of a **superior planet**. For an **inferior planet**, the term is taken to mean the interval between successive conjunctions with the Sun.

syzygy The position of the Moon in its orbit when at new or full phase.

T

tektites Small, glassy objects which are aerodynamically shaped, and seem to have been heated twice. It has been suggested that they are meteorite, but it is now generally believed that they are of terrestrial origin.

telemetry The technique of transmitting the results of measurements and observations made on instruments in inaccessible positions (such as unmanned probes in orbit, in deep space or on other planets) to a point where they can be used and analysed.

telescope The main instrument used to collect the light from celestial bodies, thereby producing an image which can be magnified. There are two main types: the reflector and the refractor. All the world's largest telescopes are reflectors, because a mirror can be supported by its back, whereas a lens has to be supported around its edge – and if it is extremely large, it will inevitably sag and distort under its own weight, thereby rendering itself useless.

terminator The boundary between the day and night hemispheres of the Moon or a planet. Since the lunar surface is mountainous, the terminator is rough and jagged, and isolated peaks may even appear to be detached from the main body of the Moon. Mercury and Venus, which also show lunar-type phases, seem to have almost smooth terminators, but this is probably because we cannot see them in such detail (at least in the case of Mercury, whose surface is likely to be as mountainous as that of the Moon). Mars also shows a smooth terminator, although it is now known that the surface of the planet is far from being smooth and level. Photographs of the Earth taken from space or from the Moon show a smooth terminator which appears much 'softer' than that of the Moon, because of the presence of atmosphere.

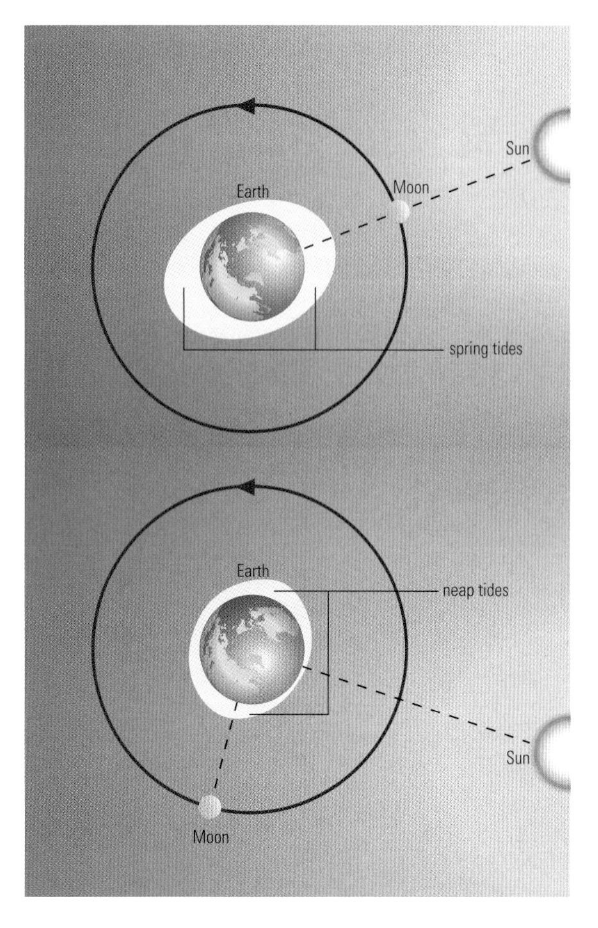

▶ **tides** The gravitational effects of the Sun and Moon combine to distort the oceans on Earth. When the Sun and Moon are lying in the same direction, high tides are at their highest and the difference between high and low tides is at its greatest magnitude. When the Sun and Moon are pulling in different directions, neap tides result – high tides are not particularly high, and the difference between high and low tide is less.

thermocouple An instrument used for measuring very small quantities of heat. When used in conjunction with a large telescope, it is capable of detecting remarkably feeble heat-sources.

tidal bulge Deformation of a celestial body in orbit around another as a result of the higher gravitational attraction on the nearside and lower attraction on the farside of the body, compared with the attraction at the centre of the body. The resulting distortion lies along the line joining the two bodies, with bulges both towards and away from the attracting body (*see* **tides**). The effect on a fluid ocean is easily understood, but even solid bodies are not completely rigid and will deform because of tidal forces. The Sun raises tidal bulges on the planets, and planets raise tidal bulges on the satellites orbiting them. Satellites also raise tidal bulges on their planets. In the case of the Earth and Moon, the tides raised on the Earth are large, because the Moon is a fairly large fraction ($\frac{1}{81}$) of the mass of the Earth. However even a tiny satellite such as Phobos raises tides on its planet, Mars. These tides are small, and have no significant effect on Mars, but the attraction of this bulge acting back on Phobos causes very significant **tidal evolution** of the orbit of Phobos.

tidal evolution Mechanism that can cause changes to the spin rates of planets and satellites, and to the orbital radii of satellites. The rates of change are very slow, but the accumulated effect can be very significant over the age of the Solar System. The attraction of one body on another raises a **tidal bulge** on that body. **Tidal friction** causes a slight delay in the rise and fall of the bulge, with the result that it does not quite lie along the line between the two bodies: it will either lead or lag, depending on whether the attracting body is orbiting faster or slower than the body is rotating. Thus the gravitational attraction between a satellite and its planet is asymmetrical, with the force having a component at right angles to the line between the bodies; this force exerts a torque. In the case of a satellite orbiting slower than the planet is rotating (that is, it is above the synchronous orbit), the nearer bulge exerts a larger torque than the farther bulge, and so there is a net torque acting in the direction of motion of the satellite, and an equal and opposite torque acting on the rotation of the planet. This causes the satellite to spiral outwards, and slows the rotation of the planet. A satellite below the synchronous orbit would spiral inwards, and the planet would rotate faster.

tidal friction Dissipation of a small fraction of the energy contained in a **tidal bulge** as it rises and falls. In the case of the Earth the main dissipation of tidal energy arises from friction at the bottom of shallow seas as the tides sweep through. In bodies without oceans the dissipation mechanisms are not so obvious, and estimates of the rate of dissipation are very uncertain. In a few cases the dissipation of tidal energy within a body can cause significant heating of the body, such as for Jupiter's satellite Io. More often the most significant effect of the tidal friction is to cause a slight delay in the rise and fall of the tidal bulge, with the result that it does not quite lie along the line between the two bodies. This slight asymmetry causes a torque between the two bodies, which can cause **tidal evolution**.

tides Distortions induced in a celestial body by the gravitational attraction of one or more others. The force a body experiences is greatest on the side nearest the attracting body, and least on the side farthest away, causing it to elongate slightly towards and away from the attracting body, acquiring a **tidal bulge** on each side. On the Earth tides are raised by the Moon and by the Sun, the effect of the Moon being about three times that of the Sun. As the Earth rotates, different parts of the surface move into the tidal bulge region, causing two high tides and two low tides in just over a day. The tides most obviously affect the fluid ocean, but even the solid crust, which is supported by a fluid mantle, is able to flex, experiencing tides of up to 25-cm (10-in.) amplitude. When the Sun and the Moon are exerting a pull in the same direction (as at new moon or full moon) their effects are additive and the high tides are higher (spring tides). When the pull of the Sun is at right angles to that of the Moon (as at first or last quarter) the high tides are lower (neap tides). **Tidal friction** caused by the tidal ebb and flow of water over the ocean floor causes a slight delay in the response of the oceans to the tidal attracting forces, with the result that the tidal bulge does not lie precisely along the line to the attracting body.

time dilation effect According to relatively theory, the 'time' experienced by two observers in motion compared with each other will not be the same. To an observer moving at near the velocity of light, time will slow down; also, the observer's mass will increase until at the actual velocity of light, time will stand still and mass will become infinite! The time and mass effects are negligible except for very high velocities, and at the speeds of modern rockets they may be ignored completely.

transfer orbit (or Hohmann orbit) The most economical orbit for a spacecraft which is sent to another planet. To carry out the journey by the shortest possible route would mean continuous expenditure of fuel, which is a practical impossibility. What has to be done is to put the probe into an orbit which will swing it inwards or outwards to the orbit of the target planet. To reach Mars, the probe is speeded up relative to the Earth, so that it moves outwards in an elliptical orbit; calculations are made so that the probe will reach the orbit of Mars and rendezvous with the planet. To reach Venus, the probe must initially be slowed down relative to the Earth, so that it will swing inwards towards the orbit of Venus. With a probe moving in a transfer orbit, almost all the journey is carried out in free fall, and no propellant is being used. On the other hand, it means that the distances covered are increased, so that the time taken for the journey is also increased.

transit (1) The passage of a celestial body, or a point on the celestial sphere, across the observer's meridian; thus the **First Point of Aries** must transit at 0 hours **sidereal time**. (2) Mercury and Venus are said to be in transit when they are seen against the disk of the Sun at inferior conjunction. Transits of Mercury are quite frequent (e.g. in 1973, 1986,1993, 1999, 2003 and 2006 and the next in 2016, 2019 and 2032). Transits of Venus are rarer; they occur in pairs (e.g., 1874 and 1882). Th last occurred in 2004, to be followed by another in 2012, then another pair in 2117 and 2125. A satellite of a planet is said to be in transit when it is seen against the planet's disk. Transits of the four large satellites of Jupiter may be seen with small telescopes; also visible are shadow transits of these satellites, when the shadows cast by the satellites are seen as black spots on the face of Jupiter.

transit instrument A **telescope** which is specially mounted; it can move only in elevation, and always points to the meridian. Its sole use is to time the moments when stars cross the meridian, so providing a means of checking the time. The transit instrument set up at Greenwich Observatory by Sir George Airy, in the nineteenth century, is taken to mark the Earth's prime meridian (longitude 0°).

trojans Groups of **asteroids** which move around the Sun at a mean distance equal to that of Jupiter. One group of Trojans keeps well ahead of Jupiter and the other group well behind. Hundreds of Trojans are now known.

troposphere The lowest part of the Earth's atmosphere, reaching to an average height of about 11 km (7 mi) above sea-level. It includes most of the mass of the atmosphere, and all the normal clouds lie within it. Above, separating the troposphere from the **stratosphere**, is the tropopause.

twilight, astronomical The state of illumination of the sky when the Sun is below the horizon, but by less than 18°.

twinkling Common term for **scintillation**.

U

ultra-violet radiation Electromagnetic radiation which has a wavelength shorter than that of violet light, and so cannot be seen with the naked eye. The ultra-violet region of the **electromagnetic spectrum** lies between visible light and X-ray radiation. The Sun is a very powerful source of ultra-violet, but most of this radiation is blocked out by layers in the Earth's upper atmosphere – which is fortunate for us, since in large quantities ultra-violet radiation is lethal. Studies of ultra-violet radiation emitted by the stars have to be carried out by means of instruments sent up in rockets or artificial satellites.

umbra (1) The dark inner portion of a sunspot. (2) The main cone of shadow cast by a planet or the Moon.

Universal Time The same as Greenwich Mean Time.

V

Van Allen belts Radiation belt regions in the terrestrial **magnetosphere** in which charged particles are trapped as they spiral along the magnetic field and bounce up and down between reflection points located towards the magnetic poles. These regions were discovered in 1958 by James Van Allen on analysis of Geiger counter data from the Explorer 1 satellite. The inner region of the Van Allen belts is located between 1000 and 5000 km (600 and 3000 mi) above the equator; it contains protons and electrons, which are either captured from the **solar wind** or originate from collisions between upper atmosphere atoms and high-energy **cosmic rays**. The outer belt region is located between 15,000 and 25,000 km (9000 and 16,000 mi) above the equator, but it curves downwards towards the magnetic poles. This region contains mainly electrons from the solar wind. The Van Allen belts are a potential hazard to Earth-orbiting spacecraft, since the radiation levels are high enough to have an adverse effect on the electronic subsystems and on-board instruments, especially following a **geomagnetic storm**. Similar radiation belt regions of trapped, charged particles have been discovered around Jupiter, Saturn, Uranus and Neptune.

variable stars Stars which fluctuate in brightness over short periods of time. In extrinsic variables, the amount of light reaching us is affected by external factors such as obscuring dust and eclipses. Intrinsic variables vary because of some process such as events in their atmosphere or pulsations.

variation (1) An inequality in the motion of the Moon, due to the fact that the Sun's pull on it throughout its orbit is not constant in strength. (2) Magnetic variation: the difference, in degrees, between magnetic north and true north. It is not the same for all places on the Earth's surface, and it changes slightly from year to year because of the wandering of the magnetic pole.

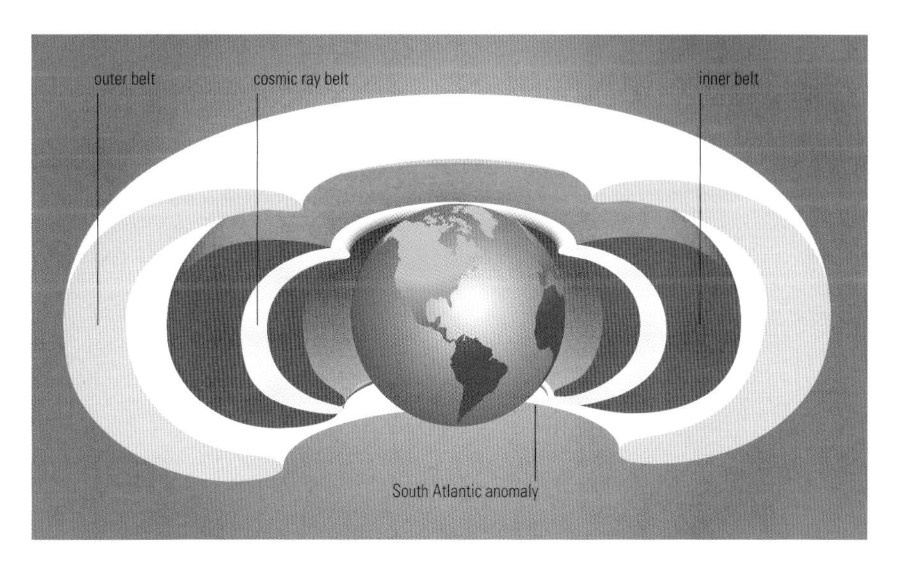

▼ Van Allen belts Earth is surrounded by radiation belts, which contain energetic particles trapped by the magnetic field. The inner belt dips low over the South Atlantic, where its particle population can present a hazard to satellites.

Vernal Equinox See **First Point of Aries**.

visual binary A **binary star** that can be resolved into separate components and for which orbital motion can be observed.

Vulcan The name given to a hypothetical planet once believed to move around the Sun at a distance less than that of Mercury. It is now certain that Vulcan does not exist.

W

white dwarf A very small, extremely dense star. The atoms in it have been broken up and the various parts packed tightly together with almost no waste space, so that the density rises to millions of times that of water; a spoonful of white dwarf material would weigh many tonnes. Evidently a white dwarf has used up all its nuclear 'fuel'; it is in the last stages of its active career, and has been aptly described as a bankrupt star. **Neutron stars** are even smaller and denser than white dwarfs.

Widmanstätten patterns If an iron **meteorite** is cut, polished and then etched with acid, characteristic figures of the iron crystals appear. These are known as Widmanstätten patterns. They are never found except in meteorites.

Wolf-Rayet star Exceptionally hot, greenish-white star whose spectrum contain bright emission lines as well as the usual dark absorption lines. The surface temperature of such stars may approach 100,000°C (180,000°F), and they seem to be surrounded by rapidly expanding envelopes of gas. Attention was first drawn to them in 1867 by the astronomers Wolf and Rayet, after whom the class is named. Recently, it has been found that many of the Wolf-Rayet stars are **spectroscopic binaries**.

wormhole Connection of two black holes either from two locations in our universe, or one from our universe and one from another, separate universe. These black holes are connected by their **singularities** and theoretically can provide a bridge or shortcut from one location to another, or from one universe to another. Wormholes are also called Einstein–Rosen bridges after the first physicists to consider the mathematical possibility of such connections as suggested by general relativity.

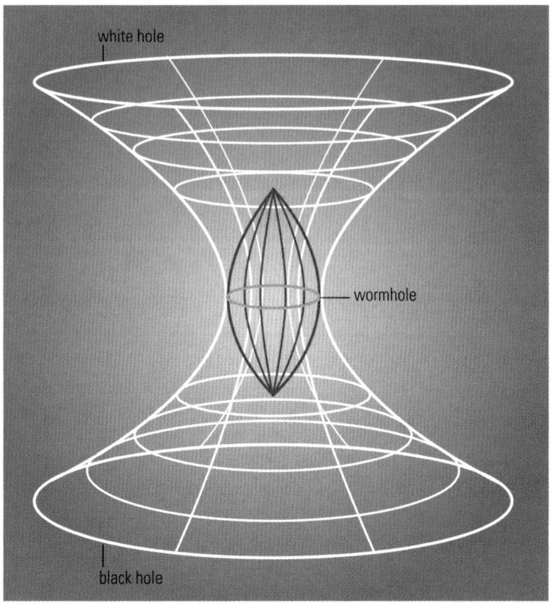

white hole
wormhole
black hole

▶ **wormhole** This is a mathematical representation of the distortion of spacetime produced by a Schwarzschild wormhole connecting two universes via a black hole and a white hole. While predicted by Einstein's equations, such wormholes cannot exist in reality, since the occurrence of white holes is forbidden by the second law of thermodynamics.

X

X-rays Electromagnetic radiations of very short wavelength. There are many X-ray sources in the sky; studies of them must be undertaken by space research methods.

X-ray astronomy X-rays are very short electromagnetic radiations, with wavelengths of from 0.01 to 10 **nanometres.** Since X-rays from space are blocked by the Earth's atmosphere, astronomical researches have to be carried out by means of instruments taken up in rockets. The Sun is a source of X-rays; the intensity of the X-radiation is greatly enhanced by solar **flares**. Sources of X-rays outside the **Solar System** were first found in 1962 by American astronomers, who located two sources, one in Scorpius and the other in Taurus; the latter has now been identified with the Crab Nebula. Since then, various other X-ray sources have been discovered, some of which are variable.

Y

year The time taken for the Earth to go once around the Sun; in everyday life it is taken to be 365 days (366 days in Leap Years). (1) Sidereal year: The true revolution period of the Earth: 365.26 days, or 365 days 6 hours 9 minutes 10 seconds. (2) Tropical year: The interval between successive passages of the Sun across the **First Point of Aries**. It is equal to 365.24 days, or 365 days 5 hours 48 minutes 45 seconds. The tropical year is about 20 minutes shorter than the sidereal year because of the effects of precession, which cause a shift in the position of the First Point of Aries. (3) Anomalistic year: The interval between successive **perihelion** passages of the Earth. It is equal to 365.26 days, or 365 days 6 hours 13 minutes 53 seconds. It is slightly longer than the sidereal year because the position of the perihelion point moves by about 11 seconds of arc annually. (4) Calendar year: The mean length of the year according to the Gregorian calendar. It is equal to 365.24 days, or 365 days 5 hours 49 minutes 12 seconds.

Z

zenith The observer's overhead point (altitude 90 degrees).

zenith distance The angular distance of celestial body from the observer's zenith.

Zodiac A belt stretching right round the sky, 8 degrees to either side of the **ecliptic**, in which the Sun, Moon and bright planets are always to be found. It passes through 13 constellations, the 12 commonly known as the Zodiacal groups plus a small part of Ophiuchus (the Serpent-bearer).

Zodiacal constellations The 12 constellations used in **astrology**. They are Aquarius, Aries, Cancer, Capricornus, Gemini, Leo, Libra, Pisces, Sagittarius, Scorpius, Taurus and Virgo.

Zodiacal light A cone of light rising from the horizon and stretching along the ecliptic. It is visible only when the Sun is a little way below the horizon, and is best seen on clear, moonless evenings or mornings. It is thought to be due to small particles scattered near the main plane of the **Solar System**. A fainter extension along the ecliptic is known as the Zodiacal band.

Index

D

E

N

T

Acknowledgements

Photographic Credits

Abbreviations used are:
t top; c centre; b bottom; l left; r right

AU	Associated Universities, Inc.
AURA	Association of Universities for Research in Astronomy, Inc.
Caltech	California Institute of Technology
DLR	Deutschen Zentrum für Luft- und Raumfahrt
ESA	European Space Agency
ESO	European Southern Observatory
GSFC	Goddard Space Flight Center
JHU	Johns Hopkins University
JPL	Jet Propulsion Laboratory
JSC	Johnson Space Center
MIT	Massachusetts Institute of Technology
NASA	National Aeronautics and Space Administration
NSO	National Solar Observatory
NOAO	National Optical Astronomy Observatory
NRAO	National Radio Astronomy Observatory
NSF	National Science Foundation
PM	Patrick Moore Collection
SOHO	Solar and Heliospheric Observatory. SOHO is a project of international cooperation between ESA and NASA
STScI	Space Telescope Science Institute
SwRI	Southwest Research Institute, Boulder, Colorado
TRACE	Transition Region and Coronal Explorer, Lockheed Martin Solar and Astrophysics Laboratories
USGS	US Geological Survey

1 ESA & NASA/ E Olszewski (University of Arizona)/HST; **2** NASA/JPL-Caltech/E Churchwell (University of Wisconsin-Madison); **6** STS-121 Shuttle Crew/NASA; **8** NASA/JPL-Caltech/R A Gutermuth (Harvard Smithsonian CfA); **10–11** NASA; **12** Patrick Moore; **13** Robin Rees; **14** Patrick Moore; **15** Nik Szymanek; **16** Patrick Moore; **17** Patrick Moore; **18** Patrick Moore; **19** NOAO/AURA/NSF; **20** Ian Morison Jodrell Bank; **21** NRAO/AU/NSF; **22** Patrick Moore; **23** Patrick Moore; **24** NASA; **25** NASA/KSC; **26** NASA/JSC; **27**t STS-115 Shuttle Crew/NASA; **27**b STS-89 Shuttle Crew/NASA; **28**l NASA, ESA, and the Hubble Heritage Team (STScI/AURA) – ESA/Hubble Collaboration; **28**r NASA; **29**t AURA/STScI/NASA; **29**b JPL-Caltech/NASA/B Brandl (Cornell & Leiden); **30–31** NASA/JPL/Space Science Institute; **34** NASA; **35**t NASA Goddard Space Flight Center, Reto Stöckli, Robert Simmon, MODIS, USGS; **35**b NASA; **37** Christine Wheeler; **38** Pete Lawrence; **40**t NASA/JPL; **40–41** Eckhard Slawik/SPL; **41**t AkiraFujii/David Malin Images; **43** Bruce Kingsley; **44** Dave Tyler; **45** Bruce Kingsley; **46** NASA/JPL; **47** NASA; **48** NASA/JSC; **49** NASA/JSC; **50**t SMART/ESA; **50**bl NASA/JPL/USGS; **50**br NASA/GSFC; **51** ESA; **61** Jamie Cooper; **62** NASA; **63** NASA/KSC; **64** NASA; **65** NASA/JPL/Mariner 10; **66** Patrick Moore; **67** Damian Peach; **68–9** NASA/JPL/USGS; **70** NASA/JPL; **71**tl ESA/VIRTIS/INAF-IASF/Observatoire de Paris-LESIA; **71**tr NASA/JPL; **71**bl NASA/JPL; **71**bc NASA; **72** Damian Peach; **73**t NASA/STSCI/AURA; **73**b Patrick Moore; **74**t G Neukum (FU Berlin) et al/DLR/ESA; **74-5** NASA/ESA; **75**t G Neukum (FU Berlin) et al/DLR/ESA; **76** G Neukum (FU Berlin) et al/DLR/ESA; **77**t NASA/JPL; **77**b NASA/JPL; **78**t NASA/JPL; **78**b NASA/JPL; **79**t NASA/JPL/MSSS; **79**b NASA/JPL; **80**t NASA/JPL; **80**b G Neukum (FU Berlin) et al/DLR/ESA; **81**t NASA/JPL/MSSS; **81**b NASA/JPL/MSSS; **82**t Mars Exploration Rover Mission/Texas A&M/Cornell/JPL/NASA; **82–3** Mars Exploration Rover Mission/Cornell/JPL/NASA; **83t** HiRISE/MRO/LPL (University of Arizona)/NASA; **84** NASA/JPL/University of Arizona; **85** NASA/JPL; **86**l NASA/JPL; **86**r NASA/USGS; **87**tl Stephen Ostro et al (JPL)/Arecibo Telescope/NSF/NASA; **87**tr Patrick Moore; **87**b USGS/NASA; **88** NASA; **89**t JHU Applied Physics Laboratory/NASA; **89**c Patrick Moore; **89**b JHU Applied Physics Laboratory/NASA; **90** Dave Tyler; **91** Dave Tyler; **92**t NASA/ESA/I de Pater and M Wong (UC Berkeley); **92**b Dave Tyler; **93** Christopher Go; **94** NASA/JPL; **95**t Caltech/NASA/JPL; **95**b NASA/JPL; **96** HST Comet Team/NASA; **97** H Weaver, T Ed Smith (STScI)/NASA; **98**tl NASA/JPL; **98**tr NASA/JPL/DLR; **98**bl NASA/JPL; **98**br NASA/JPL/DLR; **99** NASA/JPL; **100–101** NASA/JPL; **102** NASA/JPL; **103** NASA/JPL; **105** NASA/Hubble Herritage Team (STScI/AURA) R G French (Wellesley College), J Cuzzi (NASA/Ames), L Dones (SwRI); **106**t NASA/Hubble Herritage Team (STScI/AURA) R G French (Wellesley College), J Cuzzi (NASA/Ames), L Dones (SwRI); **106c** NASA/Hubble Herritage Team (STScI/AURA) R G French (Wellesley College), J Cuzzi (NASA/Ames), L Dones (SwRI); **106–7** NASA/Hubble Herritage Team (STScI/AURA) R G French (Wellesley College), J Cuzzi (NASA/Ames), L Dones (SwRI); **107**t Damian Peach; **108** NASA/JPL; **109**t NASA/JPL/SSSI; **109**b NASA/JPL; **110–11** NASA/JPL/SSI; **112** NASA/JPL/SSI; **113** NASA/JPL/SSI; **114** NASA/JPL/SSI; **115** NASA/JPL/SSI; **116**t ESA/NASA/JPL/University of Arizona; **116**bl NASA/JPL/SSI; **116**br NASA/JPL/SSI; **117** ESA/NASA/JPL/University of Arizona; **118** E Lellouch/T Encrenaz (Observatoire de Paris)/J Cuby/A Jaunsen (ESO-Chile)/VLT ANTU (ESO); **119**tl Patrick Moore; **119**tr Kenneth Seidelmann, US Naval Observatory/NASA; **120** NASA/JPL; **121**t Erich Karkoshka (University of Arizona)/NASA; **121**b NASA/JPL; **122–3** NASA/JPL; **124** NASA/JPL; **125**t L Sromovsky (University of Wisconsin-Madison)/NASA; **125**b NASA/JPL; **126** NASA/JPL; **127** NASA/JPL; **128** Patrick Moore; **130** Chad Trujillo and Mike Brown (Caltech); **131**b ESA/NASA; **132**l Akira Fujii/David Malin Images; **132**r Gordon Rogers; **133** NASA/JPL/Caltech; **134** Mount Wilson and Las Campanas Observatories of the Carnegie Institution of Washington; **135**t Patrick Moore; **135**b ESA; **136–7** Patrick Moore; **138** Akira Fujii/David Malin Images; **139**l Patrick Moore; **139**tr Pete Lawrence; **139**br Akira Fujii/David Malin Images; **140**t NASA/JPL; **140**b JASA/JPL-Caltech/UMD; **141** NASA/JPL; **142**tl; **142**b Patrick Moore; **142–3** Jimmy Westlake; **143**r Paul Doherty; **143**b Patrick Moore; **144-5** Patrick Moore; **146**tl D F Trombino; **146**bl Patrick Moore; **146-7** Patrick Moore; **147**t David Parker/SPL; **147**c Gerry Gerrard; **148-9** Royal Swedish Academy of Sciences; **150** Hinode JAXA/NASA; **151** Courtesy SOHO/EIT Consortium; **152** Patrick Moore; **153** Pete Lawrence; **154** Bill Livingston/NOAO/AURA/NSF; **155** National Solar Observatory, Sacramento Peak; **156**t Institute of Space and Astronomical Scence, Japan; **156–7** Pete Lawrence; **157**t TRACE - a mission of the Stanford Lockheed Institute for Space Research/NASA; **157** inset Courtesy SOHO/LASCO Consortium; **158** Chris Doherty; **159**t Henry Brinton; **159**bl Bill Livingston; **159**br Patrick Moore; **160–1** Nik Szymanek; **162–3** Nik Szymanek; **165** Nik Szymanek; **166** Patrick Moore; **167** Akira Fujii/David Malin Images; **169** Sloan Digital Sky Survey; **170** ESO; **173** ESO; **175** H J P Arnold; **176** Jon Morse (University of Colorado)/NASA; **177** N A Sharp/WIYN/NOAO/NSF; **178** Patrick Moore; **179** ESO; **180** Anglo-Austrailian Observatory, photograph by David Malin; **181**t Hubble Heritage Team (AURA/STScI/NASA); **182** Anglo-Australian Observatory/Royal Observatory Edinburgh, photograph by David Malin; **183**t Rebecca Elson and Richard Swird (Cambridge and NASA); original J Westphal (Caltech); **183**b Brad Moor, William Optics; **184**t Raghvendra Sahal and John Trauger (JPL)/the WFPC2 Science Team/NASA; **184**b NASA/Jeff Hester (Arizona State University); **185**t ESA/STScI/J Hester and P Scowen (Arizona State University); **185**b R Barba, N Morell et al (UNLP)/CTIO/NOAO/NSF; **186–7** NASA/JPL-Caltech/E Churchwell (University of Wisconsin); **187**t ESO; **188–9** Hubble Heritage Team/NASA; **190**l ESO; **190**r T A Rector/B A Wolpa/NOAO/AURA/NSF; **191** NASA/H Ford (JHU)/ G Illingworth (UCSC/LO)/ M Clampin (STScI)/ G Hartig (STScI)/the ACS Team/ESA; **194** T A Rector (NRAO/AUI/NSF and NOAO/AURA/NSF) and M Hanna (NOAO/AURA/NSF); **194–5** Local Group Galaxies Survey Team, NOAO/AURA/NSF; **195** NOAO/AURA/NSF; **196** ESA/INT/DSS2; **196–7** Hubble Herritage Team/ESA/NASA; **197**b NASA, Andrew S Wilson (University of Maryland); Patrick L Shopbell (Caltech); Chris Simpson (Subaru Telescope); Thaisa Storchi-Bergmann and F K B Barbosa (UFRGS, Brazil); and Martin J Ward (University of Leicester); **198**r NASA/N Benitez (JHU)/T Broadhurst (Racal Institute of Physics/The Hebrew University)/H Ford (JHU)/M Clampin (STScI)/G Hartig (STScI)/G Illingworth (UCO/Lick Obs)/the ACS Science Team/ESA; **199**t NASA; **199** inset NASA and the Hubble Heritage Team (STScI/AURA); **199**b ESO; **200** Duncan A Forbes (Swinburne University, Australia); **201** Max-Planck-Institüt; **202** NOAA/National Undersea Research Program; **203** NASA/JPL/MSSS; **204–5** NASA/ESA/S Beckwith and the HUDF Team (STScI) and B Mobasher (STScI); **206** NASA/JPL-Caltech/J Bally (Univ. of Colorado); **207**t NASA/JPL-Caltech/STScI/D Elmegreen (Vassar); **207**b NASA/JPL-Caltech/K Su (University of Arizona); **207** inset NASA and the Hubble Heritage Team (STScI); **208–9** The Art Archive/Palazzo Farnese Caprarola/Dagli Orti; **223** T A Rector and Monica Ramirez/NOAO/AURA/NSF; **225** J P Harrington and K J Borkowski (University of Maryland)/HST/NASA; **233** ESO; **235** B Whitmore (STScI)/NASA; **240–41** John Fletcher; **247** Bernard Abrams; **249** Doug Williams, N A Sharp/NOAO/AURA/NSF; **251** NOAO/AURA/NSF; **255** Bernard Abrams; **257** T A Rector and B A Wolpa/NOAO/AURA/NSF; **265** W Keel/NOAO/AURA/NSF; **266–7** Patrick Moore; **268** Pete Lawrence; **271** PeteLawrence; **272** Pete Lawrence; **273** Derek St Romaine; **274–5** Pete Lawrence; **276** Pete Lawrence; **278** Patrick Moore; **279**t Robin Rees; **279**b Jerry Gunn/Galaxy; endpapers NASA/JPL-Caltech/T Megeath (University of Toledo)/M Robberto (STScI)

Artwork Credits (© Philip's)
Paul Doherty, Raymond Turvey and Julian Baum

Mapping Credits (© Philip's)
MOON MAPS John Murray
WHOLE SKY MAPS AND SEASONAL CHARTS Wil Tirion
STAR MAPS prepared by Paul Doherty and produced by Louise Griffiths